KU-714-180

MCBU
Molecular and Cell Biology Updates

Free Radicals: From Basic Science to Medicine

Edited by G. Poli
E. Albano
M.U. Dianzani

Birkhäuser Verlag
Basel · Boston · Berlin

Editors

Professor Dr. G. Poli
Professor Dr. E. Albano
Professor Dr. M.U. Dianzani

Department of Experimental Medicine and Oncology
University of Torino
I-101 25 Torino
ITALY

A CIP catalogue record for this book is available from the Library of Congress, Washington D.C., USA

Deutsche Bibliothek Cataloging-in-Publication Data

Free radicals: from basic science to medicine / ed. by G. Poli
...– Basel; Boston; Berlin: Birkhäuser, 1993
(Molecular and cell biology updates)
ISBN 3-7643-2763-4 (Basel...) Gb.
ISBN 0-8176-2763-4 (Boston) Gb.
NE: Poli, Giuseppe [Hrsg.]

Camera-ready copy prepared by the author
Printed on acid-free paper produced from chlorine-free pulp
Printed in Germany

ISBN 3-7643-2763-4
ISBN 0-8176-2763-4

9 8 7 6 5 4 3 2 1

CONTENTS

Aging

Cancer

Metabolic disorders

Antioxidants

Preface

Free radical-mediated reactions have been well known in chemistry and physical chemistry for many years. Applying this knowledge to living organisms, biochemists have shown that reactive free radicals are formed at many intracellular sites during normal metabolism, and they have started to suggest possible roles in various pathological processes and conditions, for example in radiation damage, in the metabolism of xenobiotics, in carcinogenesis and in metabolic disorders.

At present, a large and relevant mass of experimental evidence supports the view that reactive free radicals are involved in the pathogenesis of several diseases and syndromes. This literature has captured the attention and interest of people involved in the biomedical field. Exciting developments in radical research are probable in the near future, establishing a greater interaction between basic science research and medicine. While the task of defining the involvement of free radicals in human pathology is difficult, it is nonetheless extremely important that such interaction be fulfilled as soon as possible.

These were the considerations motivating us during the organization of the VI Biennial Meeting of the International Society for Free Radical Research held in Torino, Italy, in June 1992, and also during the preparation of this book. Experts in the various aspects of free radical research were invited to participate in the Torino Meeting and to contribute chapters for this volume.

Reported here are current opinions on the role of free radical reactions during ageing and carcinogenesis and in the pathogenesis of various metabolic disorders as well as the first attempts to characterize free radical-mediated damage directly in humans. A whole section is dedicated to pharmaceutical intervention with antioxidants. We would like to emphasize that all these practical approaches have been achieved with the direct involvement and the supervision of so called 'basic' scientists!

The book is dedicated to a great scientist who was able to create consistent and fruitful links among chemists, biochemists and medical doctors engaged in the common effort to investigate "free radical pathology" in relation to humans: Trevor Slater.

Among Trevor's many unforgettable achievements, there is one which we would like to mention in particular: the formation of the "free radical task force" (one of the first and most effective of its kind) at Brunel University, West London. Working in conjunction with various groups within and outside of Europe, the force also formed a most productive union with Mario Dianzani's school in Italy.

We thank Dr. Martina Di Paolo, Dr. Rita Carini and Dr. Maurizio Parola for their invaluable help in organizing the material for the book, and the staff of Birkhauser for their kind cooperation.

The sponsorship of UNESCO, the University of Torino and the C.N.R. Centre for Immunogenetics and Experimental Oncology, Torino, is also acknowledged.

G. Poli
E. Albano
M.U. Dianzani

The Turin and Brunel collaboration. Mario U. Dianzani and Trevor Slater when they first met in 1961.

The Berlin, Brunel and New Zealand collaboration. Outside the linear accelerator building "just before the cricket season" 1979. Left to right: Robin Willson, John Packer, Dieter Asmus, Trevor Slater, Detlef Bahnemann.

Free Radicals: Generation and Mechanisms of Damage

Free Radicals: From Basic Science to Medicine
G. Poli, E. Albano & M. U. Dianzani (eds.)

TREVOR SLATER, FREE RADICAL REDOX CHEMISTRY AND ANTIOXIDANTS: FROM NAD^+ AND VITAMIN C TO CCl_4 AND VITAMIN E, TO THIOLS, MYOGLOBIN AND VITAMINS A AND D.

R.L. Willson

Biology and Biochemistry Department, Brunel University, Uxbridge, UB8 3PH, U.K.

SUMMARY: Free radicals, once considered the esoteric property of chemists, are now ardently discussed by biochemists, biologists, nutritionalists, toxicologists, physicians and surgeons alike. No-one has contributed more overall to bridging these disciplines than Trevor Slater who died earlier this year. In this paper some of the advances he catalysed concerning our knowledge of antioxidant and oxygen free radical redox reactions in solution will be described, together with some recent fascinating findings concerning the reactions of sulphur free radicals with vitamins A and D.

Potentially damaging oxygen free radicals can be generated within the body, by a variety of processes. The superoxide free radical, in particular, is thought to be formed enzymically during oxygen reperfusion following ischaemia, during phagocytosis or through the action of toxic chemicals.
A scheme of some of the reactions which may subsequently occur, with the possible eventual formation of sulphur free radicals, as it is shown in Fig 1.

PULSE RADIOLYSIS STUDIES OF ONE-ELECTRON TRANSFER REACTIONS IN WATER

Radiation chemical and biochemical methods, particularly using the fast reaction technique of pulse radiolysis have been especially valuable in characterising, quantitatively *in vitro*, many

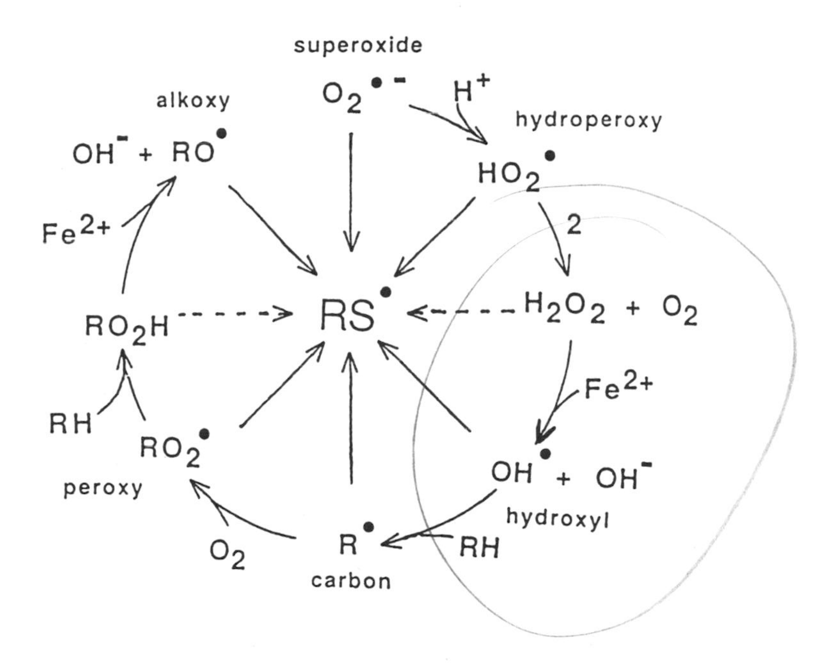

Fig. 1. Free radicals stemming from the superoxide radical, O_2^-.

of the reactions outlined above. In any biological system the fate of a free radical is determined by its own unimolecular natural lifetime in the absence of other reactants and by the rates of its reaction with other free radicals and solutes present. These bimolecular rates are in turn determined by the concentrations of the radicals and solutes and the respective absolute reaction rate constants.

Before the advent of pulse radiolysis, information concerning the rate constants of most free radical reactions could only be obtained indirectly by competition studies. The yield of formation of a relatively long-lived product or the consumption of a reference reactant was determined in the absence and presence of the reactant of interest. From these studies only relative rate constants could be obtained. A few absolute rate constants had been determined using flash photolysis. This technique, unlike pulse radiolyis, however, requires a light absorbing chromophore and is not nearly as versatile.

In retrospect it was fortunate that the advent of pulse radiolysis in the late fifties and early sixties occurred similtaneously with a growing interest in the development of drugs that might sensitise hypoxic cells to the lethal effects of radiation. A number of mechanisms involving free radical redox reactions were proposed for the ways such drugs might operate, and this gave a considerable impetus for the study of such reactions in aqueous solution (Adams and Dewey, 1963; Emmerson and Howard-Flanders, 1965). A large number of electron transfer reactions in which a free radical was the reducing species became the focus of considerable attention (Adams et al. 1967). Few of the reactions had been studied before and those that had, had usually only been studied in organic solvents. Many of the reactions could now be followed directly for the first time by observing the increase or decrease in the optical absorptions of a product or reactant. In the case of NAD^+ for example it was shown that the reaction of organic radicals with the coenzyme, a reaction which had been indicated in some early X-ray studies (Swallow, 1955), occurred rapidly with the formation of $NAD^\cdot$ (Land and Swallow 1968). When similar systems were pulse irradiated in the presence of benzoquinone the product semiquinone free radical was observed even when oxygen was present, indicating that the superoxide free radical was the free radical product of the reaction of $NAD^\cdot$ with oxygen (Willson, 1970). The overall free radical reactions occurring were subsequently published using the biochemical format of interacting circular arrows (Fig. 2).

In parallel to this research, related pulse radiolysis experiments in which a free radical was the oxidant rather than the reductant were also being undertaken (Aldrich et al. 1969; Land and Swallow, 1971;.Adams et al., 1972; Redpath and Willson, 1973).

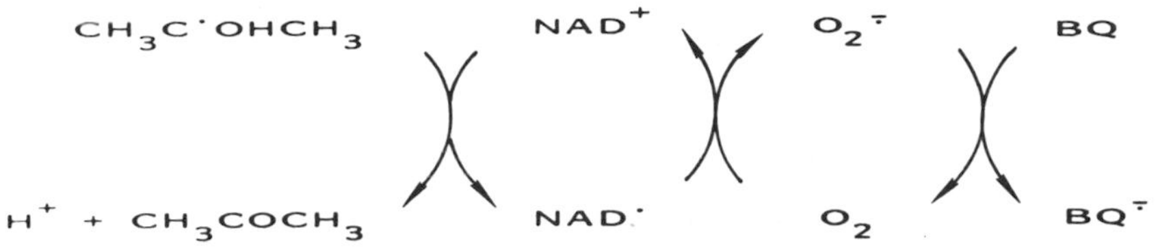

$CH_3C^{\cdot}OHCH_3$ NAD^+ $O_2^{\cdot -}$ BQ

$H^+ + CH_3COCH_3$ $NAD^{\cdot}$ O_2 $BQ^{\cdot -}$

Fig. 2. Sequential one-electron transfer in aerobic solution.

The following reactions of NADH and vitamin C (vitCH) are examples of the many reactions studied.

$$Br_2^{\cdot -} + NADH \longrightarrow 2Br^- + NAD^{\cdot} + H^+$$

$$Br_2^{\cdot -} + VitCH \longrightarrow 2Br^- + VitC^{\cdot} + H^+$$

In a number of pulse radiolysis studies free radical equilibria have been observed. From measurements of the radical concentrations at equilibrium or of the rates of the forward and reverse reactions, equilibium constants have been obtained and relative one-electron oxidation-reduction potentials subsequently calculated (Patel and Willson, 1973; Wardman, 1989). In other instances dissociative electron capture occurs simultaneously with reduction. Any reverse reaction can not then take place. Futhermore the radical resulting from fragmentation can have completely different redox properties to the initial reacting radical. The one-electron reduction of bromouracil, a well known radiosensitiser, is one such example. In this instance bromide ions are liberated. When the compound was reduced by the isopropanol radical in the presence of the radiosensitiser p-nitroacetophenone, simple competition was observed and from studies with various bromouracil concentrations the relative rate constants of the following reactions were obtained (Adams and Willson, 1972).

$$CH_3C^{\cdot}OHCH_3 \quad + \quad PNAP \longrightarrow CH_3COCH_3 \quad + \quad PNAP^{\cdot -} \quad + \quad H^+$$
$$CH_3C^{\cdot}OHCH_3 \quad + \quad BrU \longrightarrow CH_3COCH_3 \quad + \quad U^{\cdot} \quad + \quad Br^- \quad + \quad H^+$$

Since the absolute rate constant for the former reaction had been measured directly by following the appearance of the absorption of $PNAP^{\cdot -}$, the rate constant of the bromuracil reaction was determined. Cleary other halogenated organics might behave similarly, and it was not long before similar rate constants for carbon tetrachloride, iodoacetamide and iodoacetate were obtained (Willson, 1973).

This then was the background to the development of pulse radiolysis at Brunel and the beginning of collaborations with Trevor Slater which were to be so stimulating.

CCl_3O^2 and the Brunel-Berlin-New Zeland collaboration

In 1966 Trevor Slater proposed that the toxic effects of CCl_4 were due to the formation of the $CCl_3{}^{\cdot}$ free radical. This was thought to initiate lipid peroxidation by abstracting a labile hydrogen atom from a polyunsaturated fatty acid. On reduction of bromouracil, the resulting uracil radical formed as a result of dissociative electron capture, had been shown to have oxidising properties and to be able to abstract a hydrogen atom from a nearby sugar (Zimbrick et al 1969). Would the $CCl_3{}^{\cdot}$ free radical have similar properties and could its reactions with biological molecules be studied by pulse radiolysis? Ascorbate, tryptophan, and tyrosine were known to be particularly sensitive to oxidative free radical attack, with tyrosine more so in alkaline solution (Adams et al 1972). Would $CCl_3{}^{\cdot}$ also react readily with these compounds or with promethazine (Phenergan) which had been shown to protect against certain aspects of liver injury

induced by CCl_4. Pulse radiolysis studies were soon undertaken: strong transient absorption spectra were observed which were ostensibly attributed to reactions of $CCl_3^{\cdot}$. In most pulse radiolysis studies the experimental solutions must be saturated with a particular gas before irradiation. In these experiments solutions were bubbled with nitrogen in order to remove the oxygen. However to avoid any loss of the halocarbon as a result of its low solubility and relative volatility, the solutions were not treated as vigorously as usual and there was a possibility that traces of oxygen were present in the irradiated solutions. It was appreciated that this may be so and that the related peroxy free radical $CCl_3O_2^{\cdot}$ may be formed to some extent. But since peroxy radicals had been found to be relatively unreactive compared to other oxidising species and the absorptions formed rapidly and exponentially, this possibility was only briefly considered (Willson and Slater, 1975).
Around this time at the Hahn Meitner Institute in Berlin, unknown to Brunel, Dieter Asmus and colleagues were also investigating the reduction of carbon tetrachloride by pulse radiolysis, but in their experiments they were following changes changes in conductivity rather than optical absorbance. The liberation of chloride ions from CCl_4 results in a measurable increase in the conductivity of the irradiated solution (Koster and Asmus, 1971). In 1973 at the LH Gray meeting in Sussex, results from the Berlin and now what had become the Brunel free radical group, were compared. This was to be the beginning of many fruitful collaborations which have continued until today. Numerous Berlin students, starting with Detlef Bahnemann who undertook much of the early pulse radiolysis studies of phenothiazines have since worked at Brunel and vice -versa (Bahnemann et al.1981, 1983) - see photograph. In 1977 the Brunel pulse radiolysis facility became operational. Very soon afterwards John Packer from the University of Auckland joined the group on a year's sabbatical and took up the study of carbon tetrachloride where the work had ended five years earlier. His initial studies were to be extre-

mely significant. The earlier results did not seem to be repeatable. The transient radical-cation from promethazine, so strongly apparent earlier, could not be observed. At first it was thought that this might be due to the loss of halocarbon during gas saturation. But no. To cut a long story short the critical experiment came when John admitted air into the remaining stock solution of the sample that had been pulse irradiated but had given no signal. On repeating the experiment with the now aerated solution, the missing transient absorption appeared, indicating strongly that the absorption was not due to the reaction of $CCl_3^{\cdot}$ but of $CCl_3O_2^{\cdot}$.(Packer et al. 1978). This was to mark the beginning of many studies on the reaction of halogenated peroxy free radicals with biochemicals during the course of which the absolute rate constant for peroxyradical formation from CCl_4 has been determined (Monig et al. 1983).

$$CCl_3^{\cdot} + O_2 \longrightarrow CCl_3O_2^{\cdot} \qquad k = 3.3 \times 10^9 M^{-1}s^{-1}.$$

Polymer chemists had long known that in organic solvents $CCl_3^{\cdot}$ could readily add to unsaturated compounds such as styrene and as a result bind covalently (Kharasch et al. 1947). The characterisation of the above reaction thus raised the possibility that in the endoplasmic reticulum the lifetime and reactions of the halocarbon radical and hence also its diffusion distance may depend strongly on the local oxygen concentration (Slater et al. 1980).

<u>Vitamin E, vitamin C, ABTS and sulphur radicals.</u>

As well as vitamin C, vitamin E (vitEOH) was also known to protect the liver against certain aspects of liver injury and it was not long before it too was investigated. The CCl_4 system with its high alcohol content was particularly suited for experiments with such compounds that were poorly soluble in water and the subsequent first measurement of the absortion spectrum of the tocopherol phenoxy free radical was particularly rewarding (Packer et al. 1979).

$$CCl_3O_2^{\cdot} + vitEOH \longrightarrow CCl_3O_2H + vitEO^{\cdot} \quad k = 5 \times 10^8 \, M^{-1}s^{-1}.$$

The effectiveness of vitamin E in scavenging peroxy radicals was reinforced by the finding that the rate constant of its reaction with $CCl_3O_2^{\cdot}$ was fifty times higher than for the corresponding reaction with p.methoxy-phenol.

In 1971 a report describing how phenoxyl radicals derived from a range of substituted water-soluble phenols reacted readily with vitamin C (vitCH), was published (Schuler, 1977). Thanks to $CCl_3O_2^{\cdot}$ the way was now open to see if the vitamin E free radical behaved similarly. It did in accordance with the following scheme.

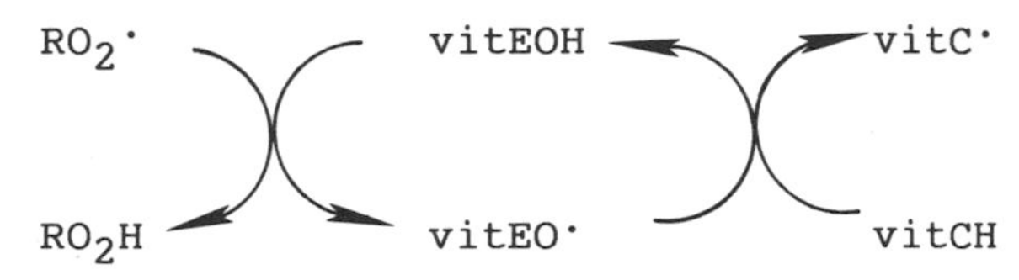

Encouraged by these findings, pulse radiolysis studies with other halocarbons and a range of biomolecules and antioxidants including beta carotene and hydroxydimethylcarbazole were undertaken (Packer et al 1980, 1981a; Monig et al. 1983a; Searle et al. 1984). Of the amino acids, it was confirmed that tryptophan was sensitive to $CCl_3O_2^{\cdot}$ attack whether as the free amino acid or when incorporated in lysozyme and parallel stationary state studies showed that the radical readily inactivated both the enzyme and the bacteriophage T_4 (Packer et al. 1981b, Hiller et al. 1983; Willson, 1985; Willson et al. 1985).

In some pulse radiolysis studies transient absorptions could not be followed, competition studies were made using the salivary peroxidase reagent ABTS, (2,2 azinobis-(3-ethylbenzthiazoline-6-sulphonate) as a reference (Wolfenden and Willson, 1982; Forni et al. 1983b). By measuring the effect of different concentrations of various polyunsaturated fatty acids, for example, on the magnitude of the $ABTS^{\cdot+}$ absorption, the rate constants of the corresponding reactions with the peroxy radicals could be determined.

$$CCl_3O_2^{\cdot} + ABTS \longrightarrow CCl_3O_2^{-} + ABTS^{\cdot+}$$
$$CCl_3O_2^{\cdot} + PUFA \longrightarrow CCl_3O_2H + PUFA(-H)^{\cdot}$$

Related studies on the vitamin E-vitamin C antioxidant system, using a variety of end points have continued. (Niki et al, 1982; Fukazawaet al. 1985). Recently the tocopherol free radical has also been generated, using a pulsed nanosecond XeCl excimer laser, in a phospholipid bilayer membrane and results indicate that the reaction with vitamin C is considerably decreased when an acidic lipid is incorporated (Bisby and Parker 1992).

Several thiols had also been shown to afford protection against some aspects of halocarbon induced liver injury and it was hoped that the reaction of the peroxy free radicals with these compounds too, could be measured by pulse radiolysis. However simple competition does not occur whether using ABTS or a phenathiazine as a reference scavenger. Thiyl free radicals formed in the reaction or others derived from them, react with the reference compounds in a so far unpredicatable manner. Whilst these complications have proved a nuisance in studies of halocarbons they are indicative of a much wider area of potential importance which until recently has received little attention; the role of sulphur in free radical biology.

SULPHUR FREE RADICAL GENERATION AND REACTIONS WITH ABTS, VITAMIN C, NADH AND OTHER ELECTRON OR HYDROGEN DONATING COMPOUNDS.

Although oxygen free radicals have been, and are likely to continue to be, the principal focus of most free radical biologists, increasing attention is being directed to a possible important role for sulphur (thiyl) free radicals in tissue function and injury (Wefers & Sies, 1983; Harman et al., 1984; Ross et al., 1985; Eling et al., 1986; Stock et al., 1986; Davies et al., 1987; Wardman, 1988; D'Aquino et al., 1989; Foureman & Eling, 1989; Schoneich, et al. 1989a, 1989b; Chatgilialoglu & Asmus, 1990; Schoneich & Asmus, 1990).

Pulse radiolysis and related studies have shown that most if not all oxygen-centred free radicals and many carbon-centred free radicals are likely to react readily with thiol compounds to form the corresponding thiyl free radicals (Fig.1). Thiyl free radicals can be formed directly or indirectly through the action of peroxides and myoglobin (Romero et al 1992), through the action of peroxidases (Ohnishi et al., 1969; Eling et al., 1986; Harman et al 1986), and during the autoxidation of cysteine (Saez et al., 1982).

In the case of myoglobin, a number of reactions may occur depending on the concentrations of the reactants present (Galaris et al 1988; Romero et al 1992). If myoglobin is represented as $HXFe^{II}$, oxymyoglobin as $HXFe^{II}O_2$, metmyoglobin as $(HXFe^{III})^+$, ferrylmyoglobin as $HXFe^{IV}O$ and the oxidised ferrylmyoglobin radical as $\cdot XFe^{IV}O$, the reactions that might take place can be written as follows.

$$HXFe^{II} + H_2O_2 \longrightarrow HXFe^{IV}O + H_2O \quad (1)$$
$$HXFe^{II}O_2 + H_2O_2 \longrightarrow HXFe^{IV}O + H_2O + O_2 \quad (2)$$
$$HXFe^{II}O_2 + HXFe^{IV}O + H_2O \longrightarrow 2(HXFe^{III})^+ + 2OH^- + O_2 \quad (3)$$

$$(HXFe^{III})^+ + H_2O_2 \longrightarrow \cdot XFe^{IV}O + H_2O + H^+ \quad (4)$$

$$(HXFe^{III})^+ + \cdot XFe^{IV}O + H_2O \longrightarrow 2HXFe^{IV}O + H^+ \quad (5)$$

$$\cdot XFe^{IV}O + RSH \longrightarrow HXFe^{IV}O + RS\cdot \quad (6)$$
$$HXFe^{IV}O + RSH \longrightarrow (HXFe^{III})^+ + RS\cdot + OH^- \quad (7)$$
$$(HXFe^{III})^+ + RSH \longrightarrow HXFe^{II} + RS\cdot + H^+ \quad (8)$$

Some of these reactions such as (2) are likely to occur only very slowly if at all, whereas others such as (1) and (4) may occur at an appreciable rate (Whitburn, 1987; George and Irvine, 1952; King and Winfield, 1963). In the comproprotionation reaction (3) electron tunneling may occur (Whitburn, 1985; Haranda et al, 1986; Giulivi and Davies, 1990). In the case of reaction (7) an alternative two-reducing eqivalent reaction leading to the formation of sulphmyoglobin (FeII) in its oxygenated or deoxygenated form may also take place (Berzofsky et al, 1972). Reaction (8) has only been shown to occur readily with dihydrolipoate: no analogous reaction has been observed with cysteine, N-acetylcysteine or ergothioneine (Romero et al, 1992). Clearly, the myoglobin and thiol free radical system is complex and the reactions predominating will be very dependent on the local conditions.

Should thiyl free radicals be formed they might readily undergo a variety of electron transfer reactions with electron donors such as ABTS (Willson and Wolfenden, 1982), ascorbate and phenothiazines (Forni and Willson, 1983; Forni et al. 1983), NADH and reduced cytochrome C (Forni and Willson, 1986a,b). They have also been shown to abstract hydrogen atoms from various organic compounds including polyunsaturated fatty acids (Akhlaq et al.1981; Schoneich et al.1989a, 1989b; Surdhar et. al.1989.)

It has been known for some time that thiyl radicals can also add to styrene, precocene and several benzanthracene dihydrodiols. In doing so, it has suggested they may have a role in the detoxification of foreign compounds (Wardman, 1988; Foureman & Eling, 1989). Recently again by chance we have found that they also react rapidly with muconic acid, a metabolite of benzene (Dunster et al. 1990). This has led to studies with other conjugated olefinic compounds. Of particular interest has been the observation that the presence of glutathione strongly promotes the oxygen radical-induced destruction of curcumin (turmeric) and of vitamins A and D (D'Aquino et al 1989, 1992; Diva et al 1992). The additional presence of vitamin C or a water soluble derivative of vitamin E (Trolox C) affords protection. In the light of published chemical studies (Burkhardt, 1934; Oswald et al., 1962;

1963) and by analogy with more recent biochemical studies with styrene, glutathione and horseradish peroxidase (Stock et al., 1986), addition of the thiyl radical to the conjugated unsaturated grouping is thought to occur. Depending on the particular molecule, bond rearrangement may then take place.

$$RS^{\cdot} + C_6H_5CH{=}CH_2 \longrightarrow C_6H_5CH^{\cdot}{-}CHSR$$

$$RS^{\cdot} + {-}CH{=}CH{-}CH{=}CH{-} \longrightarrow {-}CH(RS){-}CH{=}CH{-}C^{\cdot}H{-}$$

The transfer of a hydrogen to the resulting adduct from another thiol molecule can then follow, either directly, or more likely indirectly, after the addition of oxygen.

$$-CH(RS){-}CH{=}CH{-}C^{\cdot}H{-} + RSH \longrightarrow -CH(RS){-}CH{=}CH{-}CH_2{-} + RS^{\cdot}$$
$$-CH(RS){-}CH{=}CH{-}C^{\cdot}H{-} + O_2 \longrightarrow -CH(RS){-}CH{=}CH{-}CH(O_2^{\cdot}){-}$$
$$-CH(RS){-}CH{=}CH{-}CH(O_2^{\cdot}){-} + RSH \longrightarrow -CH(RS){-}CH{=}CH{-}CH(O_2H){-} + RS^{\cdot}$$

The protection of curcumin observed with vitamin C and Trolox C (TxOH) are in agreement with previous observations of their high reactivity with thiyl radicals (Forni et al., 1983a; Davies et al. 1988).

$$GS^{\cdot} + \text{vit CH}^- \longrightarrow GS^- + \text{vitC}^{\cdot -} + H^+$$
$$(k = 6.0 \times 10^8\ M^{-1}\ s^{-1})$$

$$\text{cysteineS}^{\cdot} + TxOH \longrightarrow \text{cysteineS}^- + TxO^{\cdot} + H^+$$
$$(k = 1.0 \times 10^8\ M^{-1}\ s^{-1})$$

The relevance of thiyl free radicals studied in _vitro_ to biological damage in _vivo_ has been questioned from time to time because of the possible occurrence of the alternative relatively rapid reactions of thiyl free radicals with oxygen. This may occur, directly, or indirectly through the reaction with the thiolate ion, RS^- (Adams et al 1967, Barton & Packer, 1970; Schaefer et al 1978).

$$RS^{\cdot} + O_2 \longrightarrow RSO_2^{\cdot}$$
$$RS^{\cdot} + RS^{-} \longrightarrow RSSR^{\cdot -}$$
$$RSSR^{\cdot -} + O_2 \longrightarrow RSSR + O_2^{-}$$

However the fact that destruction of conjugated olefins can occur even at relatively high oxygen concentrations means that damaging reactions of thiyl free radicals cannot be dismissed out of hand. The possibility that reaction with oxygen is very much an equilibrium reaction, with the reverse reaction having a significant role, must also be considered (Tamba et al 1986, Monig et al 1987).

Conjugated diene formation is a well known feature of lipid peroxidation. This and the fact that with the notable exception of vitamins A and D, the leukotrienes, and related compounds few conjugated unsaturated compounds are present in healthy mammalian cells (especially when compared to plant cells), is particularly intriguing. Could this relate in some way to the susceptibility of these molecules to thiyl free radical reactions?

ACKNOWLEDGEMENTS

Radiation studies at Brunel would not have been possible without the vision and support of Trevor Slater. All who have used the Brunel pulse radiolysis equipment and associated facilities will always remember the many exhilirating times shared with him. We are forever grateful.
I am indebted to Enrique Cadenas and Cecelia Giulivi for stimulating discussions regarding the myoglobin system.

REFERENCES

Adams, G.E. and Dewey, D.L. (1963) Biochem. Biophys. Res. Comm. 12 473.

Adams, G.E., McNaughton, G.S. & Michael, B.D (1967) in Chemistry of Ionization and Excitation" eds Johnson, G.R.A. and Scholes, G. Taylor and Francis, London 1967 pp 281-293.

Adams, G.E., Michael, B.D. and Richards, J.T. (1967) Nature 215 1248.

Adams, G.E., Aldrich, J.E., Bisby, R.H., Cundall, R.B., Redpath, J.L. and Willson, R.L. (1972). Radiat. Res., 49, 278-289.

Adams, G.E., and Willson, R.L. (1972) Int.J. Radiat.Biol., 22, 589-597.

Akhlaq, S., Schuchmann, H-P. and von Sonntag, C. (1987) Int. J.

Radiat. Biol. 51 91-102.
Aldrich, J.E., Cundall, R.B., Adams, G.E. and Willson, R.L. (1969) Nature, 221, 1049-1050.
Bahnemann, D., Asmus, K-D. and Willson, R.L. (1981) J.Chem.Soc. Perkin II, 890-895.
Bahnemann, D., Asmus, K-D. and Willson, R.L. (1983) J.Chem.Soc. Perkin II. 1661-1668.
Barton, J.P. & Packer, J.E. (1970) Int. J. Radiat. Phys. Chem. 2 159-166.
Berzofsky, J.A., Peisach, J. and Horecker, B.L. (1972) J. Biol. Chem. 247 3783-3791.
Bisby R.H. and Parker (1992) this symposium.
Burkhardt, G.N. (1934) Trans. Farad. Soc. 30 18.
Chatgilialoglu C. & Asmus K-D ed.(1990) "Sulfur-centered Reactive Intermediates in Chemistry and Biology". ASI Series Plenum Press New York and London.
D'Aquino, M., Dunster, C. & Willson, R.L. (1989) Biochem. Biophys. Res. Comm. 161 1199-1203.
D'Aquino, M. and Willson, R.L. (1992) Int. J. Radiat. Biol. 62 103.
Davies, M.J. (1990) Free Radical Res. Comm. 10 361-370.
Davies, M.J., Forni, L.G. & Shuter, S.L. (1987) Chem. Biol. Interactions. 61 177-188.
Davies, M.J., Forni, L.G. & Willson, R.L. (1988) Biochem. J. 255 513-512.
Diva, D.M.,Diva, H.S., Sharma, G.J., James, G., Tipping, I. and Willson, R.L. (1992) Int. J. Radiat. Biol. 62 105.
Eling, T.E., Curtis, J.F., Harman, L.S. & Mason, R.P. (1986) J. Biol. Chem. 261 5023-5028.
Forni, L.G., Monig, J., Mora-Arellano, V.O. & Willson, R.L. (1983a) J. Chem. Soc. Perkin Trans. II 961-965.
Forni, L.G., Packer, J.E., Slater, T.F. and Willson,R.L. (1983b) Chem.Biol. Interactions. 45, 171-177.
Forni, L.G. and Willson, R.L. (1983) In: 'Protective agents in cancer', (D.C.H. McBrien and T.F. Slater, eds.), Academic Press, London/New York. p.159-173.
Forni, L.G. and Willson, R L (1986a) Biochem. J. 240 897-903.
Forni, L.G. and Willson, R L (1986b) Biochem. J. 240 905-907.
Foureman, G.L. & Eling, T.E. (1989) Arch. Biochem. Biophys. 269 55-68.
Fukuzawa, K., Takase, S. and Tsukatani, H. (1985) Arch. Biochem. Biophys. 240 117-120.
Galaris, D., Cadenas, E., and Hochstein, P. (1988) Free Rad. Biol. Med. 6 473-478.
George, P. and Irvine, D.H. (1952) 52 511-517.
Giulivi, C. and Davies, K.J.A. (1990) J. Biol. Chem. 265 19453-19460.
Harada, K., Tamura, M., and Yamazaki, I. (1986) J. Biochem. (Tokyo)100 499-504.
Harman, L.S., Mottley, C. & Mason, R.P. (1984) J. Biol. Chem. 259, 5606-5611.
Harman, L.S., Carver, D.K., Schreiber, J. & Mason, R.P. (1986) J.

Biol. Chem. 261 1642-1648.
Hiller, K.-O., Hodd, P.L. and Willson, R.L. (1983) Chem. Biol. Interact. 47 293-305
King, N.K. and Winfield, M.E. (1963) J. Biol. Chem. 238 1520-1528.
Kharasch, M.S., Jensen, E.V. and Urry, W.H. J. Amer. Chem. Soc. 69 1100-1105.
Koster, R. and Asmus, K-D. (1971) Z. Natuforsch 266 1104.
Land, E.J. and Swallow, A.J. (1968) Biochim. Biophys. Acta 162 327-337.
Land, E.J. and Swallow, A.J. (1971) Biochim. Biophys. Acta 234 34-42.
Milia, A. Cheeseman, K.H. Forni, L. Willson, R.L. Corongiu, P. and Slater,T.F. (1985) Biochem.Pharmacol. 34, 437-38.
Monig, J., Asmus, K-D., Schaeffer, M., Slater, T.F., and Willson,R.L. (1983a) J.Chem.Soc. Perkin II, 1133-1137.
Monig, J., Bahnemann, D. and Asmus, K-D. (1983b) Chem. Biol. Interact. 47 15-27.
Monig J., Asmus K-D., Forni L.G. & Willson R.L. (1987) Int. J. Radiat. Biol. 52 589-602.
Niki, E., Tsuchiya, J., Tanimura, R. and Kamiya, Y. (1982) 789-792.
Ohnishi, T., Yamazaki, H., Iyanagi, T., Nakamura, T. & Yamazaki, I. (1969) Biochim. Biophys. Acta 172 357-369.
Oswald, A.A., Griesbau, K. & Hudson, B.E. (1963) J. Org. Chem. 28 2355-2361.
Oswald, A.A., Hudson, B.E., Rodgers, G. & Noel, F. (1962) J. Org. Chem. 27 2439 - 2448.
Packer,J .E., Slater, T.F., and Willson, R.L. (1978). Life Sciences, 23, 2617-2620.
Packer, J.E., Slater, T.F., and Willson, R.L. (1979) Nature 278, 737-738.
Packer, J.E., Willson, R.L., Bahnemann, D., and Asmus, K.-D. (1980) J.Chem.Soc., Perkin II, 296-299.
Packer, J.E., Mahood, J.S., Mora-Arellano, V.O., Slater, T.F., Willson, R.L. and Wolfenden, B.S.(1981a) Biochem. Biophys. Res.Comm., 98, 901-906.
Packer,J.E., Mahood,J.S., Willson,R.L. and Wolfenden, B.S. (1981b) Int. J. Radiat. Biol., 39, 135-141.
Patel,K.B., and Willson,R.L. (1973) J.Chem.Soc. Faraday Trans. I. 69 814-825.
Redpath,J.L. and Willson,R.L. (1973).Int.J.Radiat.Biol., 23, 51-65.
Romero, F.J.,Ordonez I., Arduini, A. and Cadenas, E. (1992) J. Biol. Chem.267 1680-1688.
Ross, D., Norbeck, K. & Moldeus, P. (1985) J. Biol.Chem. 260, 15028 -15032.
Saez,G., Thornalley, P.J.,Hill, H.A.O., Hems, R. & Bannister, J.V. (1982) Biochim. Biophys. Acta 719 24-31.
Schaefer, K., Bonifacic, M., Bahnemann, D. & Asmus K-D (1978) J. Phys. Chem. 82 2777.
Schoneich, C., Bonifacic, M. & Asmus, K-D (1989a) Free Radical

Res. Comm. 6 393-405.

Schoneich, C., Asmus, K-D., Dillinger, U. & Bruchhausen, F. (1989b) Biochem. Biophys. Res. Comm. 161 113-120.

Schoneich, C. & Asmus, K-D. (1990) Radiat. Environ. Biophys. 29 263-271.

Schuler, R.H. (1977) Radiat. Res. 69 417-433

Searle, A.J.F. Gee,C. and Willson,R.L. (1984) in 'Oxygen Radicals in Chemistry and Biology'. Pub.Walter de Gruyter, Berlin. pp. 377-381.

Slater, T.F.(1966) Nature 209 36-40.

Slater, T.F., Ahmed, M., Benedetto, C., Cheeseman,K., Packer,J.E. and Willson, R.L. and Dianzani, M.U.(1980). Int.J. Quantum Chem., Quantum Biol. Symp.7. pp.347-356.

Stock, B.H., Schreiber, J., Guenat, C., Mason, R.P., Bend, J.R. & Eling, T.E. (1986) J. Biol. Chem. 261 15915-15922.Surdhar, P.S., Mezyk, S.P. and Armstrong, D.A. (1989) J. Phys. Chem. 93 3360-3363.

Swallow, A.J. (1955) Biochem. J. 61 197-203.

Tamba, M., Simone, G. & Quintiliani, M.(1986) Int. J. Radiat. Biol. 50 595-600.

Wardman, P. (1988) in "Glutathione Conjugation: Mechanisms and Biological Significance" Sies, H. & Ketterer, B., eds. Academic Press, London pp. 44-72.

Wardman, P. (1989) J. Phys. Chem. Ref. Data 18 1637-1755.

Wefers, H. & Sies, H. (1983) Eur. J. Biochem. 137 29-36.

Whitburn, K.D. (1985) J. Inorg. Biochem. 24 35-46.

Whitburn, K.D. (1987) Arch. Biochem. Biophys. 253 419-430.

Willson, R.L. (1970) Chem. Comm. 1005.

Willson, R.L. (1973) Trans. Biochem. Soc. 1, 929-931.

Willson, R.L. (1985) in 'Oxidative Stress' ed. H. Sies, Academic Press, Lonon and New York pp.41-72.

Willson, R.L., and Slater,T.F. (1975) in 'Fast processes in radiation, chemistry and biology' Eds. Adams, Fielden & Michael. The Institute of Physics. John Wiley & Sons. 147-161.

Willson, R.L., Dunster, C.A., Forni, L.G., Gee, C.A. and Kitteridge, K.J.(1985) Phil. Trans. R.Soc. Lond. B 311 545-563.

Wolfenden, B.S. & Willson, R.L. (1982) J. Chem. Soc. Perkin Trans. II 805-812.

Zimbrick, J.D., Ward, J.F. and Myers, L.J. (1969) Int. J. Radiat. Biol.16 525-534.

Free Radicals: From Basic Science to Medicine
G. Poli, E. Albano & M. U. Dianzani (eds.)

REGULATION OF GENE EXPRESSION IN ADAPTATION TO OXIDATIVE STRESS

Kelvin J. A. Davies[1,2], Anne G. Wiese[2†], Robert E. Pacifici[2††], and Joanna M. S. Davies[3]

[1]Department of Biochemistry & Molecular Biology, The Albany Medical College, Albany, New York, U.S.A.; [2]Institute for Toxicology and Department of Biochemistry, University of Southern California, Los Angeles, California, U.S.A.; and [3]Division of Rheumatology, Department of Medicine, The Albany Medical College, Albany, New York, U.S.A.

SUMMARY:Transient adaptation to an acute oxidative stress is a genetically regulated property of facultative bacteria, yeast, and mammalian cells. Experimentally, an initial exposure to sub-lethal concentrations of hydrogen peroxide or superoxide is found to provide significant protection against a subsequent exposure to concentrations that would normally be lethal. Such adaptation is short-term (disappearing within one cell division) and relies upon RNA synthesis and protein synthesis, but does not require DNA replication. Some 30-40 proteins are overexpressed during the bacterial response to hydrogen peroxide or superoxide, and some 20-30 proteins exhibit increased expression in eucaryotes: At least some of these shock or stress proteins are thought to impart the induced resistance to oxidative stress. Four oxidative stress regulons have so far been described in bacteria and at least one antioxidant responsive element is operative in the mammalian genome. Oxidative stress signals appear to be transduced via altered interactions of specific DNA binding proteins with nucleotide target sequences or loci within stress-inducible genes. Efforts are underway to further explain the mechanisms of signal transduction involved in activation of transcription and translation during oxidative stress.

INTRODUCTION

The "Oxygen Paradox" informs us that an oxygen environment is both a major advantage in energy conservation for those organisms able to conduct aerobic metabolism and a major toxic threat to life

---

† Current affiliation: Allergan Inc., Irvine, CA, U.S.A.
†† Current affiliation: Amgen Corporation, Thousand Oaks, CA, U.S.A.

in general. To deal with this problem aerobic organisms appear to have evolved a series of antioxidant enzymes whose sole purpose is apparently to minimize cellular damage by reactive oxygen species. In addition, all aerobic cells appear to either manufacture or sequester a variety of antioxidant compounds for the same purpose. In the past few years it has become apparent that a series of damage removal and/or repair enzymes for oxidatively modified proteins, lipids, and DNA are used to deal with the damage that escapes the antioxidant defensive network (Davies, 1986; Ryter *et al.*, 1990; Davies *et al.*, 1990; Pacifici and Davies, 1991). Recently significant interest has centered on the possible inducibility of both antioxidant enzymes and repair enzymes. In particular, several groups have been testing the hypothesis that cellular adaptation to oxidative stress is conferred by elevated expression of antioxidant and repair enzymes. Current knowledge of this subject, including several new findings, is reviewed in this paper.

ADAPTIVE RESPONSES TO SHOCK OR STRESS

An adaptive response occurs when a relatively small amount of an agent confers resistance to a greater amount of the same or a related agent. In this way a slight change in the environment can trigger protective mechanisms in an organism so that it is better able to survive in that altered environment. Instances of adaptive response to a variety of agents have been documented. These include adaptive response to alkylating agents (Lindahl *et al.*, 1988), heat stress (Lindquist and Craig, 1988), cold stress (Jones *et al.*, 1987), low pH (Stonczewski *et al.*, 1987), thiols (Hitz *et al.*, 1975), and detergents (Adamowicz *et al.*, 1991) as well as the SOS response (Walker, 1984). The best characterized examples are 1) the SOS response, 2) adaptive response to alkylating agents, and 3) adaptive response to heat stress. Brief descriptions of these examples follows.

In the bacterial SOS response, damage to DNA by agents such as ultraviolet radiation results in an increased capacity to repair damaged DNA (Walker, 1984). The DNA damage activates the protease activity of RecA protein. Activated RecA recognizes and cleaves LexA repressor. (LexA repressor is normally bound to many operons repressing the transcription of at least 14 genes.) When LexA is cleaved by RecA, transcription of the normally repressed genes occurs. Many of the gene products induced during the SOS response are involved in a long-patch excision repair system and the Rec recombination-repair pathways of DNA repair.

Pre-exposure of *E. coli* to alkylating agents such as MNNG causes a resistance to the mutagenic and lethal effects of a subsequent higher dose (Lindahl *et al.*, 1988). The adaptive response to alkylating agents causes induction of at least 4 genes. The products of 2 of these genes are DNA repair enzymes. One of the gene products, O^6-alkylguanine-DNA alkyltransferase, repairs the potentially mutagenic lesion O^6-alkylguanine. The other DNA repair enzyme induced is 3-methyladenine-DNA glycosylase II, which specifically removes methylated DNA bases.

The heat shock response involves induction of a small family of heat shock proteins (hsp) following an increase in temperature (Lindquist and Craig, 1988). The hsp are believed to protect cells from the toxic effects of heat and other stresses. Other inducers of hsp include ethanol, anoxia, viral infection and DNA damage. Regulation of these proteins is rapid with synthesis increasing during a rise in temperature and decreasing with a decrease in temperature. Genetic regulation plays an important role in the heat shock response. In *E. coli* transcription of hsp is regulated by the sigma factor $\sigma 32$ which binds to RNA polymerase and redirects it to heat shock promoters. Heat shock causes an immediate increase in the translational efficiency of the $\sigma 32$ message and stabilization of its normally very unstable protein product. In eucaryotes the heat shock factor is constitutively produced but in an inactive form. Heat causes activation of the heat shock factor by a post-translational modification; this results in a rapid increase in hsp production.

All three of the responses described above involve global

response at the genetic level to a specific stress. The global response results in the induction of a set of proteins which enable cells to resist later, possibly more severe, stress.

ADAPTIVE RESPONSES TO OXIDATIVE STRESS IN PROCARYOTES

Demple and Halbrook have demonstrated that in *E. coli* a sublethal challenge of H_2O_2 confers resistance to a subsequent, more severe challenge with H_2O_2 or ionizing radiation (Demple and Halbrook, 1983). Christman *et al.* have shown that a similar adaptive response to H_2O_2 is evident in *S. typhimurium* (Christman *et al.*, 1985). Similarly, O_2^- generating compounds, such as paraquat, impart an adaptive response in *E. coli* (Hassan and Fridovich, 1977; Farr *et al.*, 1985).

Two-dimensional polyacrylamide gel electrophoresis (2D PAGE) analysis shows that the adaptive response to H_2O_2 in procaryotes is accompanied by the synthesis of 30 to 40 polypeptides as well as an increase in DNA repair capabilities (Demple and Halbrook, 1983; Christman *et al.*, 1985). Similar analysis in the adaptive response to O_2^- shows that approximately 40 new polypeptides are synthesized in addition to the 30 to 40 polypeptides induced by H_2O_2. (Because H_2O_2 is produced from the dismutation of O_2^- one would expect to see the appearance of H_2O_2 induced polypeptides in the O_2^- response). As with the H_2O_2 response DNA repair capabilities are increased during the O_2^- induced adaptive response (Farr *et al.*, 1985). Further studies of the adaptive responses to H_2O_2 and O_2^- in procaryotes have shown separate responses to these agents which are regulated at the genetic level. (For reviews see Demple, 1991; Storz *et al.*, 1990). These responses and their regulation are described below.

The Hydrogen Peroxide Response in Procaryotes

Of the 30 to 40 polypeptides induced by H_2O_2 nine are under the control of the *oxyR* gene (Christman *et al.*, 1985; Greenberg *et al.*, 1989). Deletion mutants of *oxyR* are unable to induce the nine proteins and have an increased sensitivity to H_2O_2 (Christman *et al.*, 1985). It should be noted, however, that these mutants can still mount an adaptive response to H_2O_2 (S. Lin and K.J.A. Davies, unpublished data). Conversely, there are H_2O_2 resistant mutants which are constitutive overexpressers of the nine *oxyR* induced polypeptides (*oxyr1* in *S. typhimurium* and *oxyR2* in *E. coli*) (Christman *et al.*, 1985).

OxyR positively regulates genes located throughout the genome and negatively regulates its own gene product. Proteins induced by *oxyR* include HPI catalase, encoded by *katG*, glutathione reductase, encoded by *gorA*, and an alkyl hydroperoxide reductase encoded by *ahpC* and *ahpF* (Christman *et al.*, 1985, Richter and Loewen, 1981). All three of these induced proteins serve to either prevent or reduce damage caused by oxidative stress. Catalase can break down H_2O_2 to oxygen and water and alkyl hydroperoxide reductase serves to reduce hydroperoxides to their corresponding alcohols. Glutathione reductase helps in maintaining a pool of reduced glutathione which can serve to keep cellular proteins in a reduced state. The remainder of the nine induced proteins are, as yet, unidentified.

The *oxyR* gene has been located at 89.6 minutes on *E. coli* and *S. typhimurium* genetic maps (Christman *et al.*, 1985). The gene has been cloned and sequenced and the predicted product, OxyR, is a 34.4 kDa protein which has strong sequence identity with a family of bacterial regulators, including LysR and NodD (Christman *et al.*, 1989, Tao *et al.*, 1989; Tartaglia *et al.*, 1989). Footprinting experiments have identified some putative binding sites of OxyR as being sequences upstream of *ahpC* and *katG* as well as a sequence within the *oxyR* gene itself (Storz *et al.*, 1990). OxyR is activated directly by oxidative stress (Storz *et al.*, 1990). It appears that oxidation of OxyR results in a conformational change. This bound, oxidized form of OxyR in some way triggers transcription of the *oxyR* induced genes and prevents its own transcription (Storz *et al.*, 1990).

Recent work has revealed the existence of a putative second regulon involved in H_2O_2 adaptation (S. Lin and K.J.A. Davies, unpublished observations). Point mutation of the *E. coli oxyR* deletion mutant TA4112 has provided us with a strain (LD115) that exhibits high constitutive resistance to H_2O_2, and high constitutive expression of some 10 of the H_2O_2 induced proteins (some 30-40 in all) that are not under *oxyR* control. In addition two of the *oxyR* regulated proteins are constitutively over-expressed by LD115. Work to date indicates that we have identified a second H_2O_2 inducible regulon which maps between 85′ and 95′: We have tentatively named this regulon *oxoR*. Our putative *oxoR* regulon when constitutively overexpressed (*oxoR*^{c}) provides extensive protection against H_2O_2 with little or no apparent induction of classical antioxidant enzymes. We are pursuing the possibility that *oxoR* may provide protection by inducing the synthesis of repair enzymes during oxidative stress. Interestingly, *oxyR* appears to regulate expression of the "early proteins" (0-20 min) following H_2O_2 treatment, whereas *oxoR* may be more responsible for several of the "late proteins" (20-50 min) induced by H_2O_2.

The Superoxide Response in Procaryotes

Generators of O_2^- such as paraquat and plumbagin induce a distinct set of polypeptides. Nine of these induced polypeptides are under the control of the *soxRS* regulon (Greenberg and Demple, 1989; Walkup and Kogoma, 1989). Proteins induced by *soxRS* include Mn–SOD, encoded by *sodA*, endonuclease IV, encoded by *nfo*, and glucose–6–phosphate dehydrogenase, encoded by *zwf,* as well as the protein products of *soi17*, *soi19*, and *soi28* (Chan and Weiss, 1987; Greenberg *et al.*, 1990; Tsaneva and Weiss, 1990; Touati, 1988; Kogama *et al.*, 1988). The outer membrane protein OmpF is negatively regulated and the small–subunit ribosomal protein S6 is altered by *soxRS* (Greenberg *et al.*, 1990).

The antioxidant activity of three of the induced proteins is well established. Mn–SOD catalyzes the dismutation of O_2^-, endonuclease IV initiates repair of oxidatively damaged DNA, and glucose–6–phosphatedehydrogenase provides NADPH for enzymes which protect the cell from oxidative damage such as glutathione reductase. Repression of *ompF* seems to be unrelated to antioxidant capacity but correlates with an increased resistance to antibiotics (Cohen *et al.*, 1988). The *soi17* and *soi28* gene products confer an increased resistance in *E. coli* to the O_2^- generator paraquat (Kogama *et al.*, 1988).

The *soxR* gene is located at 92.2 minutes on the genetic map of *E. coli* (Tsaneva and Weiss, 1990; Greenberg *et al.*, 1991). Both *soxR* and *soxS* are found within the *soxRS* locus and are involved in regulation of the *soxRS* regulon (Tsaneva and Weiss, 1990). Treatment with O_2^- generators results in induction of SoxS (Wu and Weiss, 1991). The predicted gene product of *soxRS* shares some sequence identity with MerR, a regulator of the resistance to mercury, and the predicted gene product of *soxS* shares sequence identity with the AraC family of regulatory proteins (Demple, 1991). The mechanism of activation of the SoxR and SoxS proteins is currently unknown. Interestingly, a second O_2^- inducible regulon, named *soxQ*, appears to duplicate the function of *soxRS* (Greenberg *et al.*, 1991). Located at 34′ on the *E. coli* genetic map, *soxQ* is triggered by O_2^- and appears to provide increased protection through the same set of inducible proteins that are regulated by the *soxRS* locus (Greenberg *et al.*, 1991).

In summary, current knowledge indicates the existence and importance of at least four bacterial regulons that respond positively to oxidative stress, the *oxyR* regulon and the *oxoR* regulon which are both induced by H_2O_2, as well as the *soxRS* regulon and the *soxQ* regulon which are both induced by superoxide. Each regulon appears to induce nine to twelve proteins in response to oxidative stress. Among the proteins induced are antioxidant enzymes, heat shock proteins, dehydrogenases, and DNA repair enzymes. Recent studies have indicated some overlapping regulation in response to oxidative stress. Some heat shock proteins are induced by H_2O_2 and O_2^-, and there is evidence for cross-resistance to heat shock during the adaptive response to oxidative stress (Christman *et al.*, 1985; Greenberg and Demple, 1989; Morgan *et al.*, 1986). Furthermore, proteins induced during the starvation response and during stationary phase overlap somewhat with H_2O_2 inducible proteins (Jenkins *et al.*, 1988). It now also appears that induction of the SOS response may result from oxidative stress (Goerlich *et al.*, 1989). Although there is a great deal known about procaryotic responses to oxidative stress and genetic regulation during oxidative stress, further work is necessary to identify all of the inducible proteins and to unravel the complex mechanisms of genetic regulation in response to oxidative stress.

ADAPTIVE RESPONSES TO OXIDATIVE STRESS IN EUCARYOTES

Pickett and his colleagues have reported a genetic control mechanism for regulating the expression of the glutathione *S*-transferase gene and the NAD(P)H:quinone reductase gene (also called DT-diaphorase) during oxidative stress (Rushmore *et al.*, 1990; Rushmore and Pickett, 1990; Favreau and Pickett, 1991; Rushmore *et al.*, 1991). This newly discovered control mechanism involves a nucleotide sequence in the 5′ flanking region of the glutathione *S*-transferase and the NAD(P)H:quinone reductase target genes, termed the "Antioxidant Responsive Element". In response to oxidative stress a specific peptide appears to bind (or to change its pattern of binding) to the antioxidant responsive element of the target genes, and thus signal transcriptional activation.

The antioxidant responsive element appears to be necessary for signal transduction in response to metabolizable planar aromatic hydrocarbons, phenols, and hydrogen peroxide. Importantly, the antioxidant responsive element has been shown to be structurally and functionally distinct from the previously known xenobiotic responsive element, which is also triggered by planar aromatic hydrocarbons and phenols, but not by hydrogen peroxide. It appears likely that other antioxidant or oxidant repair genes may contain similar responsive elements or loci.

Interestingly, the work of Pickett and his colleagues (Rushmore *et al.*, 1990; Rushmore and Pickett, 1990; Favreau and Pickett, 1991; Rushmore *et al.*, 1991) in eucaryotes is highly reminiscent of current knowledge about the bacterial *oxyR* regulon, where each of the inducible target genes contains an OxyR protein binding domain or locus, upstream of a functional promotor that transduces the transcriptional activation signal (Storz *et al.*, 1990). Recently, H_2O_2 has also been implicated as a second messenger which can activate the NF−κB transcription factor (Schreck *et al.*, 1991). The NF−κB transcription factor is thought to influence expression and replication of several species.

An adaptive response to oxidative stress in eucaryotic cells has been demonstrated (Spitz *et al.*, 1987; Laval, 1988). Spitz *et al.* have shown that Chinese hamster ovary cells (HA−1) exhibit an adaptive response to H_2O_2 (Spitz *et al.*, 1987). HA−1 cells when pretreated with 3−5 μmoles $H_2O_2/10^7$ cells for one hour become resistant to a challenge of H_2O_2 administered 16−36 hours after pretreatment. Pretreatment with H_2O_2 also confers a very slight resistance to a later heat stress. Conversely, heat stressed cells exhibit a strong resistance to a later heat stress and a lesser, though significant, resistance to H_2O_2. One dimensional SDS−polyacrylamide gel electrophoresis (1D SDS-PAGE) reveals a two−fold increase in synthesis of a 70 kD protein (perhaps hsp70) during the adaptive response to H_2O_2.

Similar results have been observed in Chinese hamster ovary (CHO) cells and in rat hepatoma (H4) cells (Laval, 1988). A pretreatment of 5 uM H_2O_2, or xanthine plus xanthine oxidase, for one hour confers a resistance to H_2O_2 and γ rays administered 24 hours later. In these cells catalase activity is unaltered by pretreatment, and levels of superoxide dismutase activity increase two−fold in xanthine/xanthine oxidase pretreated cells but not at all in H_2O_2 pretreated cells. Pretreatment with xanthine/xanthine oxidase also decreases the mutagenic effect of γ rays.

In further work with HA−1 cells we have detected (2D SDS-PAGE) the overexpression of some 21 proteins in response to H_2O_2 pretreatment (A.G. Wiese, R.E. Pacifici, and K.J.A. Davies, unpublished observations). These H_2O_2 inducible proteins appear to be responsible for the adaptive response to H_2O_2 since inhibitors of either protein synthesis or RNA synthesis block both the induction of H_2O_2 resistance and expression of the stress proteins: In contrast, inhibitors of DNA replication have no effect on either adaptation or expression of the stress proteins. Our new work with HA−1 cells further reveals that the adaptation to relatively low concentrations of H_2O_2 is an acute and transient response that is lost within one generation and appears to be independent of cell cycle.

Pretreatment of HA−1 cells with low concentrations of H_2O_2 appears to induce the expression of stress proteins in two distinct phases or waves (A.G. Wiese, R.E. Pacifici, and K.J.A. Davies, unpublished observations). This pattern is highly reminiscent of the H_2O_2 effects in bacteria described above. The first wave of H_2O_2 induced protein synthesis in HA−1 cells occurs 1-4 hours after exposure, at which point only minor protection is observed. A second phase of protein synthesis occurs between 12-16 hours following H_2O_2 exposure, and H_2O_2 resistance reaches a maximum at approximately 18 hours. Interestingly, H_2O_2 adaptation is accompanied by a prolongation of division time from 18 hours to 36 hours, and H_2O_2 resistance exhibits a clear decline towards pre-treatment levels by 24 hours following exposure.

In our typical growth medium (Minimal Essential Medium containing 15% fetal calf serum) HA−1 cells exhibit very low levels of catalase, superoxide dismutase, glutathione peroxidase, and glutathione reductase activities; rates of transcription of the respective mRNA's for each of these enzymes are correspondingly low. Interestingly, we observe little or no increase in either transcription or translation of any of these classical antioxidant enzymes during H_2O_2 adaptation (A.G. Wiese, R.E. Pacifici, and K.J.A. Davies, unpublished observations). Therefore, although HA−1 cells may be considered deficient in antioxidant enzymes, our results point to key roles for non-classical proteins (perhaps repair enzymes for DNA, proteins, and lipids) in the adaptive response to H_2O_2. This feature of adaptation to oxidant stress also appears important in bacteria where the "late proteins" induced by H_2O_2 are largely non-antioxidant enzymes.

We have observed similar transient adaptive responses to H_2O_2 in other mammalian cell lines including V79 cells, CHO cells, and C3H 10T1/2 cells (A.G. Wiese, R.E. Pacifici, and K.J.A. Davies, unpublished observations). Furthermore, the yeast *Sacharomyces cerevisiae* exhibits the adaptive ability to survive normally lethal doses of H_2O_2 if first exposed to low concentration of H_2O_2 (J.M.S. Davies, C.V. Lowry, and K.J.A. Davies, unpublished observations). Such observations give us substantial confidence that inducible resistance to oxidative stress is a general genetic property of eucaryotic cells.

CONCLUSIONS

In summary, both procaryotes and eucaryotes exhibit effective transient adaptive responses to non-lethal oxidative stress. Such adaptive responses appear to be mediated via regulation of gene expression (both transcription and translation) and provide a significant selective benefit. Although several laboratories have made important advances in this area, we are really just beginning to unravel the complex array of genetic regulatory elements involved in ensuring survival in an oxygen-rich environment.

Acknowledgements

This work was supported by a grant, ES03598, from the National Institutes of Health/National Institute of Environmental Health Sciences to K.J.A.D.

REFERENCES

Adamowicz, M., Kelley, P.M., and Nickerson, K.W. (1991) J. Bacteriol. 173, 229-233.

Chan, E. and Weiss, B. (1987) Proc. Natl. Acad. Sci. USA 84, 3189-3193.

Christman, M.F., Storz, G., and Ames, B.N. (1989) Proc. Natl. Acad. Sci. USA 86, 3484-3488.

Christman, M.F., Morgan, R.W., Jacobson, F.S., and Ames, B.N. (1985) Cell 41, 753-762.

Cohen, S.F., McMurray, L.M., and Levy, S.B. (1988) J. Bacteriol. 170 5416-5422.

Davies, K.J.A. (1986) J. Free Radical Biol. Med. 2, 153-173.

Davies, K.J.A., Wiese, A.G., Sevanian, A., and Kim, E.H. (1990) In: *Molecular Biology of Aging*, (Finch, C.E. and Johnson, T.E., eds.) Alan R. Liss, Inc., New York. Pp. 123-141.
Demple, B. (1991) Ann. Rev. Genet. 25, 315- 337.
Demple, B. and Halbrook, J. (1983) Nature 304, 466-468.
Doetsch, P.W., Henner, W.D., Cunningham, R.P., Toney, J.H., Helland, D.E. (1987) Mol. Cell Biol. 7, 26-32.
Favreau, L.V. and Pickett, C.B. (1991) J. Biol. Chem. 266, 4556-4561.
Farr, S.B., Natvig, D.O., Kogoma, T. (1985) J. Bacteriol. 164, 1309-1316.
Goerlich, O., Quillardet, P., and Hofnung, M. (1989) J. Bacteriol. 171, 6141-6147.
Greenberg, J.T. and Demple, B. (1989) J. Bacteriol. 171, 3933-3939.
Greenberg, J.T., Monach, P., Chou, J.H., Josephy, D.P., and Demple, B. (1990) Proc. Natl. Acad. Sci. USA 87, 6181-6185.
Greenberg, J.T., Chou, J.H., Monach, P.A., Demple, B. (1991) J. Bacteriol. 173, 443-4439.
Hassan, H.M., and Fridovich, I. (1977) J. Biol. Chem. 252, 7667-7672.
Hitz, H., Schafer, D., and Wittmann-Liebold, B. (1975) FEBS Lett. 56, 259-262.
Jenkins, D.E., Schultz, J.E., and Matin, A. (1988) J. Bacteriol 170, 3910-3914.
Jones, P.G., vanBogelen, R.A., and Neidhardt, F.C. (1987) J. Bacteriol. 169, 2092-2095.
Kogama, T., Farr, S.B., Joyce, K.M., Natvig, D.O. (1988) Proc. Natl. Acad. Sci. USA 85, 4799-4803.
Laval, F. (1988) J. Cell Physiol. 201, 73-79.
Lindahl, T., Sedgwick, B., Sekiguchi, M., and Nakabeppu, Y. (1988) Ann. Rev. Biochem. 57, 133-157.
Lindquist, S. and Craig, E.A. (1988) Ann. Rev. Genet. 22, 631-677.
Morgan, R.W., Christman, M.F., Jacobson, F.S., Storz, G., and Ames, B.N. (1986) Proc. Natl. Acad. Sci. USA 83, 8059-8063.
Pacifici, R.E. and Davies, K.J.A. (1991) Gerontology 37, 166-180.
Richter, H.E. and Loewen, P.C. (1981) Biochem. Biophys. Res. Commun. 100, 1039-1046.
Rushmore, T.H. and Pickett, C.B. (1990) J. Biol. Chem. 265, 14648-14653.
Rushmore, T.H., King, R.G., Paulson, K.E., and Pickett, C.B. (1990) Proc. Natl. Acad. Sci. USA 87, 3826-3830.
Rushmore, T.H., Morton, M.R., and Pickett, C.G. (1991) J. Biol. Chem. 266, 11632-11639.
Ryter, W.S., Pacifici, R.E., and Davies, K.J.A. (1990) In: *Biological Oxidation Systems*, Vol.2 (Reddy, C.C., Hamilton, G.A., and Madyastha, K.M., eds.) Academic Press, Inc., San Diego. Pp. 929-952.
Schreck, R., Rieber, P., Baeuerle, P.A. (1991) EMBO J. 16, 2247-2258.
Spitz, D.R., Dewey, W.C., and Li, G.C. (1987) J. Cell Physiol. 131, 364-373.
Stonczewski, J.L., Gonzalez, T.N., Bartholomew, F.M., and Holt, N.J. (1987) J. Bacteriol. 169, 3001-3006.

Storz, G, Tartaglia, L.A., and Ames, B.N. (1990) Science 248, 189-194.
Storz, G., Tartaglia, L.A., Farr, S.B., and Ames, B.N. (1990) Trends Gen. 6, 363-368.
Tao, K., Makino, K., Yonei, S., Nakata, A., Shinagawa, H. (1989) Mol. Gen. Genet. 218, 371-376.
Tartaglia, L.A., Storz, G., and Ames, B.N. (1989) J. Mol. Biol. 210, 709-719.
Touati, D. (1988) J. Bacteriol. 170, 2511-2520.
Tsaneva, I.R. and Weiss, B. (1990) J. Bacteriol. 172, 4197-4205.
Walker, G.C. (1984) Microbiol. Rev. 48, 60-93.
Walkup, L.K.B. and Kogoma, T. (1989) J. Bacteriol 171, 1476-1484.
Wu, J. and Weiss, B. (1991) J. Bacteriol. 173, 2864-2871.

Free Radicals: From Basic Science to Medicine
G. Poli, E. Albano & M. U. Dianzani (eds.)

RADIATION-INDUCED FREE RADICAL REACTIONS

Roger H Bisby[1] and Anthony W Parker[2]

[1]Department of Biological Sciences, University of Salford, Salford M5 4WT, UK; and [2]Laser Support Facility, Rutherford Appleton Laboratory, Science and Engineering Council, Chilton, Didcot, Oxon OX11 0QX, UK.

The irradiation of aqueous systems gives rise to the formation of both transient radical species ($OH^{\bullet}$, e_{aq}^{-} , $H^{\bullet}$, $O_2^{\bullet-}$ etc, Spinks and Woods, 1990) and more persistent products (Czapski et al., 1992) whose chemical reactivity is of great importance within the area of free radical chemistry and biology. Indeed, for the *chemical* study of free radicals the technique of radiolytic generation such as in pulse radiolysis is probably the method of choice, since the identity and yield of individual radical types formed are extremely well known. Oxygen radicals such as $^{\bullet}OH$ and $O_2^{\bullet-}$ have been intensively studied and both kinetic and thermodynamic aspects of their reactions have been extensively tabulated (Wardman and Ross, 1991). A further aspect of this approach is that a great many relevant organic radicals may also be produced and studied by radiolysis (von Sonntag, 1987) and this provides a basis for the understanding of more complex free radical reactions in biological systems such as DNA strand breakage, lipid peroxidation and protein damage. Indeed, now that the initial reactions of oxygen radicals with biological targets are becoming well characterised, the emphasis is turning to the study of subsequent free radical transformations such as those that allow radical migration in proteins (Prutz et al., 1982), the formation of cross-links between macromolecules such as proteins and DNA (Schuessler and Jung, 1989), the formation of long-lived reductive moieties in proteins (Simpson et al., 1992) and the mechanisms of antioxidant action (Neta et al., 1989).

CONCERTED ELECTRON AND PROTON TRANSFER REACTIONS

One point of particular interest is in the mechanism of one-electron oxidation-reduction reactions. In many cases including, for example, the reduction of deleterious free radicals by low molecular weight antioxidants, these reactions may be formally written as either a hydrogen atom transfer (equation 1) or as an electron transfer accompanied by associated protonation and/or deprotonation steps (equation 2).

$$AH + B^{\bullet} \longrightarrow A^{\bullet} + BH \qquad (1)$$

$$AH + B^{\bullet} \longrightarrow AH^{+} + B^{-} \xrightarrow{(H_2O)} A^{\bullet} + BH \qquad (2)$$

Antioxidants such as ascorbate must have the capacity to rapidly reduce organic free radicals (such as α-tocopheroxyl) but must not react with molecular oxygen (i.e. autoxidise) at an appreciable rate, otherwise they become pro-oxidant. The latter process appears to occur only appreciably in the presence of transition metal ions (Miller et al., 1990). It has recently been argued (Njus and Kelley, 1991) that this difference in chemical reactivity is due to a difference in mechanism: reaction of an antioxidant with an organic radical might occur by a facile hydrogen atom transfer whereas oxidation of the antioxidant by molecular oxygen requires an initial electron transfer step which is thermodynamically unfavourable and therefore slow.

Neta et al. (1989) have studied the oxidation of several types of organic reductants (including the water soluble α-tocopherol analogue Trolox C) by peroxyl radicals. It was found that the rates of these reactions are strongly solvent dependent and decrease by over an order of magnitude on changing from aqueous to organic solvents. In the extreme, it is suggested that this might reflect a change in mechanism from electron to hydrogen atom transfer. The ratio of rate constants for reaction of hydrogen and deuterium substituted molecules would be expected to be unity for a pure electron transfer reaction. Neta et al. (1989) have studied the reaction of the methyl peroxyl radical ($CH_3OO^{\cdot}$) with tetramethylphenylenediamine (TMPD) for which the mechanism can only be electron transfer and not hydrogen atom transfer (equation 3). Their finding of a $k(H_2O)/k(D_2O)$ ratio of 2.6 is

$$CH_3OO^{\bullet} + \mathrm{TMPD} \longrightarrow CH_3OO^{-} \xrightarrow{H^+} CH_3OOH \;+\; \mathrm{TMPD}^{\bullet+} \qquad (3)$$

offered as strong support for the reaction mechanism being an electron transfer accompanied by concerted protonation of the hydroperoxide product. Prutz et al. (1984) find a ratio of approximately 3 for the kinetic isotope effect in the reaction of tryptophanyl radicals with tyrosine in L-Trp-L-Tyr and also favour an electron transfer mechanism. Very recently Nagaoka et al. (1992) have studied the oxidation of α-tocopherol by stable phenoxyl radicals and report kinetic isotope (H/D) ratios of over 20. This is ascribed to hydrogen atom tunelling. Further evidence for the mechanism of radical reactions is obtained from values of the activation parameters. The entropy of activation for electron transfer reactions with concerted protonation is found to be a negative quantity reflecting the ordering of water molecule(s) in the transition state (Neta et al., 1989; Jovanovic, 1992). In contrast, hydrogen atom transfer reactions have a positive entropy of activation and are expected to be important in cases were the bond strength of the hydrogen atom donor is low (Jovanovic, 1992). According to Neta et al. (1989) the transition state for a proton-transfer-mediated electron transfer reaction from a donor (D) is as depicted below.

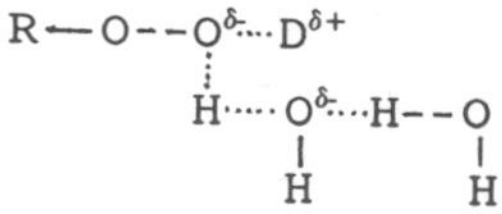

$$R{-}O{-}{-}O^{\delta-}\cdots D^{\delta+} \quad \vdots \quad H\cdots O^{\delta-}\cdots H{-}{-}O \quad (H)\quad (H) \tag{4}$$

THE REGENERATION OF α-TOCOPHEROL BY ASCORBATE IN BILAYER SYSTEMS

In contrast to the experimental findings described above for peroxyl radical reactions, Njus and Kelley (1991) suggest on theoretical grounds that the reaction of ascorbate with α-tocopherol (equation 5) is most likely to involve a hydrogen

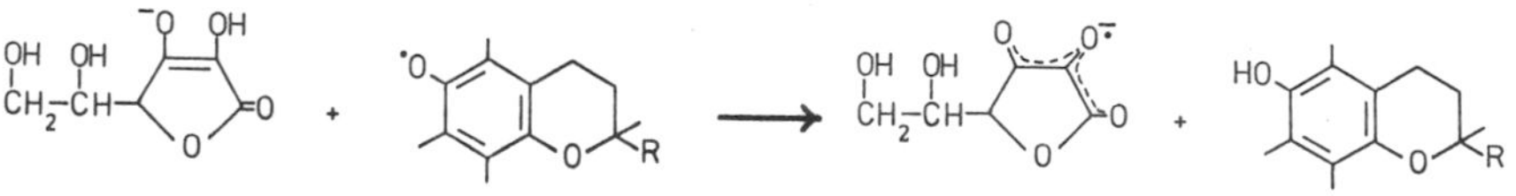

(5)

atom transfer. The rate of reaction (5) in water (for Trolox C) is 8.3×10^{6} dm^{3} mol^{-1} s^{-1} (Davies et al., 1988) compared with a value of 1.55×10^{6} dm^{3} mol^{-1} s^{-1} in a 50% isopropanol/40% water/10% acetone mixture (Packer et al., 1979) and 2×10^{5} dm^{3} mol^{-1} s^{-1} in lecithin bilayers inferred from the results of Scarpa et al. (1984). The Table contains some recent results obtained using nanosecond flash photolysis employing a XeCl excimer laser producing 308nm light to produce the α-tocopheroxyl radical by photoionisation (Bisby and Parker 1991) within micellar or membrane systems. The rate constant obtained in DMPC bilayers confirms the result of Scarpa et al. (1984) and shows that as in the case of an α-tocopherol derivative (Mukai et al. 1989,1991) reaction (5) is not significantly impeded by the intervening micellar or membrane interface and must have a low activation energy as expected for a fast reaction. As in those reactions studied by Neta et al. (1989), the rate constant for reaction (5) decreases with

TABLE Kinetic and thermodynamic parameters for reaction of ascorbate with the α-tocopheroxyl radical in various micellar and bilayer systems.

System	$\Delta H^{\ddagger}$ (kJ mol^{-1})	$\Delta S^{\ddagger}$ (J K^{-1} mol^{-1})	$k(H_20)/k(D_20)$	k_2* (dm^3 mol^{-1} s^{-1})
SDS micelles	16.4 ± 0.5	-103.4 ± 0.5	7.97 ± 0.17	$(4.97 \pm 0.08) \times 10^4$ [25°C]
DMPC bilayers	26.3 ± 1.3	-57.9 ± 4.1	3.11 ± 0.16	$(3.05 \pm 0.14) \times 10^5$ [35°C]
DMPC + DDAB bilayers (25 : 75 mole %)	14.8 ± 0.5	-49.8 ± 1.6	3.44 ± 0.25	$(3.53 \pm 0.21) \times 10^7$ [35°C]
DMPC + DPPA bilayers (80 : 20 mole %)	- - -	- - -	3.47 ± 0.34	$(1.77 \pm 0.16) \times 10^5$ [45°C]

* in 0.1 mol dm^{-3} sodium phosphate buffer, pH 7.0 ± 0.5

$\Delta S^{\ddagger}$ and $\Delta H^{\ddagger}$ values calculated at 298K

SDS = sodium dodecyl sulphate

DMPC = dimyristoylphosphatidylcholine

DPPA = dipalmitoylphosphatidic acid

DDAB = diodecyldimethylammonium bromide.

decreasing solvent polarity, the ratios of rate constants in D_2O and H_2O are substantially greater than unity, and the activation entropies are quite negative. In the membrane bilayers the rate constants show the anticipated response to change in surface charge of the bilayer, but the deuterium isotope effect (approximately 3.3 ± 0.3) is unchanged by altering the membrane charge. Overall the results appear to indicate that the mechanism of (5) under these conditions is electron transfer accompanied in this case by two concerted proton transfers - deprotonation of the product ascorbate radical and protonation of the α-tocopherolate anion. With α-tocopherol in SDS micelles the rate constant is lower, the kinetic isotope effect is much larger (about 8, pointing towards a hydrogen transfer reaction) but the activation entropy is extremely negative. The latter may reflect the decrease in entropy needed to bring the negatively charged ascorbate anion to the surface of the negatively charged SDS micelle. It is anticipated that the transition state for an electron transfer to the α-tocopheroxyl radical in the negatively charged SDS micelle would be extremely unfavourable. Therefore in this special case a slower hydrogen atom transfer may be the dominant mechanism.

CONCLUSIONS

In bilayer membranes the evidence points to a mechanism for the reaction of the α-tocopheroxyl radical with ascorbate involving concerted electron and proton transfers, equivalent to an overall hydrogen atom transfer. Despite the fact that the α-tocopheroxyl radical is located within the bilayer throughout the reaction, due to its extremely high hydrophobicity (Castle and Perkins, 1986), the involvement of water molecules is possible because it is expected that the radical would be oriented with the oxyl radical site at, or very near, the aqueous interface.

Acknowledgement

This work was supported by the Science and Engineering Research Council through the Laser Support Facility, Rutherford Appleton Laboratory, Chilton, Didcot, Oxon OX11 0QX, UK.

REFERENCES

Bisby, R.H. and Parker, A.W. (1991) FEBS Letters 290, 205-208.
Castle, L. and Perkins, M.J. (1986) JACS 108, 6381-6382.
Czapski,G., Goldstein,S., Andorn,N. and Aronovitch,J. (1992) Free Rad. Biol. Med. 12, 353-364.
Davies, M.J., Forni, L.G. and Willson,R.L. (1988) Biochem. J. 255, 513-522.
Jovanovic, S.V. (1992) in Oxidative Damage and Repair (Ed. K.J.A.Davies) Pergamon Press, Oxford, pp. 93-97.
Miller, D.M., Buettner,G.R. and Aust,S.D. (1990) Free Rad. Biol. Med. 8, 95-108.
Mukai,K., Nishimura,M., Ishizu,K. and Kitamura,Y. (1989) Biochim. Biophys. Acta 991, 276-279.
Mukai,K., Nishimura,M. and Kikuchi,S. (1991) J. Biol. Chem. 266, 274-278.
Nagaoka,S., Kuranaka,A., Tsuboi,H., Nagashima,U. and Mukai,K. (1992) J. Phys. Chem. 96, 2754-2761.
Neta, P., Huie, R.E., Maruthamuthu, P., and Steenken, S. (1989) J. Phys. Chem. 93, 7654-7659.
Njus, D. and Kelly, P.M. (1991) FEBS Letters 284, 147-151.
Packer, J.E., Slater, T.F. and Willson, R.L. (1979) Nature 278, 737-738.
Prutz,W.A., Siebert,F., Butler,J., Land,E.J., Menez, A. and Garestier, T.M. (1982) Biochim. Biophys. Acta 705, 139-149.
Scarpa, M., Rigo,A., Maiorino,M., Ursini,F. and Gregolin,C. (1984) Biochim. Biophys. Acta 801, 215-219.
Schuessler, H. and Jung, E. (1989) Int. J. Radiat. Biol. 56, 423-435.
Simpson,J.A., Narita,S., Gieseg,S., Gebicki,S., Gebicki,J.M. and Dean, R.T. (1992) Biochem. J. 282, 621-624.
Spinks and Woods "An Introduction to Radiation Chemistry" 3rd Edn., Wiley, New York, 1990.
Von Sonntag, C. "The Chemical Basis of Radiation Biology" Taylor and Francis, London, 1987.
Wardman, P. and Ross,A.B. (1991) Free Rad. Biol. Med. 10, 243-247.

Free Radicals: From Basic Science to Medicine
G. Poli, E. Albano & M. U. Dianzani (eds.)

NITRIC OXIDE AND RELATED RADICALS

H. Nohl

Institute of Pharmacology and Toxicology, Veterinary University of Vienna, Linke Bahngasse 11, A-1030, Vienna, Austria.

SUMMARY: The endothelium derived relaxing factor (EDRF), that was recognized to have vasodilatatory effects in vascular smooth muscle, prevents platelet aggregation and adhesion of platelets as well as polymorphonuclear leukocytes to endothelium. EDRF is most likely synonymous with nitric oxide (NO). This simple chemical compound exerts most of its vital physiological functions by stimulation of guanylate-cyclase. NO was also found to play a significant role in various physiological functions and pathophysiological disorders. The basis for these activities may be the deleterious decrease of the various physiological functions by rapid decomposition of NO, especially in the presence of superoxide radicals ($O_2^{\cdot-}$). In $O_2^{\cdot-}$ generating tissues, NO rapidly converts to peroxynitrite ($ONOO^-$), a compound with properties similar to hydroxyl radicals. Macrophages generating both NO and $O_2^{\cdot-}$ during their activities in inflammation were proposed to contribute to tissue injury via the formation of $ONOO^-$. Endothelium cells appear also capable of generating NO and $O_2^{\cdot-}$ radicals simultaneously. The subsequent formation of an $ONOO^-$ may cause oxidative degradation of low density lipoproteins. The protection of the oxidant-induced damage by hydroxyl radical scavengers revealed also the possibility of $OH^\cdot$ formation via homolitic cleavage of $ONOO^-$. It is becoming more clear that SOD exerts important functions in oxidant/antioxidant homeostasis of the tissue. SOD not only prevents conversion of NO to $ONOO^-$ by removing $O_2^{\cdot-}$ from the reaction,it also seems to be involved in converting NO^- from an intermediate form of EDRF to NO. These functions appear to be important in the control of ischemia/reperfusion damage as demonstrated by the protecting effect of SOD in gastric mucosa alterations. On the other hand raising tissue levels of NO above normal (which was shown to occur following application of butylated hydroxytoluene) may cause increased gene expression (including oncogenes) via disturbance of cellular c-GMP levels. Another example of the genotoxic effect of exogenous nitrocompounds is the DNA-damaging activity of a biotransformation product of 2-nitropropane.
The present chapter reports on the existence and tissue regulation of NO and NO-related compounds and their involvement in various physiological disorders.

There is increasing evidence that the endothelium-derived-relaxing factor (EDRF) is synonymous with nitric oxide (NO) or closely related to an NO-releasing compound (Palmer et al., 1987; Ignarro, et al., 1987; Vanin, 1991). Apart from endothelium cells, NO was also found to be released from macrophages (Marletta et al., 1988) and brain tissue (Drummond, 1983). It is unequivocally accepted that EDRF is synthetized from L-arginine by an NO-synthase. Several isoforms of nitric oxide synthase exist (Förstermann et al., 1991), differing in their tissue localization and the requirement of cofactors. The endothelium-related NO-synthase requires Ca^{++}-calmodulin and NADPH (Förstermann et al., 1991). Acetylcholine, ATP, Ca^{++}-ionophores and bradykinine initiate a receptor-mediated influx of Ca^{++} thereby triggering the production and extracellular release of NO. Nitroxide radicals bind to the heme moiety of guanylate cyclase which cause increasing c-GMP levels in vascular smooth muscle and platelets. Increased c-GMP promotes relaxation in vascular smooth muscle and inhibits platelet aggregation as well as adhesion of platelets and polimorphonuclear leucocytes to endothelium. The half life of NO is determined by a great variety of concurrent pathways all competing with NO-binding to guanylate-cyclase. The mean half life of NO *in vitro* ranges from 4-50 sec depending on the availability of respective reaction partners. Apart from reactions with other hemoproteins such as hemoglobin, NO may also react with dioxygen or more efficiently with superoxide ($O_2^{\bar{\cdot}}$). The latter reaction is prevented by SOD, suggesting an unexpected physiological role of this enzyme in modulating vascular blood flow.

Hereafter are reviewed the most recent aspects of NO-related physiopathology as discussed at the VI Biennal Meeting of International Society for Free Radical Research (Torino, June 16-20, 1992).

M. Murphy (Department of Pharmacology, University of Vermont) presented a cascade of reaction pathways following interactions between nitric oxide and $O_2^{\bar{\cdot}}$ radicals both of which are supposed to be of significance as multifunctional physiological signals.

Unexpectedly SOD appeared to have a catalytic role other than dismutation of $O_2^{\cdot-}$. Cupric copper of SOD was suggested to act as an e^- acceptor from the nitroxyl anion (NO^-) thereby reversibly converting NO^- to NO.

$$1. \quad NO^- + SOD(Cu^{++}) \rightleftharpoons NO + SOD(Cu^+)$$

The equilibrium of reaction 1. is thermodynamically in favour of NO. This novel activity of SOD may affect tissue concentration of NO in a variety of way. For example, NO-synthesizing enzymes might actually convert L-arginine to NO^-, while SOD supports the final one e^- oxidation to NO, eliminating the need for such a step in the reaction sequence of NO-synthase. Thus, it is conceivable that SOD may control the levels of NO and/or EDRF by a mechanism in addition to $O_2^{\cdot-}$ scavenging. Other reaction of NO and NO^- were reported to proceed with heme proteins such as catalase and cytochrome c and by analogy might also play a role in the regulation of guanylate cyclase activities. The author presented evidence that SOD and NO affect the oxidation state of catalase according to reaction 2-6.

$$2. \quad Cat + H_2O_2 \longrightarrow Cat\text{-}Compound\ I$$

$$3. \quad Cat\text{-}Compound\ I + NO \longrightarrow Cat\text{-}Compound\ II$$

$$4. \quad NO + SOD(Cu^+) \rightleftharpoons NO^- + SOD(Cu^{++})$$

$$5. \quad NO^- + Cat\text{-}Compound\ II \longrightarrow NO + Cat$$

$$6. \quad NO + Cat \rightleftharpoons Cat\text{-}NO$$

Since in some of the other heme proteins, compound I (Fe^{5+}) may drive a one e^- oxidation of protein damage (Davies et al., 1990; Nohl et al., 1992), NO might prevent this damage by converting compound I to compound II as in reaction 3. Evidence has been found that NO/NO^- undergoes a reversible one e^- exchange with cytochrome c (reaction 7) (Doyle et al., 1988).

$$7. \quad NO^- + cyt\ c\ (Fe^{3+}) \rightleftharpoons NO + cyt\ c\ (Fe^{2+})$$

The addition of SOD to such a system would provide an alternate pathway favouring NO^- to NO conversion and shift the equilibrium of reaction 7 towards the left (i.e. the oxidation of cyt c (Fe^{2+}) by NO). In the presence of both NO and reductants (other than superoxide), this SOD effect might be indistinguishable from the SOD inhibitable cyt c-reduction of $O_2^{\bar{\cdot}}$.
Guanylate cyclase is activated upon binding of NO to the ferrous iron. NO^- could combine with the otherwise unresponsive ferric state to yield the active complex.
Furthermore Beckman and coworkers reported on a catalytic role of SOD in the heterolytic cleavage of the otherwise toxic peroxinitrite ($ONOO^-$) normally formed from NO in $O_2^{\bar{\cdot}}$ generating tissue (Beckman et al., 1989).
The question remained open as to what extent these many pathways of NO decomposition will proceed *in vivo*. Rate constants and steady state concentrations are required to evaluate the existence and the significance of the various suggested pathways in the tissue.
J. Beckman et al. (Dept. of Anestesiology Physics, and the Center of Macromol. Crystallogr., Univ. Alabama, Birmingham, USA) reported on the formation of $ONOO^-$ by activated alveolar macrophages. The authors used phorbol 12-myristate 13 acetate to activate $O_2^{\bar{\cdot}}$ and NO generation in macrophages. Convertion of NO to peroxynitrite ($ONOO^-$) was quantitatively followed by SOD-catalyzed nitration of 3-nitro-4-hydroxyphenylacetic acid. The generation of this dye required the presence of intact SOD; Cu-depleted Zn-SOD was ineffective. Peroxynitrite decomposition generates a strong oxidant with the properties similar to hydroxyl radical. The strong oxidizing property of peroxynitrite does not implicate $OH^{\cdot}$ radicals from homolytic cleavage of $ONOO^-$. The metal chelator diethyltriaminepentaacetic acid had no effect on the formation of this strong oxidizing species while desferrioxamine was found to be a potent competitive inhibitor because of a direct reaction between desferrioxamine and $ONOO^-$ rather than by iron chelation.

The generation of the secondary cytotoxic species via a reaction of NO with $O_2^{\cdot-}$ following the release of this radicals from activated macrophages was proposed to contribute to inflammatory cells-mediated tissue injury. SOD prevents the formation of $ONOO^-$ by removing $O_2^{\cdot-}$ from a reaction with NO. Consequently the existence of the powerful oxidant resulting from the decomposition of $ONOO^-$ is also prevented. This highlights a further beneficial role of SOD where this enzyme exerts a protection of the tissue against a cytotoxic compound which is suggested to be involved in tissue damage during inflammation.

The involvement of NO-derived prooxidant on the pathogenesis of atherosclerosis was investigated by V.N. Darley-Usmar et al. from Wellcome Research Laboratories Beckenham, Kent, UK. Starting from the concept that the oxidative modification of low density lipoprotein (LDL) trapped within the intima of the artery wall is a key event in the formation of early atherosclerotic lesions, the sensitivity of LDL towards NO in the presence of $O_2^{\cdot-}$ was studied. Oxidation of LDL is thought to occur first in the lipid phase of the lipoprotein and to subsequently cause modification of the apo-B-protein. This modification is established in the later stages of the reaction when the endogenous antioxidants, like β-carotene, α-tocopherol and ubiquinol, are depleted. Modified apo-B-protein is recognized by macrophages forming foam cells which ultimately became deposited in the vascular wall. Macrophages also generate superoxide anions and the rate of generations is greatly enhanced upon activation. On the other hand, NO is generated by the vascular endothelium (and macrophages). The authors assumed, therefore, that the generation of NO and $O_2^{\cdot-}$ in/or around the artery wall leads to the production of $ONOO^-$ which, as J. Beckman and coworkers have shown, gives rise to the existence of a powerful oxidant in the close proximity of LDL.

Darley-Usmar et al. have tested part of the hypothesis by exposing human LDL to the simultaneous generation of $O_2^{\cdot-}$ and NO, formed during the oxidation of the vasodilatatory drug sydnonimine (SIN-1). This molecule was shown by the same authors to form an oxidant that will degradate deoxyribose to malondialdehyde and can be scavenged by OH· scavengers such as ethanol. Upon incubation of LDL with SIN-1 the liprotein undergoes a lipid peroxidation reaction as evidenced by an increase in lipid peroxide content, depletion of α-tocopherol and an increase in the negative charge of the lipoprotein. The peroxidation reaction and protein modification was found to be inhibited by both the peroxyl radical scavenger butylated hydroxy-toluene and SOD, but not by catalase. The addition of exogenous peroxide to the LDL particle did not alter the rate of SIN-1-dependent LDL oxidation. From this observation it was concluded that peroxidation is initiated from hydrogen atom extraction rather then the breakdown of existing lipid hydroperoxides. In summary, the authors concluded that LDL is oxidized as result of the simultaneous generation of NO and $O_2^{\cdot-}$ in a SOD sensitive but catalase insensitive mechanism caracteristic of the reactions mediated by peroxynitrite.

Another regulatory function of SOD was presented by Y. Naito et al. (First Department of Medicine, Kyoto Prefectural Univ. of Medicine, Kyoto, Japan). The authors reported on the synergistic action of L-arginine and SOD in protecting the gastric mucosa from ischemia/reperfusion injury. The idea behind this investigation was the assumption that the EDRF-mediated microvascular blood supply of the gastric mucosa is affected by ischemia/reperfusion causing gastric erosions and dysfunction. L-Arginine was assumed to increase steady state concentration of NO via activation of its synthesis, while SOD should prolong the activity of NO by preventing its conversion to $ONOO^-$. The hypothesis was first suggested by the inhibition of NO biosynthesis using N^G-nitro-L-arginine, a competitive inhibitor of NO-synthase. The use of this compound revealed that endogenous NO in fact plays a significant role in the manteinance of gastric mucosal integrity. Stimulation of NO synthesis upon the application of L-arginine was not very efficient in preventing ischemia/reperfusion-induced injuries of the gastric mucosa. SOD alone exibited stronger protecting effects. Protection was however still more pronounced

when the aminoacid was applied in combination with SOD. In that case all parameters indicating structural and functional damage of the mucosa were significantly reduced. In the presence of SOD the amount of serum nitrite and nitrate was above normal. The authors interpretated this observation in terms of a SOD-dependent inhibition of NO decomposition to $ONOO^-$ by removing $O_2^{\cdot-}$ from the reaction. The higher steady state concentration of NO may help to maintain sufficient blood supply and simultaneously prevent thrombosis. The synergistic effect of SOD and L-arginine may result from an additional increase of NO-serum levels via activation of NO biosynthesis (which was however not proved experimentally). These interpretations implicate the existence of increasing amounts of superoxide radicals in the endothelium cell during reperfusion and assume that the control of $O_2^{\cdot-}$ by tissue SOD is more important in regulating NO levels than stimulation of NO *de novo* synthesis by the availability of the precursor molecule. Another aspect is the direct implication of $ONOO^-$ in the development of gastric mucosa injury. Prevention of the formation of this cytotoxic compound by SOD together with the mantainance of regular vascular blood supply via NO may establish two independent mechanisms to protect the gastric mucosa from ischemia/reperfusion induced injuries.
V. Koltover (Institute of Chemical Physics of the Russian Academy of Sciences, Moscow district) reported parenteral administration of hydroxylamine to mice gives rise to the generation of abnormal levels of NO radicals. The detection of the normally unstable NO species was possible upon the formation of a nitrosyl complex with heme-iron in the presence of thioether donors. Using diethyldithiocarbamate as a source for thioether, the caracteristic ESR-triplet-spectrum of the nitrosyl-heme-complex could be observed in liver homogenates following administration of hydroxylamine. Observation of other laboratories upon a positive correlation between the amount of nitrate-uptake with food and the incidence of cancer made Koltover to study the effect of high NO-levels on the transcription rates of liver cells. Hydroxylamine-induced NO

generation caused a significant transcription increase measured via incorporation rate of tritium labelled uridine into cellular RNA. Comparison of steady state levels of liver NO-radicals with transcription rate revealed a clear positive correlation. From this observation the author concluded that NO-radicals may be implicated in carcinogenesis. As a possible mechanism he suggested an abnormal NO-induced activation of guanylate cyclase with a subsequent disturbance of c-GMP which in turn affects gene expression including oncogenes. These assumptions were discussed in relation to carcinogenesis and aging. The observation of a reciprocal correlation between guanylate cyclase activities in various animal tissue and life-span (Cutler, 1984) was taken as indirect evidence supporting the validity of Koltover's interpretation.

M. Saran et al. (GSF-Institut fur Strahlenbiologie Neuherberg, Munich) investigated DNA damaging mechanism of another nitrocompound 2-nitropropane. The compound is known to be hepatotoxic and induces liver carcinoma via alteration of liver DNA. The DNA damaging effect requires the metabolic alteration of 2-nitropropane by the hepatic cytochrome P450 system. The genotoxic metabolite is however unknown. One candidate is the 2-nitropropane radical. The formation of this radical from the nitronate anion can be assumed to initiate radical chain reactions with the implication of superoxide and peroxyl-radicals. The authors used kinetic spectroscopy in combination with competition studies to estimate the possible reactivity of the 2-nitropropane radicals with selected nucleosides and nucleotides. The radical formation was initiated either in a pulse radiolysis system or with horseradish peroxidase plus H_2O_2, and its existence was followed with ESR-spin trapping-technique and optical spectroscopy. The failure to detect any 2-nitropropane radicals with this method was explained by the low concentration of this radical due to its extreme instability. However, it remained opened whether or not the radical was formed in the physical and biochemical system. Instead, the authors observed transient absorbtion spectra which were identified as nitrogendioxy-radicals (NO_2). Kinetics of the latter compound in the absence and presence of deoxyguanosine and deoxyguanosine monophosphate showed changes which strongly supported a reactivity of the NO_2 radical with DNA. Additional evidence came from alteration of the rate constants for this type of reaction in the presence of nitroguaiacol and kaempferol, both of which are known to react with NO_2. NO_2 were reported to be one split product of biotransformation of 2-nitropropane besides acetone. The authors, therefore, concluded that the NO_2 radical is responsible for genotoxic effects of 2-nitropropane.

REFERENCES.

Beckman, J.S., Beckman, T.W., Chen, J., Marshall, I.A., and Freeman, B.A. (1989) Proc. Natl. Acad. Sci. USA 87, 1620-1624.
Cutler, R.G. (1984) In: Free Radicals in Biology (Pryor, W.A., ed.) Academic Press, pp. 371-437.
Davies, M.J. (1990) Free Rad. Res. Comm. 10, 361-370.
Doyle, M.P., Mahapatro, S.M., Broene, R.D., and Guy, J.K. (1988) J. Am. Chem. Soc. 110, 593-599.
Drummond, G.I. (1983) In: Advances in Cyclic Nucleotide Research (Greengard, P., and Robinson, G.A., eds.) Raven Press, New York, pp. 373-494.
Forstermann, U., Schmidt, H.W., Pollock, J.S., Sheng, H., Mitchell, J.A., Warner, T.D., Nakane, M., and Murad, F. (1991) Biochem. Pharmacol. 42, 1849-1857.
Forstermann, U., Pollock, J.S., Schmidt, H.W., Heller, M., and Murad, F. (1991) Proc. Natl. Acad. Sci. USA 88, 1788-1792.
Ignarro, L.J., Buga, G.M., Wood, K.S., Byrns, R.R., and Chaudhuri, G. (1987) Proc. Natl. Acad. Sci. USA 84, 9265-9269.
Marletta, M.A., Yoon, P.S., Iyengard, R., Leaf, C.D., and Wishnok, J.S. (1988) Biochemistry 27, 8706-8711.
Nohl, H., and Stolze, K. (1992) Free Rad. Biol. Res. Comm., submitted.
Palmer, R.M.J., Ferrige, A.J., and Moncada, S. (1987) Nature 327, 524-526.
Vanin, A.F. (1991) F.E.B.S. Lett. 289, 1-3.

Free Radicals: From Basic Science to Medicine
G. Poli, E. Albano & M. U. Dianzani (eds.)

MECHANISMS OF OXIDATIVE CELL DAMAGE

S. Orrenius

Department of Toxicology, Karolinska Institutet, Box 60400,
S-104 01 Stockholm, Sweden

SUMMARY: Exposure of mammalian cells to oxidative stress induced by redox-active quinones and other prooxidants results in glutathione and NAD(P)H oxidation, followed by the modification of protein thiols, ATP depletion and the loss of cell viability. Protein thiol modification is normally associated with the impairment of various cell functions, including inhibition of agonist-stimulated phosphoinositide metabolism, disruption of intracellular Ca^{2+} homeostasis, and perturbation of cytoskeletal organization. The latter effect appears to be responsible for the formation of the numerous plasma membrane blebs, typically seen in cells exposed to cytotoxic concentrations of prooxidants. Following the disruption of thiol homeostasis in prooxidant-treated cells, there is a perturbation of intracellular Ca^{2+} homeostasis with a sustained increase in intracellular Ca^{2+} concentration. This Ca^{2+} overload can cause activation of various Ca^{2+}-dependent degradative enzymes (phospholipases, proteases, endonucleases) and may contribute to the mitochondrial damage seen in oxidative stress. Severe oxidative stress is also associated with extensive DNA damage which, in turn, may lead to excessive stimulation of poly(ADP-ribose)polymerase activity and subsequent NAD^+ and ATP depletion which may contribute to cell killing. In contrast with the cytotoxic effects of severe oxidative stress, low levels of oxidative stress can lead to the activation of enzymes involved in cell signaling. In particular, the activity of protein kinase C is markedly increased by redox-cycling quinones through a thiol/disulfide exchange mechanism, and this may represent a mechanism by which prooxidants can modulate cell growth and differentiation.

It is now well established that oxidative cell injury generated by chemicals, during reoxygenation of hypoxic tissue, or as a result of acute or chronic inflammatory processes is associated with multiple alterations of cell structure and function. Among these perturbations are oxidation of intracellular thiols and pyridine nucleotides, impairment of signal transduction and ion homeostasis, modification of cytoskeletal organization, inhibition of glycolysis, infliction of DNA damage and activation of poly(ADP-ribose)polymerase, NAD^+ depletion, collapse of the mitochondrial membrane potential, ATP depletion, etc. Although the relative contribution of these alterations to the development of cell injury during oxidative stress is not known, several of them seem to be toxicologically relevant. It also appears that alternate toxic mechanisms may be recruited under different pathophysiological conditions and that perturbations of different cell functions may be responsible for oxidative cell killing in different experimental systems.

Early studies in our laboratory focussed on the role of glutathione (GSH) depletion and protein thiol modification in the development of oxidative cell damage and emphasized the importance of the cytoskeleton as a target in oxidative stress. More recently, we have devoted particular attention to prooxidant-induced alterations in intracellular signal transduction and Ca^{2+} homeostasis and their contribution to the development of acute and delayed cytotoxicity. It is the purpose of this overview to recapitulate some of our findings and discuss their significance for the development of oxidative cell and tissue damage.

ROLE OF GLUTATHIONE AND PROTEIN THIOL MODIFICATION DURING OXIDATIVE STRESS

Glutathione plays a unique role in the cellular defense against active oxygen species and reactive intermediates. GSH functions both as a reductant in the metabolism of hydrogen peroxide and organic hydroperoxides and as a nucleophile which can conjugate electrophilic molecules (Cotgreave et al., 1988). During glutathione peroxidase-catalyzed metabolism of hydroperoxides, GSH serves as an electron donor, and the glutathione disulfide (GSSG) formed in the reaction is subsequently reduced back to GSH by glutathione reductase, at the expense of NADPH. Under conditions of oxidative stress, when the cell must cope with large amounts of H_2O_2 or organic hydroperoxides, the rate of glutathione oxidation exceeds the slower rate of GSSG reduction by glutathione reductase, and GSSG accumulates. To avoid the detrimental effects of increased intracellular levels of GSSG (e.g. formation of mixed disulfides with protein thiols), the cell excretes GSSG, which can lead to depletion of the intracellular glutathione pool (Fig. 1).

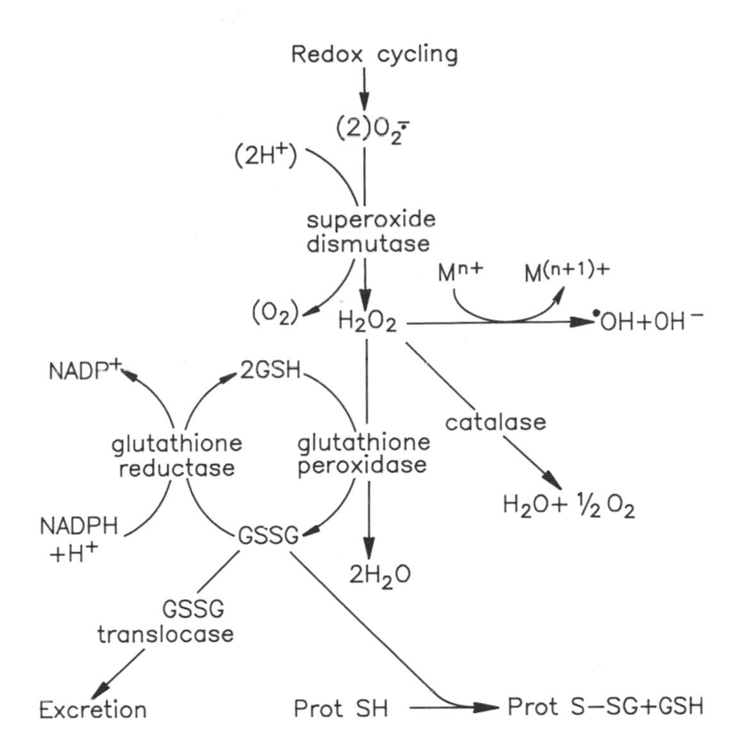

Fig. 1. Formation and metabolism of active oxygen species. M, metal; Prot SH, protein thiol.

Although it is now well established that cell killing caused by conditions of oxidative stress is preceded by depletion of intracellular GSH (Thor et al., 1982), the exact relationship between GSH depletion and cell death has not yet been clarified. Sulfhydryl groups are, in general, highly reactive and cellular protein thiols may represent critical targets in oxidative stress. Indeed, the generation of oxidative stress during the metabolism of menadione (2-methyl-1,4-naphthoquinone) in isolated rat hepatocytes results in the loss of protein thiols (Di Monte et al., 1984). This loss of protein thiols follows glutathione depletion and precedes the onset of cell death. Moreover, pretreatment of hepatocytes with agents which deplete intracellular GSH potentiate menadione-induced protein thiol modification and cytotoxicity.

ALTERATIONS IN INTRACELLULAR Ca^{2+} HOMEOSTASIS DURING OXIDATIVE STRESS

Studies using permeant indicators that are selective for calcium ions have shown that the free Ca^{2+} concentration in the cytosol is maintained between 0.1 μM and 0.2 μM (Carafoli, 1989). Thus, there is a concentration difference of approximately four orders of magnitude between the extracellular Ca^{2+} level (~ 1.3 mM) and the cytosolic free Ca^{2+} concentration, resulting in a large electrochemical driving force in favor of net Ca^{2+} accumulation by the cells. This tendency to take up Ca^{2+} is balanced primarily by active Ca^{2+} extrusion systems located at the plasma membrane and by the coordinated activities of Ca^{2+} sequestering systems located in the endoplasmic reticular, mitochondrial and nuclear membranes (Fig. 2).

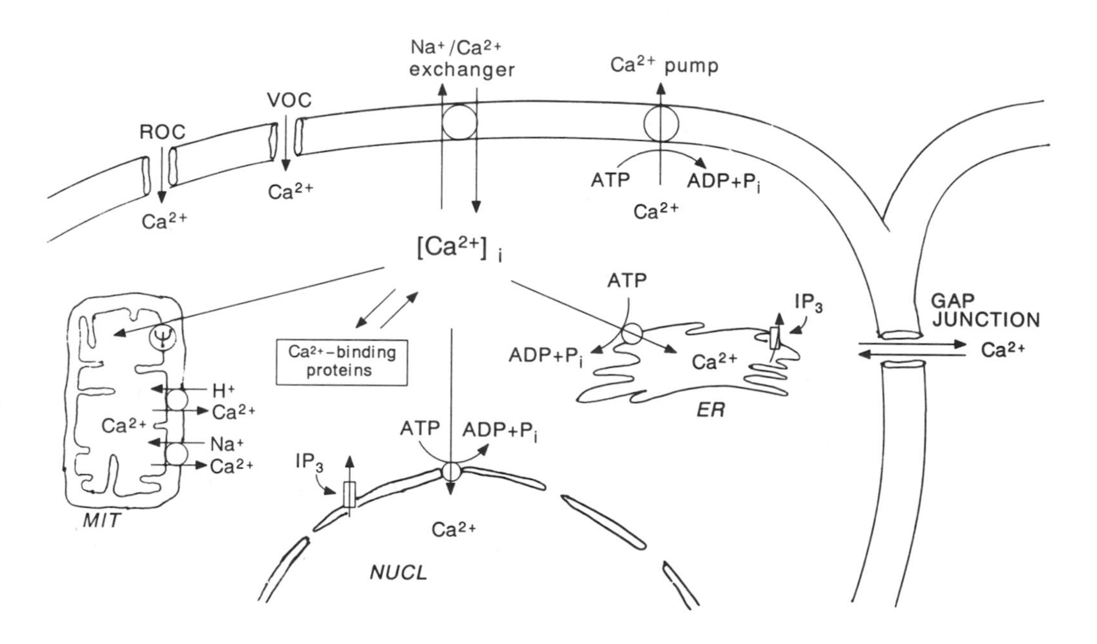

Fig. 2. Intracellular Ca^{2+} transport processes. ROC, receptor-operated channel; VOC, voltage-operated channel; IP_3, inositol 1,4,5-trisphosphate; ER, endoplasmic reticulum; MIT, mitochondrion; NUCL, nucleus.

With the knowledge that Ca^{2+} can operate as an intracellular signal for various hormones and growth factors, it has also become clear that disruption of intracellular Ca^{2+} homeostasis and subsequent Ca^{2+} overload may result in alteration of cell function and ultimately in cell death. Cellular Ca^{2+} overload can be the result of either an enhanced influx of extracellular Ca^{2+} or an impairment of Ca^{2+} extrusion from the cell. In addition, interference with individual Ca^{2+} translocases can compromise the ability of the cell to buffer cytosolic Ca^{2+} changes and thereby contribute to intracellular Ca^{2+} overload.

Ca^{2+} Sequestration by the Endoplasmic Reticulum and its Role in Oxidant Injury

Over a decade ago, Moore and coworkers showed that Ca^{2+} sequestration by liver microsomes isolated from carbon tetrachloride-intoxicated rats was substantially inhibited (Moore et al., 1976). Since then, a number of chemical toxins, including several prooxidants, have been found to impair Ca^{2+} sequestration by isolated microsomal fractions (Jones et al., 1983; Thor et al., 1985). The thiol-reducing agents, dithiothreitol and GSH, protected against the inhibition by the oxidants. It therefore appeared that the oxidation of essential sulfhydryl groups(s) critical for Ca^{2+}-ATPase activity may be involved in the mechanism of oxidative inactivation of the ER Ca^{2+} translocase. However, recent work also suggests that the inhibitory effect on Ca^{2+} sequestration observed with low concentrations of oxidants may be due to the stimulation of a specific release pathway (Missiaen, et al., 1991).

The findings of Moore and coworkers also suggest that an impairment of Ca^{2+} sequestration may be the mechanism by which chemical toxins can cause liver cell death. However, recent work in our laboratory has shown that the selective inhibitor of the microsomal Ca^{2+}-ATPase, 2,5-di-(tert-butyl)-1,4-benzohydroquinone (tBuBHQ) (Kass et al., 1989a), rapidly releases ER Ca^{2+} without correspondingly producing hepatotoxicity in isolated hepatocytes or in the isolated, perfused rat liver (Farrell et al., 1990). Hence, at least short term interference with Ca^{2+} sequestration by the endoplasmic reticulum does not appear to play a major role in the development of acute hepatotoxicity.

Ca^{2+} Sequestration by Mitochondria and Its Role in Oxidant Injury

Mitochondria contain little Ca^{2+} under physiological conditions, although they have the capacity to sequester large quantities of Ca^{2+} and could therefore act as efficient buffers of cytosolic Ca^{2+} under toxic conditions. However, this potentially important line of defense may not be operational under conditions of oxidative stress because many oxidants have been found to stimulate the rapid release of Ca^{2+} from isolated liver mitochondria (Richter and Kass, 1991). The mechanism by which oxidants cause mitochondrial Ca^{2+} efflux has been a matter of intense debate over the past fifteen years. Several groups have suggested that oxidants, in the presence of Ca^{2+}, stimulate the reversible opening of a pore (Crompton et al., 1987) that show many of the characteristics of the adenine nucleotide carrier of the mitochondrial inner membrane (Halestrap and Davidson, 1990). Other investigators have implicated a non-selective damage to the inner membrane, which would result in the loss of the transmembrane potential and reversal of the uniport Ca^{2+} uptake route, in the mechanism of oxidant-induced Ca^{2+} release (Beatrice, et al., 1980). In apparent contrast, work from Richter's and our laboratories has demonstrated that (i) the initial phase of Ca^{2+} release occurs from intact mitochondria under conditions of high transmembrane potential, and the observed loss of transmembrane potential is the result of the continuous re-uptake of Ca^{2+} by the uniporter, and that (ii) Ca^{2+} release following exposure to oxidants seems to be regulated by mitochondrial pyridine nucleotides. Addition of oxidants such as menadione, _tert_-butyl hydroperoxide or 3,5-dimethyl-N-acetyl-_p_-benzoquinone imine (3,5-Me_2-NAPQI) results in the rapid oxidation of NADH and NADPH followed by their hydrolysis to nicotinamide and ADP-ribose. Pyridine nucleotide oxidation is necessary, although not sufficient, to cause Ca^{2+} release and requires further hydrolysis to nicotinamide and ADP-ribose. Richter and co-workers have postulated that the oxidant-sensitive Ca^{2+} release mechanism involves the mono-ADP-ribosylation of a target protein (possibly the Ca^{2+}/H^{+} antiporter), regulating Ca^{2+} efflux (Richter and Kass, 1991). Evidence in support of this mechanism is the prevention of oxidant-induced Ca^{2+} release by cyclosporin A, an inhibitor of mitochondrial pyridine nucleotide hydrolysis (Richter et al., 1990; Weis et al., 1992), and by m-iodo-benzylguanidine, a competitive inhibitor of protein mono-ADP-ribose formation (Weis et al., 1992; Richter, 1990). Furthermore, we have found evidence that during oxidant-induced Ca^{2+} cycling (i.e. in the presence of Ca^{2+}) there is no release of ATP from the mitochondrial matrix. This demonstrates that in liver mitochondria, the mechanism of oxidant-induced Ca^{2+} efflux does not initially involve the opening of a pore or modification of the adenine nucleotide carrier into a non-selective channel.

Ca^{2+} Fluxes Across the Plasma Membrane

There is compelling evidence that many oxidants interfere with Ca^{2+} uptake and extrusion mechanisms at the plasma membrane level. Inhibition of Ca^{2+} efflux will result in the net accumulation of Ca^{2+} and in a pathological elevation of $[Ca^{2+}]_i$. In addition, it has become clear that chemical toxins can stimulate Ca^{2+} entry by interacting with existing Ca^{2+} channels or by increasing the plasma membrane permeability to Ca^{2+}. For instance, we have observed recently that tributyltin, a highly immunotoxic environmental pollutant, stimulates Ca^{2+} influx in immature rat thymocytes, in addition to inhibiting their plasma membrane Ca^{2+} translocase (Aw et al., 1990). Interestingly, tributyltin also releases intracellular Ca^{2+} stores, including the agonist-sensitive pool located within the ER (Chow et al., 1992). Hence, the stimulation of Ca^{2+} entry by tributyltin may involve a capacitative type of mechanism similar to that suggested for Ca^{2+}-mobilizing hormones and growth factors.

MECHANISMS OF Ca^{2+}-MEDIATED KILLING

The main evidence for the importance of Ca^{2+} overload in cell killing comes from experiments in which removal of extracellular Ca^{2+}, or loading of cells with intracellular Ca^{2+} chelators, such as quin-2 or BAPTA, have been found to prevent, or delay, cell killing induced by various agents (Nicotera et al., 1990). In addition, Ca^{2+} channel blockers have also been used to successfully prevent Ca^{2+} overload and cell death in several experimental systems.

Both the duration and the extent of the increase in $[Ca^{2+}]_i$ appear to be critical for the development of cytotoxicity. Even moderate increases in cytosolic Ca^{2+} can impair the ability of the cell to respond correctly to agonist stimulation and thereby interfere with cell control by hormones and growth factors. Another early effect of a sustained elevation of the cytosolic Ca^{2+} concentration is the impairment of mitochondrial function. In addition, more prolonged and intense increases in cytosolic Ca^{2+} will result in the disruption of cytoskeletal organization and in the activation of a number of Ca^{2+}-stimulated catabolic processes, such as proteolysis, membrane degradation and chromatin fragmentation (Fig. 3). The involvement of these Ca^{2+}-dependent alterations in cell killing will be briefly discussed in the following sections.

Alterations of Cell Signaling by Oxidants

Calcium ions are required for many physiological functions, including the control of metabolic processes, cell differentiation and proliferation, and secretory functions. These Ca^{2+}-dependent processes are tightly controlled by hormones and growth factors. The loss of the ability of the cell to respond to such hormones and growth factors will not only deprive the cell of a trophic stimulus but may also result in the activation of a suicide process that is characteristic of apoptotic cell death (Barnes, 1988).

The inability of cells to respond to Ca^{2+}-mobilizing hormones can be the consequence of the depletion of the intracellular agonist-sensitive Ca^{2+} pool by compounds such as tBuBHQ (Kass et al., 1989a) or bromotrichloromethane (Benedetti et al., 1989). Also, G_p, i.e. is the transducing G protein for inositol 1,4,5-trisphosphate-generating receptors, has been found to be susceptible to inactivation by oxidants (Bellomo et al., 1987). Finally,

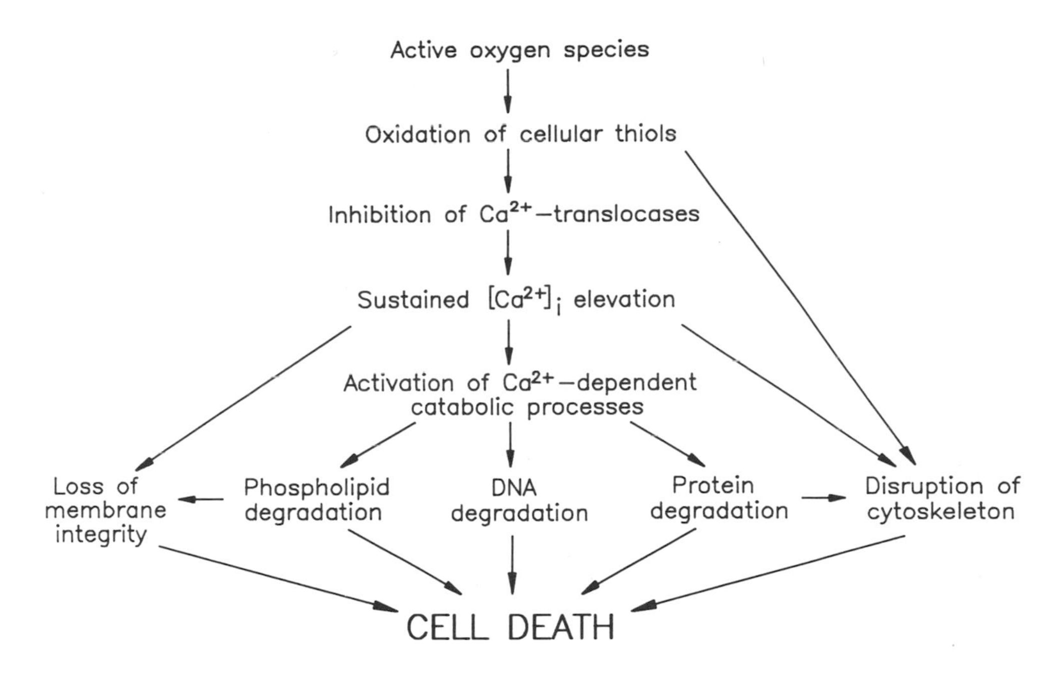

Fig. 3. Ca^{2+}-dependent mechanisms involved in oxidant-induced cell injury.

prolonged increases in $[Ca^{2+}]_i$ may obliterate the Ca^{2+} transients normally evoked by physiological agonists, thereby resulting in an impairment of cell signaling.

In contrast to the inhibition of cell signaling by high concentrations of oxidants, low levels of the same oxidants can have stimulatory effects (Burdon and Rice-Evans, 1989). For example, non-cytotoxic levels of prooxidants can lead to tumor promotion (Cerutti, 1985). Since the Ca^{2+}-stimulated, phospholipid-dependent protein kinase (protein kinase C) plays a crucial role in cell proliferation and the activation of this enzyme by phorbol esters has been associated with tumor promotion, we investigated the effects of oxidants on protein kinase C. Exposure of rat hepatocytes to low levels of oxidants resulted in a rapid increase in the specific activity of protein kinase C (Kass et al., 1989b). This increase was due to the oxidative modification of the protein, most likely through a modification of the thiol/disulfide balance of the enzyme. The involvement of thiol residues in the activation phenomenon was confirmed when we found that partially purified protein kinase C from rat brain could be activated using low concentrations of GSSG in a glutathione redox buffer (Kass et al., 1989b).

The selective, dose-dependent recruitment of different mechanisms for proliferation and cell death caused by oxidants is exemplified by our recent studies on the effects of the redox-cycling quinone, 2,3-dimethoxy-1,4-naphthoquinone (DMNQ) in RINm5F cells, a rat insulinoma cell line. At a concentration of 10 µM, DMNQ stimulated cell proliferation, whereas at only marginally higher concentrations of DMNQ (30 µM), cell growth was inhibited and a portion of the cell population underwent apoptosis (see below). Finally, at 100 µM concentration, DMNQ caused GSH and ATP depletion, Ca^{2+} overload and necrotic cell death. Interestingly, in our studies with DMNQ, we observed an induction of ornithine decarboxylase (ODC) activity in connection with the stimulation of cell proliferation. Since it is well established that protein kinase C stimulation by growth factors mediates ODC induction, it is conceivable that the effects of the lowest DMNQ concentrations described above, may be the result of protein kinase C activation caused by oxidative stress.

Cytoskeletal Alterations

One of the early signs of cell injury caused by oxidants and a variety of other toxic agents is the appearance of multiple protrusions (blebs) on the cell surface (Jewell et al., 1982). The events leading to bleb formation have not yet been fully elucidated and several mechanisms may independently contribute to their formation. However, it is generally accepted that a perturbation of cytoskeletal organization and of the interaction between the cytoskeleton and the plasma membrane plays an important role. The finding that Ca^{2+} ionophores induce similar blebbing, and that this is prevented by the omission of Ca^{2+} from the incubation medium, led to the proposal that Ca^{2+} is involved in the cytoskeletal alterations associated with the formation of surface blebs during cell injury. Many cytoskeletal constituents, including actin-binding proteins, such as caldesmon, gelsolin and villin, require Ca^{2+} to be able to interact with other cytoskeletal constituents. Moreover, Ca^{2+} regulates the function of three other actin-binding proteins which are directly involved in the association of microfilaments with the plasma membrane. Among these proteins, alpha-actinin is involved in the normal organization of actin filaments into regular, parallel arrays. However, in the presence of micromolar Ca^{2+} concentrations, alpha-actinin dissociates from the actin filaments (Rosenberg et al., 1991). The two other actin-binding proteins, vinculin and actin-binding protein, are substrates for Ca^{2+}-dependent proteases (Dayton et al., 1991). Thus, an increase in the cytosolic Ca^{2+} concentration to micromolar levels will result in the proteolysis of these two polypeptides.

Recent work has provided evidence for the involvement of Ca^{2+} in the toxic alterations of actin microfilaments and actin-binding proteins. For example, the incubation of human platelets with the redox active quinone, menadione, resulted in the dissociation of alpha-actinin from the whole cytoskeleton and in the proteolysis of actin-binding protein (Mirabelli,, et al., 1989). These changes were largely prevented in cells preloaded with the intracellular Ca^{2+} chelator, quin-2. Further, immunocytochemical investigations have revealed that dissociation of alpha-actinin from the actin filaments may contribute to bleb formation during oxidant injury (Bellomo et al., 1990). Other studies on canine heart in reperfusion injury, have shown a progressive loss of vinculin staining along the lateral margin of myocytes (Steenbergen et al., 1987). This loss was associated with the appearance of subsarcolemmal blebs and breaks in the plasma membrane. Since vinculin is a substrate for Ca^{2+}-dependent proteases and since the cytosolic Ca^{2+} concentration during ischemia and reperfusion rises well above the level necessary for protease activation, it appears that Ca^{2+}-activated proteases may be responsible for the loss of vinculin.

Ca^{2+}-Dependent Enzymes

The catabolism of phospholipids, proteins and nucleic acids involves enzymes, most of which require Ca^{2+} for activity. Ca^{2+} overload can result in a sustained activation of these enzymes and in the degradation of cell constituents which may ultimately lead to cell death.

Phospholipases catalyze the hydrolysis of membrane phospholipids. They are widely distributed in biological membranes and generally require Ca^{2+} for activation. A specific subset of phospholipases, collectively known as phospholipase A_2, have been proposed to participate in the detoxication of phospholipid hydroperoxides by releasing fatty acids from peroxidized membranes (Van Kuijk et al., 1991). However, phospholipase activation can also mediate pathophysiological reactions by stimulating membrane breakdown or by generating toxic metabolites. Hence, it has been suggested that a sustained increase in cytosolic Ca^{2+} can result in enhanced breakdown of membrane phospholipids and, in turn, in mitochondrial and cell damage. However, although a number of studies have indicated that accelerated phospholipid turnover occurs during anoxic or toxic cell injury (Chien et al., 1979), the importance of phospholipase activation in the development of cell damage by oxidants remains to be established.

Proteases During the past decade, the involvement of non-lysosomal proteolysis in various cell processes has become progressively clear. Proteases which have a neutral pH optimum include the ATP and ubiquitin-dependent proteases and the calcium-dependent proteases, or calpains. Calpains are present in virtually all mammalian cells and appear to be largely associated with membranes in conjunction with a specific inhibitory protein (calpastatin). The extra-lysosomal localization of this proteolytic system allows the proteases to participate in specialized cell functions, including cytoskeletal and cell membrane remodelling, receptor cleavage and turnover, enzyme activation, and modulation of cell mitosis.

Cellular targets for these enzymes include cytoskeletal elements and membrane integral proteins. Thus, the activation of Ca^{2+}

proteases has been shown to cause modification of microfilaments in platelets and to be involved in cell degeneration during muscle dystrophy and in the development of ischemic injury in nervous tissue. Studies from our laboratory have suggested the involvement of Ca^{2+}-activated proteases in oxidant injury in liver (Nicotera et al., 1986; Nicotera et al., 1989). Although the substrates for protease activity during cell injury remain largely unidentified, it appears that cytoskeletal proteins may be a major target.

Endonucleases During physiological cell killing a suicide process is activated in affected cells which is known as "apoptosis" or programmed cell death. Typical morphological changes occur in apoptotic cells, including widespread plasma and nuclear membrane blebbing, compacting of organelles and chromatin condensation (Arends et al., 1990). A characteristic marker of apoptotic cell death is the activation of a Ca^{2+}-dependent endonuclease which results in the cleavage of cell chromatin into oligonucleosome-length fragments. Endonuclease activation has been implicated in the killing of target cells by cytotoxic T lymphocytes and natural killer cells, and in thymocytes exposed to glucocorticoid hormones and other immunotoxicants (Nicotera et al., 1992).

The results of several recent studies have shown that Ca^{2+} overload can trigger endonuclease activation in certain cells. The Ca^{2+} ionophore A23187 stimulates DNA fragmentation in thymocytes, and characteristic endonuclease activity in isolated nuclei is dependent on Ca^{2+} (Jones et al., 1989). In addition, Ca^{2+}-mediated endonuclease activation appears to be involved in the cytotoxicity of 2,3,7,8-tetrachlorodibenzo-p-dioxin (TCDD) and of tributyltin in thymocytes (McConkey et al., 1988; Aw et al., 1990). Endonuclease activation has also been implicated in damage to macrophages caused by oxidative stress (Waring et al., 1988). Although the responsible endonuclease requires Ca^{2+} for activity, its regulation appears to be more complex and probably involves additional signals. Thus, recent work in this and other laboratories has indicated that the ability of the endonuclease to cleave DNA is dependent upon the chromatin superstructure (Brüne et al., 1991).

The chemical mechanisms by which active oxygen species, such as the hydroxyl radical, can cause DNA strand-breaks are relatively well established. For example, electron spin resonance spectroscopy, coupled to continuous flow systems, has allowed the direct observation of radicals formed on DNA model substrates during the oxidative fragmentation of the nucleic acid sugar-phosphate backbone by the hydroxyl radical (Udvardy et al., 1986). However, although it might be expected that similar mechanisms of DNA strand breakage are responsible for the DNA fragmentation observed in cells exposed to oxidative stress, there exists a growing body of evidence to suggest that during oxidative stress, DNA fragmentation can also be brought about via the activation of calcium-dependent mechanisms (Dypbukt et al., 1990; Cantoni et al., 1990).

The role of Ca^{2+} in DNA damage during oxidative stress has recently been the subject of intense investigation. The currently accepted hallmark of endonuclease-mediated DNA fragmentation, and hence apoptosis, is the appearance of a characteristic "ladder" of discrete oligonucleosome-sized bands when the DNA fragments are separated on an agarose gel. This is believed to result from the enzyme's preference for DNA cleavage at the internucleosomal linker DNA sites. In contrast, fragmentation brought about by the hydroxyl radical might be expected to occur at random sites on the sugar-phosphate backbone, resulting in the generation of DNA fragments of a continuous, rather than discrete, size range. However, much of the evidence suggesting that•OH attacks DNA at random sites is from radiation studies in which the radical is generated, at relatively high concentrations, in the free, bulk solution via the radiolysis of water molecules. In contrast, the results from studies employing transition-metal-catalysed •OH formation suggest that DNA damage may occur specifically at metal binding sites (Sagripanti et al., 1989). Whilst many metal ions, including iron, are known to form complexes with DNA, it appears that, at least quantitatively, copper is the most important transition metal ion associated with DNA in cell nuclei (Lewis and

Laemmli, 1982). Copper, along with calcium, appears to play a key role in maintaining the higher order structure of chromatin, resulting in the organization of DNA (along with its associated histones) into large loops, anchored to matrix proteins at specific sites. The findings from studies employing model systems demonstrate that copper bound to DNA can indeed interact with hydrogen peroxide and induce oxidative damage on the DNA (Prutz et al., 1990); and since the binding of copper ions to DNA is known to occur preferentially at guanine residues, then damage is expected to be site-specific. However, the biological significance of copper-mediated, site-specific DNA damage has yet to be fully evaluated, as well as the relative importance of free radical versus enzymatic mechanisms of strand breakage. Thus, further investigations are required before the relative significance of the enzymatic and free radical mechanisms of DNA fragmentation during oxidative stress can be evaluated.

CONCLUDING REMARKS

Thus, it appears safe to conclude that calcium ions play an important role in the development of oxidative cell injury. Recent research has revealed some of the biochemical mechanisms by which intracellular Ca^{2+} overload can cause cytotoxicity. However, the relative importance of the various Ca^{2+}-dependent processes in cell killing needs to be further clarified. Finally, it should be emphasized that different levels of oxidative stress may exert different effects on the cell, i.e. proliferation at low levels and death by apoptosis or necrosis at considerably higher levels.

ACKNOWLEDGEMENT

Work reported from the author's laboratory was supported by grants from the Swedish Medical Research Council (Proj. No. 03X-2471).

REFERENCES

Arends, M.J., Morris, R.G. and Wyllie, A.H. (1990) Amer. J. Pathol. 136, 593-608.

Aw, T.Y., Nicotera, P., Manzo, L. and Orrenius, S. (1990) Arch. Biochem. Biophys. 283, 46-50.

Barnes, D.M. (1988) Science 242, 1510-1511.

Beatrice, M.C., Stiers, D.L. and Pfeiffer, D.R. (1980) J. Biol. Chem. 255, 8663-8671.

Bellomo, G., Mirabelli, F., Richelmi, P. and Orrenius, S. (1983) FEBS Lett. 163, 136-139.

Bellomo, G., Thor, H. and Orrenius, S. (1987) J. Biol. Chem. 262, 1530-1534.

Bellomo, G., Mirabelli, F., Richelmi, P., Malorni, W., Iosi, F. and Orrenius, S. (1990) Free Rad. Res. Comms. 8, 391-399.

Benedetti, A., Graf, P., Fulceri, R., Romani, A. and Sies, H. (1989) Biochem. Pharmacol. 38, 1799-1805.

Brune, B., Hartzell, P., Nicotera, P. and Orrenius, S. (1991) Exp. Cell Res. 195, 323-329.

Burdon, R.H. and Rice-Evans, C. (1989) Free Rad. Res. Comms. 6, 345-358.

Cantoni, O., Sestil, P. and Catabeni, F. (1990) Eur. J. Biochem. 182, 209-212.

Carafoli, E. (1989) Annu. Rev. Biochem. 56, 395-343.

Cerutti, P.A. (1985) Science 227, 375-381.

Chien, K.R., Pfau, R.G. and Farber, J.L. (1979) Amer. J. Pathol. 97, 505-530.

Chow, S.C., Kass, G.E.N., McCabe, M.J. Jr. and Orrenius, S. (1992) Arch. Biochem. Biophys. In Press.

Cotgreave, I.A., Moldéus, P. and Orrenius, S. (1988) Ann. Rev. Pharmacol. 28, 189-212.

Crompton, M., Costi, A. and Hayat, L. (1987) Biochem. J. 245, 915-918.

Dayton, W.R., Shollmayer, J., Lepley, R.A. and Cortes, L.R. (1991) Biochem. Biophys. Res. Commun. 659, 48-61.

Di Monte, D., Bellomo, G., Thor, H., Nicotera, P. and Orrenius, S. (1984) Arch. Biochem. Biophys. 235, 343-350.

Dypbukt, J.M., Thor, H., Nicotera, P. (1990) Free Rad. Res. Comms. 8, 347-354.

Farrell, G.C., Duddy, S.K., Kass, G.E., Llopis, J., Gahm, A. and Orrenius, S. (1990) J. Clin. Invest. 85, 1255-1259.

Halestrap, A.P. and Davidson, A.M. (1990) Biochem. J. 268, 153-160.

Jewell, S.A., Bellomo, G., Thor, H., Orrenius, S. and Smith, M.T. (1982) Science 217, 1257-1259.

Jones, D.P., Thor, H., Smith, M.T., Jewell, S.A. and Orrenius, S. (1983) J. Biol. Chem. 258, 6390-6393.

Jones, D.P., McConkey, D.J., Nicotera, P. and Orrenius, S. (1989) J. Biol. Chem. 264, 6398-6403.

Kass, G.E., Duddy, S.K., Moore, G.A. and Orrenius, S. (1989a) J. Biol. Chem. 264, 15192-15198.

Kass, G.E., Duddy, S.K. and Orrenius, S. (1989b) Biochem. J. 260, 499-507.

McConkey, D.J., Hartzell, P., Duddy, S.K., Håkansson, H. and Orrenius, S. (1988) Science 242, 256-259.

Lewis, C.D. and Laemmli, U.K. (1982) Cell 29, 171-181.

Mirabelli, F., Salis, A., Vairetti, M., Bellomo, G., Thor, H. and Orrenius, S. (1989) Arch. Biochem. Biophys. 270, 478-488.

Missiaen, L., Taylor, C.W. and Berridge, M.J. (1991) Nature 352, 241-244.

Moore, L., Davenport, G.R. and Landon, E.J. (1976) J. Biol. Chem. 251, 1197-1201.

Nicotera, P., Hartzell, P., Baldi, C., Svensson, S.-Å, Bellomo, G. and Orrenius, S. (1986) J. Biol. Chem. 26, 14628-14635.

Nicotera, P., Rundgren, M., Porubek, D.J., Cotgreave, I., Moldeus, P., Orrenius, S. and Nelson, S.D. (1989) Chem. Res. Toxicol. 2, 46-50.

Nicotera, P., Bellomo, G. and Orrenius, S. (1990) Chem. Res. Toxicol. 3, 484-494.

Nicotera, P., Bellomo, G. and Orrenius, S. (1992) Annu. Rev. Pharmacol. Toxicol. 32, 449-470.

Prutz, W.A., Butler, J. and Land, E.J. (1990) Int. J. Radiat. Biol. 58, 215-234.

Richter, C., Theus, M. and Schlegel, J. (1990) Biochem. Pharmacol. 40, 779-786.

Richter, C. (1990) Free Rad. Res. Comms. 8, 329-334.

Richter, C. and Kass, G.E.N. (1991) Chem.-Biol. Interactions 77, 1-23.

Rosenberg, S., Stracher, A. and Lucas, R.C. (1991) J. Cell Biol. 91, 201-211.

Sagripanti, J.-L. and Kraemer, K.H. (1989) J. Biol. Chem. 264, 1729-1734.

Steenbergen, C., Hill, M.L. and Jennings, R.B. (1987) Circ. Res. 60, 478-486.

Thor, H., Smith, M.T., Hartzell, P., Bellomo, G., Jewell, S. and Orrenius, S. (1982) J. Biol. Chem. 257, 12419-12425.

Thor, H., Hartzell, P., Svensson, S.-Å, Orrenius, S., Mirabelli, F., Martino, A. and Bellomo, G. (1985) Biochem. Pharmacol. 34, 3717-3723.

Udvardy, A., Schedl, P., Sander, M. and Hsieh, T. (1986) J. Mol. Biol. 191, 231-246.

Van Kuijk, F.J.G.M., Sevanian, A., Handleman, G.J. and Dratz, E.A. (1991) Trends Biochem. Sci. 12, 31-34.

Waring, P., Eichner, R.D., Mullbacher, A. and Sjaarda, A. (1988) Science 246, 1165-1168.

Weis, M., Kass, G.E.N., Orrenius, S. and Moldeus, P. (1992) J. Biol. Chem. 267, 804-809.

Free Radicals: From Basic Science to Medicine
G. Poli, E. Albano & M. U. Dianzani (eds.)

LIPID PEROXIDATION. AN OVERVIEW

M. Comporti

Istituto di Patologia Generale, Università di Siena, Siena, Italy

Some decades ago lipid peroxidation was only known in the chemistry of oil and fat rancidicity, and its interest was mainly confined to the field of food technology. The possible importance of lipid peroxidation in biology as a damaging process for cellular membranes was first suggested by the studies of Tappel (1962). However, at that time, any possible implication in biopathology appeared restricted to conditions of deficiencies of vitamin E or other antioxidants, which are mainly limited to animals maintained on particular diets. The knowledge that lipid peroxidation can be linked to the electron transport chain of drug metabolism (Hochstein and Ernster, 1963), the recognition that the metabolism of carbon tetrachloride yields aloalkane free radicals (see Slater, 1972, for a review), and the observation that CCl_4 greatly stimulates the peroxidation of liver microsomal lipids (Comporti et al., 1965; Recknagel and Ghoshal, 1966), led to the assumption that lipid peroxidation could be a basic mechanism of toxicity for a wide variety of chemicals.
Today it is well established that lipid peroxidation is one of the reactions set into motion as a consequence of the formation of free radicals in cells and tissues. The mechanisms of free-radical induced cell injury have recently been reviewed by several authors (Slater, 1984; Halliwell and Gutteridge, 1989). Briefly, these mechanisms include (i) reactions with nucleic acids, nucleotides, polysaccharides, protein and non-protein thiols (thiol oxidation); (ii) covalent binding to membrane components (proteins, lipids, enzymes, receptors and transport systems); and (iii) initiation of lipid peroxidation. Thus the production of free radicals triggers off an expanding network of

multifarious disturbances. The question is, therefore, how free radicals are generated in the living cell.

Except for certain anaerobic microorganisms, oxygen is indispensible for life, and acts as the terminal oxidant in cell respiration, the main source of energy in aerobic organisms. Iron is necessary for life, too, being an essential component of vital enzymes such as cytochromes, cytochrome oxidase and catalase, and iron complexes such as hemoglobin, myoglobin, etc. Yet oxygen and iron possess properties which make them potentially damaging to biological structures. During the reduction of oxygen, reactive oxygen species can be generated. Iron can react with these species or with molecular oxygen itself to yield additional oxy radicals or reactive iron-oxygen complexes. Similar considerations hold for cytochrome P-450. This enzyme is a key component of the mixed function oxidase system, especially expressed in liver endoplasmic reticulum, and is involved in a wide variety of reactions concerned with detoxification of xenobiotics. Yet there are a large number of xenobiotics which, while not toxic per se, elicit their toxic effects after their activation to radical species or to electrophilic intermediates by the cytochrome P-450 mixed function oxidase system.

Several defense mechanisms against free radical attack have been developed by aerobic organisms . One line of defense appears to be against oxygen species, and includes superoxide dismutase (SOD), catalase and glutathione (GSH) peroxidase. A second line involves coumpounds which minimize peroxidation of membrane lipids. It includes mainly α-tocopherol (vitamin E) and the systems involved in its continuous regeneration, the membrane-bound form of GSH peroxidase (Ursini et al., 1982) and phospholipase A_2 (Sevanian et al., 1981). An additional safeguard is represented by the sequestration of iron in iron complexes, such as ferritin, hemosiderin, lactoferrin, etc. Finally, many electrophilic intermediates are conjugated with GSH by the activity of a broad class of cytosolic enzymes known as GSH-transferases.

These general notions sketch out only briefly the very complex network in which lipid peroxidation takes place. Lipid peroxidation was one of the first aspects of oxidative reactions to be recognized in cellular pathology, probably because it represents the most prominent phenomenon of uncontrolled oxidative stress. With the discovery of SOD (McCord and Fridovich, 1969) and with the knowledge that oxy radicals can easily be produced in biological systems, a much more complex spectrum of pathological oxidations has progressively been recognized, so that the term "oxidative stress" has been introduced, to signify any condition in which the pro-oxidant/antioxidant balance is disturbed in favour of pro-oxidation (Sies, 1985).

Lipid Peroxidation in Cellular Membranes

Lipid peroxidation is commonly divided into three phases, namely, initiation, propagation and termination. The initiation step is the interaction of free radicals with polyenoic fatty acids of membrane phospholipids. Such an attack occurs at the allylic hydrogens on the carbon atom between two double bonds, because of the relatively low bond dissociation energy (75 Kcal/mole), which renders these hydrogens particularly susceptible to attack. A fatty acid radical is thus formed. Carbon-centered radicals can initiate the process, e.g., the trichloromethyl radical resulting from CCl_4 metabolism. Oxygen-centered radicals can also initiate lipid peroxidation, but it seems very probable that the presence of catalytically active iron (or other transition metals) is also required. The one-electron reduction products of O_2, superoxide anion ($O_2^{\cdot-}$), hydrogen peroxide (H_2O_2) and hydroxyl radical ($OH^{\cdot}$), participate in the initiation of lipid peroxidation, but the exact mechanisms are still uncertain. Superoxide is insufficiently reactive to abstract H from lipids, while its protonated form $HO_2^{\cdot}$ is more reactive, and could initiate lipid peroxidation (in addition, $HO_2^{\cdot}$, being uncharged, should enter membranes fairly easily).

Hydroxyl radical certainly has the capacity of abstracting H from polyenoic fatty acids, but, because of its extreme reactivity, it is unlikely that it could migrate from its site of origin to fatty acid molecules in cellular membranes. Iron(II) ions (Fe^{2+}) can take part in electron transfer reactions with molecular oxygen

$$Fe^{2+} + O_2 \rightleftharpoons (Fe^{2+}-O_2 \longleftrightarrow Fe^{3+}-O_2^{\cdot}) \rightleftharpoons Fe^{3+} + O_2^{\cdot}$$

Superoxide can dismutate to form H_2O_2, and $OH^{\cdot}$ is produced via the Fenton reaction.

It is well known that the addition of iron(II) salts to unsaturated fatty acids, or to liver microsomes, or liposomes does initiate lipid peroxidation, but in most studies carried out with such systems, catalase or $OH^{\cdot}$ scavengers do not inhibit the reaction. Yet $OH^{\cdot}$ radicals can be detected in these systems and their formation is inhibited by the scavenging enzymes. It seems therefore that $OH^{\cdot}$ formation is not required for the initiation of lipid peroxidation, even if the possibility exists that, as suggested by Halliwell and Gutteridge (1989), membrane bound iron ions catalyze "site specific" formation of $OH^{\cdot}$; the latter would react immediately with membrane lipids before being available to scavenging enzymes. However, a number of studies (see Aust et al., 1985, for a review) led to the hypothesis that an iron-oxygen complex, rather than $OH^{\cdot}$, is required for initiation. Such a complex could be ferryl ion (FeO^{2+}) (Gutteridge, 1982) or perferryl ion ($Fe^{2+}O2 \longleftrightarrow Fe^{3+}O_2^{\cdot}$) (Aust and Svingen, 1982; Kornbrust and Mavis, 1980), but according to some chemical investigations it seems that the perferryl ion is not sufficiently reactive to initiate the process. It has recently been proposed (Minotti and Aust, 1989) that the active oxidant is a $Fe^{3+}-O_2-Fe^{2+}$ complex (Fe^{3+}/Fe^{2+} ratio 1/1), but its precise nature has not been established.

In any case, in the cell environment, iron must be released in a free form from iron complexes to be available for redox reactions. The problem of iron delocalization, initially proposed by Aust and coworkers (1985) has received much attention. It has recently been shown (Ferrali et al., 1992) that, in erythrocytes subjected to oxidative stress iron is released from haemoglobin in a desferrioxamine-chelatable form, and is capable of inducing oxidative reactions, including peroxidation of cell membrane, accompanied by cell lysis.

Carbon-centered radicals, such as the fatty acid radicals, easily react with oxygen to form oxygen-centered lipid peroxy radicals (LOO·). The latter can abstract hydrogen from a neighbouring polyunsaturated fatty acid (PUFA) and form the corresponding lipid hydroperoxide. Propagation is the reaction of a peroxyl radical with another PUFA to yield hydroperoxide and a new lipid radical, thus conserving the number of radicals in the reaction sequence. Decomposition of hydroperoxides, catalyzed by iron or other transition metals, yields alkoxy radicals and peroxy radicals and amplifies propagation, since such a secondary generation of free radicals can initiate new chains of lipid hydroperoxide formation.

$LOOH + Fe^{2+} \longrightarrow LO\cdot + Fe^{3+} + OH^-$ reductive decomposition

$LOOH + Fe^{3+} \longrightarrow LOO\cdot + Fe^{2+} + H^+$ oxidative decomposition

Decomposition of the peroxidized fatty acid, involving a breakdown of the molecule (scission of C-C bonds), yields a variety of end products. Decomposition of fatty acid endoperoxides yields malonic dialdehyde (MDA), the most commonly measured lipid peroxidation product. A large number of other aldehydes (alkanals, alkenals, 4-hydroxyalkenals, alkadienals, etc.) are also formed (see Esterbauer et al., 1991, for a review). Other decomposition products are small-chain hydrocarbons, such as ethane (from omega-6 fatty acids) and penthane (from omega-3 fatty acids).

The reaction chain can be terminated through the combination of two radicals to yield a nonradical product. In biological membranes, lipid-lipid interactions can occur. The latter is due to the fact that lipid radicals can abstract hydrogen from proteins which results in lipid-protein cross linking.

The consequence of lipid peroxidation in cellular membranes can

be summarized as follows. Damage to PUFA tends to reduce membrane fluidity which is essential for proper functioning. Membrane functions are also affected by the fact that the production of hydroperoxides and of carbonyl groups in the hydrophobic regions of phospholipids leads to the formation of hydrophilic centers, which may approach the external acqueous phase. This would increase the permeability of the membrane, thus explaining passive swelling of mitochondria, vesiculation of endoplasmic reticulum, leakage of enzymes and coenzymes, lysis of red blood cells, etc.
Alterations of protein structure and function also result from lipid peroxidation. Several membrane proteins derive some of their structure from closely associated lipids. During lipid peroxidation, two adjacent fatty acids can be joined by abnormal bonds, so that the enzyme structure may be altered and the activity affected. In addition, as mentioned above, lipid peroxy radicals can abstract hydrogen atoms from neighboring proteins, resulting in lipid-protein and protein-protein cross linking. The binding of aldehydes, such as 4-hydroxyalkenals, to functional -SH groups of proteins, is another aspect of protein damage induced by lipid peroxidation. Aldehydes can also react with amino groups of amino acids, proteins, phospholipids, and nucleic acids, to yield Schiff bases. The bifunctional aldehyde MDA can cross-link two amino compounds, to produce N,N'-disubstituted 1 amino-3-iminopropenes, which are fluorescent conjugated Schiff bases (Tappel, 1980).

Lipid Peroxidation and Toxic Injury

Lipid peroxidation has been proposed as the toxicity mechanism of a wide and increasing variety of compounds. Since most chemicals are metabolized in the liver, the hepatocyte is the cell where a free radical attack resulting in lipid peroxidation is most likely to occur. Before mentioning the most-studied examples of chemical induced lipid peroxidation, it is necessary to consider what evidence is necessary, in the opinion of this reviewer, to assess the occurrence of lipid peroxidation and its role in the mechanism of toxicity of a given compound. With *in vitro* systems, lipid peroxidation can be assessed using a variety of methods (thiobarbituric acid, TBA test for MDA, detection of conjugated dienes, measurement of fluorescent products, chemiluminescence, appearance of carbonyl functions in phospholipids, ethane and penthane evolution, etc.), whose results in general substantially agree. In addition, there generally good correlation between the extent of lipid peroxidation and the degree of functional membrane damage (e.g., alterations of enzymatic activities). The widespread skepticism about similar methods used to assess lipid peroxidation *in vivo* can be obviated if: (i) the various methods used give similar indications (Pompella et al., 1987) (the use of one method only, e.g. TBA test, may be unreliable for many reasons (Girotti, 1985)); and (ii) the results of *in vivo* studies are in agreement with those of *in vitro* studies carried out with appropriate systems.

Another fact which has generated confusion over the assessment of lipid peroxidation *in vivo* is the assumption that lipid peroxidation occurs whenever its toxic effects are ameliorated by prior administration of antioxidants. The effectiveness of antioxidants *in vivo* may only be a circumstantial indication of the involvement of free radical processes - not necessarily lipid peroxidation - under certain conditions. Induction of lipid peroxidation in cells and tissues must be demonstrated directly rather than indirectly deduced. Finally, the claim that lipid peroxidation may in some instances be the effect rather the cause of cell death has been discussed elsewhere (Comporti, 1987).

Another consideration that must be added is that, in many instances, lipid peroxidation is probably not the sole cause of cell death. Rather, the initial peroxidative damage in cell membranes is the trigger for a subsequent cascade of pathological events culminating in cell death. Nevertheless, lipid peroxidation may play a critical role as a link between the metabolism of a certain toxin and subsequent events. In other cases, lipid peroxidation can be induced as the consequence of particular imbalances and disorders created in the cell by the toxic insult. It could be argued that such disturbances can produce cell death even in the absence of lipid peroxidation, but, in our experience, the prevention of lipid peroxidation in such cases, largely prevents cell death.

The toxins which can induce lipid peroxidation in the cell fall broadly into the following groups: (i) those which are metabolized to reactive free radicals which alkylate cellular macromolecules and promote lipid peroxidation; (ii) those which are metabolized to electrophilic intermediates which readily conjugate with GSH, thus producing extensive GSH depletion (removal of one of the key factors in antioxidant systems); and (iii) those which are converted to intermediates generating reactive oxygen species by redox cycling. As will be discussed, many redox cycling compounds can also produce GSH depletion as a consequence of their metabolism, but their mechanism of action is more directly connected with their ability to produce reactive oxygen species.

With regard to the first group of toxins, carbon tetrachloride and the more potent analog monobromotrichloromethane must be considered as prototypes. They are metabolized by the microsomal mixed function oxidase system, utilizing the NADHP-cytochrome P-450 electron transport chain. It is generally accepted (see Slater, 1972; and Recknagel and Glende, 1973, for reviews) that the reaction involved in the metabolism is a one-electron reduction catalyzed by cytochrome P-450, the latter being maintained in the reduced form by NADPH:

$$CCl_4 + e^- \longrightarrow CCl_3^{\cdot} + Cl^-$$

The trichloromethyl-radical is thus generated, and, in the presence of molecular oxygen, can form the trichloromethyl peroxy radical. These free radicals can act both directly, by covalent binding to membrane lipids and proteins, and indirectly, through interactions of membrane unsaturated fatty acids, and the consequent promotion of lipid peroxidation. Both possibilities have been demonstrated, and occur few minutes after the *in vivo* intoxication. However, studies from our and other laboratories (see Comporti, 1985, for a review) have shown that at least some effect of CCl_4, such as inactivation of microsomal glucose-6-phosphatase, are mediated by lipid peroxidation rather than by covalent binding. Furthermore, in addition to loss of membrane structure, lipid peroxidation also induces the release of a number of toxic aldehydes from membrane phospholipids. The detection of such metastable products, in particular of 4-hydroxy-nonenal (Benedetti et al., 1980) and other 4-hydroxyalkenals (Esterbauer et al. 1982), which readily bind to -SH groups, represented an approach to the understanding of the mechanisms by which CCl_4 metabolism at a discrete site in the endoplasmic reticulum could result in alterations at distant loci. It was in fact difficult to imagine that the short-lived CCl_4 radicals could travel from their site of origin and act at a distance. The cytopathological effects of aldehydes derived from lipid peroxidation have been extensively reviewed (Dianzani, 1982; Esterbauer et al., 1991).

Additionals studies (Recknagel, 1983) have implied a disturbance in cellular calcium homeostasis as a link between CCl_4 metabolism and the subsequent cytopathological effects. This alteration would represent the trigger for the final common pathway leading to cell death in many types of cell injury. According to this hypothesis, which was put forth by Orrenius and coworkers (see Orrenius and Bellomo, 1986, for a review) for some redox cycling compounds, the initial pathological events in the membrane (lipid peroxidation, covalent binding, loss of protein thiols, etc.) lead to a disturbance of the mechanisms which regulate the concentration of free cytosolic calcium. In particular, it has been shown (Moore et al., 1976) that CCl_4 and $BrCCl_3$, as a consequence of their metabolism *in vitro* and *in vivo*, cause a marked inhibition of the microsomal calcium- sequestering activity. When the microsomal calcium pump is impaired, a sustained rise of cytosolic Ca^{2+} can be anticipated. This would cause activation of lytic enzymes (Nicotera et al., 1992) and impairment of mitochondrial functions, with collapse of membrane potential and consequent ATP depletion. The critical level of cytosolic Ca^{2+} capable of producing mitochondrial membrane potential collapse would be reached only when, in addition to an inhibition of the microsomal Ca^{2+} sequestration activity, a severe disturbance in the plasma membrane Ca^{2+} pumps is also produced (Bànhegyi et al., in press).

Many other halogenated aliphatic hydrocarbons, such as CCl_3Br, $CHBr_3$, CHI_3, alothane, etc., have been reported to promote lipid peroxidation. In general, the toxicity is inversely related to the dissociation energy of the carbon-halogen bond.

The second group of toxins which can promote lipid peroxidation in cellular membranes is represented by chemicals generally known as GSH depleting agents. They include aryl halide derivatives, such as bromobenzene and acetaminophen, alcohols, such as allyl alcohol, some halogenated aliphatic hydrocarbons, such as 1,2 dibromoethane and 1,1 dichloroethylene, diethylmaleate and many other substances. On the basis of a number of studies, carried out in particular with bromobenzene and acetaminophen, it was generally accepted that liver injury was due to the covalent binding of their reactive metabolites to cellular macromolecules (see Gillette et al., 1974, for a review). However, it was subsequently shown, (Devalia et al., 1982) that the treatment of animals or isolated hepatocytes with substances which prevent liver cell death do not substantially modify the extent of covalent binding. As an alternative toxic mechanism, lipid peroxidation was detected by several groups (see Comporti, 1987, for a review), using different methods. Lipid peroxidation starts only when the hepatic GSH depletion has reached critical values (less than 20% of the initial level). Treatment of the animals with antioxidants, even after bromobenzene administration, prevents both lipid peroxidation and liver necrosis, while not affecting the extent of covalent binding nor is GSH depletion substantially affected. Lipid peroxidation could be the consequence of constitutive oxidative stress in a cell depleted of GSH. Alternatively, free radicals derived from the toxin, or oxy radicals originating from the cytochrome P-450 system during mixed function oxidation, could be the active species initiating lipid peroxidation in a cell whose antioxidant defences are compromised by a dramatic loss of GSH. In some cases, active oxygen species could be produced by redox cycling, as has been proposed (Rosen, 1983) for the reactive metabolite of acetaminophen, N-acetyl-p-benzoquinoneimine.

In summary, the studies carried out with GSH depleting agents have made a significant contribution to understanding the importance of the antioxidant potential for cell life. In addition to GSH, ascorbate seems to be a very important factor in controlling oxidative stress. The enzymatic systems capable of regenerating reduced ascorbate from the oxidized forms are increasingly under investigation (Coassin et al., 1991; Casini et al., 1992).

The third group of toxins to be considered here are those generally known as redox cycling compounds. They give non-alkylating metabolites which readily react with oxygen and generate active oxygen species by intracellular redox cycling (see Bellomo and Orrenius, 1985, for a review). Paraquat, diquat, menadione, alloxan, 6-hydroxydopamine, together with some anthracyclines and nitrofurans are the most studied molecules (see Horton and Fairhurst, 1987, for a review). With many of them, the one-electron reduction pathway, catalyzed by NADPH-cytochrome P-450 reductase, yields the corresponding free radicals, which, in the presence of dioxygen, are rapidly reoxidized to the parent compounds, with concomitant formation of superoxide anions. Dismutation of O_2 produces H_2O_2; more potent oxidants ($OH^{\cdot}$) can be produced by the Fenton reaction. In the opinion of this reviewer, the oxy radicals generated by these compounds can directly produce cell and membrane damage, all the more so with severely decreased GSH levels. Lipid peroxidation, even if occurring, is probably not the primary nor the necessary mechanism of cell death. For instance it is well documented (Orrenius et al., 1985) that menadione produces cell death in the absence of lipid peroxidation. Evidence for lipid peroxidation in paraquat-induced lung toxicity is extremely controversial (Horton and Faihurst, 1987). Conflicting results have also been obtained with adriamycin in seeking to explain its heart toxicity. It has been suggested, however, that adriamycin can form complexes with iron and ADP-iron and that the ferrous forms of these complexes may initiate lipid peroxidation.

Lipid Peroxidation in other Pathological Oxidative Reactions

The possibility that lipid peroxidation is involved in pathological processes other than those induced by toxic compounds has been known for many years and has been the subject of many reports to this meeting. Thus, oxidation of low density lipoproteins (LDL) appears to be involved in the pathogenetic mechanisms of atherosclerosis (see Steinbrecher et al., 1990, for a review). Oxidized lipoproteins are also formed in diabetic rats and have the capacity to injure cells *in vitro* (Chisolm, 1992). Promising studies aimed to demonstrate by immunological means that oxidized LDL are indeed present in plasma of many atherosclerotic patients have been reported (Tatzber et al., 1992). Other studies have successfully attempted to detect lipid peroxidation by histochemical techniques (Pompella and Comporti, 1992).

Lipid peroxidation is also implicated in ischemia-reperfusion injury. It is likely that such injury results from an overall production of oxy radicals and that the generation of oxy radicals by endothelial cells and neutrophils rather than by parenchymal cells, plays an important role (Jaeschke, 1991). In the liver, the contribution of Kupffer cells may be significant.

Of particular interest is the demonstration that, in hypoxic hepatocytes, an acidotic intracellular pH retards the onset of cell death and that, during reperfusion, pH rises sharply and

precipitates cell death ("pH paradox") (Lemasters, 1992).
Evidence has been put forth suggesting the possible involvement of lipid peroxidation in several other conditions leading to clinical disease in humans. For instance, lipid peroxidation products have been detected in retinal tissue during experimental uveitis, an autoimmune condition in which infiltrating neutrophils are presumably the source of oxygen radicals.
It has been shown that conjugated dienes and TBA-reactivity increase in the synovial fluid of rheumatoid arthritis patients, and that the concentration of TBA-reacting substances indeed correlates with disease severity, as measured both by clinical and laboratory parameters (Rowley et al., 1984). However, the relevance of peroxidative processes in rheumatoid arthritis is still controversial (Halliwell et al., 1988).
The possible involvement of lipid peroxidation in Parkinson's disease has received much attention recently, following the discovery that 1-methyl-4-phenyl-1,2,3,6-tetrahydropyridine (MPTP) - a synthetic toxin structurally related to environmental pollutants - is able to induce lipid peroxidation and neuronal loss in the midbrain of mice (Adams and Odunze, 1991).
Less convincing evidence has been provided for lipid peroxidative processes in muscular dystrophy (Murphy and Kehrer, 1989) and central nervous system trauma and stroke (Hall and Braughler, 1989). Elevated indices of lipid peroxidation have also been reported in adult respiratory distress syndrome (Frei et al., 1988), as well as in traumatic shock (Lieners et al., 1989), burn shock (Demling et al., 1989) and endotoxaemic shock (Peavy and Fairchild, 1986).

REFERENCES

Adams, J.D., and Odunze, I.N. (1991) Biochem. Pharmacol. 41, 1099-1105.
Aust, S.D., and Svingen, B.A. (1982) in Free Radicals in Biology (Pryor, W.A., ed.), vol. V, pp. 91-113, Academic Press, New York and London.
Aust, S.D., Morehouse, L.A., and Thomas, C.E. (1985) J. Free Rad. Biol. Med. 1, 3-25.
Bànhegji, G., Bellomo, G., Fulceri, R., Mandl, J., and Benedetti,

A., Biochem. J., in press.
Bellomo, G., and Orrenius, S. (1985) Hepatology 5, 876-882.
Benedetti, A., Comporti, M., and Esterbauer, H. (1980) Biochem. Biophys. Acta 620, 281-296.
Casini, A.F., Maellaro, E., Del Bello, B., Sugherini, L., and Comporti, M. (1992) Free Rad. Res. Commun. 16, Suppl. 1, Abstr. 16.6.
Chisolm, G.M. (1992) Free Rad. Res. Commun. 16, Suppl. 1, Abstr. C.2.
Coassin, M.,Tomasi, A., Vannini, V., and Ursini, F. (1991) Arch. Biochem. Biophys. 290, 458-462.
Comporti, M., Saccocci, C., and Dianzani, M.U. (1965) Enzymologia 29, 185-204.
Comporti, M. (1985) Lab. Invest. 53, 599-623.
Comporti, M. (1987) Chem. Phys. Lipids 45, 143-169.
Comporti, M. (1989) Chem. Biol. Interactions 72, 1-56.
Demling, R.H., Lalonde, C., Fogt, F., Zhu, D., and Liu, Y. (1989) Crit. Care Med. 17, 1025-1030.
Devalia, J.L., Ogilvie, R.C., and McLean, A.E.M. (1982) Biochem. Pharmacol. 31, 3745-3749.
Dianzani, M.U. (1982) in Free Radicals, Lipid Peroxidation and Cancer (McBrien, D.C.H., and Slater, T.F., eds.), pp. 129-158, Academic Press, New York and London.
Esterbauer, H., Cheeseman, K.H., Dianzani, M.U., Poli, G., and Slater, T.F. (1982) Biochem. J. 208, 129-140.
Esterbauer, H., Shaur, R.J., and Zollner, H. (1991) Free Rad. Biol. Med. 11, 81-128.
Ferrali, M., Signorini, C., Ciccoli, L., and Comporti, M. (1992) Biochem. J. 285, 295-301.
Frei, B., Stocker, R., and Ames, B. (1988) Proc. Natl. Acad. Sci. USA 85, 9748-9752.
Gillette, J.R., Mitchell, J.R., and Brodie, B.B. (1974) Annu. Rev. Pharmacol. 14, 271-288.
Girotti, A.W. (1985) J. Free Rad. Biol. Med. 1, 87-95.
Gutteridge, J.M.C. (1982) FEBS Lett. 150, 454-458.
Hall, E.D., and Braughler, J.M. (1989) Free Rad. Biol. Med. 6, 303-313.
Halliwell, B., Hoult, J.R., and Blake, D.R. (1988) FASEB J. 2, 2867-2873.
Halliwell, B., and Gutteridge, J.M.C. (1989) Free Radicals in Biology and Medicine, Second Edition, Clarendon Press, Oxford.
Hochstein, P., and Ernster, L.(1963) Biochem. Biophys. Res. Commun. 12, 388-394.
Horton, A.A., and Fairhurst, S. (1987) CRC Crit. Rev. Toxicol. 18, 27-79.
Jaeschke, H. (1991) Chem. Biol. Interactions 79, 115-136.
Kornbrust, D.L., and Mavis, R.D. (1980) Mol. Pharmacol. 17, 400-407.
Lemasters, J.J. (1992) Free Rad. Res. Commun. 16, Suppl. 1, Abstr. 9.1.
Lieners, C., Redl, H., Molnar, H., Furst, W., Hallstrm, S., and Schlag, G. (1989) Prog. Clin. Biol. Res. 308, 345-350.

McCord, J.M., and Fridovich, I. (1969) J. Biol. Chem. 244, 6049-6057.
Minotti, G., and Aust, S.D. (1989) Chem. Biol. Interactions 71, 1-19.
Moore, L., Davenport, G.R., and Landon, E.J. (1976) J. Biol. Chem. 251, 1197-1201.
Murphy, M.E., and Kehrer, J.P. (1989) Chem. Biol. Interactions 69, 101-173.
Nicotera, P., Bellomo, G., and Orrenius, S. (1992) Annu. Rev. Pharmacol. Toxicol. 32, 449-470.
Orrenius, S., Rossi, L., Eklwow, L., and Thor, H. (1985) in Free Radicals in Liver Injury (Poli, G., Cheeseman, K.H., Dianzani, M.U., and Slater, T.F., eds.), pp. 99-105, IRL Press, Oxford.
Orrenius, S., and Bellomo, G. (1986) in Calcium and Cell Functions (Cheung, W.J., ed.), vol. VI, pp. 186-208, Academic Press, New York and London.
Peavy, D.L., and Fairchild, E.J. (1986) Infect. Immunity 52, 613-616.
Pompella, A., Maellaro, E., Casini, A.F., Ferrali, M., Ciccoli, L., and Comporti, M. (1987) Lipids 22, 206-211.
Pompella, A., and Comporti, M. (1992) Free Rad. Res. Commun. 16, Suppl. 1, Abstr. 12.15.
Recknagel, R.O., and Ghoshal, A.K. (1966) Lab. Invest. 15, 132-146.
Recknagel, R.O., and Glende, E.A., Jr. (1973) CRC Crit. Rev. Toxicol. 2, 263-297.
Recknagel, R.O. (1983) Trends Pharmacol. Sci. 4, 129-131.
Rosen, G.M., Singletary, W.V., Jr., Rauckman, E.J., and Killenberg, P.G. (1983) Biochem. Pharmacol. 32, 2053-2059.
Rowley, D., Gutteridge, J.M.C., Blake, D., Farr, M., and Halliwell, B. (1984) Clin. Sci. 66, 691-695.
Sevanian, A., Stein, R.A., and Mead, J.F. (1981) Lipids 16, 781-789.
Sies, H. (1985) Oxidative Stress, Academic Press, London.
Slater, T.F. (1972) Free Radical Mechanisms in Tissue Injury, Pion Limited, London.
Slater, T.F. (1984) Biochem. J. 222, 1-15.
Steinbrecher, U.P., Zhang, H., and Longheed, M. (1990) Free Rad. Biol. Med. 9, 115-168.
Tappel, A.L. (1962) Vitamins and Hormones 20, 493-510.
Tappel, A.L. (1980) in Free Radicals in Biology (Pryor, W.A., ed.), vol. IV, pp. 2-47, Academic Press, New York and London.
Tatzber, F., Waeg, G., Puhl, H., Salonen, J., and Esterbauer, H. (1992) Free Rad. Res. Commun. 16, Suppl. 1, Abstr. 2.18.
Ursini, F., Maiorino, M., Valente, N., Ferri, L., and Gregolin, C. (1982) Biochim. Biophys. Acta 710, 197-211.
Wu, G.-S., Sevanian, A., and Rao, N.A. (1992) 12, 19-27.

Free Radicals: From Basic Science to Medicine
G. Poli, E. Albano & M. U. Dianzani (eds.)

LIPID PEROXIDATION IN DIVIDING CELLS

Kevin H.Cheeseman

Department of Biology and Biochemistry,Brunel University
Uxbridge, UB8 3PH (U.K).

SUMMARY: For many years it has been established that tumour tissue exhibits a strong resistance to lipid peroxidation. Other rapidly dividing cells were also shown to be resistant and it was suggested that low lipid peroxidation activity is a common feature of cell division and that there is actually a biochemical connection between the two processes. The crux of this hypothesis was that possibly lipid peroxidation is a normal cellular process that somehow acts as a coarse regulator of cell division. The mechanisms of resistance of tumours to lipid peroxidation and the relative roles of pro-oxidants such as cytochrome P-450 and antioxidants such as α-tocopherol have been studied. The rapidly dividing non-transformed regenerating liver has provided interesting observations on the tight temporal connection between cell division and resistance and the role of α-tocopherol in this phenomenon. Foetal liver and intestinal epithelial cells have also been investigated and the overall impression from all these cell types is that there is no common mechanism of resistance.

INTRODUCTION

It has long been established that dividing cells display resistance to lipid peroxidation. Examples are cancer cells, regenerating liver, intestinal epithelial cells and foetal liver and observations in this area go back to the 1950s. Eminent figures in the field (e.g.Burlakova, 1975; Szent-Györgi, 1979; Slater, 1988) have suggested that this may mean that lipid peroxidation can act as a coarse control of cell division. In this proposal it is envisaged that continuous low-level lipid peroxidation produces biologically active products (e.g. hydroxyalkenals) that inhibit cell division. This is discussed below after examples of peroxidation-resistant dividing cells.

RESISTANCE TO LIPID PEROXIDATION IN CANCER CELLS

Pioneering work in this field was by Donnan et al. (1950) who described a low rate of autoxidation in hepatoma tissue and by Shuster (1955) who found Ehrlich ascites tumour cells were resistant to spontaneous and UV-induced lipid peroxidation. These reports were closely followed by those of Thiele and Huff (1960) and Utsumi et al. (1965) who reported the resistance of hepatoma mitochondria to lipid peroxidation.

In the ensuing years these observations were confirmed and expanded by many other researchers (Lash, 1966; Burlakova, 1975; Player et al., 1977a; Bartoli et al., 1979; Player et al. 1979; Dianzani et al., 1984; Cheeseman et al., 1986a; Minotti et al., 1986; Cheeseman et al., 1988a). This field of study has been the subject of reviews in recent years (Masotti et al., 1988; Dianzani, 1989; Galeotti et al., 1991). The majority of this work was carried out on experimental hepatomas since liver is by far the best characterised tissue with regard to lipid peroxidation and it is always important to compare the tumour tissue to the tissue of origin. The general features of resistance to lipid peroxidation found in hepatomas can be summarised as follows:

(i) lipid peroxidation, both spontaneous and forced, is much lower than in the liver

(ii) the degree of resistance correlates with the degree of dedifferentiation

(iii) hepatoma membranes have a very abnormal lipid composition: low amounts of phospholipid, high amounts of cholesterol, low amounts of PUFAs

(iv) cytochrome P-450 and associated enzymes, often important as pro-oxidants, are usually undetectable in hepatomas

(v) hepatomas contain increased levels of certain 'antioxidants', that were only detected empirically in early studies but characterised in more recent ones.

These properties of hepatomas can be illustrated with reference to the fast growing, undifferentiated Novikoff and Yoshida ascites hepatoma cells (Cheeseman et al., 1986a; Cheeseman et al., 1988a). As can be seen in Table I, these hepatomas are extremely resistant to lipid peroxidation whether it be enzyme catalysed (NADPH/ADP-Fe dependent) or non-enzymic (ascorbate-Fe dependent).It can be seen that both of these hepatomas are characterised by low concentrations of the PUFAs that are the most susceptible substrates for lipid peroxidation: arachidonic acid and docosahexaenoic acid.Moreover they contain an elevated level of α-tocopherol in their membranes. For this reason we postulated that one of the principle reasons for the resistance of these hepatoma cell membranes to lipid peroxidation was the increased ratio of antioxidant:substrate, or more specifically, α-tocopherol:PUFA.

TABLE I: LIPID PEROXIDATION AND ITS DETERMINING FACTORS IN THE MICROSOMES OF NOVIKOFF AND YOSHIDA HEPATOMAS RELATIVE TO NORMAL LIVER

PARAMETER	NORMAL LIVER	NOVIKOFF	YOSHIDA
NADPH/ADP-Fe lipid peroxidation (nmolsO_2/min/mg)	82.4	nd	nd
Ascorbate-Fe lipid peroxidation (nmolsMDA/min/mg)	1.29	0.06	0.24
Fatty acids (%)			
16:0	18.9	13.9	18.5
18:0	22.0	19.5	13.7
18:1	8.6	20.2	18.1
18:2	17.6	25.1	21.6
20.4	19.1	10.6	9.6
22:6	6.1	1.4	5.3
α-Tocopherol (nmol/mg lipid)	1.67	4.87	3.06
α-Tocopherol per bisallylic methylene ($x10^4$)	5.2	33.0	38.6
Cytochrome P-450 (nmol/mg)	0.71	nd	nd

The shortfall in arachidonic acid and docosahexaenoic acid is to some extent countered by increased levels in linoleic acid (18:2) but this is much less sensitive to peroxidation. Since the susceptibility of a PUFA to lipid peroxidation is determined by its number of bis-allylic methylene groups we have expressed this increased antioxidant protection of the hepatoma membranes as the ratio of α-tocopherol to bis-allylic methylene on a molar basis. As can be seen it is very much elevated. α-Tocopherol accounted for all of the lipid-soluble antioxidant activity. Another important factor is probably the complete absence of cytochrome P-450 that is usually important for propagating the lipid peroxidation reactions. Taken together these factors may explain the extremely high resistance.

The group of Galeotti (see reviews by Masotti et al., 1988 and Galeotti et al., 1991) have contributed important observations to this area and have come to somewhat different conclusions as to the mechanism of resistance. They have always stressed the prime importance of the absence of the pro-oxidant haemoprotein cytochrome P-450 and they give much less emphasis to the role of vitamin E. The basis of these assertions is that induction of cytochrome P-450 restored some lipid peroxidation activity to a Morris hepatoma and that vitamin E deficiency did not. Moreover, they have viewed the whole issue from a completely different point of view in their assertion that the low content of PUFAs found in hepatoma membranes is due to their having already undergone lipid peroxidation *in vivo* and thereby having lost the PUFAs during the growth of the tumour. It should be mentioned at this point that although hepatomas contain elevated levels of α-tocopherol, they are paradoxically deficient in enzymic antioxidant defences such as superoxide dismutase and GSH-peroxidase and also in GSH itself. It is probably not surprising that we have failed to explain, to the complete satisfaction of all concerned, the extreme resistance of hepatoma membranes to lipid peroxidation: this situation actually reflects upon how much we still have to learn about the process of lipid peroxidation in biological membranes *per se*.

RESISTANCE TO LIPID PEROXIDATION IN REGENERATING LIVER

The low level of lipid peroxidation during liver regeneration was first noted and associated with rapid cell division by Wolfson et al. (1956). They found less lipofuscin in regenerating liver and suggested a possible involvement of lipoperoxides in the control of cell division. Our own group turned to regenerating liver as a model of rapid cell division in order to compare it to the situation in hepatoma cells, it being our aim to determine whether the resistance to lipid peroxidation described above was a feature of cell division or a feature of transformation (Cheeseman et al., 1986b; Cheeseman et al., 1988b; Slater et al., 1990). Regenerating liver is much more resistant to lipid peroxidation than is normal liver but far less resistant than hepatoma. It is important to stress that in our particular model of liver regeneration we used an entrainment regime for the rats that was designed to bring about a certain degree of synchrony of cell division; this is an important feature lacking in many other such studies. The success of this approach can be seen in the cycling of DNA synthesis indicated by the cycling in activity of thymidine kinase in the liver cytosol. The peaks of activity at 24h intervals indicate the peaks of cell division. Remarkably, using NADPH/ADP-Fe dependent lipid peroxidation measurements *in vitro*, the resistance to lipid peroxidation can also be found to exhibit cyclic behaviour with maxima of resistance coinciding with maxima of thymidine kinase activity. The basis of this resistance seems to be a flux in the membrane concentration of α-tocopherol; other important factors such as PUFA composition and P-450 content did not change significantly. If the regenerating liver is deficient in vitamin E then the resistance to lipid peroxidation disappears. The regenerating liver and hepatoma models can be seen to be rather different in that (i) the degree of resistance is much less in the regenerating liver and cycles with cell division; (ii) regenerating liver microsomes have normal PUFA.compositions: (iii) cytochrome P-450 decreases but is never completely absent; (iv) α-tocopherol is elevated only at certain times. The degree of resistance does not seem to be completely, quantitatively proportional to the concentration of α-tocopherol in the membrane so other factors may be also involved but at the time of writing these remain elusive.

These observations offer some support to the theory that lipid peroxidation may act as a control of cell division, bearing in mind the close temporal association between the fluxes in antioxidant activity and thymidine kinase activity. However, it could also be seen from the point of view that the cells have evolved a mechanism to increase their protection against oxidative damage at the very sensitive time of mitosis. Moreover we have found (Cheeseman, Emery and Slater, unpublished results) that when rats are made vitamin E deficient the rate of regeneration of the liver is unchanged suggesting that endogenous lipid peroxidation in the absence of vitamin E is not capable of inhibiting cell division.

RESISTANCE TO LIPID PEROXIDATION IN FOETAL LIVER

The foetal liver undergoes rapid cell division, much faster than most tumours. This has long been associated with low rates of lipid peroxidation (Cole 1956; Player et al., 1977b). Using the guinea pig foetal liver as an experimental model we have examined the lipid peroxidation activity at various stages of gestation (Cheeseman, Safavi and Kelly, manuscript in preparation). At 55d of gestation (full term is 68d) lipid peroxidation activity, measured *in vitro* in the foetal liver microsomes is indeed very low. By full term, lipid peroxidation is essentially at the normal adult level.It should be pointed out that cell division is still rapid at this time. Over the period of gestation, the changes in PUFAs and in α-tocopherol in the microsomal membranes are not sufficient to explain the change in lipid peroxidation activity. It is striking, however, that lipid peroxidation (even 'non-enzymic' ascorbate/iron dependent lipid peroxidation) is directly proportional to the concentration of cytochrome P-450 in the membrane, which may point to the importance of this haemoprotein in propagating reactions.

RESISTANCE TO LIPID PEROXIDATION IN INTESTINAL EPITHELIAL CELLS

A final example of dividing cells that are resistant to lipid peroxidation are epithelial cells of the gastrointestinal tract. Thanks to studies by Balasubramanian and colleagues, these have been the subject of revived interest in recent years. Balasubramanian et al. (1988) showed that not only are intestinal epithelial cell membranes resistant to lipid peroxidation but that these membrane preparations inhibited iron-dependent lipid peroxidation in liver membranes *in vitro*. The inhibitory substance was found to be a mixture of fatty acids, principally oleic acid, released by phospholipase activity during the preparation of the intestinal cell membranes (Diplock et al., 1988). This was a remarkable, unexpected discovery.It was subsequently demonstrated that these fatty acids were responsible for conferring resistance only to iron-dependent lipid peroxidation systems, as they are able to sequester the iron (Balasubramanian et al., 1989). Intestinal cell membranes are also resistant to cumene hydroperoxide-induced and to carbon tetrachloride-induced lipid peroxidation and in this case the resistance is due to the very low level of cytochrome P-450. These membranes are, however, almost as susceptible to lipid peroxidation induced by the 'azo-initiator' ABAP and by gamma-irradiation as are liver microsomes.Interestingly, the latter observation seems to contradict one of the earliest observations that actually stimulated this whole field of research (Barber and Wilbur, 1959).

CONCLUSIONS

Overall, looking at these different models of resistance to lipid peroxidation exhibited by rapidly dividing cells, it is evident that the nature of the resistance differs markedly and the biochemical mechanisms involved are often quite different (see Table II).

TABLE II: SUMMARY OF CHARACTERISTICS AND PUTATIVE MECHANISMS OF RESISTANCE TO LIPID PEROXIDATION IN EXAMPLES OF DIVIDING CELLS

CELL TYPE	RESISTANCE CHARACTERISTICS	RESISTANCE MECHANISM
Fast-growing hepatoma	complete	hightocopherol:PUFA ratio absent P-450
Regenerating liver	intermittent & incomplete	intermittent high tocopherol
Foetal liver	only at early gestation	low P-450?
Intestinal-epithelium	only resistant to Fe-dependent peroxidation	Fe-sequestration by free fatty acids

Although nature has often found different ways of achieving the same ends, the divergent characteristics and mechanisms of resistance may suggest that the resistance to lipid peroxidation observed in dividing cells may actually be coincidental rather than the manifestation of a biochemical mechanism.

The proposed mechanisms for lipid peroxidation to influence cell division are based on the biological activities of certain products of lipid hydroperoxide breakdown. The most well-known candidate in this context is certainly 4-hydroxynonenal (HNE) that was originally isolated during investigations into the cytostatic activity of autoxidised fatty acids (Schauenstein et al., 1977). Indeed, it can be justifiably said that one of the most significant 'spinoffs' from this line of research has been the study of HNE, now considered to be a most important product of lipid peroxidation and implicated in a wide range of tissue injuries and even in the pathogenesis of atherosclerosis (see Esterbauer et al., 1991 for a review).HNE and other hydroxyalkenals are strong inhibitors of many enzymes, notably those involved in DNA synthesis and protein synthesis; this has been used as further justification for the proposal that on-going lipid peroxidation acts as a coarse control of cell division. Counterarguments raised against this suggestion are that HNE is not produced at sufficiently high concentrations under normal conditions and that it is very rapidly removed by a number of different metabolic pathways. Further, lipid peroxidation is such an uncontrolled process in itself that it seems ill-suited to the role of controlling the important and sophisticated process of cell division. Nevertheless, HNE has been shown to be active in whole cells even at low concentrations and in the presence of enzymes that metabolise it and. Moreover, a short half-life does not rule out its having important cellular effects (cf. established second messengers). HNE is known to affect gene expression: it decreases the expression of c-myc proto-oncogene (Barrera et al., 1987), causes transcription of shock proteins (Cajone et al., 1989) and blocks the cell cycle (Poot et al., 1988). It is of great interest to note that recent investigations in another field have led to the hypothesis that lipid peroxidation is involved in controlling expression of the gene for collagen (Chojkier et al., 1989), so there is still the possibility that lipid peroxidation, perhaps through the intermediacy of biologically active products, may influence gene expression and cell division.

ACKNOWLEDGEMENT

I am grateful to the Association for International Cancer Research for funding.

REFERENCES

Balasubramanian, K. A., Manohar, M. and Mathan, V. I. (1988) Biochim.Biophys.Acta 962 51-58.

Balasubramanian, K. A., Nalini, S., Diplock, A. T., Cheeseman, K.H. and Slater, T. F.(1989) Biochem.Biophys.Acta 1003 232-237.

Barber, A. A. and Wilbur, K. M. (1959) Radiat.Res. 10 165-175.

Barrera, G., Martinotti, S., Fazio, V., Manzari, V., Paradisi, L.,Parola, M., Frati, L. and Dianzani, M. U. (1987) Toxicol.Pathol. 15 (2), 238-240.

Bartoli, G. and Galeotti, T. (1979) Biochim.Biophys.Acta 574 537-541.

Burlakova, E. B. (1975) Russ.Chem.Rev. 44 871-880.

Cajone, F., Salina, M. and Bernelli-Zazzera, A. (1989) Biochem.J.262 977-979.

Cheeseman, K. H.,Collins, M..,Maddix, S. P., Milia, A., Proudfoot, K.,Slater, T. F., Burton, G. W., Webb, A. and Ingold, K. U (1986b) FEBS Lett. 209 191-196.
Cheeseman, K. H., Collins, M. M., Proudfoot, K., Slater,T. F., Burton, G. W., Webb, A. C. and Ingold, K. U. (1986a) Biochem.J.235 507-514.
Cheeseman, K. H.,Emery, S., Maddix, S.,Proudfoot, K.,Slater, T.F., Burton, G. W., Webb, A. and Ingold, K. U. (1988b).Lipid peroxidation,antioxidants and cell division . In: Eicosanoids, Lipid peroxidation and Cancer. (Ed:S. Nigam, D. C. H. McBrien and T. F. Slater) pp195-202, Springer-Verlag, Berlin.
Cheeseman, K. H., Emery, S., Maddix, S., Slater, T. F., Burton, G. W. and Ingold, K. U. (1988a) Biochem.J. 250 247-252.
Chojkier, M., Houglum, K., Solis-Herruzo, J. and Brenner, D. (1989) J.Biol.Chem. 264 16957-16962.
Cole, B. T. (1956) Proc.Soc.Exp.Biol.Med. 93 290-294.
Dianzani, M. U. (1989) Tumori 75 351-357.
Dianzani, M. U., Canuto, R. A., Rossi, M. A., Poli, G., Garcea, R., Biocca, M.,Cecchini, G., Biasi, F., Ferro, M. and Bassi, A. (1984)Toxicol.Pathol. 12 189-199.
Diplock, A. T., Balasubramanian, K. A., Manohar, M., Mathan, V. I. and Ashton, D. (1988) Biochim.Biophys.Acta 962 42-50.
Donnan, S. K. (1950) J.Biol.Chem. 182 415-419.
Esterbauer, H., Schaur, R. J. and Zollner, H. (1991) FreeRad.Biol.Med. 11 81-128.
Galeotti, T., Masotti, L., Borrello, S. and Casali, E. (1991) Xenobiotica 21 (8), 1041-1051.
Lash, E. D. (1966) Arch. Biochem.Biophys. 115 332-336.
Masotti, L., Casali, E. and Galeotti, T. (1988) Free Rad. Biol.Med. 4 377-386.
Minotti, G., Borello, S., Palombini, G. and Galeotti, T. (1986) Biochim.Biophys,Acta 876 220-225.
Player, T. J., Mills, D. J. and Horton, A. A. (1977a) Biochem.Soc.Transact. 5 1506-1508.
Player, T. J., Mills, D. J. and Horton, A. A. (1977b) Biochem.Biophys.Res.Commun. 781397-1402
Player, T. J., Mills, D. J. and Horton, A. A. (1979) BrJ.Cancer 39 773-778.
Poot, M., Esterbauer, H., Rabinovitch, P. S. and Hoehn, H. (1988) J.Cell.Physiol. 137 421-429.
Schauenstein, E., Esterbauer, H. and Zollner, H. (1977). Aldehydes in Biological Systems. Pion Ltd., London.
Shuster, C. W. (1955) Proc.Soc. Exp. Biol.Med. 90 423-426.
Slater, T. F. (1988). Lipid peroxidation and cell division in normal and tumour tissues. In: Eicosanoids, Lipid peroxidation and Cancer. (Ed: S.Nigam, D. C. H. McBrien and T. F. Slater) pp137-142, Springer-Verlag, Berlin.
Slater,T. F., Cheeseman, K. H., Benedetto, C., Collins, M, Emery,S., Maddix, S. P., Nodes, J. T., Proudfoot, K. P., Burton, G. W. and Ingold, K. U. (1990) Biochem.J. 265 51-59.
Szent-Györgi, A. (1979). The living state and cancer. In:Sub-molecular Biology and Cancer. (Ciba Foundation Symposium 67) pp3-18, Excerpta Medica, New York.
Thiele, E. H. and Huff, J. W. (1960) Arch. Biochem. Biophys. 88 208-211.
Utsumi, K., Yamamoto, G. and Inaba, K. (1965) Biochim.Biophys.Acta 105 368-371.
Wolfson,N.,Wilbur, K.and Bernheim, F.(1956)Exp.Cell Res.10 556-558

Free Radicals: From Basic Science to Medicine
G. Poli, E. Albano & M. U. Dianzani (eds.)

FORMATION AND METABOLISM OF THE LIPID PEROXIDATION PRODUCT 4-HYDROXYNONENAL IN LIVER AND SMALL INTESTINE

W.G. Siems[1], T. Grune[1], H. Zollner[2] and H. Esterbauer[2]

[1]Institute of Biochemistry, Medical Faculty (Charite'), Humboldt University, O-1040 Berlin, Germany and [2]Institute of Biochemistry, University of Graz, A-8010 Graz, Austria

SUMMARY: During reoxygenation of isolated rat hepatocytes or rat small intestine following 60 min of anoxia/ischemia an increase of the concentration of the cytotoxic lipid peroxidation product 4-hydroxynonenal was observed. This increase was short-termed. The rapid disappearance of 4-hydroxynonenal is due to its immediate reactions with cellular compounds and its metabolism. The capacity of eucaryotic cells to metabolize 4-hydroxynonenal was investigated. As primary and secondary metabolic products the glutathione-HNE conjugate, the 4-hydroxynonenoic acid, the corresponding alcohol 1,4-dihydroxynonene, the glutathione-dihydroxynonene conjugate and water were identified. It is postulated that the rapid metabolism of 4-hydroxynonenal and other aldehydic products of lipid peroxidation could be an important part of the intracellular antioxidative defense system.

Abbreviations: GSH - glutathione (reduced form); HNE - 4-hydroxynonenal; DHN - 1,4-dihydroxynonene; HNA - 4-hydroxy-2-nonenoic acid; PUFA - polyunsaturated fatty acids; DNPH - dinitrophenylhydrazine; TLC - thin-layer chromatography; HPLC - high-performance liquid chromatography.

NUCLEOTIDE DEGRADATION AND OXYGEN RADICAL FORMATION DURING POSTISCHEMIC REOXYGENATION

There occurs a fast decrease of ATP level in hepatocytes, small intestine and other cells and tissues after the onset of oxygen deficiency. This decrease of ATP level is followed by a wave-like behaviour of the degradation products of ATP. That means the decrease of ATP level is followed by contemporary increases of the levels of ADP, AMP, adenosine, IMP and finally by the accumulation of hypoxanthine. The conversion of hypoxanthine to uric acid in presence of oxygen, i.e. during the reoxygenation, is connected with the formation of superoxide radicals. In liver and small intestine the xanthine oxidase reactions seem to contribute to a high degree to the formation of free radicals, subsequent lipid peroxidation and to cell damages during the postischemic phase (Granger, 1988; Gerber et al., 1991). The oxidative loading of the cells was demonstrated in those experiments by means of the thiobarbituric acid-reactive material, which was accumulated during the early reperfusion phase (Siems et al., 1989; Kowalewski et al., 1991). The thiobarbituric acid-reactive substances in the small intestine reached their maximal level at the time point 70 min of the experiments, which is equal to 10 min after the onset of reperfusion following 60 min of ischemia (Fig. 1). The measurement of thiobarbituric acid-reactive substances is an unspecific parameter, and it was interesting to measure more specific metabolic changes due to postischemic lipid peroxidation. The most interesting group of lipid peroxidation products, we are interested in, are the 4-hydroxyalkenals and especially its main representative 4-hydroxynonenal (HNE).

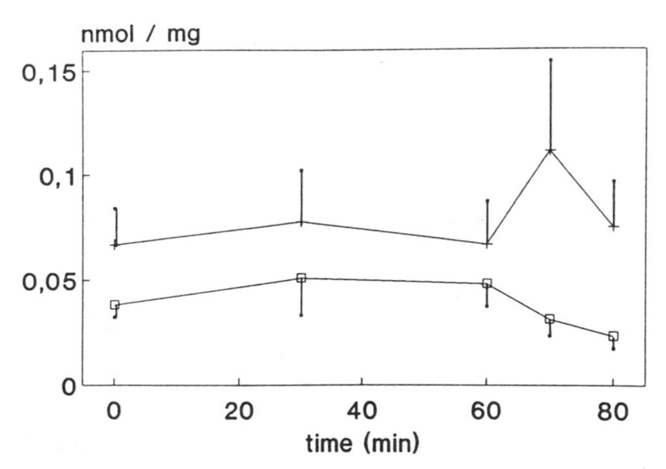

Fig. 1. Content of thiobarbituric acid-reactive substances of rat small intestine at ischemia (60 min) and reperfusion. Data from 6 control animals (-+-) and 6 animals treated with xanthine oxidoreductase inhibitor (-□-). Values as mean ± S.D.

PROPERTIES OF THE LIPID PEROXIDATION PRODUCT 4-HYDROXYNONENAL

The biological effects of 4-hydroxyalkenals have been studied by a number of laboratories. Such effects of HNE are the inhibition of DNA synthesis, inhibition of various enzyme activities, the increase of capillary permeability, stimulation of chemotoxis etc. (Esterbauer, 1982; Poli et al., 1987; Esterbauer et al., 1988; Schaur et al., 1991). It was postulated that the hydroxyalkenals can be an important factor for the progression of the radical induced cell damage (Esterbauer et al., 1988 and 1989; Schaur et al., 1991). 4-Hydroxynonenal as the main representative of 4-hydroxyalkenals is a major product formed from omega 6-PUFA. It exhibits in the range of μM concentrations cytotoxic, mutagenic and genotoxic properties. In the submikromolar range HNE is chemotactic, and it was also shown that it stimulates adenylate cyclase, guanylate cyclase and particularly phospholipase C at submikromolar concentrations (Schaur et al., 1991).

ACCUMULATION OF 4-HYDROXYNONENAL DURING REOXYGENATION FOLLOWING OXYGEN DEFICIENCY

Experiments on the measurement of 4-hydroxynonenal during oxygen deficiency and reoxygenation were carried out with liver cells and with the small intestine under in vivo conditions of the animals. For the measurements the classical HNE analysis was used, which was developed by the groups in Austria and Italy (Esterbauer et al., 1982; Poli et al., 1985 and 1986). The procedure includes the derivatization of carbonyls with dinitrophenylhydrazine, the TLC separation into three groups of DNPH carbonyls and the final HPLC analysis of dinitrophenylhydrazones of 4-hydroxyalkenals which are detected at 378 nm.
The experimental results were the following: During reoxygenation both in hepatocyte suspension and in small intestine an increase of the HNE concentration was demonstrated. In the intestinal mucasa the peak of HNE accumulation was observed at 10 min of reperfusion

following the ischemic period of 60 min (Kowalewski et al., 1991; Siems et al., 1991). There occur increased HNE levels also in the first minutes of postischemic reperfusion of isolated rat hearts which was investigated together with I.E. Blasig's group (Institute of Molecular Pharmacology, Berlin-Friedrichsfelde, Germany). And the HNE increase was also demonstrated in hepatocytes experiments at 75 min which corresponds to 15 min of reoxygenation after 60 min of anoxia. The highest values of HNE level were measured in the reperfused small intestine with about 6 μM. The increase of HNE concentration in hepatocytes and in small intestine was prevented in presence of the xanthine oxidoreductase inhibitor oxypurinol. In all experimental models the increase of the HNE level was very short-termed. It seems to reflect only a part of the acceleration of HNE formation during postischemic reoxygenation. We assumed that the rapid disappearance of HNE which was formed in the cells is due to immediate reactions of HNE with cellular compounds and to a rapid metabolism of HNE. Therefore, experiments on investigation of HNE metabolism were carried out. Those experiments included i) the developement of methods for the identification and measurement of HNE metabolites, ii) the appliction of such methods to biological samples and iii) the quantification of HNE product formation, that means the flux rate analysis by means of direct measurements of HNE products, also by means of tracerkinetic experiments with radioactive labeled hydroxynonenal and by means of inhibitors of pathways of HNE metabolism.

DEVELOPEMENT OF METHODS FOR THE IDENTIFICATION AND QUANTITATIVE MEASUREMENT OF 4-HYDROXYNONENAL METABOLITES

If one adds 4-hydroxynonenal to cell suspensions, this HNE is rapidly degraded by the cells. Table I gives the values of the HNE utilization rat in different eucaryotic cell types under comparable incubation conditions (pH 7.4, 37°C, initial HNE concentration of 100 μM).

Table I. 4-Hydroxynonenal utilization rates in different eucaryotic cell types. The incubations were carried out at 37°C, pH 7.4 and initial HNE concentration of 100 μM. Values are given as nmoles/mg wet weight/min.

Cell type	Species	HNE utilization rate
Enterocytes	Rat	25.2 ± 2.7
Hepatocytes	Rat	28.4 ± 1.8
Ehrlich ascites tumor cells	Mouse	8.7 ± 1.5

Several metabolites of HNE were postulated to exist (Esterbauer et al., 1985; Fauler, 1987; Ferro et al., 1988; Schaur et al., 1991). HNE metabolites should be the corresponding alcohol of the HNE, the 1,4-dihydroxynonene (DHN), the 1:1-glutathione-HNE conjugate (GSH-HNE) and the 4-hydroxynonenoic acid (HNA). Enzymes for the formation of those metabolites of HNE like alcohol dehydrogenase or aldehyde reductases, aldehyde dehydrogenases and glutathione S transferases are available with high capacities in liver and in small intestine. It was necessary to develope sensitive methods for the identification and quantitative measurement of HNE metabolites (Fauler, 1987; Siems et al., 1992a and 1992b). These methods include HPLC separations with precolumn derivatization, HPLC with UV detection, HPLC of the ^{3}H-labeled compounds. The use of tritium-labeled (2-^{3}H) HNE enabled us to quantify the formation of metabolites by TLC without prior derivatization. Two different solvent systems were used for the TLC separation of HNE metabolites: one separates the HNE, the 4-hydroxynonoic acid and the 1,4-dihydroxynonene, the second allows the determination of glutathione conjugates of HNE and of 1,4-dihydroxynonene. The position of a particular metabolite on the TLC-plate was determined by separating the standards on the same plate. The concentration was determined by measuring the radioactivity associated with the TLC-spots. For that purpose the TLC-plates were scanned with an automatic TLC linear analyzer (Berthold, Wildbach, Germany). Based on the recovered radioactivity, HPLC and TLC separations gave concurrent results for the three HNE metabolites mentioned above.

The determination of DHN by TLC received preference over its determination by HPLC, since the TLC method is less tedious and much faster.

FLUX RATES OF THE PATHWAYS OF 4-HYDROXYNONENAL METABOLISM IN HEPATOCYTES AND ENTEROCYTES OF THE SMALL INTESTINE

It could be demonstrated that the three primary HNE products (GSH-HNE, HNA, DHN) are formed in hepatocytes and in enterocytes of the small intestine after the addition of HNE to the cell suspension. In hepatocyte suspensions the main product of HNE from the quantitative point of view was the GSH-HNE conjugate followed by the HNE and the DHN which was accumulated only to a minor extent. Fig. 2 demonstrates the accumulation of GSH-HNE conjugate in hepatocyte or enterocyte suspensions after addition of 100 μM HNE.

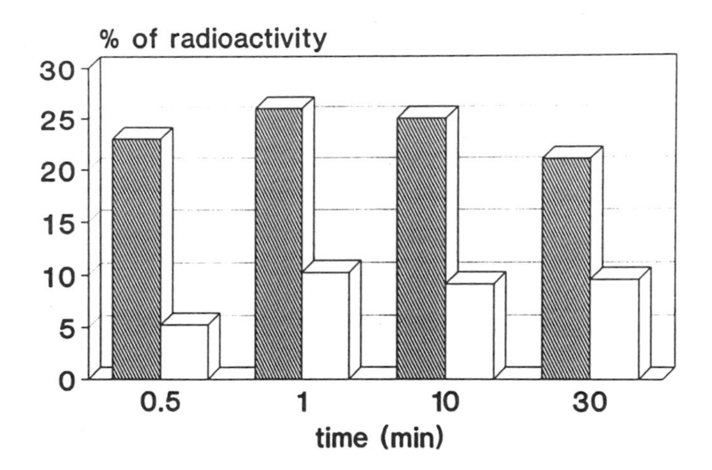

Fig. 2. Accumulation of glutathione-HNE conjugate in hepatocyte suspensions (left columns) and enterocyte suspensions (right columns) after addition of 100 μM HNE. Values are given as nmol/ml which is equal to % of initial HNE concentration.

If one compares the total disappearance of HNE in hepatocytes with the sum of the three primary HNE metabolites one can notice a gap which is increasing with the duration of incubation. Furthermore, there were observed at least seven different compounds which were separated by HPLC of extracts from hepatocyte suspensions treated with ^{3}H-labeled HNE. That was a further argument for the formation of secondary products from the primary HNE intermediates. As secondary products of HNE metabolism up to now GSH-DHN conjugate and water, which is generated through the beta-oxidation of the 4-hydroxynonenoic acid, could be identified. In 10 min of incubation at 37°C about 10 to 15% of the HNE in 10^6 hepatocytes/ml suspension were degraded to water from HNA. The water formation from the acid could be inhibited by 4-pentenoic acid as inhibitor of beta-oxidation of fatty acids, particularly as inhibitor af acyl CoA dehydrogenase. In presence of this inhibitor the time course of HNA concentration was modulated, too. That means, in presence of the beta-oxidation inhibitor the accumulation of HNA was increased.
Ofcourse HNE reacts with sulfhydryl groups and amino groups of proteins, too. Therefore, the binding of HNE to proteins was also measured. The binding of HNE to proteins accounted only for about 3% of the initial HNE concentration. But the protein binding is assumed to be of much higher functional importance because of the inhibition of enzymes.
The HNE metabolism was balanced both in hepatocyte and in enterocyte suspensions. Table II shows the data on the concentration of primary and secondary HNE intermediates in these cells after 3 min of incubation in presence of 100 μM HNE. In Fig. 3 the main pathways of HNE metabolism in hepatocytes (Siems et al., 1990) and in enterocytes (Grune et al., 1992) are presented.
HNE metabolites were also analyzed in other cell types and tissues: in the heart (Ishikawa et al., 1986), in MH_1C_1 cells, a cell line derived from the minimal deviation 7795 Morris hepatoma (Ferro et al., 1988), in HTC cells (Canuto et al., 1990) and in HA1 Chinese hamster fibroblasts and H_2O_2-resitant cells derived from HA1 cells (Spitz et al., 1991). We investigated the HNE metabolism in further

Table II. Concentration of HNE intermediates in rat hepatocyte suspensions and rat enterocyte suspensions 3 min after the addition of 100 μM HNE. Incubations were carried out at pH 7.4, 37°C and with 10^6 cells/ml suspension. Values as nmol/ml suspension which is equal to % of the initial HNE concentration.

Metabolite	Hepatocytes	Enterocytes
1,4-Dihydroxynonene	8.1 ± 2.1	5.4 ± 0.6
GSH-HNE conjugate	27.5 ± 2.5	11.0 ± 0.5
4-Hydroxynonenoic acid	25.3 ± 5.6	4.2 ± 0.6
HNE-modified proteins	3.0 ± 0.6	1.3 ± 0.2

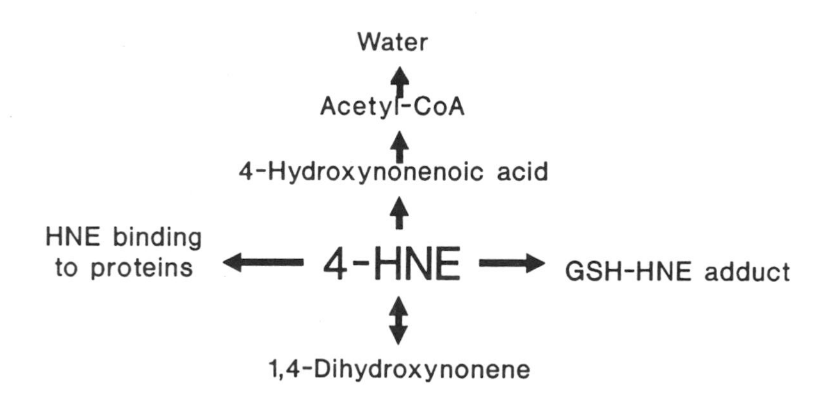

Fig. 3. Main pathways of HNE metabolism in rat hepatocytes and enterocytes of rat small intestine (according to Siems et al., 1990 and Grune et al., 1992).

eucaryotic cell types including ascites tumor cells in dependence on the growth phase of the tumor (Siems et al., 1992). In the proliferating phase of tumor growth the formation of glutathione conjugates of HNE and DHN and the protein binding are increased, and in the resting phase the oxidative metabolism of HNE to HNA and water is dominant. Interestingly, in agreement with data on decreased lipid peroxidation rate in tumor cells (Cheeseman et al., 1984 and 1986; Masotti et al., 1988) the maximal HNE degradation rate in tumor cells was markedly lower than in hepatocytes and enterocytes (see also Table I).

HNE FORMATION AND HNE METABOLISM IN THE PERFUSED SMALL INTESTINE

What about the meaning of the experiments with cell suspensions for the reals situation in organs? The reality in the organism is the solid perfused organ, not the cell suspension. In the organ under in vivo conditions HNE does not only undergo metabolic conversions, but also the washout with the blood stream. Therefore, experiments on the kinetics of HNE formation, HNE metabolism and HNE washout in the perfused small intestine were carried out. After loading of small intestine with different HNE concentrations the rate of HNE disappearance in the tissue and the HNE wahout in the perfusate outflow were measured. It was found that the washout of HNE in relation to the rate of HNE metabolism is very low. The washout in all experiments was less than 1% of the HNE metabolism in the tissue. Additionally, the rate of HNE metabolism in different periods of ischemia-reperfusion experiments was compared. The rate of HNE metabolism was markedly reduced during the phase of oxygen deficiency. The real flux rate of intracellular HNE formation during ischemia and reperfusion of rat small intestine was estimated on the basis of tissue HNE concentration, HNE utilization rate at defined HNE tissue concentration and HNE washout by perfusate circulation. Such estimation led to the values of HNE formation rate which are shown in Fig. 4 and which reach a maximum in the first minutes after the onset of reperfusion. The parameter shown in Fig. 4 is the HNE formation rate as metabolic flux rate, not the HNE tissue concentration.

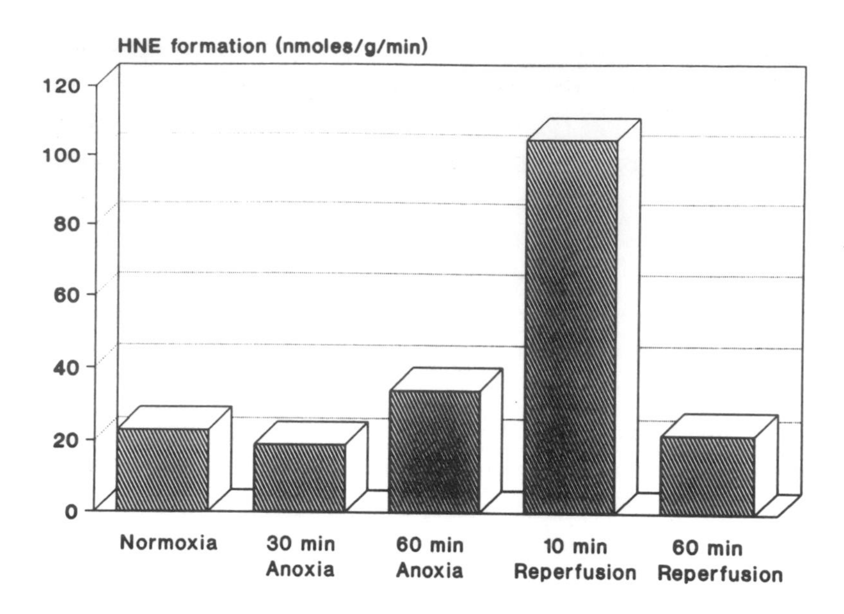

Fig. 4. Formation of HNE in rat small intestinal mucosa during 60 min of ischemia and following reperfusion under in vivo conditions estimated from HNE tissue levels, HNE utilization rate and HNE washout by perfusate circulation. Values are given as nmoles/g wet weight/min (6 experiments).

PATHOPHYSIOLOGICAL IMPORTANCE OF HNE METABOLISM

Concerning the pathophysiological meaning of our experiments we postulate that the rapid metabolism of HNE and other aldehydic products of lipid peroxidation could be an important part of the intracellular antioxidative defense system. Such a role of HNE metabolism is underlined if one takes into account that HNE probably is a natural compound which is always present (in very low concentrations) in tissues and body fluids. Certainly, HNE and other aldehydes which are formed continuously in the organism may possess important physiological functions, e.g. within the multitude of factors which regulate directly or indirectly the proliferation of cells. Only an HNE accumulation will be dangerous due to the initiation of structural and functional changes of cells. The pathways of HNE metabolism could be responsible for the maintenance of low intra- and extracellular HNE levels and for the detoxification of HNE generated during increased lipid peroxidation processes. Fig. 5 shows, that there are always more questions than results, also concerning the physiological and pathophysiological role of HNE. It is interesting to identify further HNE intermediates, to know more on the toxicity and stability of HNE products, on the interorgan relationships, on the intracellular compartmentation, on the transport through biological membranes and on the mechanisms of protein binding of HNE leading to enzyme inactivation.

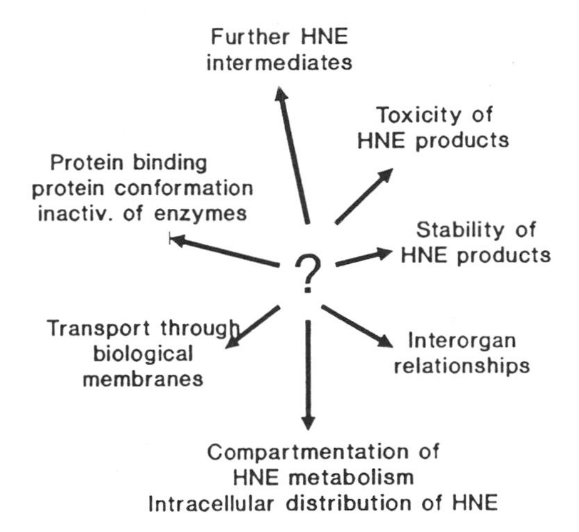

Fig. 5. Strategies for further investigations of the metabolism of 4-hydroxynonenal in eucaryotic cells.

Acknowledgements

The work has been supported by grants of the Deutsche Forschungsgemeinschaft, Bonn, F.R.G. for WGS (Ge 669/1-1/284/91), by the University Agreement between the University of Graz/Austria and the Humboldt University of Berlin/F.R.G., by grants of the Association for International Cancer Research (AICR), U.K. for HE and by grants from the Charite' Hospital Research Commission, Berlin/Germany, to WGS and TG.

REFERENCES

Canuto, R.A., Muzio, G., Bassi, A.M., Biocca, M.E., Poli, G., Esterbauer, H., and Ferro, M. (1990) in: Enzymology and Molecular Biology of Carbonyl Metabolism, vol. 3. Ed. H. Weiner. Plenum Press, New York.

Cheeseman, K.H., Burton, G.W., Ingold, K.U., and Slater, T.F. (1984) Toxicol. Pathol. 12, 552-557.

Cheeseman, K.H., Collins, M. Proudfoot, K., Slater, T.F., Burton, G.W., Webb, A.C., and Ingold, K.U. (1986) Biochem. J. 235, 507-514.

Esterbauer, H. (1982) in: Free Radicals, Lipid Peroxidation and Cancer, pp. 101-128. Eds D.C.H. McBrien and T.F. Slater. Academic Press, New York.

Esterbauer, H., Zollner, H., and Lang, J. (1985) Biochem. J. 228, 363-373.

Esterbauer, H. Zollner, H., and Schaur, R.J. (1988) ISI Atlas Sci. Biochem. 1, 311-317.

Esterbauer, H., Zollner, H., and Schaur, R.J. (1989) in: Membrane Lipid Oxidation, vol. 1, pp. 239-283. Ed. C. Vigo-Pelfrey. CRC Press, Boca Raton, FL.

Fauler, G. (1987) Investigations on the metabolism of 4-hydroxy-alkenals. Austria: Univ. Graz; Ph.D. Thesis.

Ferro, M., Marinari, U. M., Poli, G., Dianzani, M. U., Fauler, G., Zollner, H., and Esterbauer, H. (1988) Cell Biochem. Funct. 6, 245-250.

Gerber, G., Siems, W., and Werner, A. (1991) in: Membrane Lipid Oxidation, vol. 3, pp. 115-140. Ed. C. Vigo-Pelfrey. CRC Press, Boca Raton, FL.

Granger, D.N. (1988) Am. J. Physiol. 255, H1269- H1275.

Grune, T., Siems, W., Kowalewski, J., Zollner, H., and Esterbauer, H. (1991) Biochem. Int. 25, 963- 971.

Ishikawa, T., Esterbauer, H., and Sies, H. (1986) J. Biol. Chem. 261, 1576-1586.

Kowalewski, J., Siems, W., Grune, T., Werner, A., Esterbauer, H., and Gerber, G. (1991) Z. Klin. Med. 46, 143-146.

Masotti, L., Gasali, E., and Galeotti, T. (1988) Free Rad. Biol. Med. 4, 377-386.

Poli, G., Biasi, F., Chiarpotto, E., Carini, R., Cecchini, G., Ramenghi, U., and Dianzani, M.U. (1987) Free Rad. Res. Comms. 3, 279-284.

Poli, G., Cecchini, G., Biasi, F., Chiarpotto, E., Canuto, R.A., Biocca, M.E. Muzio, G., Esterbauer, H., and Dianzani, M.U. (1986) Biochim. Biophys. Acta 882, 207-214.

Poli, G., Dianzani, M.U., Cheeseman, K.H., Slater, T.F., Lang, J., and Esterbauer, H. (1985) Biochem. J. 227, 629-638.

Schaur, R.J., Zollner, H., and Esterbauer, (1991) in: Membrane Lipid Oxidation, vol. 3, pp. 141-163. Ed. C. Vigo-Pelfrey. CRC Press, Boca Raton, FL.

Siems, W., Kowalewski, J., Werner, A., Schimke, I., and Gerber, G. (1989) Free Rad. Res. Comms. 7, 347-353.

Siems, W.G., Zollner, H., and Esterbauer, H. (1990) Free Rad. Biol. Med. 9, 110.
Siems, W., Kowalewski, J., David, H., Grune, T., and Bimmler, M. (1991) Cell. Molec. Biol. 37, 213-226.
Siems, W.G., Grune, T., Beierl, B., Zollner, H., and Esterbauer, H. (1992a) in: Free Radicals and Aging, pp. 124-135. Eds. I. Emerit and B. Chance. Birkhaeuser Verlag, Basel.
Siems, W.G., Zollner, H., Grune, T., and Esterbauer, H. (1992b) Fresenius J. Anal. Chem. 343, 75-76.
Spitz, D.R., Sullivan, S.J., Malcolm, R.R., and Roberts, R.J. (1991) Free Rad. Biol. Med. 11, 415-423.

Free Radicals: From Basic Science to Medicine
G. Poli, E. Albano & M. U. Dianzani (eds.)

DNA DAMAGE BY REACTIVE OXYGEN SPECIES. THE ROLE OF METALS

R. Meneghini, E.A.L. Martins and M. Calderaro

Institute of Chemistry, Department of Biochemistry, University of Sao Paulo, Brazil.

SUMMARY: Several examples of DNA base damage induced by hydroxyl radicals have been identified recently. It is notheworthy that this damage is detected in DNA of cells submitted to oxidative stress, showing that hydroxyl radicals are formed close to DNA, by metal-induced Fenton reactions. Some of these lesions have already been shown to be mutagenic. The presence of Fe and Cu in the nucleus, probably bound to DNA, is paradoxical if we consider the potential hazard imposed on such a critical target as DNA. However, it is possible that the intracellular homeostasis of Fe and Cu is altered by the pro-oxidative status of the cell. Evidence to support this view was obtained in cells exposed to menadione. The levels of "free" Fe and Cu seem to increase substantially during menedione metabolism, and it is possible that part of the released Fe and Cu migrates to the nucleus.

INTRODUCTION: In recent years we have learned that genomic alterations are important triggers of cell malignant transformation (Weinberg, 1989). It has also been established that reactive oxygen species (ROSs) are normally produced in the cell and that numerous chemical species present in the enviroment, including in the normal diet, are capable of increasing the cellular concentrations of ROSs, above the basal level (Hallywell and Gutteridge, 1989; Ames, 1983). Because some ROSs are sufficiently reactive to produce DNA chemical modifications they must constitute a major source of malignant transformation (Cerutti et al., 1990; Borek and Troll, 1983; Ames et al., 1987) and of gene mutation (McBride et al., 1991; Shibutani et al., 1991; Nassi-Calo et al., 1989). Therefore, it has become of great interest to know which ROSs are capable of altering the DNA structure and to understand the mechanisms involved.

IN VITRO DNA DAMAGE: It is well established that O_2^-, lipid peroxides and H_2O_2 are sufficiently reactive per se to attack DNA (Meneghini and Hoffmann, 1980; Brawn and Fridovich, 1981; Lesko et al., 1980). However, OH radicals, alkoxy radicals and singlet oxygen are quite reactive towards DNA. From radiobiological studies it has long been known that OH radicals produce several types of DNA damage (Ward, 1988). the molecular mechanisms involved in the generation of these altered structures have reviewed (Cadet and Berger, 1985). More recently, with two improved analytical techniques, namely, HPLC associated to electrochemical detection and GC-MS, the possibility of identifying DNA lesions at biologically significant levels has become feasible. Over 10 base lesions have identified upon DNA exposure to sources of OH radicals generation. The most relevant were 8-hydroxy-guanine, 4,6-diamino-5-formamido pyrimidine (Fapy Gua), thymine glycol, cytosine glycol, 8-hydroxy-adenine and dihydroxy-cytidine (Aruoma, et al., 1989). An other significant base lesion seems to be 5-hydroxylmethyl-uracil, which is formed in the DNA of polymorphonuclear leukocytes (Frenkel and Chrzan, 1987). When chromatin is exposed to a source of OH radicals, a thymine-tyrosine cross-link is detected, in addition to base damage (Nackerdien et al., 1991).

The sugar moiety is also a target for OH radical attack in DNA, although to a lesser extent than the bases. The preferential center of attack is at C-4, and the carbon-centered radical produced by hydrogen abstraction undergoes rearrangements which eventually lead to base loss and strand-break with two types of termini, phosphoryl and phosphoglycolate (Demple et al., 1986; Ward, 1988; Henner et al., 1982).

Singlet oxygen is also reactive towards DNA. However, the spectrum of lesions produced is much shorter than that originating from the OH radical attack; the products identified so far are 8-hydroxy-guanine, Fapy purines and strand breaks (Di Mascio et al., 1990; Devasagayam et al., 1991; Boiteux et al., 1992). However, the biological significance of this fact depends on obtaining evidences that singlet oxygen may be generated in the cell nucleus.

MUTAGENITY AND REPAIR OF LESIONS: The mutagenicity of some DNA lesions produced by oxyradicals has been investigated. Recently it has been shown that 8-hydroxy-guanine led to the insertion of dA, dT, dG and dC in the opposite strand with almost equal frequency (Kuchino et al., 1987). Interestingly, the 3' vicinal base to 8-hydroxy-guanine was also frequently misread. Other investigation have shown that G-T transversion is the predominant point mutation induced by 8-hydroxy-guanine (Shibutani et al., 1991; Wood et al., 1990; Cheng et al., 1992); 5-hydroxylmethyl-uracil has also been shown to be mutagenic, whereas thymine glycol seems to operate more as a block for DNA replication (Shirname-More, et al., 1987). Several other forms of oxyradicals-induced DNA damage are likely to be mutagenic, and it has been shown that phage DNA exposed to Fe^{2+} and transfected into bacteria induced a ratio of mutational event to lethal event higher than for any other agent tested (McBride et al., 1991). Therefore, it is not surprising that the cells operate with distinct mechanisms to prevent the effects of these lesions, or to prevent the lesions themselves. For instance, ascorbate in the seminal fluid has been shown to prevent 8-hydroxy-guanine formation in sperm DNA (Fraga et al., 1991). In *E. coli* lacking FeSOD and MnSOD the mutation rate under aerobic growth is greatly enhanced in relation to the parental strain (Farr et al., 1989). Once the lesion is formed, DNA repair mechanisms are called into actions. DNA strand-breaks formed by attack of ROSs usually have a phosphoryl or a phosphoglycolate group at the 3' terminus (Henner et al., 1982), both constituting a block for DNA replication. In *E. coli*, exonuclease III or endonuclease IV removes these blocks, generating a 3'OH primer for DNA polymerase (Demple et al., 1986). In mammalian cells, an enzyme with the same properties of exonuclease III has been identified (Robson et al., 1991). Both 5-hydroxymethyl-uracil and thymine glycol are removed from DNA by specific glycosylases (Hollstein et al., 1984; Higgins et al., 1987). More recently, an enzyme that removes 8-hydroxyguanine and Fapy purine lesions from DNA has been identified in *E. coli* ; associated with it is an endonuclease activity which cleaves DNA at the resulting abasic site, thus creating the appropriate substrate for further DNA repair (Boiteux et al., 1992). It is likely that several more enzymes involved in the repair of DNA damaged by ROSs will be discovered in the near future. After all, there are many different DNA lesions produced by ROSs and the repair enzymes have to be relatively specific. It is interesting that, even with such elaborate machinery devoted to the prevention and repair of ROSs-induced DNA lesions, normal oxidative damage to nuclear DNA is extensive, reaching one 8-hydroxy-guanine per 130,000 bases (Richter et al., 1988).

INVOLVEMENT OF METALS IN DNA DAMAGE BY ROSs: It seems clear now that Fe is involved in the production of DNA damage by ROSs. This was first proposed when it was shown that both 1,10-phenantroline and 2,2'-dipyridyl prevented production of DNA strand breaks in mammalian fibroblasts exposed to H_2O_2 (Mello-Filho et al., 1984; Mello-Filho and Meneghini, 1984). These two compounds are strong Fe chelators which prevent the metal from participating in the Fenton reaction: $Fe^{2+} + H_2O_2 \rightarrow Fe^{3+} + OH^{\cdot} + OH^-$
Because they have a lipophilic structure, they can easily enter the cell and bind Fe(II), which is not the case for desferrioxamine, DTPA and bathophenanthroline (Mello-Filho and Meneghini, 1985). Scavengers of OH radicals also block strand-break formation, and kinetic studies indicate that the OH radical travels, on average, 15 angstroms from the site of generation to the DNA target (Ward et al., 1985). Collectively, these results suggest that H_2O_2 can reach the nucleus and react with Fe(II), which is closed to or, more likely, bound to DNA. This would generate OH radicals which react practically at the site of formation (Mello-Filho et al., 1984). Other genotoxic effects of

H_2O_2 were prevented by these chelators, such as sister chromatide exchanges (Larramendy et al., 1987), mutations at the HGPRT locus and cell transformation (Nassi-Calo et al., 1989), and most of these results have been reproduced in other laboratories (Keyse and Tyrrell, 1990; Larramendi et al., 1989; Birnboim and Kanabus-Kaminska, 1985). The recent identification of several typical types of OH radical-induced base damage in cells exposed to a source of H_2O_2 clearly demonstrates that OH· radicals are participating in site-specific Fenton reactions on DNA (Dizdaroglu et al., 1991). It also seem clear that, if OH· radicals attack DNA bases, they should be able to attack the sugar moiety as well, giving origin to strand breaks. However, some evidence has been produced to suggest that the increase in cytosolic calcium, caused by oxidative stress, might be due to activated nucleases, which would be responsible for the DNA breaks (McConkey et al., 1989; Cantoni et al., 1989; Ochi and Cerruti, 1987). It is quite possible that both types of DNA strand breaks are formed upon oxidative stress, i.e., those produced by OH radicals attack, and those produced by Ca-activated nucleases, and that, depending on the cell, one or another will prevail. However, it is unlikely that the use of Ca and Fe chelators, neither exhibiting absolute specificity, will clarify this issue. To gain knowkedge on this topic, the best approach is to look for the structure of DNA at the site of the break, since this is expected to be different, depending on the causative agent (see above).

DOES OXIDATIVE STRESS ALTER THE HOMEOSTASIS OF Fe AND Cu IN THE CELL?.

One intriguing aspect of the model proposed above is the assumption of the existence of Fe bound to DNA. In fact, Fe-DNA is a Fenton reagent (Floyd, 1981; Netto and Augusto, 1991; Aruoma et al., 1989). Therefore one would not expect Fe to be so closely associated with such a critical biological target. How much Fe or Cu is closely associated to DNA is still an open question, although some investigations have indicated the presence of these ions in the nucleus (Thorstensen and Romslo, 1984). Another possibility is that, under oxidative stress, Fe and/or Cu intracellular homeostasis is altered, as in the case of Calcium. For instance, Fe and Cu could be released from their storing sites under pro-oxidative conditions, and some of these ions could leak into the nucleus. We have recently addressed the possibility of changes in the homeostasis of Fe and Cu in cells treated with the quinone menadione. In the cell, menadione is enzymatically reduced, either partially, to the semiquinone form, or fully, to the diphenolic form. Both forms can react with dioxygen to generate superoxide, which by dismutation gives rise to H_2O_2. In fact, menadione has long been known to generate a strong pro-oxidative condition in the cell (Nohl et al., 1986). We and others have previously shown that menadione induces DNA strand breaks in the cells (Morrison et al., 1984; Ngo et al., 1991; Cantoni et al., 1991; Martins and Meneghini, 1990). One interesting aspect is that menadione-induced strand-breaks are much less susceptible to inhibition by 1,10-phenanthroline than H_2O_2-induced DNA strand-breaks. In fact, the most conspicuous effect of this chelator is to enhance the production of strand-breaks, and this is the response that is dependent on menadione concentration (Figure 1). At low menadione concentrations, even some protection by 1,10-phenanthroline is observed. We reasoned that menadione could be inducing a release of copper from some storing source. Copper is known to form a clastogenic compound with 1,10-phenanthroline, which binds DNA and causes extensive degradation; this action is known to be prevented by neocuproine, a strong Cu chelator (Que et al., 1980). In fact, neocuproine present a strong inhibitory effect in the production of strand-breaks by menadione plus 1,10-phenanthroline (Figure 1). It is interesting that neocuproine shows no inhibitory effect against the strand-breaks caused by menadione alone (not shown); therefore these latter effects are not mediated by Cu. It seems plausible to consider the following situation: the breaks induced

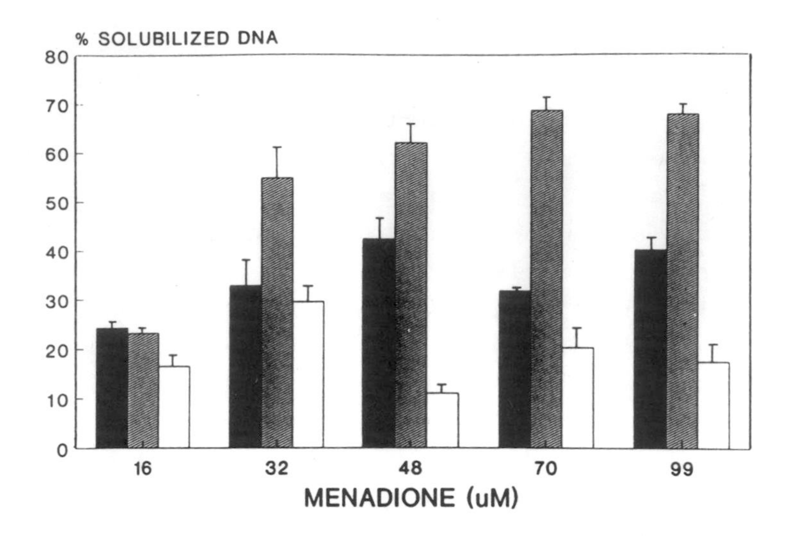

Figure 1. Menadione induced DNA strand breaks and the effect of 1,10-phenanthroline and neocuproine. V-79 Chinese hamster cells were submitted to the treatments described below. DNA strand breaks were visualized by the DNA precipitation assay (Olive, 1988) according to a modification (Martins et al., 1991). In this assay the frequency of DNA strand breaks is proportional to the amount of DNA solubilised from a K-SDS-DNA precipitate. Solid bar: breaks produced by the indicated menadione concentrations; hatched bar: breaks produced by menadione + 1 mM 1,10-phenanthroline; white bar: breaks produced by menadione + 1 mM 1,10-phenanthroline + 1 mM neocuproine.

by menadione alone are caused by H_2O_2, generated by menadione metabolism. Therefore they must presumably be produced by a Fe-induced Fenton reaction in the nucleus (Mello-Filho and Meneghini, 1991). The reason why 1,10-phenanthroline is not inhibitory in this case is because Cu ions are released by menadione metabolism, and the predominant effect becomes the clastogenic action of the Cu-phenanthroline complex.

As possible sources of Cu donors, under the effect of menadione metabolism, we might consider glutathione-Cu and metallothioneine-Cu complexes (Freedman et al., 1989). If this model is correct, then it should be possible to prevent the Fe-mediated, menadione-induced, strand breaks by an iron chelator which does not form clastogenic complexes with Cu. We have already shown that 2,2'-dipyridyl, which form a Fe complex analogous to that of

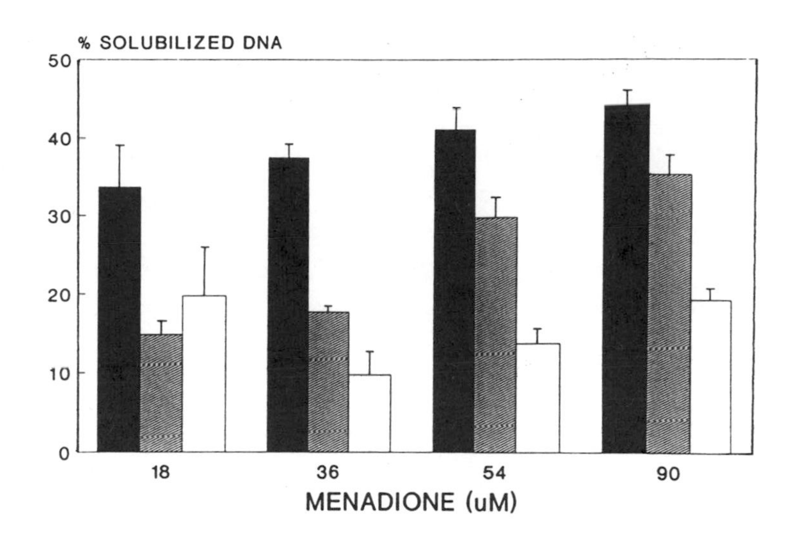

Figure 2. Menadione-induced DNA strand breaks V-79 cells and the effect of 2,2'-dipyridyl. See legend of Figure 1 for methods. Solid bar: breaks produced by the indicated menadione concentrations; hatched bar: breaks produced by menadione + 100 µM 2,2'-dipyridyl; white bar: breaks produced by menadione + 1 mM 2,2'-dipyridyl.

1,10-phenanthroline, prevents Fe-mediated Fenton reactions (Mello-Filho et al., 1984; Mello-Filho and Meneghini, 1984). However, unlike the 1,10-phenanthroline molecule, the 2,2'-dipyridyl molecule is not planar; therefore its Cu complex should not be capable of intercalating in DNA, a pre-requisite for the clastogenic action (Que et al., 1980). In fact, Figure 2 shows the inhibitory action of two different concentrations of 2,2'-dipyridyl on strand-breaks induced by menadione. Clearly, protection is strong. Because neocuproine, a strong inhibitor of Cu-mediated (but not Fe-mediated) Fenton reaction (Mello-Filho and Meneghini, 1991), has no effect on the menadione-induced breaks, it is clear that 2,2'-dipyridyl is inhibiting Fe-mediated strands. Also clear is that, as menadione concentration is increased, the inhibitory capacity is proportionally decreased.

This favour the idea that Fe ions are also being released from storing sources by menadione metabolites. In fact, it has been shown that two products of menadione metabolism, superoxide anion (Thomas and Aust, 1986) and menadione semiquinone (Monteiro et al., 1989) do promote iron release from ferritin. In conclusion, it seems clear that, during menadione metabolism, conspicuous alterations in Fe and Cu homeostasis can occur in the cell, and this might have important consequences on the genotoxic effects of oxygen radicals.

Acknowledments: This work was supported in part by the Council for Tobacco Research, USA, and in part by FAPESP, Brazil.

REFERENCES:

Ames, B.N. (1983) Science 221, 1256-1264.
Ames, B.N., Magaw, R. and Gold, L.S. (1987) Science 236, 271-280.
Aruoma, O.I., Halliwell, B. and Dizdaroglu, M. (1989) J.Biol.Chem. 264, 13024-13028.
Birnboim, H.C. and Kanabus-Kaminska, M. (1985) Proc.Natl.Acad.Sci.USA 82, 6820-6824.
Boiteux, S., Gajewiski, E., Laval, J. and Dizdaroglu, M. (1992) Biochemistry 31, 106-110.
Borek, C. and Troll, W. (1983) Proc.Natl.Acad.Sci.USA 80, 1304-1307.
Brawn, K. and Fridovich, I. (1981) Arch.Biochem.Biophys. 206, 414-419.
Cadet, J. and Berger, M. (1985) Int.J.Radiat.Biol. 47, 127-143.
Cantoni, O., Sestili, P., Cattabeni, F., Bellomo, G., Pou, S., Cohen, M. and Cerutti, P. (1989) Eur.J.Biochem. 182, 209-212.
Cantoni, O., Fiorani, M., Cattabeni, F. and Bellomo, G. (1991) Biochem.Pharmacol. 42 Suppl., S220-S222.
Cerutti, P., Amstad, P., Larsson, R., Shah, G. and Krupitza, G. (1990) Prog.Clin.Biol.Res. 347, 183-186.
Cheng, K.C., Cahill, D.S., Kasai, H., Nishimura, S. and Loeb, L.A. (1992) J.Biol.Chem. 267, 166-172.
Demple, B., Johnson, A. and Fung, D. (1986) Proc.Natl.Acad.Sci.USA 83, 7731-7735.
Devasagayam, T.P.A., Steenken, S., Obendorf, M.S.W., Schulz, W.A. and Sies, H. (1991) Biochemistry 30, 6283-6289.
Di Mascio, P., Kaiser, S.P., Thomas, , Devasagayam, P.A. and Sies, H. (1990) Adv.Exp.Med.Biol. 283, 71-77.
Dizdaroglu, M., Nackerdien, Z., Chao, B.-C., Gajewski, E. and Rao, G. (1991) Arch.Biochem.Biophys. 285, 388-390.
Farr, S.B., D'Ari, R. and Touati, D. (1989) Proc.Natl.Acad.Sci.USA 86, 8268-8272.
Floyd, R.A. (1981) Biochem.Biophys.Res.Commun. 99, 1209-1215.
Fraga, C.G., Motchnik, P.A., Shigenaga, M.K., Helbock, H.J., Jacob, R.A. and Ames, B.N. (1991) Proc.Natl.Acad.Sci.USA 88,

11003-11006.
Freedman, J.H., Ciriolo, M.R. and Peisach, J. (1989) J.Biol.Chem. 264, 5598-5605.
Frenkel, K. and Chrzan, K.C. (1987) Carcinogenesis 8, 455-460.
Halliwell, B. and Gutteridge, J.M.C. Free radicals in biology and medicin, Oxford:Clarendon press, 1989. Ed. 2nd
Henner, W.D., Grunberg, S.M. and Haseltine, W.A. (1982) J.Biol.Chem. 257, 11750-11754.
Higgins, S.A., Frenkel, K., Cummings, A. and Teebor, G.W. (1987) Biochemistry 26, 1683-1688.
Hollstein, M.C., Brooks, P., Linn, S. and Ames, B.N. (1984) Proc.Natl.Acad.Sci.USA 81, 4003-4007.
Keyse, S.M. and Tyrrell, R.M. (1990) Carcinogenesis 11, 787-791.
Kuchino, Y., Mori, F., Kaszi, H., Inoue, H., Iwai, S., Miura, K., Ohtsuka, E. and Nishimura, S. (1987) Nature 327, 77-79.
Larramendy, M., Mello-Filho, A.C., Martins, E.A.L. and Meneghini, R. (1987) Mutation Res. 178, 57-63.
Larramendy, M., Lopez-Larraza, D., Vidal-Rioja, L. and Bianchi, N.O. (1989) Cancer Res. 49, 6583-6586.
Lesko, S.A., Lorentzen, R.J. and Ts'o, P.O.P. (1980) Biochemistry 19, 3027-3028.
Martins, E.A.L., Chubatsu, L.S. and Meneghini, R. (1991) Mutation Res. 250, 95-101.
Martins, E.A.L. and Meneghini, R. (1990) Free Radical Biol.Med. 8, 433-440.
McBride, T.J., Preston, B.D. and Loeb, L.A. (1991) Biochemistry 30, 207-213.
McConkey, D.J., Hartzell, P., Jondal, M. and Orrenius, S. (1989) J.Biol.Chem. 264, 13399-13402.
Mello-Filho, A.C., Hoffmann, M.E. and Meneghini, R. (1984) Biochem.J. 218, 273-275.
Mello-Filho, A.C. and Meneghini, R. (1984) Biochim.Biophys.Acta 781, 56-63.
Mello-Filho, A.C. and Meneghini, R. (1985) Biochim.Biophys.Acta 847, 82-89.
Mello-Filho, A.C. and Meneghini, R. (1991) Mutation Res. 251, 109-113.
Meneghini, R. and Hoffmann, M.E. (1980) Biochim.Biophys.Acta 608, 167-173.
Monteiro, H.P., Ville, G.F. and Winterbourn, C.C. (1989) Free Radic.Biol.Med. 6, 587-591.
Morrison, H., Jernstrom, B., Nordenskjold, M., Thor, H. and Orrenius, S. (1984) Biochem.Pharmacol. 33, 1763-1769.
Nackerdien, Z., Rao, G., Cacciuttolo, M.A., Gajewski, E. and Dizdaroglu, M. (1991) Biochemistry 30, 4873-4879.
Nassi-Calo, L., Mello-Filho, A.C. and Meneghini, R. (1989) Carcinogenesis 10, 1055-1057.
Netto, L.E.S., Ferreira,A.M. and Augusto, O. (1991) Chem.Biol.Interact. 79, 1-14.
Ngo, E.O., Sun, T.-P., Chang, J.-Y., Wang, C.-C., Chi, K.-H., Cheng, A.-L. and Nutter, L.M. (1991) Biochem.Pharmacol. 42, 1961-1968.
Nohl, H., Jordan, W. and Youngman, R.J. (1986) Adv.Free Rad.Biol.Med. 2, 211-279.

Ochi, T. and Cerutti, P. (1987) Proc.Natl.Acad.Sci.USA 84, 990-994.
Olive, P.L. (1988) Environ.Mol.Mutag. 11, 487-495.
Que, B.G., Downey, K.M. and So, A.G. (1980) Biochemistry 19, 5987-5991.
Richter, C., Park, J.W. and Ames, B.N. (1988) Proc.Natl.Acad.Sci.USA 85, 6465-6467.
Robson, C.N., Milne, A.M., Pappin, D.J.C. and Hickson, I.D. (1991) Nucleic Acids Res. 19, 1087-1092.
Shibutani, S., Takeshita, M. and Grollman, A.P. (1991) Nature 349, 431-434.
Shirname-More, L., Ressmann, T.G., Troll, W., Teebor, G.W. and Frenkel, K. (1987) Mutation Res. 178, 177-186.
Thomas, C.E. and Aust, S.D. (1986) J.Biol.Chem. 261, 13064-13070.
Thorstensen, K. and Romslo, I. (1984) Biochim.Biophys.Acta 804, 200-208.
Ward, J.F., Blakely, W.F. and Joner, E.I. (1985) Radiation Res. 103, 383-392.
Ward, J.F. (1988) Prog.Nucl.Ac.Res. 35, 96-125.
Weinberg, R.a. (1989) Cancer Res. 49, 3713-3721.
Wood, M.L., Dizdaroglu, M., Gajewski, E. and Essigmann, J.M. (1990) Biochemistry 29, 7024-7032.

Free Radicals: From Basic Science to Medicine
G. Poli, E. Albano & M. U. Dianzani (eds.)

INFLAMMATION AND A MECHANISM OF HYDROGEN PEROXIDE CYTOTOXICITY

E.M. Link

Department of Chemical Pathology, Division of Molecular Pathology, University College and Middlesex School of Medicine, Cleveland Street, London W1P 6DB, U.K.

SUMMARY

The joint appearance of hydrogen peroxide and prostanoids at a site of inflammation, as well as H_2O_2 capacity of initiating inflammatory symptoms suggest that the involvement of both in inflammation is causative. H_2O_2 can either augment or diminish prostanoid synthesis depending on its concentration. On the other hand, indomethacin, an anti-inflammatory drug inhibiting cyclooxygenase - the enzyme crucial in prostanoid production - sensitizes or protects mammalian cells against hydrogen peroxide toxicity at H_2O_2 concentration range similar to that stimulating/inhibiting prostanoid synthesis. The survival of cells exposed to H_2O_2 in the absence or presence of indomethacin differs by up to 80%. It is, therefore, obvious that H_2O_2-dependent increase in prostanoid synthesis protects against H_2O_2 cytotoxicity. The putative mechanism involved and possible consequences of preventing inflammatory symptoms by inhibiting cyclooxygenase activity are discussed.

INTRODUCTION

Erythema, edema, local hyperthermia and pain are infallible first symptoms of inflammation (Fig.1). Their appearance is induced by locally elevated amounts of prostanoids. This is why non-steroidal anti-inflammatory drugs (NSAIDs) such as indomethacin or aspirin which inhibit cyclooxygenase - an enzyme crucial in prostanoid production - are administered to control the symptoms [Vane,1987]. The other characteristic species at a site of inflammation is hydrogen peroxide [Halliwell & Gatteridge, 1989].

Inflammatory symptoms are the first side-effect following radiotherapy [Polgar, 1988] or direct application of hydrogen peroxide [Weitzman et al., 1986; Rees & Orth, 1986]. Since ionizing radiation generates H_2O_2 [Casarett, 1968] and all forms of inflammatory reactions are associated with increased prostanoid synthesis [Willoughby, 1987], the joint appearance of hydrogen peroxide and prostanoids is suggestive of a causative link between them. A knowledge of the mechanism involved in H_2O_2 cytotoxicity might, therefore, help to elucidate such an association.

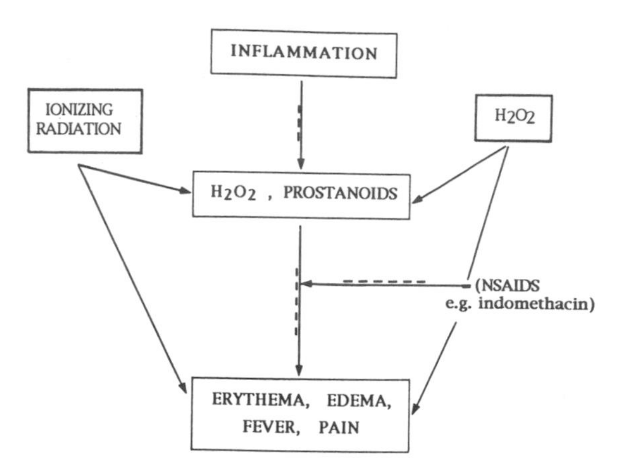

Fig.1 Schematic illustration of hydrogen peroxide and prostanoid involvement in inflammatory reactions.

SOURCE OF HYDROGEN PEROXIDE AT A SITE OF INFLAMMATORY REACTION

1. Phagocytic Cells:

Phagocytic cells such as neutrophils and macrophages accumulate at a site of inflammation. Once activated, they undergo the so called respiratory burst which results in $O_2^{\cdot -}$ and H_2O_2 production [Halliwell & Gutteridge, 1989].

2. Ionizing Radiation:

Hydrogen peroxide is one of the main products of water radiolysis. 5.5 Gy of γ-rays from ^{137}Cs delivered in 60 sec. generates 1.2-1.5 μM H_2O_2 in physiological saline [Link, 1987], 50 Gy of X-rays forms 10 μM H_2O_2 in PBS or serum-free Eagle's medium [unpublished data]. Most biological systems consist of approximately 80% of water and a single fractional dose applied during routine radiotherapy with low LET ionizing radiation amounts to 2 - 8 Gy, whereas a total dose varies from 30 to 60 Gy.

Survival of mammalian cells can significantly differ depending on whether the cells are irradiated in catalase-free medium or a medium supplemented with the enzyme (Fig.2 and [Link, 1987]). While the dose-dependent survival curve of epithelial cells exposed to γ-rays in phosphate-buffered saline

(PBS) has a typical shape with the shoulder (Fig.2), surviving fraction of the cells irradiated in PBS supplemented with catalase diminishes linearly with increasing radiation dose (Fig.2). The slope of the latter curve is significantly shallower than of that with the shoulder showing that H_2O_2 generated in the medium during irradiation contributes substantially to the overall cell damage caused by γ-rays.

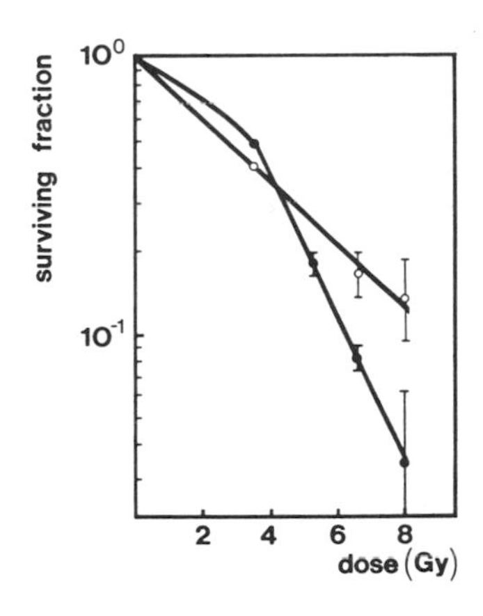

Fig.2 Survival of $2x10^5$ epithelial cells irradiated with γ-rays from ^{137}Cs source (dose rate: 5 Gy min^{-1}) in 10 ml PBS (●) or PBS supplemented with 330 units ml^{-1} of catalase (o) as assessed by clonogenic assay.

INVOLVEMENT OF HYDROGEN PEROXIDE IN PROSTANOID FORMATION

Prostaglandin endoperoxide synthase (PES) consists of two enzymic activities: cyclooxygenase and peroxidase. Cyclooxygenase catalyses fatty acid oxidation. Initially, the reaction proceeds slowly since the enzyme is using its own product - a hydroperoxide PGG_2 for autoacceleration [Lands et al., 1971; Hemler et al., 1978; Needleman et al., 1986; Lands, 1988]. However, since exogenous hydroperoxides, including H_2O_2, eliminate such kinetic lag phase [Lands et al., 1971; Hemler et al., 1978; Needleman et al., 1986; Lands, 1988], cyclooxygenase in the presence of H_2O_2 should both increase synthesis of prostanoids and protect cells against hydrogen peroxide cytotoxicity. Indeed, C.Deby and G.Deby-Dupoint [Deby & Deby-Dupoint, 1980] showed a significant influence of H_2O_2 on prostaglandin production measured as a conversion of ^{14}C-arachidonic acid to PGE_2 and $PGF_{2\alpha}$ by bull seminal vesicles (Fig.3). Hydrogen peroxide at the concentration range of 0.1 μM to 10 μM continuosly stimulated prostaglandin formation with the maximum of 185% physiological level obtained for 5-10 μM H_2O_2. Further increase in hydrogen peroxide concentration resulted in a sharp inhibition of prostaglandin production leading to 40% of the maximum value at 100 μM H_2O_2.

A similar biphasic curve illustrating PGI_2 formation by endothelial cells exposed to H_2O_2 at a comparable concentration range was obtained recently by D.J. Boswell et al. [Boswell et al., 1992].

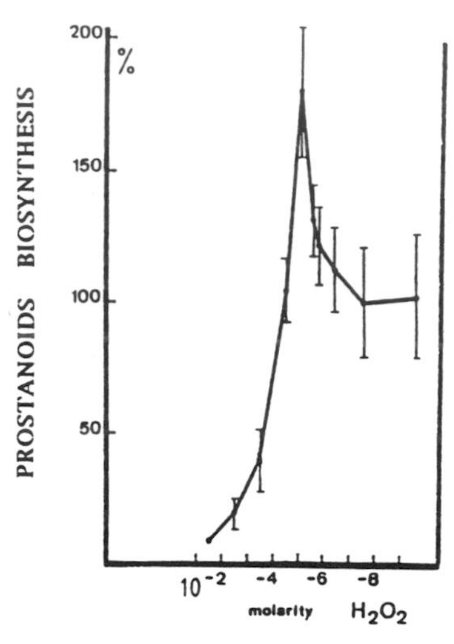

Fig.3 H_2O_2 dependence of prostaglandin E_2 and $F_{2\alpha}$ formation by bull seminal vesicles. (By C.Deby and G.Deby-Dupont, 1980; with the authors' permission)

INTERACTION OF CYCLOOXYGENASE AND GLUTATHIONE PEROXIDASE IN PRESENCE OF HYDROGEN PEROXIDE

Glutathione peroxidase is a crucial enzyme protecting cells exposed to exogenous hydrogen peroxide [Link, 1988a; Link, 1988b]. However, most enzymes are constituents of more than one biochemical pathway and do not act independently. Thus, a maximum cell protection by glutathione peroxidase from H_2O_2 cytotoxicity is conditioned by the availability of glucose for the pentose phosphate pathway which, by supplying NADPH to glutathione reductase, maintains the glutathione redox cycle and, consequently, the activity of glutathione peroxidase [Link, 1988a; Link, 1988b]. However, a maximal activation of this biochemical pathway might alter the efficiency of different chain(s) of reactions in which the enzymes in question or their substrates/products are also indispensable. Indeed, activated glutathione peroxidase in the presence of GSH interferes with the activity of cyclooxygenase by decomposing PGG_2 produced by the latter enzyme and used by it in a positive feedback reaction for autoacceleration (Fig.4). An interaction of both enzymic pathways could, therefore, modify the efficiency of hydrogen peroxide decomposition by glutathione peroxidase due to a competition between H_2O_2 and PGG_2.This, however, would inhibit cyclooxygenase

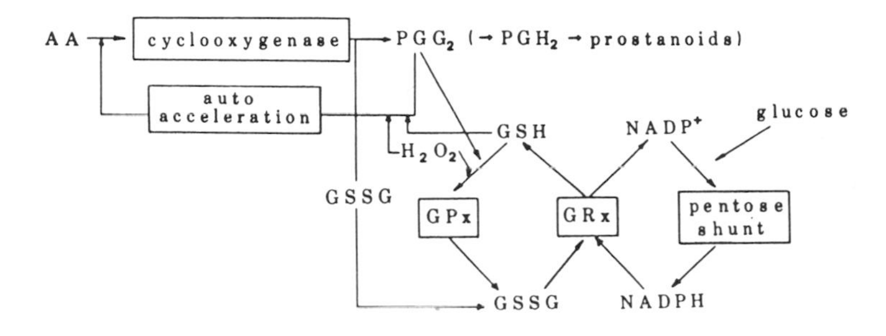

Fig.4 Interaction between glutathione peroxidase and cyclooxygenase enzymic pathways.

unless exogenous hydrogen peroxide replaces PGG_2 in the autoacceleration of the enzyme. Consequently, production of prostanoids will be altered as described in the previous section and an overall decomposition of H_2O_2 by cells will differ from that obtained if only one of the enzymes remained active. This, in turn, could result in a different level of survival of cells exposed to exogenous hydrogen peroxide.

The influence of such interaction of both enzymic pathways was investigated by inhibiting either glutathione peroxidase and pentose shunt or cyclooxygenase, or both enzymes simultaneously [Link, 1990]. Experiments were carried out using epithelial cells in the exponential growth phase. Cells were exposed to H_2O_2 whilst suspended in phosphate-buffered saline (PBS) for 1 h at 37°C. Their survival was determined by a clonogenic assay. Cellular glutathione peroxidase activity was modified by using PBS with or without glucose (4.7 mg/100ml) for a 1 h preincubation and during cell exposure to H_2O_2 carried out at either pH 6.5 or 7.5 [Link, 1988a; Link, 1988b]. (The pH-dependence of glutathione peroxidase activity is characterised by a maximum at pH 8.5; there is no activity at pH 6.0 [Chaudiere et al., 1984; Maddipati et al., 1987].) Indomethacin (0.25mg/100 ml) was used as cyclooxygenase inhibitor and was added to PBS for 1 h before and during cell incubation with H_2O_2 [Link, 1990].

At pH 6.5 and in the absence of glucose when glutathione peroxidase activity was residual, survival of cells exposed to hydrogen peroxide decreased linearly with increasing hydrogen peroxide concentration (Fig.5A).

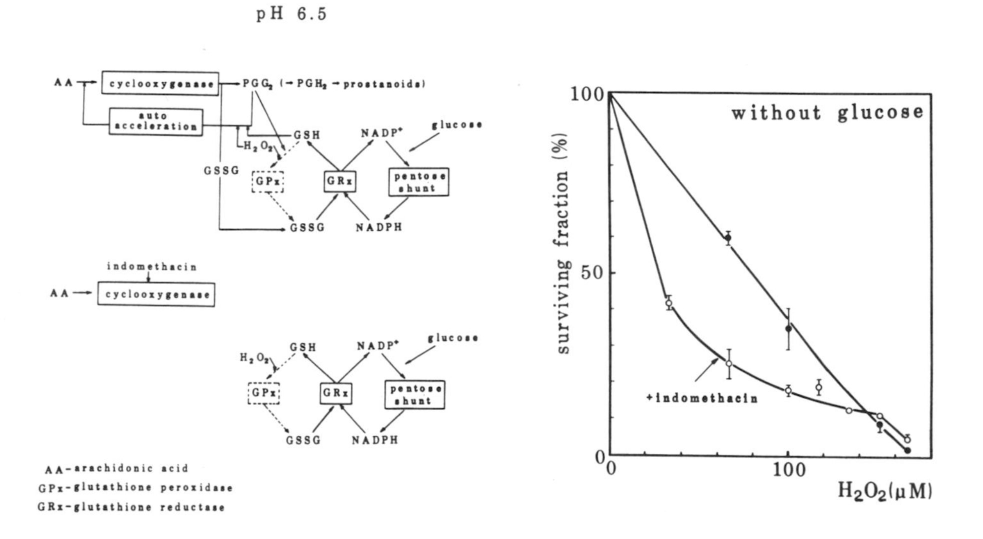

Fig.5A Dose-survival curves of cells exposed to H_2O_2 at pH 6.5 after preincubation with (o) or without (●) a cyclooxygenase inhibitor, indomethacin, in the absence of glucose and an illustration of the possible mechanism involved in cell response to hydrogen peroxide under such conditions [Link, 1990].

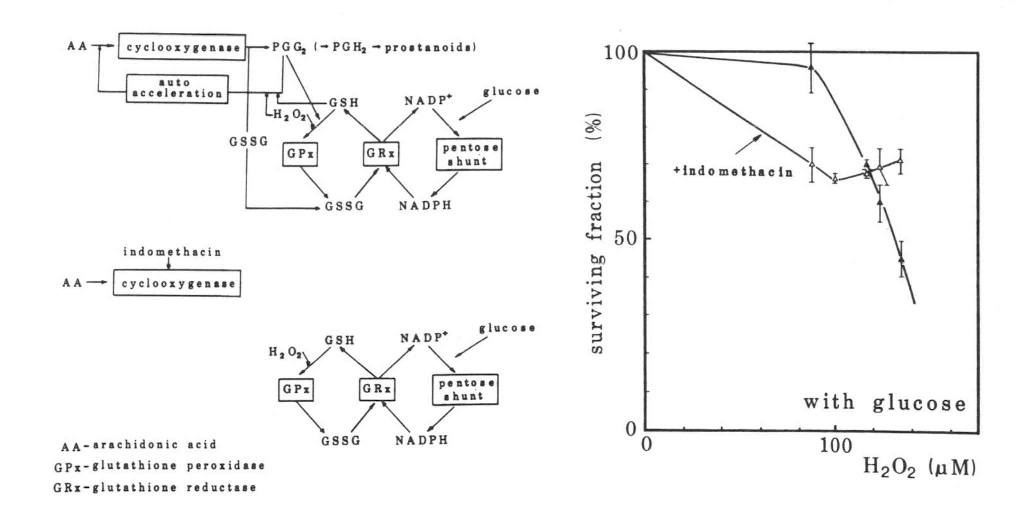

Fig.5B Dose-survival curves of cells exposed to H_2O_2 at pH 7.5 after preincubation with (△) or without (▲) a cyclooxygenase inhibitor, indomethacin, in the presence of glucose and an illustration of a possible mechanism involved in cell response to hydrogen peroxide under such conditions [Link, 1990].

It was evident that cyclooxygenase activity (as determined by the effect of indomethacin on the survival of cells exposed to H_2O_2) was essential for the protection of cells with glutathione peroxidase activity reduced to its minimum when hydrogen peroxide concentration did not exceed 35 μM (Fig. 5A). The protection was gradually less significant between 35 and 145 μM. Above this concentration range indomethacin did not affect cell response to H_2O_2.

A gradual supply of glucose and, subsequently, an increase of pH to 7.5 with a varying availability of glucose from its initial absence to the excess at the final stage, enabled a detection of the progressive involvement of glutathione peroxidase in the cellular response to hydrogen peroxide and the influence of interaction of both enzymic pathways on the cell survival after exposure to H_2O_2 [Link, 1990].

Restored activity of glutathione peroxidase (pH 7.5) and a full availability of GSH, unlimited by insufficient GSSG reduction (additional supply of glucose) led to maximal decomposition of H_2O_2 by both enzymes

(Fig.5B). This resulted in a significant protection against hydrogen peroxide cytotoxicity: up to 90 μM H_2O_2 the surviving fraction diminished only by 3%. Afterwards, however, a sharp decrease in cell survival was observed. The turning point corresponded with that of survival curve obtained for cells exposed to H_2O_2 in the presence of indomethacin when inhibition of cyclooxygenase resulted initially in lower survival but gradually protected cells above 100 μM H_2O_2 ultimately leading to the overall surviving fraction higher than that in the presence of both enzyme activities.

CONCLUSIONS

Elevated amounts of prostanoids observed at a site of all forms of inflammatory reactions are associated with the appearance of hydrogen peroxide. The latter is generated by phagocytic cells or ionizing radiation which trigger the inflammatory reaction. Alternatively, H_2O_2 causes inflammatory symptoms by itself if applied directly. A uniform pattern of changes in prostanoid synthesis by various cells depending on hydrogen peroxide concentration and illustrated by a characteristic biphasic curve, as well as a complementary survival pattern of the cells exposed to H_2O_2 at similar concentrations but in the presence of cyclooxygenase inhibitor, strongly argue for a causative link between hydrogen peroxide and prostanoid formation at a site of inflammation.

A significant interaction of glutathione peroxidase enzymic pathway with that of cyclooxygenase which results in a full cell protection against H_2O_2 cytotoxicity under conditions when both enzyme activities are at their optimum (Fig. 5B), as well as a low survival of the cells (approximately 20%) while both enzyme activities are inhibited, suggest a major role of the enzymes in defending mammalian cells against toxic action of H_2O_2. Since a turning point from the induction to the inhibition of prostanoid synthesis occurs between $5x10^{-6}$ and 10^{-5} M H_2O_2 which is in the range of K_m for peroxidase activity of PES (=10^{-5} M H_2O_2) and well above concentrations of exogenous hydroperoxides eliminating the kinetic lag phase of cyclooxygenase, it seems that H_2O_2 gradually becomes a competitor to PGG_2 - the usual substrate of the enzyme - and thus prevents a conversion of PGG_2 to PGH_2. This may gradually inhibit prostanoid production. Furthermore, such use of H_2O_2 for peroxidase activity of PES should protect cells against its cytotoxicity and, since H_2O_2 is competing with PGG_2, inhibition of cyclooxygenase by indomethacin should enable even more efficient decomposition of H_2O_2 and further cell protection. This seems to be the case (see Fig.5B).

In consequence, if the proposed mechanism is correct, we are facing a clinical paradox in that hydrogen peroxide at concentrations stimulating prostanoid synthesis induces very unpleasent inflammatory symptoms. However, the process is associated with cell protection against cytotoxic action of H_2O_2. By inhibiting cyclooxygenase activity with NSAIDs, the symptoms are alleviated but the tissue suffers a damage caused by H_2O_2. On the other hand, when an inhibition of cyclooxygenase would be beneficial, i.e. would protect the tissue against H_2O_2 toxicity (above 10^{-4} M H_2O_2), NSAIDs are probably not introduced since significantly lower production of prostanoids results in negligible inflammatory symptoms and, therefore, the treatment is either withdrawn or not applied at all.

REFERENCES

Boswell, D.J., Greenfield, S.M., Thompson, R.P.H. and Punchard, N.A. The effect of hydrogen peroxide (H_2O_2) on endothelial cell prostacyclin (PGI_2) production. Int. J. Radiat. Biol., 1992, in press

Casarett, A.P., (ed.), Radiochemistry of water. In: Radiation Biology. Prentice-Hall, Inc., Englewood Cliffs, New Jersey, 1968, pp. 62-68

Chaudiere, J., Wilhelmsen, E.C. and Tappel, Al L. Mechanism of selenium-glutathione peroxidase and its inhibition by mercaptocarboxylic acids and other mercaptans. J.Biol. Chem., 259, 1043-1050, 1984

Deby, C. and Deby-Dupoint, G. Oxygen species in prostaglandin biosynthesis *in vitro* and *in vivo*. In: Biological and Clinical Aspects of Superoxide Dismutase, (Bannister, W.H. and Bannister, J.V., eds.), Elsevier, Amsterdam, 1980, pp.84-97

Halliwell, B. and Gutteridge, J.M.C.(eds.), Phagocytosis. In: Free Radicals in Biology and Medicine. 2nd ed., Clarendon Press, Oxford, 1989, pp. 372-390

Hemler, M.E., Graff, G. and Lands, W.E.M. Accelerative autoactivation of prostaglandin biosynthesis by PGG_2. Biochem. and Biophys. Res. Commun., 85, 1325-1331, 1978

Lands, W.E.M. Enzymic assay of hydroperoxides. In: Free Radicals, Methodology and Concepts (Rice-Evans, C. and Halliwell, B., eds.), Richelieu Press, London, 1988, pp. 213-224

Lands, W.E.M., Lee, R. and Smith, W. Factors regulating the biosynthesis of various prostaglandins. Ann.N.York Acad.Sc., 180, 107-122, 1971

Link, E.M. Enzymic pathways involved in cell response to H_2O_2. Free Rad. Res.Comms., 11, 89-97, 1990

Link, E.M. In: Free Radical Biochemistry and Radiation Injury. (Wardman, P. and Willson, R.L., eds.), Br.J.Cancer, 55 (Suppl.VIII), 110-112, 1987

Link, E.M. The mechanism of pH-dependent hydrogen peroxide cytotoxicity. Arch. Biochem. Biophys., 265, 362-372, 1988a

Link, E.M. Why is H_2O_2 cytotoxicity pH-dependent? In: Free Radicals, Methodology and Concepts (Rice-Evans, C. and Halliwell, B., eds.), Richelieu Press, London, 1988b, pp. 539-548

Maddipati, K.R., Gasparski, C. and Marnett, L.T. Characterization of the hydroperoxide-reducing activity of human plasma. Arch. Biochem. Biophys., 254, 9-17, 1987

Needleman, P., Turk, J., Jakschik, B.A., Morrison, B.A. and Lefkowitch, J.B. Arachidonic acid metabolism. Ann. Rev. Biochem., 55, 69-102, 1986

Polgar, P., (ed.), Eicosanoids and Radiation. Kluwer Academic Publishers, Boston, 1988

Rees, T.D. and Orth, C. Oral ulcerations with use of hydrogen peroxide. J. Periodontol., 57, 689-692, 1986

Vane, J.R. Anti-inflammatory drugs and the many mediators of inflammation. Int. J. Tiss. Reac., IX, 1-14, 1987

Weitzman,S.A., Weitberg, A.B., Stossel, T.P., Schwartz, J. and Shklar, G. Effects of hydrogen peroxide on oral carcinogenesis in hamsters. J. Periodontol., 57, 685-688, 1986

Willoughby, D.A., (ed.), Inflammation - mediators and mechanisms. Br.Med.Bull., 43, 247-477, 1987

Aging

Free Radicals: From Basic Science to Medicine
G. Poli, E. Albano & M. U. Dianzani (eds.)

FREE RADICAL THEORY OF AGING

D. Harman

University of Nebraska College of Medicine, 600 South 42nd Street, Omaha, Nebraska 68198-4635, U.S.A.

SUMMARY: Aging is the accumulation of changes responsible for both the sequential alterations that accompany advancing age and the associated progressive increases in the chance of disease and death. The production of these changes can be attributed to the environment and disease and to an inborn aging process(es). Past improvements in general living conditions have decreased the chances for death so that they are now near limiting values in the developed countries. In these countries the aging process is the major cause of disease and death after about age 28. The free radical theory of aging postulates that aging changes are caused by free radical reactions. Support for this theory is extensive; it includes: 1) studies on the origin and evolution of life, 2) studies of the effect of ionizing radiation on living things, 3) dietary manipulation of endogenous free radical reactions, 4) the plausible explanation it provides for aging phenomena, and 5) the growing number of studies that implicate free radical reactions in the pathogenesis of specific diseases. On the basis of present data the healthy active life span may be increased by 5-10 or more years by keeping body weight down, at a level compatible with a sense of well-being while ingesting diets adequate in essential nutrients but designed to minimize random free radical reactions in the body.

INTRODUCTION

Aging is the accumulation of changes responsible for both the sequential alterations (Kohn, 1985; Upton, 1977) that accompany advancing age and the associated progressive increases in the chance of disease and death. The chance of death serves as a measure of the number of such accumulated changes, i.e., of physiologic age, while the rate of change of this parameter with time measures the rate of accumulation, i.e., the rate of aging. The production of these changes can

be attributed to the environment and disease and to an inborn aging process(es).
The chance of death for man (Harman, 1991) - readily obtained from vital statistics data - drops precipitously after birth to a minimum figure around puberty and then increases with age to a value beyond which it rises almost exponentially (Kohn, 1985; Upton, 1977; National Center for Health Statistics, 1988; Jones, 1955; Dublin et al., 1949) at a rate characteristic of man, so that few individuals reach age 100, and none live beyond about 115 years (Comfort, 1979). Improvements in general living conditions - better nutrition, medical care, etc. - decrease the chance of death (Jones, 1955; Dublin et al, 1949) in the young more than in the old; illustrated in Figure 1 by the curves of the logarithm of the chance of death versus age for Swedish females for various periods from 1751 to 1988.

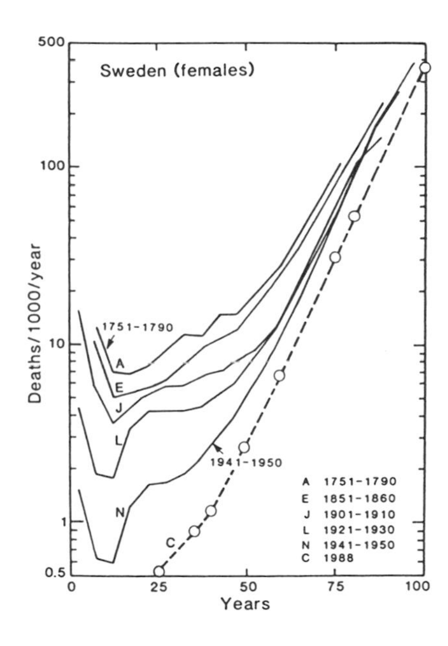

Age-Specific death rates of Swedish females in various periods from 1751 to 1950 (Adapted from Jones, H. R.: The relation of human health to age, place and time, in Handbook of Aging and the Individual, edited by Birren, J.E., Chicago Univ., Chicago Press, 1959, pp. 336-363) and for 1988 (Sveriges officiella statistik. Statistiska centralbyran, Stockholm)

Fig. 1. Age-specific death rates of Swedish females from 1751 to 1950 - adapted from Jones, H.R. (1955); data for 1988 are from Sveriges Officiella Statistik (1988).

Today in the developed countries, the chance of death rises almost exponentially after about age 28 (National Center for

Health Statistics, 1988; Sveriges Officiella Statistik, 1988; Office Federal de la Statistique, 1988). These chances are now near limiting values; only 2-3 percent of a cohort die before age 28 while average life expectancies at birth - determined by the chances for death - in the United States (National Center for Health Statistics, 1988, 1989) (Figure 2) as well as in other developed countries, approach plateau values of around 75 years for males and 80 years for females.

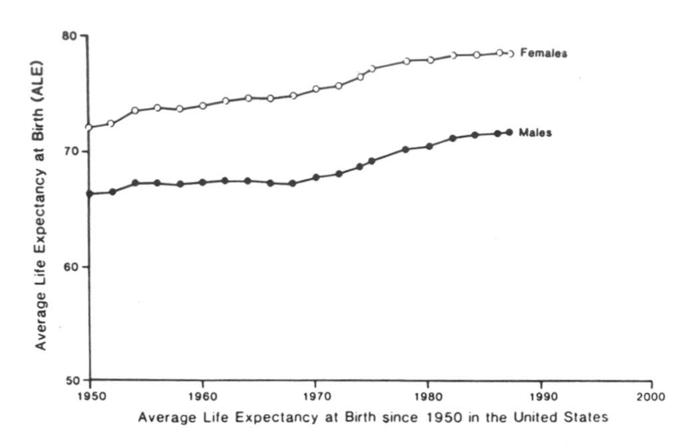

Fig. 2. Average life expectancy at birth since 1950 in the United States.

The aging process is now the major risk factor for disease and death after around age 28 in the developed countries. The importance of this process to our health and well-being is obscured by the protean nature of its contributions to non-specific change and to disease pathogenesis. As "risk factors" for diseases are detected and minimized the chance of death decreases toward that determined by the aging process, while the associated average life expectancy at birth approaches a maximum of about 85 years (Woodhall et al., 1957; Fries, 1980; Olshansky et al., 1990). Average life expectancy at birth in the developed countries is today about 10 years less than the potential maximum.

Significant increases in average life expectancy at birth in the developed countries are now likely to be achieved only by slowing the aging process. Because the average period of senescence is not known, average life span at birth serves as a measure of the span of healthy, productive life, i.e., the functional life span.

THEORIES OF AGING

Many theories have been advanced to account for the aging process (Harman, 1991). No one theory is generally accepted: "this remarkable process remains a mystery" (Rothstein, 1986); "it is doubtful that a single theory will explain all the mechanisms of aging" (Schneider, 1987; Vijg, 1990).

The importance attached to increasing the functional life span of man dictates that the aging process hypotheses be explored for practical means of achieving this goal, while continuing to work toward a consensus. The free radical theory of aging (Harman 1956, 1962, 1981, 1986) shows promise of application today.

FREE RADICAL THEORY OF AGING

Statement of the Theory

The free radical theory of aging arose in 1954 from a consideration of aging phenomenon from the premise (Harman, 1956, 1992) that a single common process, modifiable by genetic and environmental factors, was responsible for the aging and death of all living things. The theory postulates that aging is caused by free radical reactions, i.e., these reactions may be involved in production of the aging changes associated with the environment, disease and the intrinsic aging process. It predicts that the life-span of an organism can be increased by slowing the rate of initiation of random free radical reactions and/or decreasing their chain lengths. The former should be achieved by decreasing ingestion of easily oxidized dietary components, caloric intake, and temperature; the latter by increasing the concentrations of free radical inhibitors in the organism or by increasing the resistance of it's constituents to free radical attack. Studies are in accord with these predictions.

Although the free radical theory of aging is applicable to all life, the following comments are directed largely to mammalian aging, in which O_2 is the main source of damaging free radical reactions.

Free Radical Reactions and the Origin and Evolution of Life

A reasonable explanation (Harman 1981) for the prominent presence of free radical reactions in living things is provided by studies on the origin and evolution of life; these are summarized in Table 1. Life apparently originated

Table I. Overview of the origin and evolution of life.

Years ago	Main events occurring
3.5 Billion	Basic chemicals of life formed by free radical reactions, largely initiated by ionizing radiation from the sun. Life begins, excision and recombinational repair becomes possible.
	Ferredoxin appears. RH or $H_2S + CO_2 \xrightarrow{(h\nu)} CH$
2.6 Billion	Blue-Green Algae $H_2O \xrightarrow{(h\nu)} 2H + O_2$
1.3 Billion	Atmospheric O_2 reaches one percent of present value. Anaerobic procaryotes disappear. Eucaryotes become dominant cells. Eucaryotes + blue-green algae $\longrightarrow$ the green leaf plants. Eucaryotes + a procaryote able to reduce O_2 to $H_2O \longrightarrow$ animal kingdom. Emergence of multicellular organisms and plants. Meiosis becomes possible.
500 Million	Atmospheric O_2 reaches 10 percent of present value. Ozone screen allows emergence of life from sea.
65 Million	Primates appear.
5 Million	Man appears.

spontaneously about 3.5 billion years ago as a result of free radical reactions involving components of the primitive atmosphere - initiated largely by ionizing radiation from the sun, selected free radical reactions to play major metabolic roles, and assured evolution by utilizing them further to provide for mutations, as well as aging and death - the latter serving, if necessary, to ensure space and nutrients for new living things.

Endogenous Free Radical Reactions

Free radical reactions arise (Altman et al., 1970) upon exposure to ionizing radiation, from nonenzymatic reactions (Mead 1976; Scott 1965), and from enzymatic reactions, particularly those of the two major energy-gaining processes employed by living things - photosynthesis (Loach et al., 1976) and the reduction of O_2 to water (Nohl et al., 1978; Chance et al., 1979). They probably also arise as well in the reduction of terminal electron acceptors employed by anaerobes (Gottschalk et al., 1979). Probably the vast majority of these free radical reactions are enzymatic ones involved in maintenance and function, while the remainder, initiated by non-enzymatic means and by "leakage" of free radicals from enzymatic reactions, cause more-or-less random change.

Antioxidants and antioxidant enzymes have evolved (Harman, 1986) to limit the rate of production of the more-or-less random free radical damage that occurs continuously throughout the cells and tissues. Tissue levels of the free radical defenses in general vary from one tissue to another, between species, and with age (Perez-Campo et al., 1990; Rao et al., 1990; Sohal et al., 1990ab). The free radical damage includes oxidative change of mitochondrial and nuclear DNA (Richter et al., 1988; Adelman et al., 1988) and formation of "advanced glycosylation end products" (Dunn et al., 1991; Wolff et al., 1991). Precisely how such free radical-induced changes collectively cause aging and death is not known.

Support for the Theory

This includes (Harman 1986): 1) studies of the origin and evolution of life, discussed briefly above, 2) studies of the effect of ionizing radiation on living things, 3) dietary manipulations of endogenous free radical reactions, 4) the plausible explanations it provides for aging phenomena, and 5) the growing number of studies that implicate free radical reactions in the pathogenesis of specific diseases. Three, 4, and 5 are discussed below.

3. Life Span Studies

It should be possible to augment the normal defenses against free radical reaction damage by at least three dietary changes: 1. Caloric reduction; this is accompanied by a proportionate decrease in the flux of O_2 through the cells and tissues, as well as in mitochondrial oxygen utilization - resulting in a decrease in $O_2^{-\bullet}$ production. 2. Minimize dietary components that tend to increase free radical reaction levels, such as copper and polyunsaturated lipids or easily oxidize amino acids, while increasing those that enhance antioxidant capacity such as β-carotene in carrots (Diplock 1991), and oltipraz and anethole dithione in cruciferous vegetables like cabbage and cauliflower (Stohs et al. 1986). 3. Addition to the diet of one or more free radical reaction inhibitors, e.g., 2-mercaptoethylamine (2-MEA), α-tocopherol, butylated hydroxytoluene (BHT), and 1,2-dihydro-6-ethoxy-2,2,4-trimethylquinoline (ethoxyquin). Life span studies are in accord with the foregoing, for example:

a. Caloric Reduction

Caloric reduction, with adequate intake of essential nutrients, increases both average and maximum life spans and decreases the incidence of diseases (Masoro et al., 1982). Total lifetime O_2 consumption is about the same for caloric restricted and unrestricted animals (Masoro et al., 1982). The increase in life spans of restricted animals is approximately inversely proportional to the decrease in the rate of O_2 consumption; for example, in one study a 38 percent reduction in O_2 (as measured by food intake) resulted in a 41 percent increase in mean life span (Yu et al. 1982). The beneficial effects of caloric restriction may be largely due to a decrease in free radical damage (Harman, 1983, 1986) owing to the lower flux of O_2 throughout the cells and tissues - particularly in the mitochondria where over 90 percent of the O_2 consumed by a mammal is utilized. Efforts to increase the "normal" maximum life span of man, while living a normal life, will probably include some acceptable degree of food restriction.

b. Dietary Protein

Increasing the dietary content of easily oxidized animo acids might increase free radical reaction damage and thereby decrease life expectancy. In accordance (Harman, 1978), when either 1%w (percent by weight) of either histidine or lysine was added to a semi-synthetic diet containing 20%w casein as a the sole source of protein, average life expectancy was decreased by 5 and 6 percent respectively. Conversely, replacing casein by a soybean protein containing a lesser amount of easily oxidized amino acids increased life expectancy by 13 percent.

c. Antioxidants

Many studies (Harman, 1986) have shown that dietary antioxidants increase the average life span. For example, addition of 1.0% by weight 2-mercaptoethylamine to the diet of male LAF_1 mice (Harman 1968), starting shortly after weaning, increased the average life span by 30%. Corresponding increases produced by 0.5% ethoxyquin in the diet of male and female C3H mice (Comfort 1971) were 18 and 20% respectively.

Increases in average life expectancy of mice receiving antioxidants may be associated with lower body weights (Harman, 1968) of up to about 10 percent. Such decreases are not likely to have a significant effect on the average life span (Silberberg et al., 1962). Only one antioxidant longevity study has been reported (Heidrick et al., 1984) in which food consumption and body weights of both control and treated mice were the same; addition of 0.25% by weight of 2-mercaptoethanol to the diet of $BC3F_1$ mice increased average life expectancy by 13%.

A study of the effect of 2-MEA and of BHT on the life span of female C57BL mice (Kohn 1971) led to the conclusion that the antioxidants did not increase the average life span, or the maximum life span, beyond values obtained for control mice surviving under optimal conditions. In contrast, BHT significantly increased the average life span of both male and female BALB/c mice under living conditions the authors claimed to be optimal (Clapp et al., 1979); the BHT mice were generally heavier than the controls. The reason for the discrepancy between the results of the above C57BL mouse study and those of other mouse lifespan experiments with antioxidants, is not clear.

Increases in average life expectancies of mice in studies such as the above that are not associated with increases in maximum life span - and, therefore, no decreases in the rate of the inborn aging process - are attributed largely to inhibition by the antioxidants of aging changes associated with the environment and disease.

d. Antioxidants and the Maximum Life Span

By the mid 1960's it had become apparent (Harman, 1972) that decreasing endogenous free radical reactions by antioxidants or altering the composition of the diet increased average life expectancy, but had little, if any, effect on the maximum life span.

Only three compounds (Harman, 1986) have been thus far been reported to increase the maximum life span of mice: 2-mercaptoethanol (Heidrick et al., 1984) and two pyridine derivatives, 2-ethyl-6-methyl-3-hydroxypyridine and 2-6-dimethyl-3,5-diethoxycarbonyl-1,4-dihydropyridine (Emanuel, 1976; Emanuel et al., 1981).

The general failure of antioxidants to increase the maximum life span may be largely because mitochondrial function is depressed by the compounds at concentrations below those needed to slow damage to the mitochondria by free radical reactions initiated during normal mitochondrial function (Harman 1972,1983).

Increasing free radical reaction damage to mitochondria-coupled with changes, if any, in nuclear DNA contributions to mitochondrial components - can reasonably account (Harman 1983) for the decreases in mitochondrial number and function with age. State 3 respiration decreases with age (Chiu et al., 1980; Nohl et al., 1978a) while formation of $O_2^{\bar{\cdot}}$ and H_2O_2 increases (Nohl et al., 1978b; Sohal et al., 1991). It is very likely that the maximum life span is determined largely by the rate of development of these changes. Progressive decreases in ATP for reductive synthesis coupled with increased H_2O_2 and $O_2^{-\bullet}$ formation should result in increasingly higher free radical reaction levels, free radical damage, i.e., aging changes, and the chance of death. In accordance, exhalation of ethane, ethylene, butane and pentane by rats, increases exponentially with age (Sagai et al, 1980); these hydrocarbons are products of lipid peroxidation. Further, the levels of oxidized proteins in cultured human fibroblasts likewise increase almost exponentially as a function of age (9-80 years) of the fibroblast donor (Stadtman 1990). In addition the serum mercaptan concentration declines with age (Harman, 1960; Leto et al., 1970). In agreement with the foregoing, $O_2^{-\bullet}$ generation by submitochondrial particles from the mouse, rat, rabbit, pig and cow (Sohal et al., 1989) is inversely related to the maximum life span.

4. Aging Phenomena

The free radical theory of aging provides plausible explanations for aging phenomena (Harman, 1983), these include: 1) the inverse relationship between the average life spans of mammalian species and their basal metabolic rates, 2) the clustering of degenerative diseases in the terminal part of the life span, 3) the exponential nature of the mortality curve, 4) the beneficial effect of caloric restriction on life span and degenerative diseases, 5) the greater longevity of females, and 6) the increases in autoimmune manifestations with age. For example:

a. Greater Longevity of Females

The greater longevity of females may be due (Harman 1979), at least in part, to the greater protection of female embryos from free radical reaction damage during a short period (about 48 hours in the mouse) of both high mitotic and metabolic activity just prior to the random inactivation of one of the two functioning X chromosomes in the late blastocyst state of development. The X chromosome codes for glucose-6-phosphate dehydrogenase, a key enzyme in the production of NADPH. NADPH acts to maintain glutathione in the reduced state.

b. Autoimmunity

The increase in autoimmune manifestations with age may be largely due (Harman, 1980) to a disproportionate decrease in the radiosensitive T-suppressor cell function in comparison to the other cells involved in immunity owing to the increasing levels of free radical reactions with age. In accordance, addition of 0.25%w of α-tocopherol acetate, 0.25%w ethoxyquin, or 1.0%w NaH_2PO_2, to the diet of male NAB mice, starting shortly after weaning, increased the average life span by 7, 32 and 1 percent, respectively.

5. The "Free Radical" Diseases

The reason for the increasing incidence of disease with advancing age has long been of interest. A plausible explanation (Harman 1984) is based on the observation that free radical reactions are implicated (Harman, 1984, 1986; Halliwell et al., 1989; Slater et al., 1991) in pathogenesis of a growing number of disorders.

The ubiquitous free radical reactions would be expected to produce progressive adverse changes that accumulate with age throughout the body. The "normal" sequential alterations with age can be attributed to those changes more-or-less common to all persons. Superimposed on this common pattern of change are patterns that should differ from individual to individual owing to genetic and environmental differences that modulate free radical reaction damage. The superimposed patterns of change may become progressively more discernable with time and some may eventually be recognized as diseases at ages influenced by genetic and environmental risk factors. Aging may also be viewed as a disease, differing from others in that the aging pattern is universal.

The above discussion indicates that the relationship between aging and diseases in which free radical reactions are involved is a direct one. The "free radical" diseases includes the two major causes of death, cancer and atherosclerosis. Some of the data (Harman, 1986) implicating free radical reactions in cancer, atherosclerosis, essential hypertension, amyloidosis, and the immune deficiency of age are presented briefly below.

a. Cancer

Endogenous free radical reactions, like those initiated by ionizing radiation, may result in tumor formation. Support for this possibility is extensive (Harman, 1986). For example, the incidence of mammary carcinoma in C3H female mice increases as the amount and/or degree of unsaturation of the dietary fat is increased (Harman, 1971), i.e., in parallel with the expected increases in endogenous free radical reactions. In women the future risk of breast cancer is lower the higher the plasma level of vitamin E (Wald et al., 1984).

The increasing incidence of cancer with age is probably due, at least in part, to the increasing level of endogenous free radical reactions with advancing age (Noy, 1985; Stadtman, 1990, Sagai et al., 1980) resulting in an increased rate of mutation in proto-oncogenes and tumor-suppressing genes (Friend et al., 1988; Hollstein et al., 1991) coupled with the progressive diminishing capacity (Kay et al., 1976) of the immune system to eliminate the altered cells. It is likely that the probability of developing many, if not all, types of human cancer can be decreased by eating diets selected to minimize adverse free radical reactions in the cells and tissues.

b. Atherosclerosis

Atherosclerotic lesions tend to form in areas of the vascular tree that have been subjected to injury (Fuster et al., 1992). A possible constant source of injury-producing compounds (Harman, 1957, 1986) is the reaction of molecular oxygen with polyunsaturated substances present in serum and arterial wall lipids.

The first steps in atherogenesis may include interaction of localized oxidized low density lipoproteins with endothelial cells (Gerrity, 1981) followed by induction of mononuclear leukocyte adhesion molecules (Cybulsky, 1991) that aid to localize monocytes to the endothelial surface. The adherent monocytes migrate through the endothelium, change their phenotypic expression to become tissue macrophages, and then take up oxidized lipoproteins to become foam cells. The foam cells constitute the fatty streak, the first manifestation of atherosclerosis.

Support for the possibility that free radical reactions are involved in atherogenesis includes: The extensive atherosclerosis seen in patients with homocystinuria and/or hyperhomocysteinemia (Clarke et al., 1991; Harker et al., 1976) may be related to the formation of superoxide radicals during the ready oxidation by O_2 of homocysteine to homocystine (Harker et al., 1976). Xanthine oxidase has been implicated in atherogenesis (Oster, 1971); if so, it is probably related to superoxide production during the action of the enzyme (McCord et al., 1982). The enhanced atherogenesis observed in chronic renal dialysis patients (Linder et al., 1974) may also be due to free radical-induced endothelial cell injury; white cells tend to aggregate under these circumstances and release H_2O_2, $O_2^{-\bullet}$, $HO^\bullet$(Jacob et al., 1980). In addition, the free radical reactions initiated by ionizing radiation also result in atherosclerosis (McCready et al, 1983; Selwign, 1983), while the plasma selenium concentration in patients with arteriographically defined coronary atherosclerosis is inversely correlated with the severity of the coronary lesions (Moore et al., 1984. Further, plasma concentrations of vitamins C and E and of carotene have been found to be significantly inversely related to the risk of angina in man (Riemersma et al., 1991; Gey et al., 1991).

Taken as a whole (Fuster et al., 1992; Steinberg, 1990; Editorial, 1992; Luc et al., 1991; Reaven et al., 1992), studies on the pathogenesis of atherosclerosis are compatible with the possibility that the disease is basically due to free radical reactions largely involving dietary-derived lipids in the arterial well and serum to yield peroxides and other substances. These compounds induce endothelial cell injury and produce changes in other components of the arterial wall that collectively initiate, and help to sustain, an inflammatory reaction in the wall that in turn interacts with serum-derived lipids.

c. Essential Hypertension

Blood pressure normally rises with age (Buchan et al., 1960; Zachariah et al., 1991). Diastolic pressure usually plateaus around 45-50 years of age whereas systolic pressure continues to increase (Manger et al., 1982). At any given age, increases in blood pressure over the "normal" for that age are associated with increased morbidity and mortality risks.

Endothelium derived $O_2^{-}\cdot$ is a vasocontracting factor (Katusic, 1989). This radical may play a significant role in essential hypertension (Nakazono et al., 1991). As expected (Nakazono et al., 1991) the blood pressure of spontaneously hypertensive rats (SHR), but not that of normal rats, was decreased by superoxide dismutase. The $O_2^{-}\cdot$ was apparently formed by xanthine oxidase (Nakazono et al., 1991). In accord, oxypurinol, an inhibitor of xanthine oxidase, decreased the blood pressure of SHR rats but not that of normal rats. Likewise in accord, the serum uric acid in the SHR rats was higher than in normal rats; hyperuricemia is frequently observed in essential hypertension (Selby et al., 1990). Further, since $O_2^{-}\cdot$ inactivates (Nakazono et al., 1991) the endothelial-dependent relaxing factor NO, it may be responsible for the observation that endothelium-mediated vasodilation is impaired in patients with essential hypertension (Panza et al., 1990).

In view of the above the pressor effect of catecholamines on blood pressure may be due to $O_2^{-}\cdot$ formed during the ready reaction of these compounds with O_2.

d. Amyloidosis

Amyloid, formed from various proteins by several different mechanisms, is associated with a small amount (about 10 percent) of a non-fibrillary glycoprotein called amyloid P component (Glenner, 1980); the fibrils are readily proteolyzed in the absence of this substance (Hind et al., 1984).

A study of the effect of antioxidants on the life span of male LAF_1 mice (Harman, 1968) shed some light on the etiology of murine amyloidosis and its associated with age. Ethoxyquin almost completely prevented spontaneous amyloidosis in these mice; butylated hydroxytoluene (BHT) was less effective. The two antioxidants also inhibited development of amyloidosis when added to the diet of casein-injected C3HeB/FeJ male mice (Harman et al., 1976). Plasma electrophoretic studies of the casein-injected mice showed that ethoxyquin, and to a lesser extent BHT, depressed the appearance of a protein fraction in the α_1-glycoprotein region. This suggests that the antioxidants inhibited the rate of oxidative breakdown of unknown substances, possibly of the connective tissues or of cell surface components such as the amyloid precursor protein (APP) of Alzheimer's disease, to form amyloid fibrils. The slower rate of formation of the fibrils may have allowed them to be proteolyzed before there was time for them to aggregate to form amyloid and to then be stabilized against further degradation by the amyloid P component.

Although the details of the inhibiting effect of antioxidants on amyloid formation in the two mouse strains remains to be clarified it is apparent that a free radical reaction(s) is involved in pathogenesis and that this can be inhibited by dietary antioxidants. Whether free radical reaction inhibitors can also inhibit formation of the amyloid observed in Alzheimer's disease and other disorders remains to be determined.

e. Immune Deficiency of Age

Cellular and humoral immune responses decline with age (Kay et al., 1976). Studies with $BC3F_1$ mice show that old mice have only 10 percent of the humoral capacity and 25 percent of the cell-mediated capacity of young animals.

A number of antioxidants have been shown to enhance both humoral and cellular immune responses (Harman et al., 1977), indicating that some endogenous free radical reactions have adverse effects on the immune system. With increasing age, the level of more-or-less random free radical reactions increases with age, as indicated by the increase in exhalation of hydrocarbons (Sagai et al., 1980) and of oxidized proteins (Stadtman, 1990) with age, and the concomitant decline in the serum concentration of mercaptans (Harman, 1960; Leto et al., 1970). Thus, part of the decline of the immune system with age may be attributed to increasing levels of free radical reactions.

COMMENT

A beautiful coherent picture is emerging from studies on the origin of life, mutation, radiation biology, aging, degenerative diseases, and free radical reactions in biological systems. Life apparently originated as a result of free radical reactions, selected free radical reactions to play major metabolic roles, and assured evolution by also employing them to provide for mutation, aging and death. It would be remarkable that life with its beautiful order should owe its origin to, and be sustained by, a class of chemical reactions whose outstanding characteristic is their unruly nature.

Aging may never have changed; in the beginning, free radical reactions were initiated primarily by ultraviolet radiation from the sun and now free radicals arise from within from enzymatic and non-enzymatic free radical reactions. In mammalian species, mitochondria are the major source of damaging free radical reactions. Because these reactions also damage mitochondria they may serve as the "biologic clock" determining the maximum life span; the damage results in progressive decreases in ATP production and enhanced O_2 formation. Antioxidants that can slow mitochondrial aging, at concentrations that do not also significantly depress function, should increase the maximum life span.

It is reasonable to expect on the basis of present data that the healthy, active life span can be increased 5 or more years by keeping body weight down, at a level compatible with a sense of well-being, while ingesting diets adequate in essential nutrients but designed to minimize random free radical reactions in the body.

Chronic disorders now decrease the quality of life of numerous older persons while the need of many of them for services and medical care from society imposes a significant and growing burden on the remainder of the population. Amelioration of these two interrelated problems should be possible by application of measures to slow the aging process.

REFERENCES

Adelman, R., Saul, R.L., and Ames, B.N. (1988) Proc. Natl. Acad. Sci., USA, 85: 2706-2708.

Altman, K.I., Gerber, G.B., and Okada, S.A. (1990) Radiation Chemistry, Vol. 1 and 2, Acad. Press, New York.

Barden, B., and Brizzee, K.R. (1987) The histochemistry of lipofuscin andneuromelanin. In Advances in Biochemistry, Vol. 64, Advances in Age Pigment Research., Tatoro, E. H., Glees, P., and Pisanti, F. A., Eds., Pergaman Press, Oxford, 339-392.

Buchan, T.W., Henderson, W.K., Walker, D E., Symington, T., and McNeil, I.H. (1960) Health Bull. Edinburgh 18 3.

Bugiani, O., Giaccone, G., Frangione, B., Ghetti, B., and Taagliavini, F. (1989) Neurosci. Lett. 103 263-268.

Chance, B., Sies, H., and Boveris, A. (1979) Physiol. Rev. 59 527-605.

Chiu, Y.J.D., and Richardson, A. (1980) Exper. Gerontol. 15 511-517.

Clapp, N.K., Satterfield, L.C., and Bowles, N.D. (1979) J. Geront. 34 497-501.

Clarke, R.C., Daly, L., Robinson, K., Naughten, E., Cahalane, S., Fowler, B., and Graham, I. (1991) N. Eng. J. Med. 324 1149-1155.

Comfort, A. (1971) Nature 229 254-255.

Comfort, A. (1979) Elsevier, New York, 3rd Ed., pp. 81-86.

Cybulsky, M.I., and Gimbrone, Jr., M.A. (1991) Science 251 788-791.

Diplock, A.T. (1991) Amer. J. Clin. Nutr. 53 Suppl. 1, 189S-193S.

Dublin, L.E., Lotha, A.J., and Spiegel, M. (1949) Length of Life. Ronald Press, New York.

Dunn, J.A., and McChance, D.R., Thrope, S.R.. Lyons, T.J., and Baynes, J.W. (1991) Biochemistry 30 1205-1210.

Editorial, Atherosclerosis goes to the wall (1992) Lancet 339 647-648.

Emanuel, N.M. (1976) Quart. Rev. Biophys. 9 283-308.

Emanual, N.M., Duburs, G., Obukhov, L.K., and Uldrikis, J. (1981) Chem. Abst. 94 9632a.

Friend, S.H., Dryja, T.P., and Weinberg, R.A. (1988) New Eng.

J. Med. 318 618- 622.
Fries, J.F. (1980) N. Engl. J. Med. 303 130-135.
Fuster, V., Badimon, L., Badimon, J.J., and Chesebro, J.H. (1992) New Engl. J. Med. 326 242-250; 310-318.
Gerrity, R.G. (1981) Amer. J. Pathol. 103 181- .
Gey, K.F., Pusha, P., Jordan, P., and Moser, U.K. (1991) Amer. J. Clin. Nutr. 53 (Suppl. 1), 326S-334S.
Glenner, G.G. (1980) N. Engl. J. Med. 302 1283; 1233.
Gottschalk, G., and Andreesen, J.R. (1979). In International Review of Biochemistry, Microbiol. Biochemistry, Vol. 2, Quale, J.R., Ed., University Park Press, Baltimore, pp. 85-115.
Halliwell, B., and Gutteridge, J.M.C. (1989) 2nd Edition, Clarendon Press, pp. 418-419, Oxford, England.
Harker, L.A., Ross, R., Slichter, S.J., and Scott, C.R. (1976) J. Clin. Invest. 58 731-741.
Harman, D. (1956) J. Gerontol. 11 298-300.
Harman, D. (1957) J. Geront. 12 199-202.
Harman, D. (1960) J. Gerontol. 15 38-40.
Harman, D. (1962) Rad. Res. 16 753-763.
Harman, D. (1968) J. Gerontol. 23 476-482.
Harman, D. (1971) J. Geront. 26 451-457.
Harman, D. (1972) J. Amer. Geriatrics Soc. 20 145-147.
Harman, D., Eddy, D.E., and Noffsinger, J. (1976) J. Amer. Geriat. Soc. 24 203-210.
Harman, D., Heidrick, M.L., and Eddy, D.E. (1977) J. Amer. Geriat. Soc. 25 400-407.
Harman, D. (1978) Age 1 145-152.
Harman, D., and Eddy, D.E. (1979) Age 2 109-122.
Harman, D. (1980) Age 3 64-73.
Harman, D. (1981) Proc. Natl. Acad. Sci. USA 78 7124-7128.
Harman, D. (1983) Age 6 86-94.
Harman, D. (1984) Age 7 111-131.
Harman, D. (1986) Free radical theory of aging: Role of free radicals in the origination and evolution of life, aging, and disease processes. In Johnson, Jr., J.E., Walford, R., Harman, D., and Miquel, J. eds., Liss, New York, pp. 3-49.
Harman, D. (1991) Proc. Natl. Acad. Sci. USA 88 5360-5363.
Harman, D. (1992) Free radical theory of aging: history. In Free Radicals and Aging, Emerit, I., and Chance, B., Eds., Birkhauser, Basel, in press.
Heidrick, M.L., Hendricks, L.C., and Cook, D.E. (1984) Mech. Ageing and Dev. 27 341-358.
Hind, C.R.K., Caspi, D., Collins, P.M., and Baltz, M.L. (1984) Lancet II 376- 378.
Hollstein, M., Sidransky, D., Vogelstein, B., and Harris, C.C. (1991) Science 253 49-53.
Jacob, H.S., Craddock, P.R., Hammerschmidt, D.E., and Moldow, C.F. (1980) N. Eng. J. Med. 302 789-794.
Jones, H.R. (1955). In Birren, J.E. ed. Handbook of Aging and the Individual, Chicago Univ. Press, Chicago, IL, pp. 333-363.
Katusic, Z.S., and Vanhoutte, P.M. (1989) Amer. J. Physiol. 297 H33-H37.

Kay, M.M.B., and Makinodan, T. (1976) Clin. Immunol. Immunopathol. 6 394-413.
Kohn, R.R. (1985) Aging and Age-related diseases: normal processes. In Johnson, H.A. ed. Relation Between Normal Aging and Disease, Raven Press, New York, pp. 1-44.
Kohn, R.R. (1971) J. Geront. 26 378-380.
Leto, S., Yiengst, M.J, and Barrows, Jr., C.H. (1970) J. Geront. 25 4-8.
Linder, A., Charra, B., Sherrard, D.J., and Scribner, A.H. (1974) N. Eng. J. Med. 209 697-701.
Loach, P.A., and Hales, B.J. (1976) Free radicals in photosynthesis. In Free Radicals in Biology, Pryor, W.A., Ed., Academic Press, New York, pp. 199-137
Luc, G., and Fruckart, J.-C. (1991) Amer. J. Clin. Nutr. 53 (Suppl. 1) 206S- 209S.
Manger, W.M., and Page, I.H. (1982) An overview of current concepts regarding the pathogenesis and pathophysiology of hypertension. In Arterial Hypertension, Rosenthal, J., Ed., Springer-Verlag, New York, pp. 1.
Masoro, E.J., Yu, B.P., and Bertrand, H.A. (1982) Proc. Natl. Acad. Sci. USA 79 4239-4241.
McCord, J.M., and Fridovich, I. (1982) Lipids 17 331-337.
McCready, R.A., Hyde, G.L., Bivins, B.A., Mattingly, S.S., and Griffen, Jr., W. O. (1983) Surgery 93 306-312.
Moore, J.A., Noiva, R., and Wells, L.C. (1984) Clin. Chem. 30 1171-1173.
Nakazono, K., Watanabe, N., Matsumo, K., Sasaki, J., and Sato, T. (1991) Proc. Natl. Acad. Sci. USA 88 10045-10048.
National Center for Health Statistics (1988) Vital Statistics of the United States 1985. (U.S. Dept. Health Human Serv., Hyattsville, MD), PHS Publ. No. 88-1104, Life Tables, Vol. 2, Sect. 6 p. 9.
National Center for Health Statistics (1989). Annual Summary of Births, Marriages, Divorces, and Deaths: United States 1988. Monthly Vital Statistics 37: No. 13, Hyattsville, MD (U.S. Dept. Health Human Serv.) PHS Publ. No. 89-1120, p. 19.
Nohl, H., and Hegner, D. (1978a) European J. Biochem. 82 863-867.
Nohl, H., Breuninger, V., and Hegner, D. (1978b) European J. Biochem. 90 385- 390.
Noy, N., Schwartz, H., and Gafni, A. (1985) Mech. Ageing and Dev. 29 63-69.
Office Federal de la Statistique (1988) Suisse - Table de Mortalite 1986-1987. Swiss Govermment, Berne, Switzerland..
Olshansky, S.J., Carnes, B.A. and Cassel, C. (1990) Science 250 634-640.
Oster, K.A. (1971) Amer. J. Clin. Res. 2 30-35.
Panza, J.A., Quyyumi, A.A., Brush, Jr., J.E., and Epstein, S.E. (1990) New Engl. J. Med. 323 22-27.
Perez-Campo, R., Lopez-Torres, M., Paton, D., Sequeros, E., and Barja de Quiroga, G. (1990) Mech. Ageing and Dev. 56 281-292.
Rao, G., Xia, E., and Richardson, A. (1990) Mech. Ageing and Dev. 53 49-60.

Reaven, P.D., Parthasarathy, S., Beltz, W.F. and Witztum, J.L. (1992) Arterioscler. Thromb. 12 312-324..
Richter, C., Park, J.W., and Ames, B.N. (1988) Proc. Natl. Acad. Sci. USA 85 6465-6467.
Riemersma, R.A., Wood, D.A., MacIntyre, C.C.A., Elton, R.A., Gey, K.F., and Oliver, M.F. (1991) Lancet 337 1-5.
Rothstein, M. (1986) Chem. Eng. News 64 (32), 26.
Sagai, M., and Ichinose, T. (1980) Life Sciences 27 731-738.
Schneider, E.L. (1987) Theories of aging: A perspective. In Modern Biological Theories of Aging, eds. Warner, M. R., Butler, R.N., Sprott, R.L. and Schneider, E., Raven Press, pp. 1-4.
Selby, J.V., Friedman, G.D., and Quesenberry, Jr., C.P. (1990) Amer. J. Epidemiol. 131 1017-1027.
Selwign, A.P. (1983) Lancet 2 152-154.
Silberberg, R.S., Jarrett, S.R., and Silberberg, M. (1962) J. Gerontol. 17 239- 244.
Slater, T.F., and Block, G., eds. (1991) Antioxidants Vitamins and -Carotene in Disease Prevention. Amer. J. Clin. Nutr. 53 (Suppl. 1), 189S-396S.
Sohal, R.S., Svensson, I., Sohal, B.H, and Brunk, W.T. (1989) Mech. Ageing and Dev. 49 129-135.
Sohal, R.S., Arnold, L.L., and Orr, W.C. (1990a) Mech. Ageing and 56 223- 235.
Sohal, R.S., Arnold, L.A, and Sohal, B.H. (1990b) Free Rad. Biol. and Med. 10 495-500.
Sohal, R.S., and Sohal, B.H. (1991) Mech. Ageing and Dev. 57 287-202.
Stadtman, E.R. (1990) Free Rad. Biol. and Med. 9 315-325.
Steinberg, D., and Witztum, J.L. (1990) JAMA 264 3047-3052.
Stohs, S.J., Lawson, T.A., Anderson, L., and Bueding, E. (1986) Mech. Ageing and Dev. 37 137-145.
Sveriges officiella statistik (1988) Befolknings-forandringar (1987). Statistiska centralbyran, Stockholm, Sweden, pp. 114-115.
Upton, A.C. (1977) Pathobiology. In Finch, C.E., and Hayflick, L., eds. The Biology of Aging, Von Nostrand Reinhold, New York, pp. 513-535.
Vijg, J. (1990) Aging 2 227-229.
Wald, N.J., Boreham, J., Hayward, J.L., and Bulbrook, R.D. (1984) British J. of Cancer 49 321-324.
Wolff, S.P., Jiang, Z.Y., and Hunt, J.V. (1991) Free Rad. Biol. and Med. 10 339-352.
Woodhall, B., and Joblon, S. (1957) Geriatrics 12 586-591.
Yu, B.P., Masoro, E.J., Murata, I., Bertrand, H.A., and Lynd, F.T. (1982) J. Geront. 37 130-141.
Zachariah, P.K., Sheps, S.G., Bailey, K.R., Wiltgen, C.M., and Moore, A.G. (1991) JAMA 265 1414-1417.

Free Radicals: From Basic Science to Medicine
G. Poli, E. Albano & M. U. Dianzani (eds.)

OXIDATIVE STRESS STATE IN AGING AND LONGEVITY MECHANISMS

R.G. Cutler

Gerontology Research Center, National Institute on Aging, 4940 Eastern Avenue, Baltimore, MD 21224

SUMMARY: Most evidence indicates that aging is a result of normal metabolic processes that are essential for life. Thus, an important approach in biogerontology is to identify specific metabolic reactions necessary for life but which could also lead to aging. A unique characteristic of this approach is an explanation of what governs aging rate or longevity of a species or even individuals within a species. These would be mechanisms that would act to reduce the long-term toxic or aging effects of the normal metabolic and developmental reactions. The reactions involving oxygen metabolism clearly fit into this model since they are essential for life yet can potentially cause many of the dysfunctions associated with aging. Such a model can also account for differences in aging rate or longevity of different animals species by differences that may exist in their innate ability to reduce oxidative stress state. Our laboratory has been testing this oxidative stress state (OSS) hypothesis of aging and longevity by determining if a positive correlation exists between OSS of an animal and its aging rate. Much of our data has found such a positive correlation, yet there is some indication that separate causative mechanisms may exist in determining aging rate as opposed to those related to age-dependent specific diseases, such as cancer or cardiovascular disease.

Nothing definite is yet known about the causes of human aging or the processes that govern aging rate of different mammalian species. Furthermore, there is the possibility that aging could be the result of many different independently-related processes and that no simple primary process exists that causes aging or governs aging rate.

Nevertheless, steady progress is being made towards answering these key questions. For example, although we do not yet have proof of a single cause of aging, we can now state with reasonable confidence that aging is the result of normal and essential products of metabolism and development (Cutler 1991a,b,c). This is an important advance, for the other view which has been unchallenged for many years is that aging is genetically programmed, where specific genes turn on or off to cause aging for the evolutionary good of the species or individual. That this is not likely to be the case is argued by the fact that animals living in their natural ecological niche rarely live long enough to grow physiologically old or senescent. Thus, why evolve a genetic program of death or senescence if an individual in a population rarely reaches such a dysfunctional condition. So, if aging is not an "active" process, it must be a "passive" process, a result of other processes that were selected to insure a minimum survival time for the organism.

So the next question we ask is, what are these hypothetical passive processes predicted to cause aging? Are they developmental or metabolically related processes - or both? Another question is, are there only a few primary processes causing aging or does essentially every gene and/or metabolic product contribute a little to aging?

Some progress has been made in answering the latter question as to the genetic/metabolic complexity of aging or the processes governing aging rate (Cutler 1975, 1976a,b). A series of studies comparing the life span of closely related species having substantial differences in aging rates such as human vs chimpanzee has led to the suggestion that, in spite of the vast complexity of aging processes, relatively few processes may exist that actually govern aging rate. These processes have since been defined as longevity-determinant mechanisms. Such studies have led to the suggestion that perhaps a few key regulatory genes play a global role in governing the duration general health is maintained.

These studies have led to a new approach in studying human aging and longevity and this is to ask what normal processes essential to life could also lead to the loss of health maintenance and consequently aging. In this regard then, we need to determine what longevity determinant mechanisms might act to reduce the effects of these aging processes. In this sense, we now ask not so much how a human ages but why do humans live so long? This question is particularly interesting since humans are clearly the longest-lived of all mammalian species yet have no great biological differences as compared to shorter-lived species. Thus, the new focus is on longevity, not aging mechanisms.

Of the many different developmental and metabolic processes we have to chose in investigating what key developmental or metabolic programs might play a role in causing aging, we have selected to study oxygen metabolism as a "passive" aging process. This appeared to be an ideal area to study since oxygen metabolism is clearly an essential part of life, providing efficient means of generating energy - but at the same time producing well known by-products called reactive oxygen species (ROS) that could conceivably cause aging. Moreover, an exciting aspect on suggesting that ROS act as a primary cause of aging is that we easily arrive at a class of mechanisms potentially important in governing aging rate. These are mechanisms acting to reduce oxidative stress state, such as in lowered metabolic rate, more efficient mitochondria, higher levels of DNA repair or concentrations of antioxidants. I have more recently called this idea the "Oxidative Stress State hypothesis of aging" or "OSS hypothesis of aging" to bring the concept more in line with current concepts of oxidative stress mechanisms.

If this hypothesis is correct, then we would predict, as a function of longer life span of different species, a less oxidative stress state in an animal's tissues. There are of course many different mechanisms that could have evolved to reduce oxidative stress, but clearly one simple means would be to increase the concentration of the same antioxidants that are commonly found in all mammalian species. Such an increase would likely be a result of simple genetic modifications of regulatory genes, a mechanism also thought to be principle genetic mechanisms in the evolution of different species (Wilson, 1976).

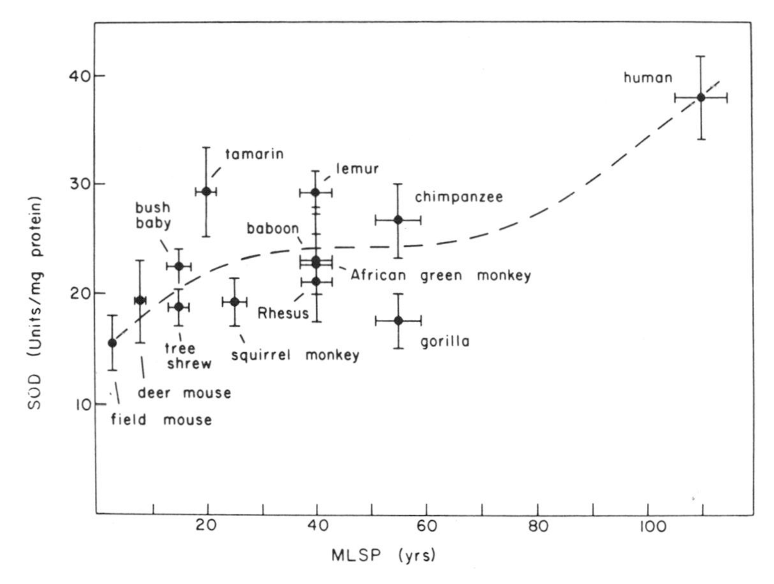

Figure 1. Superoxide dismutase (SOD) concentration in liver of mammals as a function of maximum life span potential (MLSP). From (Cutler, 1985b).

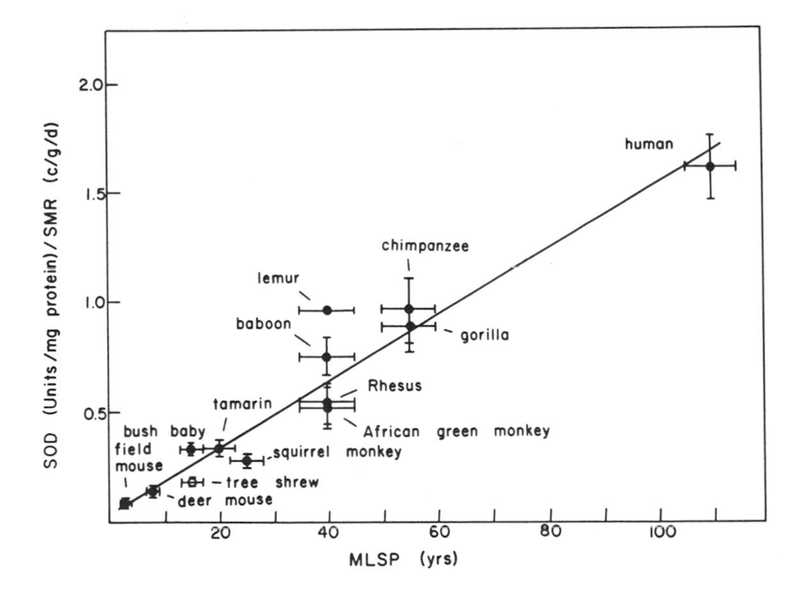

Figure 2. Superoxide dismutase (SOD) concentration per specific metabolic rate (SMR) in liver of mammals as a function of maximum life span potential (MSLP). Linear correlation coefficient: $r = 0.961$, $P \leq 0.001$, where $n = 13$. From (Cutler, 1985b).

The first study undertaken to test the OSS hypothesis of aging was to measure the specific activity of superoxide dismutase (SOD) in different tissues of mammalian species having different life spans. Since, of course, we were really interested in measuring net oxidative stress state in a cell, it is important to realize that it was argued that antioxidant concentration reflects indirectly the steady state levels of oxyradical concentration or the OSS of the cell (Cutler, 1982). To make this possible, it is necessary to normalize the antioxidant activity by dividing SOD activity by the specific metabolic rate. For example, Fig. 1 shows typical data of SOD activity per mg protein of liver tissue as a function of life span of different species. These data show a general increase with life span in SOD activity with human clearly having the greatest amount.

However, as shown in Fig. 2, by normalizing the data on a per specific metabolic rate basis, this correlation becomes remarkably linear.
Thus, we find that:

$$SOD/SMR - kMLSP$$

where SOD is superoxide dismutase (units/mg protein), SMR is specific metabolic rate (c//g/d) and MLSP is maximum life span potential (yrs). Now we cal also estimate oxidative stress state (OSS) within a cell, as:

$$[OSS] \propto SMR/[antioxidant] \propto SMR/[SOD]$$

Thus, our correlation of SOD/SMR with MLSP does relate to MSLP $\propto$ 1/OSS, as expected. Further experiments were then carried out to determine if other antioxidants showed a similar positive correlation with life span.

This was found to be the case for many antioxidants, as shown for vitamin E, urate and carotenoids in Figures 3, 4 and 5, respectively.

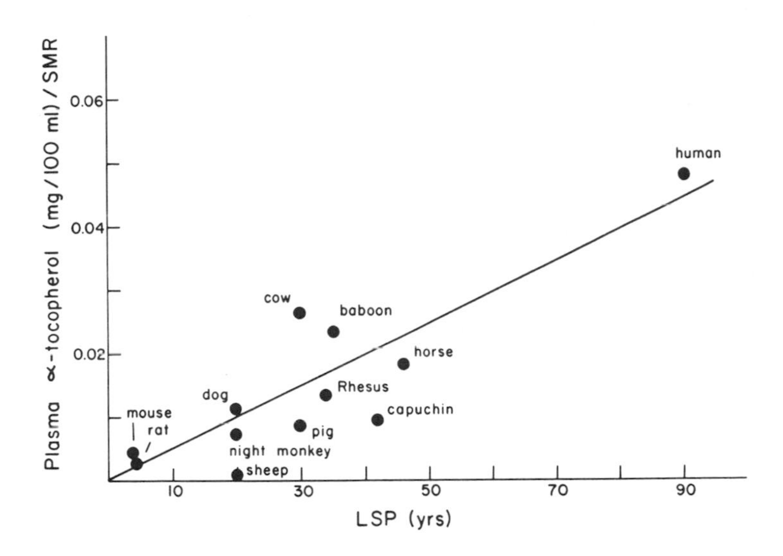

Figure 3. Plasma levels of vitamin E per specific metabolic rate (SMR) as a function of maximum life span potential (MLSP) in mammalian species. Correlation coefficient r = 0.864, P ≤ 0.001. non-life span data from Altman and Dittmer (1961) and Bernischke et al. (1978).

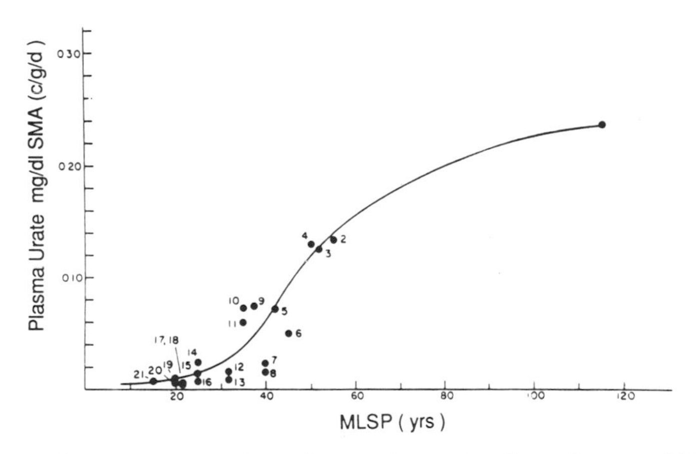

Figure 4. Plasma urate levels and urate level per SMR in primates as a function of MLSP. Values taken from the literature. Species:
1, human; 2, chimpanzee; 3, orangutan; 4, gorilla; 5, gibbon; 6, capuchin; 7, macaque; 8, baboon; 9, spider monkey; 10, Siamang gibbon; 11, woolly monkey; 12, langur; 13, grivet; 14, tamarin; 15, squirrel monkey; 16, night monkey; 17, potto; 18, patas; 19, galago; 20, howler monkey; 21, tree shrew. From (Cutler, 1986).

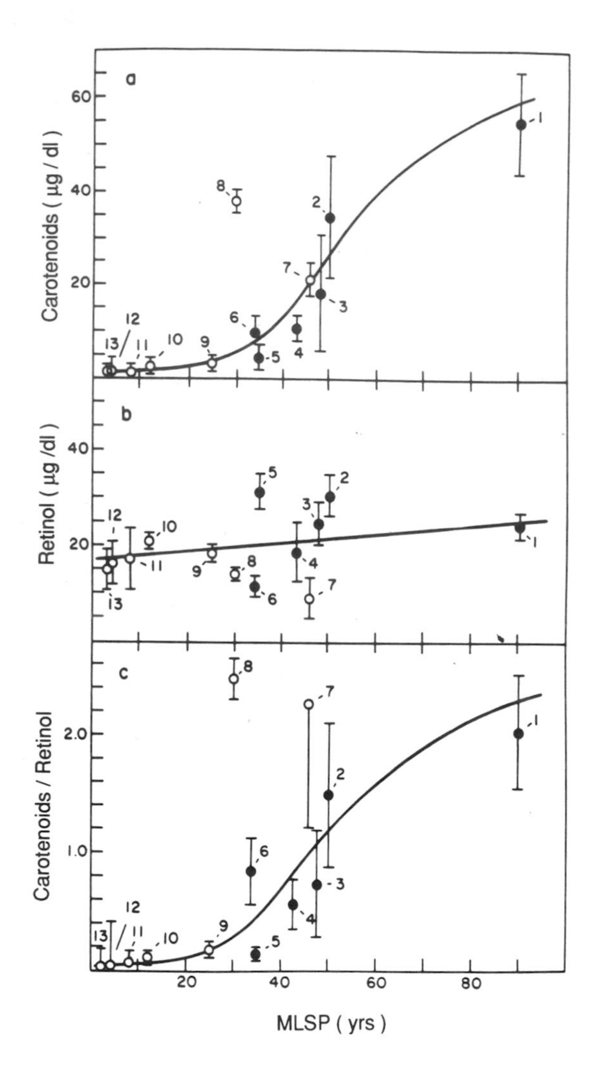

Figure 5. Serum carotenoid and retinol concentration as a function of MLSP. primates. , non primate species. Error bars represent SD. For primates only, linear correlation coefficients are: a. r = 0.926 (P < 0.010) b. r = 0.163 (P > 0.1) and c. r = 0.905 (P < 0.020). 1, human; 2, orangutan; 3, chimpanzee; 4, gorilla; 5, gibbon; 6, Rhesus; 7, horse; 8, cow; 9, goat; 10, rabbit; 11, deer mouse; 12, rat; and 13, field mouse. From (Cutler, 1984).

Some antioxidants, however, had a negative correlation, such as catalase (Figure 6), but we have reason to believe this may actually represent a different strategy to lower OSS. This is by lowering cytochrome P-450 activity (Ayala & Cutler, 1991). Other antioxidants such as ascorbate showed no correlation with life span. All of these data are summarized in Table 1.

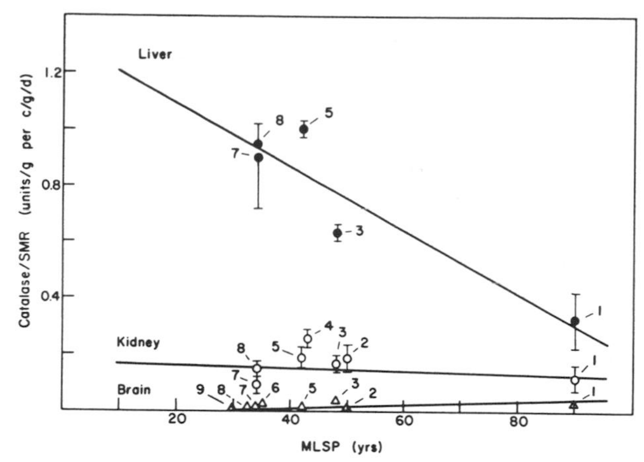

Figure 6. Catalase activity in liver, kidney, and brain of primate species per SMR as a function of MLSP. , liver; , kidney; , brain. Species: 1, human; 2, orangutan; 3, chimpanzee; 4, gorilla; 5, baboon; 6, lemur; 7, rhesus; 8, pig-tailed macaque; 9, African green monkey. From (Cutler, 1986).

Table 1. Summary of Antioxidant Comparison Results

Positive Correlation	No Correlation	Negative Correlation
1. Cu/Zn SOD	1. Ascorbate	1. Catalase
2. Mn SOD	2. Retinol	2. Glutathione
3. Carotenoids		3. Glutathione peroxidase
4. Alpha tocopherol		
5. Urate		
6. Ceruloplasmin		

Another rather simple experiment to test the hypothesis that longer-lived mammalian species are under a lesser oxidative stress state is to measure the net autoxidation sensitivity of whole tissue homogenates in an oxygen atmosphere (Cutler, 1985a). Such an experiment is shown in Figure 7, indicating a remarkable rank order correlation of increasing resistance to autoxidation reaction-producing TBARM products with increased life span.

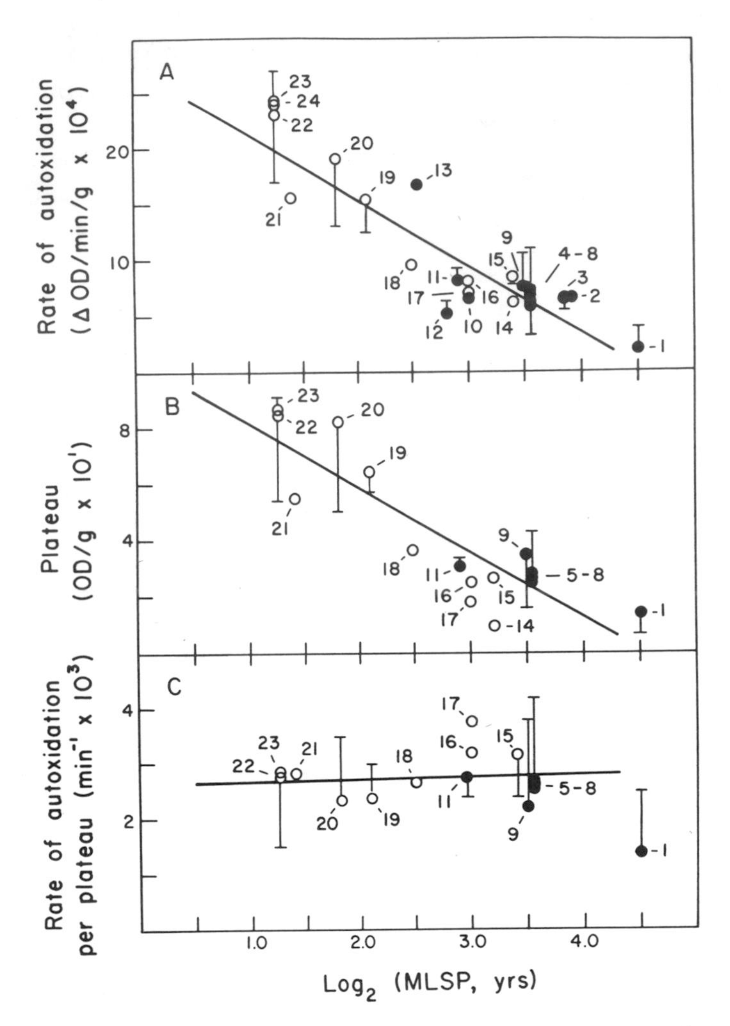

Figure 7. Characteristics of autoxidation of brain homogenate as a function of MLSP. , primates; , non-primate mammalian species. Species identification: 1, human; 2, orangutan; 3, chimpanzee; 4, gibbon; 5, baboon; 6, rhesus monkey; 7, pig-tailed macaque; 8, African green monkey; 9, rhesus monkey; 10, marmoset; 11, squirrel monkey; 12, galago; 13, tree shrew; 14, cow, 15, pig; 16, dog; 17, sheep; 18, rabbit; 19, white-footed mouse; 20, deer mouse; 21, rat; 22, field mouse; 23, C57BL/6J mouse; 24, mouse strains. (Cutler, 1985a).

These data are consistent with OSS contributing to the aging process and mechanisms acting to reduce OSS as potential longevity determinants. Thus, a general scheme of how OSS may cause aging can be drawn, as shown in Figure 8.

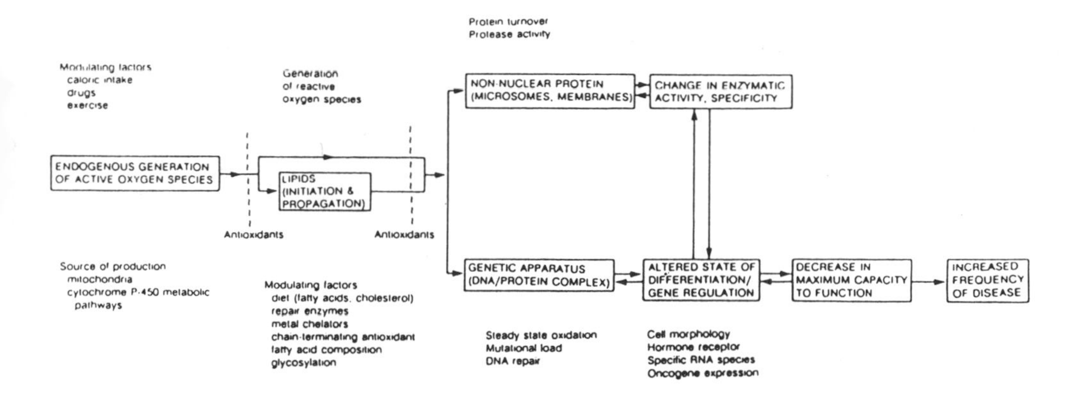

Figure 8. Oxidative initiation model of aging. From (Cutler, 1991b).

But, if OSS causes aging then how does it do it? One possibility of course is that so much damage is eventually accumulated within a cell that, like a "wear and tear" phenomena, the cell simply becomes less efficient and eventually even dies. The problem with this idea is that cells from old animals do not appear to have accumulated sufficient amounts of damage or oxidized products to cause a decrease in function. In fact, gross lack of sufficient enzymatic activity is not evident in old cells. However, better studies of this nature are certainly needed.

Another possibility is that critical targets in a cell are modified, such as nuclear DNA (regulatory genes) or mitochondrial DNA - lowering efficiency of ATP production. Another possibility is that OSS could modify signal transduction pathways through peroxidation of the cellular membrane. The idea here is that small areas of damage in critical areas within a cell or critical cell populations can lead to aging. This is the concept of the dysdifferentiation hypothesis of aging (Cutler, 1991c).

One interesting aspect of the OSS aging hypothesis and the related dysdifferentiation hypothesis of aging is that they are remarkably similar to the genetic instability hypothesis of cancer. Indeed, there is some evidence suggesting that not only does the onset frequency of cancer increase with age but also that this rate of increase is positively correlated to a species' aging rate
(Semsei & Cutler, 1989; Cutler & Semsei, 1989). That is, cancer rate appears to be roughly proportional to aging rate or inversely related to the life span of a mammalian species. This correlation suggests that perhaps the mechanism(s) causing aging is common to or even the same as those causing the types of cancer strongly related to aging (such as colon or prostate cancer). And the best candidate for this common mechanism appears to be genetic instability as it is related to oxidative DNA damage or mutations. These concepts have led to the idea that perhaps both cancer and aging processes proceed by initiation and progression steps, where the initiation step would be related to OSS.

There is now considerable data indicating that oxidative DNA damage does indeed increase with age, both in nuclear DNA as well as in mitochondrial DNA (Richter et al., 1988; Wallace, 1992). There is also data, more of a preliminary nature, to indicate that the steady state level of oxidative DNA damage also decreases with the increased life span of a mammalian species (Adelman et al., 1988; Cutler, 1991).

Thus, the OSS hypothesis of aging appears consistent with both antioxidant data as well as oxidative DNA damage data. What is now critically needed is direct intervention-types of experiments where OSS is decreased and resulting life span observed. A number of such experiments are now underway in various laboratories throughout the world with major emphasis being placed on SOD/catalase transgenic mice and Drosophila strains (Rose et al., 1988; Przedborski et al., 1992).

There are, however, some rather old data that are not consistent with the OSS hypothesis of aging (Cutler, 1991). For example, although ionizing radiation is well known to decrease life span, it does not apparently do so by increasing aging rate but rather by increasing the rate of cancer. Similar results are obtained by giving experimental animals mutagenic chemicals. Indeed, no mutagenic agent has ever been shown yet to accelerate uniformly the aging process. Only specific diseases such as cancer, cataracts and diabetes appear to be accelerated.

Such data suggest the possibility that during the evolution of longer life span perhaps at least two different strategies were involved (Cutler, 1976b; Cutler 1978). One strategy could be to reduce the long term aging effect of developmentally-linked biosenescent processes and the other was to reduce the long term aging effect of metabolically-linked biosenescent processes. Examples of potential developmentally-linked aging processes are growth control factors such as the many different hormones and various peptide factors now recently being discovered, as well as many hormones related to onset of sexual maturity.

How long-term toxic effects of such developmentally-related biosenescent processes could be reduced during the evolution of longer life span appears to be largely by extending the time when they would be expressed during development. That is, longer-lived species would simply progress through all the same stages of development - but at a slower rate then effectively postponing in time the expression of essential genes that also cause aging later in life.

Decreasing the rate of development would be expected to affect a different class of age-related diseases such as those generally associated with degenerative or chronic diseases. Examples are general loss of homeostatic control as in temperature regulation or osteoporosis.

On the other hand, a second strategy would be to decrease toxic effects of metabolic processes not related to development. Decreasing the toxic by-products of metabolism appears to be of a more straight-forward matter and can be achieved, for example, by reduction of SMR, increased efficiency of mitochondrial ATP production, and by higher levels of tissue antioxidants. These protective mechanisms against toxic by-products of metabolism may be most important in decreasing onset frequency of specific diseases related to genetic instability. Cancer would represent an ideal example.

Thus, to significantly increase life span, it may be necessary to both lower OSS, resulting in a higher genetic stability of cells, and to also decrease rate of development. Both of these mechanisms appear to have occurred coordinately during the evolution of longevity in mammalian species.

To test the role developmentally-linked biosenescent processes may have in aging, it would be necessary to intervene in the normal control mechanisms governing developmental rate. The prediction would be that any non-toxic mechanism that could lead to a slower rate of development leading to delayed appearance of sexual maturity would increase life span significantly. Ideally, we would like to genetically engineer a mouse with higher levels of expression of antioxidants, lower OSS, plus genes slowing down the rate of development.

How to decrease the rate of development of an animal appears to be quite difficult, especially without affecting other major physiological systems. For example, food restriction in early life of an animal is effective and does indeed lengthen life span. But whether this increase in life span is due to other effects of food restriction, such as lowering OSS, is not clear (Chung t al., 1992). In any case, it is becoming increasingly clear that OSS is likely to be an important factor governing long term health status and the many complex pathological processes associated with aging.

Acknowledgements

Thanks are expressed to Edith Cutler for secretarial and technical assistance. Her help was made possible by support from the Paul Glenn Foundation for Medical Research.

REFERENCES

Adelman, R., Saul, R.L. and Ames, B.N. (1988) Proc. Natl. Acad. Sci. USA 85, 2706-2708.
Altman, P. and Dittmer, D., eds. (1962) Biological Handbook, Growth. Fed. Amer. Soc. Exp. Biol., Bethesda, MD.
Ayala, A. and Cutler, R.G. (1991) in: Liver and Aging - 1991. K. Kitani, ed. Elsevier, 337-352.
Bernirschke, K., Garner, F.M. and Jones, T.C. (1978) Pathology of Laboratory Animals. Springer-Verlag, Basel.
Chung, M.H., Kasai, H., Nishimura, S. and Yu, B.P. (1992) Free Rad. Biol. Med. 12, 523-525.
Cutler, R.G. (1975) Proc. Natl. Acad. Sci. USA 72 , 4664-4668.
Cutler, R.G. (1976a) J. Human Evolution 5, 169-204.
Cutler, R.G. (1976b) in: Interdisciplinary Topics in Gerontology, Vol. 9. R.G. Cutler, ed. S. Karger, Basle, 83-133.
Cutler, R.G. (1978) in: The Biology of Aging. J.A. Behnke, C.E. Finch and G.B. Moment, eds. Plenum Press, New York, 311-360.
Cutler, R.G. (1982) in: Testing the Theories of Aging. R. Adelman and G. Roth, eds. CRC Press, Boca Raton, FLA., 25-114.
Cutler, R.G. (1984) Proc. Natl. Acad. Sci. USA 81, 7627-7631.
Cutler, R.G. (1985a) Proc. Natl. Acad. Sci. USA 82, 4798-4802.
Cutler, R.G. (1985b) in: Molecular Biology of Aging. A.D. Woodhead, A.D. Blackett and A. Hollaender, eds. Plenum Press, New York. 15-74.
Cutler, R.G. (1986) in Physiology of Oxygen Radicals. Amer. Physiological Soc., Bethesda, MD. 251-285.
Cutler, R.G. (1991a) Amer. J. Clin. Nutr. 53, 373s-379s.
Cutler, R.G. (1991b) New York Acad. Sci. 621, 1-28.
Cutler, R.G. (1991c) Arch. Geront. Geriatr. 12, 75-98.
Cutler, R.G. (1992) in press, I. Emerit, eds. Birkhauser Press.
Cutler, R.G. and Semsei, I. (1989) J. Gerontology 44, 25-34.
Przedborski, S., Jackson-Lewis, V., Kostic, V., Carlson, E., Epstein, C.J. and Cadet, J.L. (1992) J. Neurochem. 58, 1760-1767.
Richter, C., Park, J.-W. and Ames, B.N. (1988) Proc. Natl. Acad. Sci. USA 85, 6465-6467.
Rose, M.R., Nusbaum, T.J. and Fleming, J.E. (1992) Lab. Animal Sci. 42, 114-118.
Semsei, I., Ma, S. and Cutler, R.G. (1989) Oncogene 4, 465-470.
Wallace, D.C. (1992) Science 256, 628-632.
Wilson, A.C. (1976) in: Gene Regulation in Evolution, F.J. Ayala, ed. Sinauer Associates, Inc., Sunderland, Mass. 225-234.

Free Radicals: From Basic Science to Medicine
G. Poli, E. Albano & M. U. Dianzani (eds.)

MAILLARD REACTION AND OXIDATIVE STRESS ARE INTERRELATED STOCHASTIC MECHANISMS OF AGING.

V.M. Monnier, D.R. Sell, R.H. Magaraj, P. Odetti*.

Institute of Pathology, Case Western Reserve University Cleveland OH 44120, USA, and * Department of Internal Medicine, University of Genova, Italy.

SUMMARY: A number of observations point toward a tight relationship between the Maillard reaction and oxidative pathways of injury to biological molecules **in vitro** and **in vivo**. These include the decreased crosslinking rate of glycated proteins and the decreased formation rate of the glycoxidation products carboxymethyl-lysine and pentosidine in absence of O_2, the catalysis of Maillard reaction mediated damage to proteins by Vitamin C oxidation products, the formation of free radicals in early steps of the glycation reaction and in lipoproteins exposed to reducing sugars **in vitro** and **in vivo**, and the polymerization or fragmentation of macromolecules exposed to sugars in presence of metal catalysts. These reactions appear to occur as part of the aging process and, at an accelerated rate, in diabetes and uremia. Two major questions arise: Is damage inflicted by glycoxidation reactions quantitatively significant, and will antioxidant therapy or therapy aimed at blocking the advanced Maillard reaction help prevent them?

BIOCHEMICAL BACKGROUND: The Maillard reaction is initiated by the nonenzymatic condensation reaction of a reducing sugar with an amine (Fig. 1). The reaction is primarily driven by the concentration of the sugar and its reactivity which follows the anomerization rate of the sugar. Thus, the condensation rate increases inversely with chain length of the sugar, being lowest for glucose and highest for trioses and glycolaldehyde. This observation has led to the concept that mammalians may have emerged with glucose as the major carrier of energy, precisely because it is a slow reacting molecule (Bunn & Higgins, 1981). The labile Schiff base that forms between the sugar and the amine undergoes a

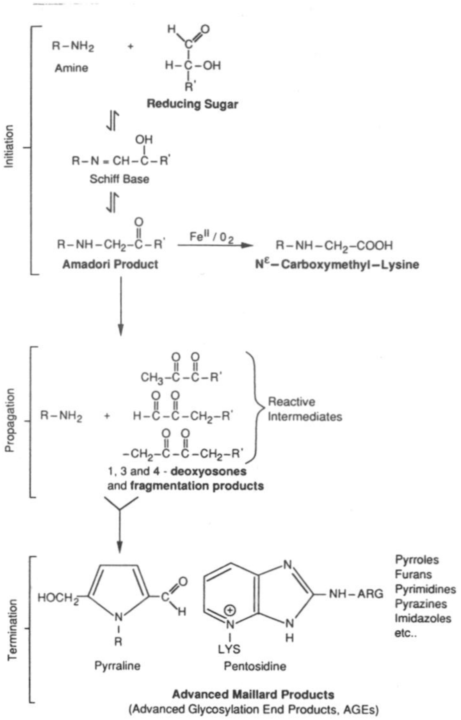

Fig. 1: General Scheme of the Maillard reaction.

rearrangement to form the relatively stable Amadori product (Fig. 1). In absence of oxidative stress, the Amadori product undergoes enolization reactions which lead to the formation of deoxyglucosones. The α-dicarbonyl compounds react again with amino acid residues to form heterocyclic protein adducts, some of which are fluorescent and can act as crosslinks. These stable end products of the Maillard/glycation reaction are also called "AGEs", i.e. advanced glycation end products (Brownlee et al., 1988) A recent review by Ledl and Schleicher (1990) describes the complexity of the chemical pathways and products of the advanced Maillard reaction.

Evidence for a modulation of the Maillard reaction by oxygen and evidence for an interface with free radicals comes from several observations. First, Baynes and his group (Ahmed et al., 1986) noted that Amadori products have a high tendency to fractionate into carboxymethyl and -lactyl lysine in presence of O_2 and metal catalyst. Interestingly, this phenomenon is not limited to proteins glycated by sugars, but includes also those modified by ascorbate (Dunn et al., 1990). Second, Hayashi et al. (1985) described an alternative pathway of aldosyl-amine reactions which leads to the formation of pyridinium free radicals very early on during the reaction (Fig. 2).

The dialkyl pyrazine radical formed is sufficiently stable to be detected by ESR. In order to form, a retroaldol condensation reaction which is catalyzed by the amine has to occur. Depending on where fragmentation occurs, C2 and C3 fragments in form of glyoxal and methylglyoxal are formed from glucose which can be detected by TLC. These fragmentation reactions are negligible at acidic pH but become quite pronounced at slightly alkaline pH. A third line of evidence comes from the work by Wolff and colleagues (1988, 1990) who described the phenomenon of autooxidation of reducing sugars undergoing Maillard reaction. These authors found that free radicals and hydrogen peroxide are produced during glucose autooxidation. The reaction leads to protein browning, conformational changes and fragmentation. Evidence for oxidative processes in these modifications comes from experiments demonstrating that catalase, and metal chelators prevent the protein changes, whereas addition of Cu^{II} makes them worse. For a full account of these reactions, the reader is referred to the article by Dean in this volume.

The tight relationship between glycation and oxidation has led Baynes (1991) to propose the term "glycoxidation" for the formation of carboxymethyl-lysine and the protein crosslink pentosidine. The biomedical relevance of glycoxidation and other advanced Maillard reaction products such as pyrraline and LM-1 are discussed below.

<u>PENTOSIDINE AS A PROBE FOR THE ADVANCED MAILLARD REACTION **IN VIVO:**</u> Pentosidine is a highly fluorescent protein crosslink involving an imidazo [4,5b] pyridinium ring formed by lysine and arginine residues. It was originally isolated from aged human dura mater following extensive proteolytic digestion and through various chromatographic steps (Sell and Monnier, 1989a, 1989b). In this

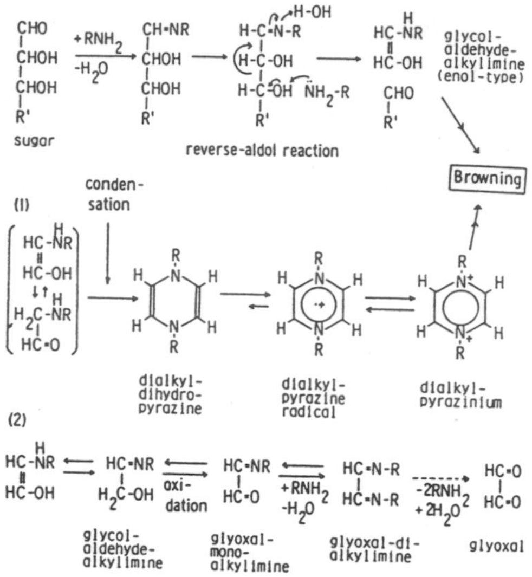

Fig.2: Proposed pathway for browning reaction involving C_2 sugar fragmentation and free radical. (M. Namiki and T. Hayashi, ACS Symp. Ser., 1983, 215, 21.)

tissue, pentosidine increases linearly with advancing age to reach approximately 250 pmol per milligram collagen at 100 years. In contrast, pentosidine formation in skin occurs at a slower rate and follows an exponential course. By the end of life-span approximately one-third of pentosidine levels present in the dura mater are found in the skin (Sell and Monnier, 1990). The slower formation of pentosidine in skin suggests that collagen turns over faster in skin than in the dura mater. On the other hand, the progressive age-related accumulation of pentosidine in normal human skin suggests the presence of collagen pools with distinct turnover, one being high, the other being very slow.

Pentosidine was also found in highly purified basement membranes where it increases in a curvilinear fashion with advancing age to reach a plateau at about 40 years of age (Sell et al., 1990). An age-realted increase in pentosidine was also found in rat tail tendons. Levels at the end of life-span, however, were much lower than in humans, being about 40 pmol/mg protein. Interestingly, dietary restriction which retards the age-related increase in thermal tendon denaturation time, a parameter of crosslinking, also retarded pentosidine formation (Sell et al., 1990).

One important question is whether pentosidine itself is of significance as a crosslinking agent **in vivo**. In human dura mater, for example, one can calculate that the presence of 250 pmol pentosidine per mg collagen translates into 3% modification of triple helical molecules of collagen. This appears to be very small compared to the physiological crosslinks such as histidinohydroxylysinonorleucine which reach one-tenth of a mole at mid-life. However, this level does not increase beyond mid-life and thus the significance of HHL as a senescence crosslink is therefore unclear. In contrast, pentosidine keeps increasing throughout life-span. When collagen gels were incubated with 100 mM ribose at 37°C and subjected to acid extraction in order to estimate collagen solubility, it was noted that acid solubility was decreased by 50% at day 6 whereas pentosidine levels were as low as 45 pmol/mg. This indicates that pentosidine itself is unlikely to be a major crosslink but that its presence reflects the tip of the iceberg in terms of overall crosslinking of the extracellular matrix by the Maillard reaction.

PENTOSIDINE AND THE SEVERITY OF DIABETIC COMPLICATIONS: Previous work from our laboratory revealed an association between collagen-linked fluorescence at 440 nm (excitation at 370 nm) and the severity of diabetic retinopathy, arterial stiffness, and joint stiffness (Monnier et al., 1986). This type of fluorescence is different from that of pentosidine which fluoresces at 385 nm upon excitation at 335 nm and is being used as an indicator for the advanced Maillard reaction. A study was undertaken to investigate the relationship between pentosidine levels, the severity of diabetic complications, and the previously measured long-wave fluorescence using the collagenase digest from the same skin biopsies that were stored at -80°C. The results showed that pentosidine is increased in all insulin-dependent diabetics above age-matched controls. Pentosidine levels corrected for age were found to correlate with the severity of diabetic retinopathy

(P<0.001) as well as joint stiffness (P<0.001). Furthermore, it correlated with a cumulative index of all four groups of complications (P<0.001) (Sell et al., in press).

The origin of skin pentosidine in this context is still unclear but the data suggest the presence of a metabolic abnormality which is expressed at a higher level in insulin-dependent diabetics who go on to develop severe complications. Interestingly there was a high correlation between the fluorescence of the age-adjusted pentosidine levels and the age-adjusted fluorescence at 440 nm, suggesting the possibility that the long-wave fluorophore is biochemically related to pentosidine.

BIOCHEMICAL ORIGIN OF PENTOSIDINE: In our initial studies, pentosidine could be synthesized through the nonenzymatic reaction of pentoses such as ribose, xylose, arabinose, and lyxose with L-lysine and L-arginine or with collagen. No pentosidine could be detected in the reaction mixture of L-lysine and L-arginine with glucose for period of one hour at 80°C. Subsequent experiments, however, revealed that prolonged incubation of bovine serum albumin with glucose at 37°C progressively led to the formation of pentosidine especially at glucose concentrations above 100 mM (Grandhee and Monnier, 1991). A detailed investigation of the mechanism of pentosidine formation from pentoses and hexoses revealed that the formation of Amadori products was a prerequisite in the synthesis of pentosidine. The highest yield and rate of pentosidine formation was obtained from the ribose Amadori product of E-amino lysine reacted with L-arginine at pH 9. The Amadori product of glucose was more reactive than glucose itself. Dyer et al. (1991) showed that other carbohydrates such as tetroses and trioses could also synthesize pentosidine. Furthermore this group showed that when protein is incubated with glucose and the reaction mixture is separated according to molecular weight, pentosidine was present in both the high and the low molecular fraction but was not a major crosslink in the high molecular weight fraction. Studies with other sugars by both groups also revealed that fructose and most interestingly ascorbate were pentosidine precursors.

PENTOSIDINE IN THE AGING HUMAN LENS: The presence of ascorbate in millimolar concentrations in the human lens as well as the

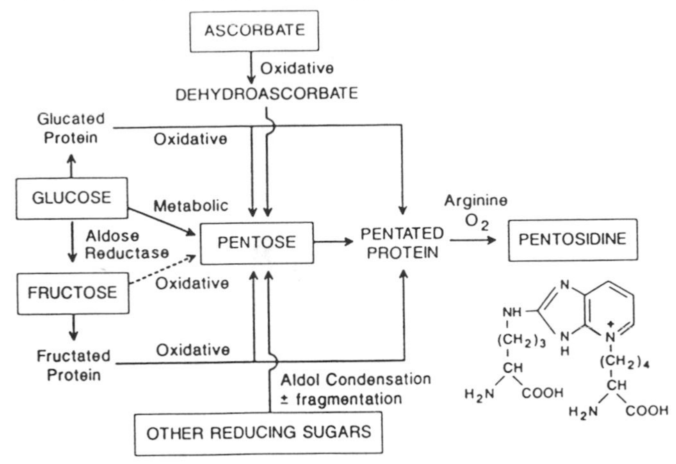

Fig. 3: Possible pathways for the biosynthesis of pentosidine.

demonstration by others that ascorbate is a potent Maillard reactant leading to browning of crystallin solutions **in vitro** led us to assay pentosidine in the human lens (Nagaraj et al., 1991). Lenses obtained at autopsy or cataract extraction surgery were classified according to their degree of pigmentation from pale yellow to severely brown. Mean pentosidine levels increased in the water soluble fraction as a function of degree of the pigmentation (Fig.4). Interestingly, normal levels were found in lenses from diabetic donors. In contrast, Lyons et al. (1991b) found elevated pentosidine levels in approximately 40% of lenses from diabetic donors. The fact that only a few diabetic subjects have elevated pentosidine levels further emphasizes that ascorbate rather than glucose or fructose is the likely precursor of pentosidine in highly pigmented lenses. Further evidence in favor of ascorbate as a major modifier of lens crystallins in aging is the discovery that a blue fluorophore called LM-1 could also be synthesized from ascorbate (Nagaraj and Monnier, 1992).

PROTECTIVE MECHANISMS AGAINST PENTOSIDINE FORMATION IN LENS: An investigation of the mechanisms by which ascorbate (ASA) can lead to pentosidine formation revealed that ascorbate first needs to be oxidized into dehydroascorbate which itself is then delactonized to 2,3-diketogulonate (2,3-DKG). No pentosidine was formed from ASA or DKA in absence of O_2, but pentosidine was able to form at a slower rate from 2,3-DKG even without O_2, (Nagaraj et al., 1991).

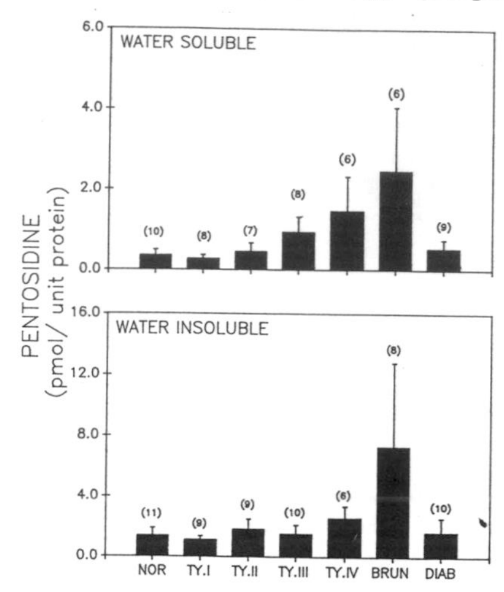

Fig.4: Pentosidine levels in cataractous lenses classified on the basis of pigmentation. Results are expressed as the mean +/- SD. Statistical significance was calculated by using Student's nonpaired *t* test. *, Significantly different compared to normal lenses (*P*<0.005). Nor, normal;Ty, type; Brun, brunescent; Diab, diabetic. (Preprinted from Nagaraj et al., 1991).

The incubation of lens crystallins with L-xylosone, the decarboxylation and rearrangement product of 2,3-DKG, led also to pentosidine formation, suggesting thereby that it is a precursor of **in vivo** pentosidine biosynthesis. Fig. 5 below is a conceptual

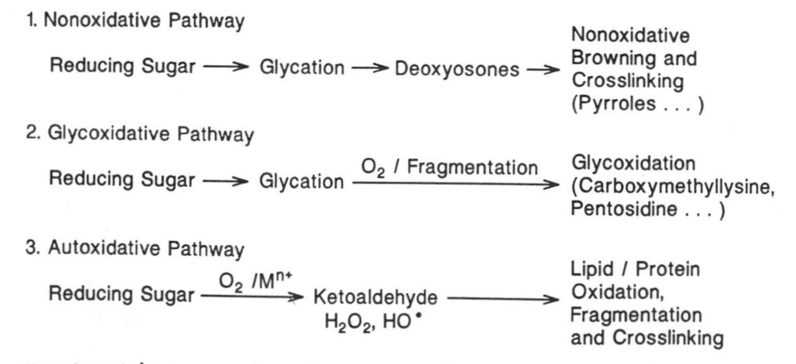

Fig. 5: Protective mechanisms against damage by the Maillard reaction in the lens.

representation of defense mechanisms against toxicity by ascorbate mediated Maillard reaction. Upon oxidation, ascorbate is immediately regenerated from DKA through glutathione and NAD(P)H dependent reductase (Bigley et al., 1981) and the ascorbate free radical (semiascorbate) reductase (Bando and Obazawa, 1990). In cataractous senile lenses, the ascorbate redox shuttle is weakened as evidenced by DKA accumulation. A second line of defense, however, is to be expected by virtue of the fact that the lens, and cells in general, contain aldose and/or aldehyde reductases which can inactivate ASA oxidation products such as L-threose, L-xylose, and L-xylosone. Recent studies by Liang and colleagues (1991) have revealed the presence of oxoaldehyde reductions in mammalian tissues. Thus, it is obvious that there are several lines of defense against ascorbate mediated toxicity which, however, depend strongly on the integrity of the NADH/NAD^+ and NADPH/$NADP^+$ redox system.

SUMMARY: The major pathways of Maillard reaction mediated damage to macromolecules can be summarized at the chemical and biological levels (Fig.6 and 7) as follows:

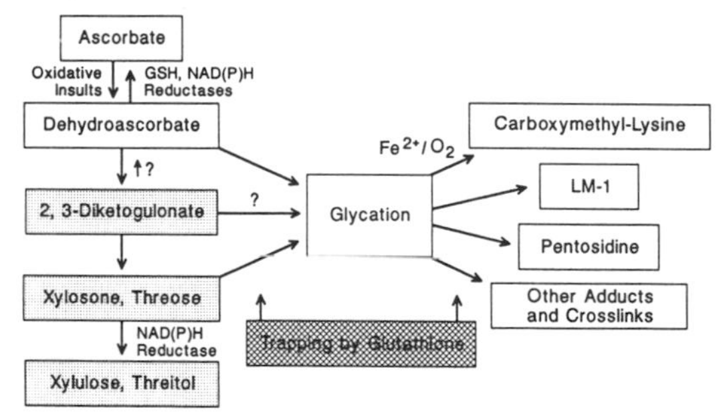

Fig.6: Biochemical pathways of Maillard reaction mediated damage to macromolecules.

First, reducing sugars can exert their damage by forming deoxysones by a nonoxidative mechanism. An example is the formation of pyrraline (Fig. 1), the presence of which was detected in tissue sections of glomeruli, arterioles, and sclerosed extracellular matrix of diabetic and non-diabetic human kidneys (Miyata and

Monnier, 1992). Second, glycated proteins can undergo glycoxidation to form carboxymethyl-lysine (CML) and pentosidine as discussed above. Preliminary data indicate that both behave similarly in tissues (Lyons et al., 1991a), although CML is in fact quantitatively more important than pentosidine. Third, reducing sugars can undergo oxidation in the presence of metals and amine catalysts to form ketoaldehydes and active oxygen radicals as proposed by Hunt et al. (1988). Although there is solid evidence for the **in vivo** occurrence of pathways 1 and 2 (Table I), the contribution of autoxidative glycation **in vivo** remains to be demonstrated. Furthermore, it is unknown whether antioxidants would be effective in preventing oxidative pathways of the Maillard reaction **in vivo**. It is likely that oxygen tension would have to be lowered below physiological levels in order to prevent glycoxidation products to form.

From a biological viewpoint, the mechanism of damage by the Maillard reaction is expected to vary depending on whether one looks at extracellular or intracellular processes (Fig.7). In the **extracellular space**, all three mechanisms of damage by the Maillard reaction are expected to occur, whereby glucose is expected to play a major role. One typical example is the glycation/oxidation of low-density lipoproteins (LDLs). There is growing evidence that oxidation of LDLs may play an important role in the pathogenesis of atherosclerosis. Recent studies suggest that LDLs incubated with glucose form free radicals suggestive of an oxidation process (Hunt et al., 1990; Mullarkey and Brownlee 1990). On the other hand, the formation of advanced Maillard reaction products may favor uptake of modified LDLs by macrophages (Vlassara et al. 1985). Thus, it will be important to separate the two mechanisms to evaluate their individual contributions. This, however, may not be possible, as glycoxidative products would automatically form in the presence of O_2, especially in hyperglycemia.

The nature of the **intracellular** damage by the Maillard reaction is expected to be minimal in unstressed cells. In metabolic diseases such as diabetes and uremia, and in conditions in which cellular homeostasis is impaired, disequilibrium in the redox potential of the cell may lead to accelerated ketoaldehyde production which, under normal circumstances, would be reduced by reductases. The latter, however, would become less efficient because of impaired dinucleotide ratio and saturating substrate concentrations. Such mechanisms appear to happen when aging human lenses become intensively pigmented. They may, in addition, play

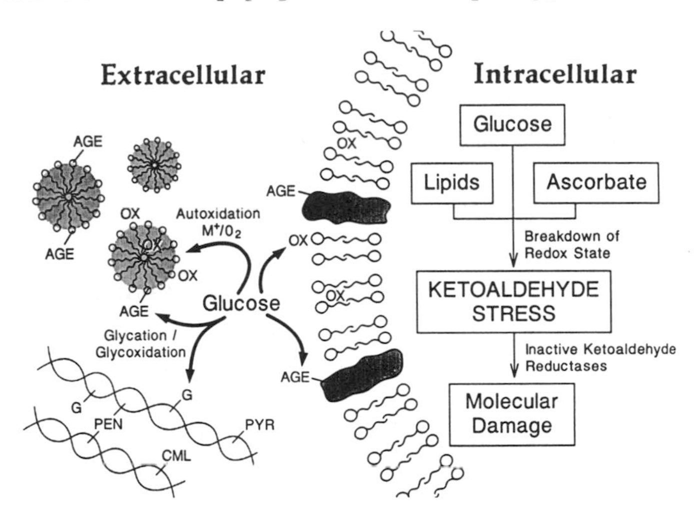

Fig. 7: Compartmentalization of damage mediated by the Maillard reaction.

an important role in apoptosis and all death.

Thus, it is becoming clear that oxidative and ketoaldehyde stress by the Maillard reaction should always be considered together when attempting to understand the biochemical nature of stochastic mechanisms of aging.

Acknowledgements

This work was supported by grants from the National Institute on Aging and the National Eye Institute, the Juvenile Diabetes Foundation International and the American Diabetes Association.

REFERENCES

Ahmed, M.U., Thorpe, S.R., Baynes, J.W. (1986) J. Biol. Chem. 261, 8816-8821.

Bando, M., and Obazawa, H. (1990) Exp. Eye Res. 50, 779-784.

Baynes, J.W. (1991) Diabetes 40, 405-411.

Bigley, R., Riddle, M., Layman, D., and Stankova, L. (1981) Biochim. Biophys. Acta 659, 15-22.
Brownlee, A., Cerami, A., and Vlassara, H. (1988) N. Engl. J. Med. 318, 1315-1321.
Bunn, H.F., and Higgins, P.J., (1981) Science 213, 222-224.
Dunn, J.A., Ahmed, M.U., Murtiashaw, M.H., Richardson, J.M., and Walla, M.D., and Thorpe, S.R., Baynes, J.W. (1990) Biochemistry 29, 10764-70.
Dyer, D.G., Blackledge, J.A., Thorpe, S.R., and Baynes, J.W. (1991) J. Biol. Chem. 266, 11654-11660.
Grandhee, S.K., and Monnier, V.M. (1991) J. Biol. Chem. 266, 11649-11653.
Hayashi, T., Mase, S., Namiki, M. (1985) Agric. Biol. Chem. 49, 3131-3135.
Hunt, J.V., Dean, R.T., and Wolff, S.P. (1988) Biochem. J. 256, 205-212.
Hunt, J.V., Smith, C.C.T., and Wolff, S.P. (1990) Diabetes 39, 1420-1424.
Ledl, F., Schleicher, E. (1990) Angewandte Chemie 6, 565-706.
Liang, Z-Q., Hayase, F., and Kato, H. (1991) Eur. J. Biochem. 197, 373-379.
Lyons, T.J., Bailie K.E., Dyer, D.G., Dunn J.A., and Baynes, J.W. (1991a) J. Clin. Invest. 87, 1910-1915
Lyons, T.J., Silverstri, G., Dunn, J.A., Dyer, D.G., and Baynes, J.W. (1991b) Diabetes 40, 1010-1015.
Miyata S. and Monnier, V.M. (1992) J. Clin. Invest. 89, 1102-1112.
Monnier, V.M., Vishwanath, V., Frank, K.E., Elmets, C.A., Dauchot, P., and Kohn, R.R. (1986) N. Engl. J. Med. 314, 403-408.
Mullarkey, J., Edelstein, D., and Brownlee, M. (1990) Biochem. Biophys. Res. Comm. 173, 932-39.
Nagaraj, R.H., and Monnier, V.M., (1992) Biochem. Biophys. Acta 1116, 34-42.
Nagaraj, R.H., Sell, D.R., Prabhakaram, M., Ortwerth, B.J., and Monnier, V.M. (1991) Proc. Natl. Acad. Sci. USA 88, 10257-10261.
Namiki, M., and Hayashi, T. ACS Symp. Ser. 215, 21.
Richard, S., Sell, D.R., Katz, M.S., Masoro, E.T., and Monnier, V.M. (1990) The Gerontol 39, 321A.
Sell, D.R., Carlson, E.C., and Monnier, V.M. (1990) Diabetes 39, 16A.
Sell, D.R., Lapolla, A., Odetti, P., Fogarty, J., and Monnier, V.M. (in press) Diabetes
Sell, D.R., and Monnier, V.M. (1989a) Conn. Tissue Res. 19, 77-92.
Sell, D.R., and Monnier, V.M. (1989b) J. Biol. Chem. 264, 21547-21602.
Sell, D.R., and Monnier, V.M. (1990) J. Clin Invest. 85, 380-384.
Vlassara, H., Brownlee, M. and Cerami, A. (1985) Proc. Natl. Acad. Sci. USA 82, 5588-5592.

Free Radicals: From Basic Science to Medicine
G. Poli, E. Albano & M. U. Dianzani (eds.)

GLYCOXIDATION: A NON-ENZYMATIC MECHANISM OF PROTEIN AGEING.

U.M. Marinari, L. Cosso, M.A. Pronzato, *G. Noberasco and P. Odetti

Institute of General Pathology and *Dept. of Internal Medicine, University of Genova, Italy

SUMMARY. Reducing sugars and reactive oxygen species are involved in the development of fluorescence in collagen connective tissue: protein modification can be generated by glycation or by reaction with lipid peroxidation products. An increase in the production of fluorescent end products and/or their defective removal system are implicated in the accumulation of such substances in the body in ageing or in various diseases. Some results suggest that the two non-enzymatic reactions are not independent, but are in fact interrelated. Our recent studies compared rat subcutaneous tissue fluorescence, and support this link, suggesting that the "glyco-xidation" hypothesis is likely to emerge as an important theory of ageing.

A large number of reactive radicals are usually produced, either enzymatically or otherwise, by exogenous and endogenous sources in our body; some of them are utilized; others might have noxious activity, but only a few cause damage, since an effective multistep defensive system is always at work neutralizing any unwanted reactive radicals.

When the body is not able to modulate these reactions, and a post-translational, often non-enzymatic, reaction overwhelms the defences, molecular modifications of cell structure and functions may occur and, if the altered protein is not removed or renewed, the damage may accumulate.

The study of the lesion mechanisms at the molecular level has been pioneered by researchers from different areas of interest (Bunn, 1981; Stadtman, 1986; Wolff and Dean, 1986), with the result that, during the 1980s, each field evolved slowly and independently.

Several studies (Wolff et al., 1986; Davies, 1987; Stadtman,

1988) have reported the most common alterations consequent on oxidation: fragmentation, formation of dityrosine or carbonyl radicals, loss of aminoacidic residues, modification of the side chain with the formation of hydroperoxides which have reactive moieties whose life is longer than that of oxygen free radicals. Moreover, besides direct modification, a cross-linking reaction takes place, increasing the detrimental effect in the proteins involved.

In many tissues, proteins are placed next to lipid moieties; consequently, a reactive radical oxidates the lipidic moiety and then the resulting lipid peroxide, or another lipid radical formed during the lipoperoxidation cascade, involves a protein-free amino-group in the reaction (Kigugawa and Ido, 1984).

The damaged protein may undergo one of two processes: i) degradation and removal by scavenger systems; ii) accumulation in the long-lived tissue with the induction of further damage, involving inter- or intra-molecular cross-linking with the development of insoluble molecules, even more difficult to remove, and the release of cytotoxic compounds (Bjorkstein, 1974).

It is well known, for example, that a slight oxidation of intracellular proteins elicits the activation of the macroxy-proteinases (M.O.P.); this enzymatic system allows a rapid and precise degradation of the intracellular protein to be renewed (Pacifici and Davies, 1991). If the free radicals and related oxidants further damage the protein, or if the side products of lipo-peroxidation modify the amino groups, some other clearance mechanism will be activated; e.g. a specific receptor mediates degradation of extracellular damaged proteins, as reported for the modified-LDL on the macrophage membranes (Fogelman et al., 1981).

Studies carried out on hepatic or connective tissue and on plasma have recently shown the presence of protein MDA- and HNE-adducts, supporting the hypothesis that a deleterious accumulation of such compounds occurs during biochemical stress, due to metabolic diseases such as hemocromatosis, diabetes mellitus, hyper-lipo-proteinemia, etc. (Houglum et al., 1990).

Another non-enzymatic reaction studied during the same period of time is glycation. At the beginning, only food chemists paid attention to this phenomenon; subsequently, biochemists, biologists and pathologists also observed protein "browning" as a "physiological and pathological event". In the catalytic sequence, a reducing sugar reacts with a free epsilon-aminogroup of a protein, producing an aldimine and then a ketoamine (the latter being more stable because of the immediate equilibrium with a cyclic hemiacetal form). Recently, research has focussed on the steps following this "early phase", because its occurrence has been observed in the long-lived tissues: the complex and almost unknown pathway generates, after many reactions, stable, irreversible and often UV-active compounds, forming intra- or intermolecular cross-links; these compounds are called advanced glycation end-products (AGEs) (Means and Chang, 1983; Monnier and Cerami, 1981). The molecular structure of some of them has been described in these last few years: pyrraline (not fluorescent), alkyl-2-formyl-3,4- diglucosyl-pyrrole and pentosidine (yellow and fluorescent), but since the greater part of these Maillard or browning products are as yet unknown, 370/440 nm fluorescence is still usually being used to evaluate their tissue level (Farmar et al., 1988; Monnier et al., 1991).

There is strong evidence that browning products are involved in, or at least related to, some pathological processes, such as cataract formation, and cross-sectional clinical studies offer solid support for their role in the genesis both of diabetic complications and the ageing processes (Sell et al., 1991).

In fact, several reports have shown that fluorescence increases exponentially with age, and very recently Sell and colleagues have shown a similar increase for pentosidine. In the human dura mater, the theoretical value of 250 pmol/mg of collagen at 100 years means that 10% of the lysine residues are irreversibly blocked, significantly slowing down protein turnover, while the percentage of insoluble collagen increases (Sell et al., 1989). If confirmed, this will stand as clear proof that ageing impairs the biological ability to replace damaged molecules.

Interestingly enough, as we have already emphasized, protein adducts from lipid peroxidation are often fluorescent, and the maximum intensity wavelength is close to the fluorescence of the Maillard products.

We have hypothesized that all the spontaneous post-translational reactions work together, and not independently. In a recent study using ageing rats, we described simultaneously evaluated fluorescence related to different final products of browning or lipoperoxidation adducts in the collagen. A close relationship between all these examples of fluorescence was thus demonstrated, although the rate of increase during ageing was different (Fig.1) (Odetti et al., 1992). In our opinion, the rate of accumulation and the final result would depend on the interaction of several factors within and outside the organism: oxidative defence status, level of blood glucose, efficiency of removal system, etc.

Our research group set up an *in vitro* study to address the uncertainty surrounding the specificity of the measurement of fluorescence, still widely used, but unavoidably imprecise. We attempted to look at this problem in greater depth, by incubating insoluble collagen with the probable precursors of the fluorescence determined: in our laboratory, the corresponding maximum of fluorescence at the expected wavelength appeared with reducing sugars and with malondialdehyde, but an overlappping area was also observed.

Interestingly, hydroxynonenal *in vitro* produced an increase of fluorescence usually associated with browning, thus suggesting the possible occurrence of interrelationships in the formation of end products (Hicks et al., 1988).

Our data agree with the experiments supporting an interrelationship between the oxidative reaction and the Maillard pathway: early glycation is able to release free radicals, autoxidation of

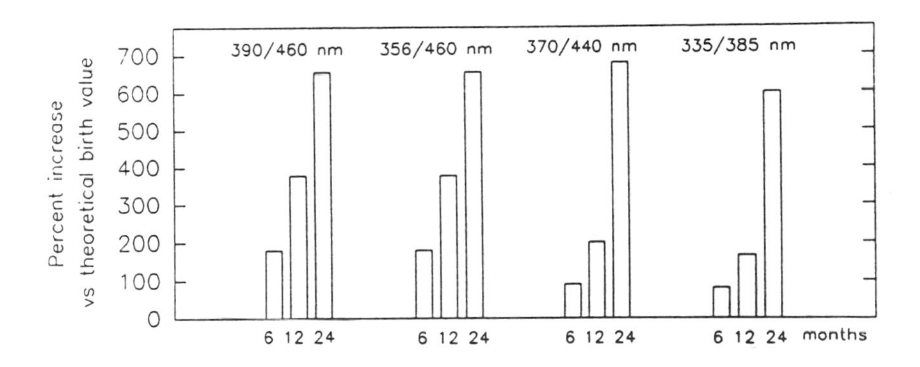

Fig.1: Percent increase of fluorescence in digests of subcutaneous collagen in Wistar rats during ageing.

glucose generates the precursors of browning and the two reactions potentiate the formation of cross-links in insoluble collagen, etc. (Hunt et al., 1988; Mullarkey et al., 1990).

In time we will probably see the consolidation of Baynes' hypothesis (Baynes, 1991) on the merging of the two reactions - with the acceptance of the term "glycoxidation" - and the involvement of all the spontaneous post-translational modifications of proteins in the pathogenesis of the major damages which occur during ageing and in many other diseases. Only more specific probes and markers can be of help in acquiring greater knowledge on the matter. Knowledge of these reactions and an understanding of their interrelationships may lead to the development of specific therapies.

REFERENCES

Baynes, J.W. (1991) Diabetes 40, 405-412.
Bjorksten, J.(1974) in Rothstein M.(ed.): Theoretical aspects of aging. Academic Press, N.Y., 43-60.
Bunn, H.F. (1981) Schweiz Med.Woch. 111,1503-1507.
Davies, K.J.A. (1987) J.Biol.Chem. 262, 9895-9920.
Farmar, J., Ulrich, P. and Cerami, A. (1988) J.Org.Chem. 53, 2346-2349.
Fogelman, A.M., Schechter, J.S., Hokom, M., Child, J.S. and Edwards, P.A. (1981) Proc. Natl. Acad. Sci. USA 77, 6499-6503.
Hicks, M., Delbridge, L., Yue, D.K. and Reeve, T.S. (1988) Biochem. Biophys. Res. Commun. 151, 649-655.
Houglum, K., Filip, M., Wiztzum, J.L. and Chojkier, M. (1990) J.Clin.Invest. 86, 1991-1998.
Hunt, J.V., Smith, C.C.T. and Wolff, S.P. (1988) Biochem. J. 256, 205-212.
Kikugawa, K. and Ido, Y. (1984) Lipids, 19, 600-608.
Means, G.E., Chang, M.K. (1984) Diabetes 31(s.3), 1-4.
Monnier, V.M. and Cerami, A. (1981) Science 211, 491-493.
Monnier, V.M., Sell, D.R., Nagaraj, R.H. and Miyata, S. (1991) Gerontology 37, 152-165.
Mullarkey, C.J., Edelstein, D. and Brownlee, M. (1990) Biochem. Biophys. Res. Commun. 173, 932-939.
Odetti, P., Noberasco, G., Pronzato, M.A., Cosso, L., Bellocchio, A., Cottalasso, D. and Marinari, U.M. (1992) Free Rad. Res. Commun. 26(s1), abs 15.3.
Pacifici, R.E. and Davies, K.J.A. (1991) Gerontology 37, 166-180.
Sell, D.R. and Monnier, V.M. (1989) J.Biol.Chem. 264, 21547-21602.
Sell, D.R., Nagaraj, R.H., Grandhee, S.K., Odetti, P., Lapolla, A., Fogarty, J. and Monnier, V.M. (1991) Diabetes/Metabolism Rev. 7(4), 239-251.
Stadtman, E.R. (1986) Trends Biochem.Sci. 11, 11-12.
Stadtman, E.R. (1988) J.Gerontol. 43, B112-120.
Wolff, S.P. and Dean, R.T. (1986) Biochem J. 234, 399-403.
Wolff, S.P., Gardner, A. and Dean, R.T. (1986) Trends Biochem. Sci. 11, 27-31.

Free Radicals: From Basic Science to Medicine
G. Poli, E. Albano & M. U. Dianzani (eds.)

MARKER OR MECHANISM : POSSIBLE PRO-OXIDANT REACTIONS OF RADICAL-DAMAGED PROTEINS IN AGING AND ATHEROSCLEROSIS, AN AGE-RELATED DISEASE

R.T. Dean, S.P. Gieseg and J.A. Simpson

Cell Biology Group, Heart Research Institute, Camperdown, Sydney, NSW 2050, Australia

SUMMARY: The accumulation during aging of unreactive stable oxidised protein functions, notably carbonyls, is clearly documented. Here we describe the identification of two novel long-lived yet reactive protein oxidation products. We suggest that a similar accumulation of an enlarged pool of these derivatives may occur during aging and atherosclerosis, for example in vascular intima as a result of low density lipoprotein oxidation. We propose that such accumulating reactive protein-bound oxidation products may be not only a marker of aging, but also part of its mechanism, by virtue of their transfer of damaging reactions to other molecules.

Protein Damage by Free Radicals and the Generation of Protein-Bound Reactive Species

The basic molecular features of free radical damage to proteins are now quite well established. They include amino acid modification, polypeptide fragmentation, and, particularly when oxygen supply is limited, cross-linking. The consequence of this complex of reactions is protein unfolding, and normally increased susceptibility to proteolytic enzymes. This seems usually to result in accelerated catabolism of the damaged protein molecules, so that they are removed efficiently (Stadtman, 1990; Pacifici and Davies, 1991; Wolff et al., 1986). However, damaged molecules accumulate in some aged systems, and we have also shown recently that this can result when oxidised albumin is endocytosed by macrophages (Grant et al., 1992). As in this particular case, an overall increased catabolism may nevertheless be associated with an increased cellular pool of damaged molecules, which would be recognised as an 'accumulation' (Fig 1).

Most of the protein oxidation products, such as protein carbonyls, dityrosine cross links, deaminated groups, and broken peptide bonds, either cannot, or do not seem to be, readily repaired. Rather, degradation is the main biological defence against such events. The occurrence of accumulations of oxidised proteins in aged cells in culture and in vivo (Carney et al., 1991) is therefore probably an example of the failure of the degradation system to overcome the rate of generation of damaged protein: it may or may not require any abnormality in the degradation system per se. It is not obvious that the limited accumulations which have been observed would have any damaging effect, since the modified protein functions are chemically relatively unreactive (e.g. the protein carbonyls). They may well be marker rather than mechanism of aging.

Our recent work has changed this picture, as we have characterised (in collaboration with the group of Dr Jan Gebicki, Macquarie University, Sydney) two reactive species on oxidised proteins which are long-lived (Simpson et al, 1992). These might transfer damage both in space and time, and thus if they too accumulate in aging, they might have significant actions. They perhaps could not only be marker, but also part of the mechanism of aging, if they inflict damage on other target molecules than themselves, as envisaged for other radical generating mechanisms.

The Nature and Properties of Protein-bound Reactive Moieties Produced by Radical Damage.

We first detected a reactive protein-bound species during studies of the influence of sugar autoxidation on protein structure. Besides observing metal catalysed radical production which could modify and fragment proteins, we found a superoxide genera-

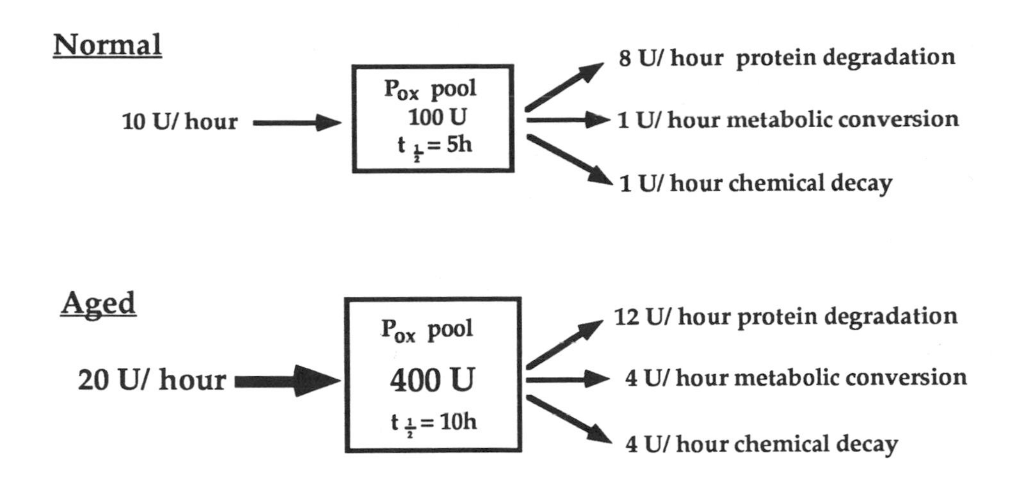

Figure I: Protein-Bound Reactive Species Could Accumulate in Aging.

Postulated steady-state fluxes into and out of a pool of protein-bound oxidation products (Pox) in 'normal' and 'aged' tissues are depicted schematically. Pox refers equally to reactive and stable protein-bound oxidation products. The influx rate may well be higher in aged (e.g. 20U/hour) than normal tissues (e.g.10U/hour). In each case, influx is probably (at least transiently) in equilibrium with the three components of efflux: chemical and metabolic decay, and proteolysis. The first two components are proposed to be unregulated, and the latter is known to be regulated. Thus chemical and metabolic decay increase in parallel with pool size, but proteolysis increases to a lesser degree. The diagram neglects the possibility that oxidised proteins may handled in different intracellular compartments from normal ones, and considers only a single pool. However, the arguments can be generalised to relate to more complex compartmentalisation.

ting protein-associated site (Wolff and Dean, 1987). This has not yet been characterised in detail, although other indications of reactive species on proteins exposed to sugars have been obtained (Hicks et al., 1988; who demonstrated the initiation of lipid peroxidation by collagen so treated), and recently by Cerami and colleagues (Bucala et al., 1991).

The first species to be chemically characterised in our work was the protein hydroperoxide group. This had been observed in very old work, but neglected since. We found upon exposure of albumin and other proteins to radiolytically generated hydroxyl radicals in the presence of oxygen, that iodometrically detectable hydroperoxides are formed. These show various chemical properties of hydroperoxides, such as reducibility by borohydride and other reductants. Individual free amino acids can also be the site of generation of hydroperoxides, and the most susceptible is valine. A surprising portion of the incident hydroxyl radicals are apparently involved in generating the hydroperoxides, on amino acid side chains: there are up to 0.4 hydroperoxides generated on bovine serum albumin per incident hydroxyl radical. It should be noted that these hydroperoxides are distinct from those which may participate in the main chain peptide bond cleavage (Garrison, 1968), and which do not accumulate.

The possible _in vivo_ reactions of protein hydroperoxides include the removal of biological reductants such as glutathione and ascorbate, which can act on them. The product of the reaction has not been rigourously characterised as yet, but is expected to be a hydroxide, analagous to those generated by the reduction of lipid hydroperoxides. While a variety of biological targets have been studied, little evidence of the transfer of damaging reactions to another molecule (e.g. the induction of lipid peroxidation) has been obtained. However, the hydroperoxides autoxidise under alkaline pH and in the presence of microperoxidase this forms the basis of a sensitive chemiluminescence detection system (Dean and Stocker, unpublished). A range of other possible reactions remain to be adequately studied.

The second species we discovered and characterised, is one which can reduce transition metals, in metalloproteins (e.g. cytochrome c) or in the form of low molecular weight chelates or complexes. The reduction of cytochrome c is independent of superoxide, as judged by lack of inhibition by superoxide dismutase, and lack of dependence on the presence of oxygen. The aromatic amino acids are the most susceptible of free amino acids to the formation of such reducing moieties. This reductant has been found to be generated also by tyrosinase treatment of proteins, which is known to generate protein bound DOPA (Ito et al., 1984; 3,4, dihydroxy-phenylalanine). To define fully the reductant produced during radical attack, we have established protein hydrolysis conditions which permit recovery of both the reducing activity and DOPA. We then analyse the hydrolysates for DOPA by HPLC with fluorescence detection, and can demonstrate parallel formation and chromatography of DOPA and reductant in a variety of radical-generating conditions, as well as with tyrosinase. In collaboration with Dr.Mark Duncan (Univ of New South Wales, Sydney) we have demonstrated by GC-MS that DOPA is the main component of the putative DOPA peak on HPLC, thus confirming that it is a major protein-bound reductant produced during radical attack (Gieseg et al., 1992).

Protein-bound DOPA is produced both in the presence and absence of oxygen, unlike the protein hydroperoxides. It too can undergo alkaline autoxidation, and be detected by chemiluminescence. However, more interestingly, it is active in the protein-quinone assay of Gallop and colleagues (Paz et al., 1991), and clearly undergoes redox cycling under these alkaline conditions (pH10), so that many molecules of Nitroblue tetrazolium (NBT) can be reduced by each protein-bound DOPA molecule, presumably cycling between the catechol and quinone forms. In recent studies we have shown that reduction of NBT can be induced by protein-DOPA at neutral pH, providing transition metal is supplied: the reduction of the metal is presumably coupled to that of NBT, and the mechanism is under study at present. Table 1 summarises some postulated actions of protein-DOPA, and also refers to its presence in the active site of some redox enzymes, and in some specialised adhesive proteins.

The protein bound reductants can thus transfer damage to other components, such as NBT, but equally important might be their influence on the local reactions of bound transition metals on the proteins themselves. Since local Fenton chemistry tends to drive the metals into the oxidised form, and hence the reactions tend to be limited by supply of reduced metal, the presence of a bound reductant pool might exaggerate some of these damaging reactions by overcoming the shortage of reduced metal. It has yet to be established whether the reductant can actually release reduced metal from protein-bound sites (as can some free radicals), but it is already known from our studies of the action of protein-bound copper that it can react to cause bulk-phase, as well as local, radical reactions, and damage adjacent as well as distal targets (Simpson and Dean, 1990). Protein-bound reductants could thus modulate vicinal and distant damaging radical reactions.

Are there biological defences against protein-bound reactive species?

Our increasing knowledge of the range of enzymes able to remove lipid hydroperoxides, or to deal with a wide variety of xenobiotics, makes one suspect that cells are endowed with systems for removing protein oxidation products. A specialised enzyme able to reduce methionine sulfoxide, one of the products of methionine oxidation, is known. We have begun to seek protein hydroperoxide removing enzymes, and have some evidence (Simpson, unpublished) that rat liver homogenates can catalyse such a reaction at neutral pH. As yet we have not found dramatic effects of putative cofactors such as GSH, and the reaction rate is not vast compared with either the quite significant rate of uncatalysed decay at physiological temperature, or the huge

Table I. Some Postulated Reactions of Protein-DOPA and its Congeners

Situation	Function/Reaction	Selected Refs.
In normal proteins		
In redox enzymes (e.g. galactose oxidase)	Donation of reducing equivalents	Nobutoshi et al., 1991
In adhesive and structural proteins of certain invertebrates and other organisms	Adhesion, wall building, etc. Cross-linking reactions	Waite et al., 1992; Dorsett et al., 1987
In hair	Related to melanogenesis	Ito et al., 1983
In radical-damaged proteins		
During physiological and pathological normoxic damage	Transition metal reduction Metalloprotein reduction	Simpson et al., 1992
	Coupled reductions via metals and/or superoxide generation	Paz et al., 1991
	Redox cycling with a co-reductant to regenerate the catechol from the quinone	O'Brien 1991; Subrahmanyam et al., 1991.
During hypoxic radical damage	Semiquinone radical-radical cross-linking?	
In other abnormal proteins		
Tyrosinase damaged proteins (e.g. melanoma?)	As for radical-damaged proteins	Simpson et al., (submitted)

metal-catalysed rate. Protein-DOPA might be expected to be a substrate for certain peroxidases, and possibly DT-Diaphorase, and this is being determined presently.

As yet we have little evidence on the degree to which the primary oxidation products are metabolised while protein bound: for example we need to determine the quantities of protein bound Valine-OH as well as Val-OOH. We have developed HPLC procedures to undertake this. It remains quite possible that the current understanding, that most protein oxidation products are removed only during degradation of the whole polypeptide on which they are present, will remain appropriate.

Proteins in lipid environments : an example

Oxidised low density lipoprotein (LDL) is claimed to accumulate within the blood vessel walls at sites of plaque, in the age-related disease atherosclerosis. However, most of the studies on LDL oxidation have entirely concentrated on oxidation of lipid, though it is known that apoB can be oxidised, and fragmented. Whether apoB is oxidised in vivo rather than simply derivatised by products of lipid oxidation is not clear.

The question obviously arises : can the reactive protein species described above be generated on apoB during LDL oxidation? Results presented at this meeting by our colleagues (Gebicki and Babiy, 1992) show that protein hydroperoxides are generated during exposure of LDL to hydroxyl radicals. Similarly, we have preliminary evidence for the generation of protein-bound reductants during copper-mediated oxidation. In both cases, delipidation of the LDL was used, and the resulting apoB was resolubilised (with the normal, extreme difficulty!) so that it could be assayed. In the case of the reductant, the quinone cycling assay involving alkaline reduction of NBT was used, and the solvent, guanidine hydrochloride had a drastic inhibitory effect on the action of protein-bound DOPA in the assay: nevertheless it seemed that at least 25% of the progressively generated

NBT reducing activity could be recovered in the protein fraction. It was also noted that there is endogenous reductant in the lipid phase of fresh LDL, presumably reflecting the action of antioxidant quinones such as ubiquinone. These studies clearly need considerable further development.

The issues canvassed above of course also relate to membrane proteins, but they have not yet been studied at all in that context.

Perspectives : Could Protein-bound Reactive Species Contribute to the Progression of Aging and Age-related Disease?

We have argued above that our view of protein oxidation must now be modified to take account of the possibility that the reactive species we have described might be active participants in ongoing oxidative stress. As illustrated in Fig 1, there is no kinetic conflict between the properties of chemical reactivity, relative long existence time, and enlarged or enlarging pools in vivo. Even in any one compartment, any biological molecule has a pool size dictated by a complex of factors, most notably the rate of biochemical generation and the rate of biochemical removal. An increased flux through the system can be coincident with an enlarged, steady, or decreased pool size.

Thus if, like protein carbonyls, protein bound reactive species accumulate enlarged pools during aging, then they may in turn inflict damage on other biological molecules and contribute to the progression of aging, and the development of age related disease. This latter problem is one of the most serious facing medicine, and the possibility of restricting age related disease is one of the most desirable preventive tasks. Understanding the possible role of each component of oxidative damage in aging is thus critical. If protein <u>mediated</u> damage is amongst them, then specific kinds of preventive approach, targetting towards the protein-reactive species themselves, need to be envisaged and developed. The lesson that antioxidants have different functions in different places, and conversely, that different antioxidants are appropriate for different defensive mechanisms, also needs to be applied to this system.

ACKNOWLEDGEMENT.

We are grateful for support from the Australian Research Council, NH&MRC of Australia and NHF of Australia.

REFERENCES.

Bucala, R., Tracey, K.J. and Cerami, A. (1991) J Clin. Invest. 87, 432-8.
Carney, J. M., Starke-Reed, P. E., Oliver, C. N., Landum, R. W., Chang, M. S., Wu, J. F. and Floyd, R. A. (1991) Proc. Natl. Acad. Sci. U.S.A. 88, 3633-3636.
Dorsett, L.C., Hawkins, C.J., Grice, J.A., Lavin, M.F., Merefield, P.M., Parry, D.L., and Ross, I.L. (1987) Biochemistry 26, 8078-8082.
Garrison, W. M. (1968) Curr. Top. Radiat. Res. 4, 43-94.
Gebicki, J. M. and Babiy, A. V. (1992) Free. Rad. Res. Comm. 16 (suppl 1), Abstract 2.14.
Gieseg, S. P., Simpson, J. A. and Dean, R. T. (1992) Free. Rad. Res. Comm. 16 (suppl 1),Abstract 6.20.
Grant, A. J, Jessup, W. and Dean, R. T. (1992) Biochem. Biophys. Acta, 1134, 203-209.
Hicks, M., Delbridge, L., Yue, D.K., and Reeve, T.S. (1988) Biochem. Biophys. Res. Comm. 151, 649-655.
Ito, S., Homma, K., Kiyota, M., Fujita, K.and Jimbow, K. (1983) J. Invest. Dermatol. 80, 207-209.
Ito, S., Kato, T., Shinpo, K. and Fujita, K. (1984) Biochem. J. 222, 407-411.
Nobutoshi, I., Phillips, S. E. V., Stevens, C.,Ogel, Z. B., McPherson, M. J., Keen, J. N., Yadav, K. D. S. and Knowles, P. F. (1991) Nature. 350, 87-90.
O'Brien, P.J. (1991) Chem. Biol. Interactions 80, 1-41
Pacifici, R. E. and Davies, K. J. A. (1991) Gerontology. 37, 166-180.
Paz, M. A., Flukiger, R., Boak, A., Kagan, H. M. and Gallop, P. M. (1991) J. Biol. Chem. 266, 689-692.
Simpson, J. A. and Dean, R. T. (1990) Free. Rad. Res. Comm. 10, 303-312.
Simpson, J.A., Narita, S., Gieseg, S., Gebicki, S., Gebicki, J. M. and Dean, R. T. (1992) Biochemical. J. 282, 621-624.
Stadtman, E. R. (1990) Free. Rad. Biol. Med.9, 315-325.
Subrahmanyam, V.V., Ross, D., Eastmond, D.A. and Smith, M.T. (1991) Free Rad. Biol. Med. 11, 495-515.
Waite, J.H., Jensen, R.A. and Morse, D.E. (1992) Biochemistry, 31, 5733-5738.
Wolff, S. P. and Dean, R. T. (1987) Bioelectrochem. Bioenerg. 18, 283-293.
Wolff, S. P., Garner, A.W. and Dean, R.T. (1986) Trends Biochem Sci. 11, 27-31.

Cancer

Free Radicals: From Basic Science to Medicine
G. Poli, E. Albano & M. U. Dianzani (eds.)

CARCINOGENESIS AND FREE RADICALS

Roy H. Burdon

Department of Bioscience & Biotecnology, Todd Centre, University of Strathclyde, Glascow G4 ONR.

SUMMARY: It appears that free radicals may be involved in the initiation, promotion and progression phase of carcinogenesis. Whilst free radicals can cause DNA alteration, they can also promote cell proliferation particularly of cells harbouring altered growth response genes. They appear to stimulate biochemical pathways common to those utilised by normal cellular growth factors. Althought inflammatory cells associated with tumors may serve as a source for such growth promoting active oxygen species, tumor cells themselves constitutively release superoxide. Not only may this released superoxide enhance tumor cell growth, but it may also be involved in the oxidative inactivation of serum antiproteases in the environment of tumor cells. This may facilitate the action of tumor proteases in growth responses and in metastasis. Consideration of the tumor phenotype also suggests various 'antioxidant' strategies aimed at tumor therapy.

INTRODUCTION

Progress toward tumor formation is slow, often taking ten or more years. Importantly, carcinogenesis involves a number of stages and can be viewed as a multistage, microevolutionary process. Most cancers derive from single abnormal cells and work with experimental systems shows carcinogenesis to be divisible into three different stages: initiation, promotion and progression. Initiation is a heritable aberration of a cell. Cells so initiated can undergo tranformation to malignancy if promotion and progression follow. Initiation appears to be irreversible and can be the outcome of DNA damage resulting in a change in the cell's DNA sequence. Promotion, on the other hand, is affected by factors that do not alter DNA sequences and involves the selection and clonal expansion of initiated cells. This process

is partly reversible and accounts for the major portion of the lenghty latent period of carcinogenesis. The final stage of tumor formation is the progression of a benign growth to a highly malignant rapidly growing neoplasm. There is a loss of growth control, an escape from the host defence mechanisms, and metastasis.

It is now clear that a single mutation is insufficient to cause cancer. From a statistical analysis of cancer as a function of age it is estimated that in humans at least, somewhere between three and seven independent mutations are required (to low figure applies to leukaemias whereas the larger relates to carcinomas). In summary cancers develop in slow stages from mildly aberrant cells and progression involves successive rounds of mutation and natural selection.

Evidence is now accumulating that support an involvement of free radicals in the complex pathways of multistep carcinogenesis. Such evidence has been accumulated from experiments which demonstrate that antioxidants which scavenge free radicals directly, or that interfere with the generation of free radical mediated events, can inhibit carcinogenesis (Sun, 1990).

FREE RADICALS & INITIATION

Certain initiators such as radiation, or chemical carcinogens, can bring about the production of various free radicals and subsequent DNA base sequence alteration (Ames, 1983; Floyd, 1982). In addition cell of the immune system such as neutrophils and macrophages produce superoxide radicals and hydrogen peroxide which have been associated with the induction of experimental cancers (Cunnigam & Lokesh, 1983). Oxygen free radicals and methyl radicals, are known to have damaging effect on DNA nucleotides. In some cases such free radicals may arise in reactions catalyzed by ferric and cupric ions localized in the vicinity of cellular DNA nucleotides. Iron may also be important in the generation of free radicals that are relevant to asbestos carcinogenicity.

Such free radical mediated damage to cellular DNA will have consequences for an organism unless "repaired". Cells are normally equipped with extensive DNA repair systems and the excretion of 10^4 oxidatively modified DNA bases per cell per day (Ames et al., 1991) attests not only to the extent of daily DNA base damage, but also to the efficiency of the human cellular DNA repair systems (Friedberg, 1986). In cases where these repair systems are congenitally defective, patients are predisposed to the development of cancer. Whilst oxygen radical effects can lead to DNA damage, they may also directly affect the protein components of the DNA repair apparatus itself. Unrepaired DNA alterations are inherited as mutations.

Despite cellular repair systems, there is normally a low rate of mutation in human cells, being of the order of $1.4x10^{-10}$ mutations/base pair/cell generation. The rate is nevertheless similar in somatic and germ line cells (Loeb, 1991). Because of the considerable number of independent mutations required for detectable tumor formation, cancers based on the normal rate of mutation are believed to be extremely rare indeed (Loeb, 1991). It is possible, however, that cancer causing genes have specific sequences that mutate at exceptional frequency but there are little data to support this. Another possibility is that single mutational events could cause multilocus deletions, however most spontaneous mutations are point mutations. Perhaps more likely is that mutations that lead to detectable cancers have multiple causal origins i.e. both spontaneous and chemically induced mutations could contribute to tumor progression. The higher incidence of cancer in the large intestine compared with the small intestine could reflect the contributions of carcinogens produced within the large intestin by the bacterial flora. Nevertheless it is argued (Loeb, 1991) that the mutation rate in normal human cells, even if exposed to carcinogens may not be sufficient to accounts for multiple mutations that accumulate during tumor progression. It is possible however that after each mutation there could occur a selective expansion of the mutant cell population. If each mutant cell expands to 10^6 cells then the cell population at risk for each relevant mutation after the first would be multiplied by 10^6. An other alternative proposed is that tumor progression involves a "mutator" phenotype (Loeb, 1991). Certainly the mutation rate of certain tumors can be 10 to 25-fold higher than that in somatic cells. It is possible that an initial mutation could occur in a gene encoding a function required for normal DNA repair. This might explain the higher mutation rates in tumor cells (or "mutator" phenotype) and be the basis for the multiple mutations that characterise many cancers. Alternatively alterations in DNA damage or repair rates could stem from metabolic derangements in cellular metabolism (see Riley, this volume). It may also be that tumor cells generate abnormal amounts of mutagenic active oxygen species. Our own observations with rodent cells transformed to the malignant state show them to have only slightly higher rates of intracellular superoxide generation. They also have slightly diminished overall levels of superoxide dismutase compared with corresponding non-transformed cells (Burdon et al., 1990). However whether such small differences would allow sufficient accumulation of intracellular superoxide or hydrogen peroxide to alter the mutation rate 25-fold remains to be ascertained.

FREE RADICALS AND PROMOTION

Promoters and oxidant radicals

Oxygen radicals and related species may also be involved in promotion. A group of well studied tumor promoters, the phorbol esters, not only can bring about changes in gene expression in cells causing them to adopt some of the phenotypic characteristics of tumor cells, but also can stimulate inflammatory leukocytes to release superoxide (Dechatelet et al., 1976). Despite the role of inflammatory cells in combating tumor development there is a paradoxical view that inflammation may contribute to tumor promotion (Cerutti, 1988). The release of superoxide by phagocytic cells following stimulation with phorbol esters is proportional to their tumor promoting activity (Goldstein et al., 1981) and inhibitors of the respiratory burst can act as antipromotors (Kensler & Trush, 1984).

Other known tumor promoters, such as benzoyl peroxide and butylated hydroxytoluene hydroperoxide, may also act through mechanisms involving free radical generation.

Oxidant and growth promotion

Normally cell growth is tightly regulated. During normal growth and differentiation there is an exact replacement of differentiated, or dying cells, without any under or over compensation. Basically four many types of protein are involved in growth control. These are serum growth factors, cellular growth factors receptors intracellular signal transducers (e.g. protein kinases, G-proteins) and nuclear transcription factors (e.g. Myc). These proteins are encoded by growth control genes, but as a result of mutation these can be inappropriately activated or altered to oncogenes which in turn give rise to inappropriately expressed or altered growth responses to various tumor promoting substances. In a way such response modification may effectively create a state of "promotability" (Cerutti, 1988). Certainly we have found that low levels of both superoxide and hydrogen peroxide, products of the "respiratory burst" will promote fibroblast growth, but particularly of fibroblasts that harbour an oncogene, or mutated protooncogene (Burdon & Rice-Evans, 1989; Burdon et al., 1990). A cellular pro-oxidant state may be important in facilitating growth responses in "initiated" cells.

Released active oxygen species as intercellular mitogenic signals

It is now clear from our own studies and studies of others, that superoxide (and hydrogen peroxide) at low levels can stimulate growth or growth responses in a variety of cell types when added exogenously to the culture medium (Burdon et al., 1981; Burdon, 1992). In particular these species stimulate the activation and the translocation of protein kinase C as well as the expression of early growth regulated genes (Amstad et al., 1991) such as the protooncogenes c-fos and c-myc. Such observations have led to suggestion that superoxide and/or hydrogen peroxide might function as mitogenic stimuli through biochemical processes common to natural growth factors. Whilst a possible source *in vivo* of either superoxide or hydrogen peroxide may be inflammatory cells in the vicinity of a tumor, it is clear that these active oxygen species are also released from a variety of non-inflammatory cells, albeit at a low level. These cells include endothelial cells, smooth muscle cells as well as primary human fibroblasts. Inhibitor studies suggest that an NADPH-oxidase like activity is involved at least in human primary fibroblasts (Meyer et al., 1991). This may be a significant aspect of a tumor phenotype.

To investigate the significance of the "constitutive" release of superoxide from oncogene transformed fibroblasts, we added exogenous superoxide dismutase or catalase to the growth medium of these cells. Surprisingly cell growth was severely reduced (Burdon et al., 1991; Burdon, 1992) suggesting that superoxide and/or hydrogen peroxide serve an important role as cell generated growth regulatory agents. In view of the many components that constitute the cellular pathways and the second messenger systems whereby growth signals are normally transduced, it is perhaps surprising that such delicately balanced multicomponent systems can be accurately activated by simple direct exposure to superoxide or hydrogen peroxide. Although oxidative modifications of various individual protein components of signal transduction pathways may have stimulatory effects on processes related to growth regulation, a key question is: how can any specificity be achieved? An alternative possibility relates to the extreme sensitivity of serum α-1 antitrypsin to oxidative inactivation. It has been demonstrated for example that fibroblasts release a growth related protease, which is believed to be involved in the local remodelling of cell coat, or glycocalyx, permitting the release, or activity, of normal growth factors (Tse & Scott, 1991). Several antiproteases are active against this growth regulated proteinase and thus block the action of growth factors, but not the intracellular signal transduction events linked to growth factor action. Whilst we find that supplementation of serum with α-1 antitrypsin will depress growth of fibroblasts, addition of hydrogen peroxide at low concentration will eliminate these effects. Thus whilst the signalling of growth responses involving released superoxide or hydrogen peroxide may be mediated through the oxidative modification of signal transduction pathway components, another possibility is the oxidative inactivation of serum protease inhibitors allowing proteases to remodel the cell surface thereby facilitating, or modulating, the action of normal growth factors. This latter mechanism would retain an important element specificity.

TUMOR PROGRESSION AND FREE RADICALS

As already mentioned tumor progression might involve successive rounds of mutation and natural selection. These may also involve free radicals and a potential source may again be associated inflammatory cells. On the other hand it has already been argued that an early step in tumour progression is one that induces a "mutator" phenotype. A significant class of as yet unrecognised "oncogenes" may be derived from genes required to maintain the normal nucleotide sequence of genes, as well as the stringent location of genes within each chromosome and the accurate segregation of chromosomes during cell division (Loeb, 1991).

A final and decisive step in carcinogenesis is the invasive and metastatic spread of the tumour to various body spaces and cavities. This appears to be facilitated by the activation of genes for the release of proteolytic enzymes (Ruddon, 1987). Paradoxically, the presence of immune cells that recognized tumour cells can enhance the colonisation of metastatic sites. Whereas high levels of immune cells appear to favour killing, the lower numbers of immune cells can favour metastasis. Again the released superoxide (from either the immune cells or the tumour cells themselves) may serve to promote metastatic growth. Alternatively they could inactivate serum antiproteases some of which are extremely sensitive to oxidative inactivation (e.g. α-1 antitrypsin), thus facilitating the proteolytic activity of the tumour released proteases. The ability of cancer cells to take advantage of the host's inflammatory response mechanisms and at the same time avoid destruction, gives cancer cells a significant selective growth advantage. In short a little immunity may be a good thing from the point of view of the cancer cells (Ruddon, 1987).

RADICALS AND THE TUMOUR PHENOTYPE

Besides the enhanced mutability of tumour cells, which has already been referred to, there are a number of other phenotypic characteristics of tumour cells which should also be emphasised in relation to free radicals.

Although a proper explanation is not yet available for the endogeous lipid peroxidation encountered in mammalian cells it is nonetheless clear that the level can vary considerably. A number of observations indicate an apparent inverse relatioship between levels of lipid peroxidation and rate of cell proliferation and extent of cell differentiation (Burdon & Rice-Evans, 1989; Begin, 1987; Barrera et al., 1991). The presence of lipid hydroperoxides within cell membranes is certainly likely to disturb membrane organisation and function due to their polarity. Whilst lipid radicals may oxidatively modify the function of membrane proteins such as ion pumps (Thomas & Reed, 1990), lipid peroxides can also yield aldehydic breakdown products through non-enzymic pathways. These aldehydes such as 4-hydroxynonenal can react with thiol and amino groups of proteins relevant to the regulation of cell proliferation, for example phospholipase C and adenyl cyclase (Curzio et al., 1986; Paradisi et al., 1985; Esterbauer, 1988). Lipid peroxidation and the level of aldehydic breakdown products is relatively low in undifferentiated, highly proliferating, tumour cells and it may be that products such as 4-hydroxynonenal play a certain role in the "down-regulation" of cell proliferation. Importantly 4-hydroxynonenal has also been shown to induce human leukaemia cells to differentiate terminally into monocyte/monophage-like cells or granulocytes (Barrera et al., 1991). Whilst a number of observations point to the potential of 4-hydroxynonenal as a regulatory molecule, little is known about factors that control its cellular concentration. Clearly a prerequisite is lipid peroxidation and this might be modulated by serum antioxidants such as α-tocopherol, but there is a need to explore mechanisms whereby its level could be more "finely-tuned". The possibility exists that aldehyde dehydrogenases may be important in this context, and by implication in the processes that relate to cell differentiation or carcinogenesis.

The low level of lipid peroxidation reported for highly proliferating cells, whilst observed in rapidly dividing tumour cells, does not appear to be specifically associated with the tumour phenotype. Some tumour cells (e.g. breast cancer) have high level of the lipid soluble antioxidant, α-tocopherol which may minimise levels of lipid peroxidation but again whether high cellular levels of α-tocopherol, are specifically associated with the tumour phenotype is not established.

Another feature of a large number of tumour cell types is low levels of Mn superoxide dismutase (MnSOD) activity (Oberley & Oberley, 1988). Tumours are also usually low in Cu/ZnSOD activity and often also low is catalase activity. Glutathione peroxidase levels are however quite variable, although high levels of glutathione reductase are observed in fast proliferating lung tumours.

As already mentioned tumour cells constitutively appear to release low levels of superoxide/and hydrogen peroxide. They also appear to generate superoxide intracellularly, possibly from mitochondrial sources (Burdon et al., 1990). Thus reduced levels of intracellular SOD or catalase may permit the accumulation of superoxide or hydrogen peroxide within a tumour. This may contribute to the mutator phenotype, but taken together with data suggesting that oncogene transformed cells respond significantly better to the growth promoting effects of low levels of superoxide or hydrogen peroxide; it may be the reduced levels of antioxidant enzymes permit achievement of a cellular prooxidant state which may facilitate advantageous growth of neoplastic cells.

THERAPEUTIC STRATEGIES

It is clear that free radical reactions are amongst the panoply of mechanisms leading human cancer and this presents opportunities for intervention to prevent the processes of initiation, or progression. Certainly there is a growing body of epidemiological evidence that antioxidant nutrients may have a preventive role in at least certain types of cancer. Those nutrients include α-tocopherol, β-carotene as well as trace minerals such as selenium.

A key problem however relates not only to the long-term nature of carcinogenesis, but to its various identifiable stages. Clearly interference with the mutagenic effects that relate to initiation will be important, but initiation events are likely to be well in the past as far as patients presenting clinically with detectable malignant tumours. To deal with such detectable tumours it may be better to consider the phenotypic properties of tumour cells as distinct from the overall process of carcinogenesis.

Rapidly proliferating cells, including malignant tumours, have low levels of lipid peroxidation. In turn this will yield only low levels of aldehydes such as 4-hydroxynonenal, which can "down-regulate" the rates of cell proliferation, or even bring about redifferentiation of certain tumor cell types. Thus any therapeutic approach which would further limit the extent of cellular lipid peroxidation, such as α-tocopherol, might not be a suitable way of attempting to limit proliferation of malignant tumour cells at a clinical levels.

An alternative approach in dealing with tumour cells is to take advantage of observations that both superoxide and hydrogen peroxide, either generated intracellularly and/or released from tumour cells, are important for proliferation as well as for tumor cell viability. Exogenously added superoxide dismutase or catalase are both significantly inhibitory towards tumour cell proliferation and viability (Burdon et al., 1991; Burdon, 1992). Similar growth inhibition of cultured tumour cells was observed when the low molecular weight superoxide dismutase mimic copper (II)-(3,5-diisopropyl salicylate)$_2$ (Burdon, 1992), was added to the culture medium, thus suggesting a basis for a novel class of antiproliferative drugs.

REFERENCES

Ames, B.N. (1983) Science 221, 1256-1262.
Ames, B.N., Shah, G., Peskin, A., and Cerutti, P. (1991) in: Oxidative Damage and Repair (Davies, K.J.A., ed.) Pergamon Press, Oxford, pp. 767-773.
Barrera, G., Di Mauro, C., Muraca, R., Ferrero, D., Cavalli, G., Fazio, V.M., Paradisi, L., and Dianzani, M.U. (1991) Exptl. Cell Res. 197, 148-159.
Begin, M.E. (1987) Chem. Phys. Lipids 45, 269-313.
Burdon, R.H. (1992) Proc. Roy. Soc. Edin. (in press).
Burdon, R.H., Gill, V., and Rice-Evans, C. (1990) Free Rad. Res. Comms. 11, 65-76.
Burdon, R.H., Gill, V., and Rice-Evans, C. (1991) in: Oxidative Damage and Repair (Davies, K.J.A., ed.) Pergamon Press, Oxford, pp. 791-794.
Burdon, R.H., and Rice-Evans, C. (1989) Free Rad. Res. Comms. 6, 345-358.

Cerutti, P. (1988) Carcinogenesis 9, 519-526.
Cerutti, P. (1988) in: Growth Factors, Tumour Promoters and Cancer Genes, Alan Liss, Inc. New York, pp. 239-247.
Cunningham, M.L. and Lockesh, B.B. (1983) Mutation Res. 121, 299-304.
Curzio, M., Esterbauer, H., Di Mauro, C., Cecchini, G., and Dianzani, M.U. (1986) Biol. Chem. Hoppe-Seyler 367, 321-329.
Dechatelet, L.R., Shirley, P.S., and Johnson, R.B. (1976) Blood 545-554.
Esterbauer, H., Zollner, H., and Schaur, R.J. (1988) ISI Atlas Sci. 1, 311-317.
Floyd, R.A. (1982) Free Radicals and Cancer, Marcel Dekker, New York.
Friedberg, E.C. (1986) DNA Repair, Freeman, New York.
Goldstein, B., Witz, G., Amoroso, M. Stone, D., and Troll, W. (1981) Cancer Lett. 11, 257-262.
Kensler, T., and Trush, M. (1984) Environ. Mutagen 6, 593-599.
Loeb, L.A. (1991) Cancer Res. 51, 3075-3078.
Meier, B., Cross, A.R., Hancock, J.J., Kaup, F., and Jones, O.T.G. (1991) Biochem. J. 275, 241-245.
Oberley, L.W., and Oberley, T.D. (1988) Mol. Cell Biochem. 84, 147-153.
Paradisi, L., Panagini, C., Parola, M., Barrera, G., and Dianzani, M.U. (1985) Chem. Biol. Interactions 53, 209-217.
Raddon, R.W. (1987) Cancer Biology, 2nd Ed. Oxford Univ. Press, Oxford.
Sun, Y. (1990) Free Rad. Biol. Med. 8, 583-599.
Thomas, C.E., and Reed, D.J. (1990) Arch. Biochem. Biophys. 281, 96-105.
Tse, C.A., and Scott, G.K. (1991) Biochem. Soc. Trans. 19, 2855.

Free Radicals: From Basic Science to Medicine
G. Poli, E. Albano & M. U. Dianzani (eds.)

DERANGEMENTS OF CELLULAR METABOLISM IN THE PRE-MALIGNANT SYNDROME

P.A. Riley

Division of Molecular Pathology, University College & Middlesex School of Medicine, The Windeyer Building, Cleveland Street, London W1P 6DB, U.K.

SUMMARY: Carcinogenesis is a multi-stage process involving several independent genetic events. It has been proposed that initiation involves the generation of a clone of cells with raised mutation rate. Thus, the essential component of the pre-malignant syndrome (PMS) consists of an elevated intrinsic mutation rate. Alteration in the steady-state level of mutagenesis is the consequence of metabolic alterations in the cells, affecting either the rate of DNA damage and/or the rate of repair. These derangements of cellular metabolism may permit early detection of premalignant (initiated) cells. The biochemical abnormalities are considered in terms of oxyradical production and may lead to the development of screening techniques and suggest possibilities for cancer prevention based on antioxidant supplementation.

INTRODUCTION

Age-specific Cancer Incidence

More than 90% of human malignancies are carcinomas and the majority of these are of relatively late onset. The age-specific incidence data for these cancers provide evidence that malignancy is the result of multiple independent somatic mutations. In the majority of cases the slope of the age-specific incidence indicates that at least four mutations are necessary to imbue a normal cell with malignant properties.

The general ideas underlying this communication were the subject of many illuminating pleistocene discussions with Professor Trevor Slater to whose memory it is dedicated in recognition of his inspiration and friendship.

Given the estimated number of cells at risk in any tissue the simplest plausible assumption is that four independent mutations at specific loci are necessary (but not necessarily sufficient) for the ultimate expression of the malignant phenotype.

GENETIC INSTABILITY IN PRE-MALIGNANT CELLS

Mutation Rate

Calculations based on these assumptions indicate that the mutation rate required in order to generate the observed age-specific cancer incidence is in the order of 10^{-3} mutations per gene per cell per year. In general terms this is much higher than the estimations of normal mutation rates in humans (approximately 10^{-6} per gene per cell per year). Such considerations lead to the conclusion that cancers arise from clones of cells that possess an elevated mutation rate (Loeb, 1974; Riley, 1982,1990; Loeb, 1991). It seems probable, therefore, that the initiating event in the classical two-stage model of carcinogenesis (Berenblum, 1974) consists of the establishment of a clone of cells with this "mutator" phenotype. Thus the initiation process becomes the rate-limiting step of carcinogenesis and the transition that is most sensitive to external carcinogens.

The nature of the "mutator" phenotype

It is well-recognized that genetic instability is a characteristic of established cancers and underlies the ideas on tumour progression by rapid clonal evolution (Nowell, 1976) and the expression of the morphological abnormalities that form the basis of the cytological diagnosis of malignancy. According to the proposal outlined here, the mutation rate is raised in pre-malignant cells. The net mutation rate in cells is the result of the balance between the rate of damage to the genome and the rate of accurate repair. Therefore, several possible abnormalities could be responsible for the mutator phenotype.

(a) Susceptibility of DNA to damaging reactions

One such change could be an increase in the proliferation rate of the population (Ames & Gold, 1990). It is known that DNA damage is more likely to occur during the S-phase of the cell cycle which would implicate growth factors and other agents responsible for increasing cellular proliferation in the enhancement of the mutation rate. However, an increased growth rate by itself is insufficient to raise the mutation rate and proliferation rates in normal cells are not related to cancer incidence. For example, the small gut epithelium exhibits a very high turnover rate in humans but has an extremely low incidence of cancer. Such considerations would seem to exclude increased growth rate as a primary mechanism for increasing the mutation rate. But if, for example, S-phase cells exhibit increased sensitivity to the action of certain specific mutagens, growth stimulation might be a contributory factor to the raised rate of damage to DNA.

(b) Repair Deficiency

It is clear that a deficiency in DNA repair would be a powerful mechanism of increasing the mutation rate in cells. There are well-established inherited abnormalities of DNA repair which are known to be associated with a raised mutation rate in the affected cells and an increased incidence of cancer, for example, xeroderma pigmentosum, ataxia telangiectasia and Bloom's syndrome (Cairns, 1981). In all these cases, however, the nature of the repair deficiency appears to be too specific to permit facile generalization. However, it is certainly possible that deficiencies in certain types of repair, such as the excision processes involved for removing oxidised bases from DNA (see below) could play a significant role in the process of carcinogenesis. Thus, a possible metabolic derangement characteristic of initiated (pre-malignant) cells may consist of one or more defective DNA repair mechanism.

(c) Diminished Detoxication or Reduced Scavenging Capacity

Initiated cells might be made more susceptible to the action of mutagens, of either extrinsic or intrinsic origin, by virtue of constitutive loss of detoxification pathways or reduced scavenging capacity. The latter possibility would be of special significance in the case of DNA damage by free radical processes associated with intrinsic production of reactive oxygen species. For example, transition metal-catalysed peroxide (R'OOH) decomposition to give hydroxyl or alkoxyl radical-mediated oxidation of DNA (RH):

$$RH + R'OOH + M^{(n-1)+} \qquad R\cdot + R'OH + HO^{-} + MP^{n+}$$

leading to base modification (Wagner et al., 1992) would be highly susceptible to alterations in intracellular metal chelation and to the action of peroxide metabolising enzymes such as catalase (for R'= H) and glutathione peroxidase.

(d) Intrinsic Mutagens

A derangement in the pre-malignant clone that may be important as a source of the raised mutation rate is the extent of intrinsic generation of mutagens in the cell.

It is possible, of course, that in reality the initiated state of pre-malignant cells consists of a combination of more than one of these factors and it would be surprising if a unique pathway were to be identified. Nevertheless, the cellular state manifested by initiated cells may be regarded as comprising a metabolic syndrome, which we name the pre-malignant syndrome (PMS), of which the most significant feature is a raised mutation rate.

Oxidative metabolism

On the basis of the age-association of cancer and the close correspondence between the age-specificity of cancer incidence in species with quite different life-expectancies it is likely that similar processes are involved in both the somatic deterioration that occurs in ageing and, at some stage, in the events leading to the emergence of malignant cells (Cutler, 1992 this volume). Many authorities have drawn attention to the correlation between ageing and oxidative metabolism and, in the light of this and the very large body of evidence suggesting that the promotional stage of carcinogenesis may involve reactive oxygen species (ROS) (Cerutti, 1985), it has been argued that any metabolic lesion resulting in increased ROS generation would be expected to increase the rate of DNA damage and thus the mutation rate in the affected cells (Batlle & Riley, 1991).

Consequently, one characteristic of the pre-malignant syndrome (PMS) could be the increased intrinsic generation of reactive oxygen species. If we further pursue this argument it is possible to propose a number of specific derangements of cellular metabolism that could give rise to this state of affairs. A proposal for such a metabolic abnormality, which appears to possess attractive features, is that pre-malignant cells exhibit an abnormality of haem synthesis (Batlle & Riley, 1991). This suggestion is based on the induction of a relative haem deficiency in cells by inhibition of the haem biosynthetic pathway. There are many steps in haem biosynthesis (Rimington, 1989) which are potentially susceptible to (partial) inhibition and which would result in similar sequelae. A generalized haem deficiency would lead to:

(i) defective electron transport in mitochondria; with resultant
(ii) increased oxy-radical generation;
(iii) decreased catalase activity, which could exacerbate the effects of raised ROS generation; and
(iv) diminished P_{450} activity;
(v) the possibility that the amount of reactive non-haem iron in the cell is increased;
(vi) accumulation of porphyrins proximal to the metabolic block.

Thus in cells with a deficiency in haem synthesis conditions would be created with the potential of increasing oxidative damage to the genome resulting in an increased mutation rate.

DIAGNOSTIC AND THERAPEUTIC IMPLICATIONS

Detection of pre-malignant cells

If the abnormalities listed above were widespread in pre-malignant cells it should be possible to use the deranged metabolism as a method for their early detection. The feasibility of this approach to diagnostic markers of pre-malignancy is currently under examination. In an initial test we examined the production of reactive oxygen species by a tetrazolium assay and demonstrated that it is possible by this means to distinguish between benign and malignant cells in cytological smears of aspirates (Lee et al., 1991). This seems to confirm that there is metabolic similarity in malignant cells with regard to their raised tetrazolium-reducing capacity which is in keeping with the proposal that malignant cells exhibit features of permanent oxidative stress (Galeotti et al., 1992, this volume) and consistent with the hypothesis of intrinsic mutagenesis by elevated oxy-radical production.

It remains to be seen whether this (and related phenoma with diagnostic potential) applies to pre-malignant cells as proposed. If it does it suggests that circumstances increasing oxidative stress (e.g. dietary effects on metabolism such as high fat intake) will be associated with an increase in the incidence of malignant disease. Of greater significance may be the implication that antioxidant therapy, by reducing the intracellular levels of ROS or diminishing the availability of transition metal ions, may provide a useful approach to cancer prevention.

Acknowledgements

I thank Professor Alcira Batlle, with whom the haem deficiency hypothesis was jointly developed, for many useful discussions. I am grateful to Miss Maruschka Malacos for typing the manuscript. The financial support of the AICR is gratefully acknowledged.

References

Ames, B.N., and Gold, L.S. (1990) Science 249, 970-971.
Batlle, A.M.C., and Riley, P.A. (1991) Cancer Journal 4, 326-331.
Berenblum, L. (1974) In: Carcinogenesis as a Biological Problem. Elsevier: New York.
Cairns, J. (1981) Nature 289, 353-357.
Cerutti, P.A. (1985) Science 227, 375-381.
Lee, J.A., Spargo, J.D., and Riley, P.A. (1991) J. Clin. Pathol. 44, 749-752.
Loeb, L.A., Springgate, C.F., and Battula, N. (1974) Cancer Res. 34, 2311-2321.
Loeb, L.A. (1991) Cancer Res. 51, 3075-3079.
Nowell, P. (1976) Science 194, 23-28.
Riley, P.A. (1982) Medical Hypotheses 9, 163-168.
Riley, P.A. (1990) Free Rad. Res. Comm. 11, 59-63.
Rimington, C. (1989) J. Clin. Chem. Clin. Biochem. 27, 473-486.
Wagner, J.R., Hu, C.-C., and Ames, B.N. (1992) Proc. Natl. Acad. Sci. USA 89, (in press).

Free Radicals: From Basic Science to Medicine
G. Poli, E. Albano & M. U. Dianzani (eds.)

MOLECULAR MECHANISMS OF OXIDANT CARCINOGENESIS

P. Cerutti, G. Shah, A. Peskin and P. Amstad

Department of Carcinogenesis, Swiss Institute for Experimental Cancer Research - 1066 Epalinges/Lausanne, Switzerland

SUMMARY: Oxidants can act at several stages of tumorigenesis. They produce heritable changes in DNA structure ad may play a role in the mutational activation or inactivation of cancer related genes. Mutations in these genes in general do not produce a selectable cellular phenotype. Therefore, we have developed the RFLP/PCR protocol (RFLP, restriction fragment lenght polymorphism; PCR, polymerase chain reaction) for the "genotypic" analysis of oxidant-induced base pair changes in hotspot codons of the human H-ras protooncogene and the p53 tumor suppressor gene.

Growth promotion by oxidants is observed with cultured human- and mouse fibroblasts as well as epidermal cells. It is expected to play a role in inflammation, fibrosis and tumorigenesis. Indeed, oxidants trigger (patho)physiological reactions which resemble those induced by growth- and differentiation factors. They activate protein kinases, cause DNA breakage and induce the growth competence related protooncogenes c-fos, c-jun and c-myc.

Mechanistic studies indicate that protein-phosphorylation and -polyADP-rybosilation are required for the transcriptional induction of c-fos by oxidants and the synthesis of protein factors, including FOS- and JUN-proteins, which bind to the fos-AP1 enhancer element. PolyADPR participates in the efficient repair of DNA breakes which otherwise may retard or block transcriptional elongation. A fine balance of the multiple components of the cellular antioxidant defence determines the growth response of cells to oxidative stress. Transfectants of mouse epidermal cells which overproduce Cu,Zn-SOD were sensitized to the toxic effects of an oxidant while overproducers of catalase (CAT) were protected.

INTRODUCTION

Oxidants can act as irritants, carcinogenic initiators, -promoters, -progressors or cytotoxic agents. They produce genetic damage in the form of DNA base damage and strand breakage which can lead to heritable sequence changes in the form of mutations. However oxidants also act as tumor promoters and cause the clonal expansion of "initiated cells" by a variety of mechanisms (Cerutti, 1985; Cerutti, 1988;m Cerutti & Trump, 1991;, Cerutti, 1991). The recognition that oxidants can activate cellular signal transduction pathways and stimulate, rather than inhibit, cell growth was of fundamental importance (Zimmerman & Cerutti, 1984; Muhlematter et al., 1988). Our discovery that they transcriptionally induce the growth-related protooncogene c-fos, c-jun and c-myc support the notion that they can act as "growth factors" (Crawford et al., 1988). Alternatively, increased resistance of initiated cells to oxidant cytotoxicity can favour their expansion at the cost of the surrounding normal tissue (Cerutti, 1988; Crawford et al., 1989). Increased resistance may result from enhanced cellular defences by antioxidant enzymes and low molecular weight antioxidants. Evidence for a protective role of antioxidants in human cancer derives from epidemiological studies in Switzerland and the US. It was found that individual with high plasma levels of the antioxidants micronutrients Vitamin C and E, β-carotene, Se and Vitamin A exhibited a lower mortality from lung- and colorectal cancer (Gey, 1987; Menkes et al., 1989; Harris et al., 1991). On the bases of these and other studies large scale cancer chemoprevention programs are underway in several countries with emphasis on micronutrient antioxidants.

Bona fide oxidants with promotional activity include H_2O_2, superoxide, ozone, hyperbaric oxygen, peroxyacetic acid, chlorobenzoic acid, benzoyl-peroxide, cumene-hydroperoxide, p-nitroperbenzoic acid, periodate. Phagocytes/istiocytes which infiltrate the tissue in response to irritants represent a major pathophysiological source of oxidants and in several instances inflammation appears to be a prerequisite for promotion (Cerutti, 1985; Cerutti, 1988; Cerutti & Trump, 1991; Cerutti, 1991). This is of particular relevance to tobacco smoke induced lung cancer since heavy smokers invariably suffer from chronic inflammation of the respiratory tract (Alexander-Williams, 1976; Argyrus & Slaga, 1981; Mass et al., 1985). However, it should be recognized that oxidants and chronic inflammatory processes have been implicated in the etiology of the other forms of human malignancies, e.g. cancer of the colorectum, bladder, liver and the skin (Alexander-Williams, 1976; Argyris & Slaga, 1981; Mass et al., 1985).

A major effort is made in our laboratory to evaluate the capacity of oxidant carcinogens to introduce activating mutations into protooncogenes and inactivating mutations into tumor suppressor genes. Such permanent genetic changes have been implicated in several stages of tumorigenesis process. In order to be able to detect base pair changes in a minute minority of cells without the need to select phenotypically altered cells we have developed the RFLP/PRC protocol (RFLP, restriction fragment lenght polymorphism; PCR, polymerase chain reaction) which measures mutations in restriction endonuclease recognition sequences (Zijlstra et al., 1990; Felley-Bosco et al., 1991; Sandy et al., 1992; Chiocca et al., 1992). This "genotypic" mutation system is being applied to oxidant mutagenesis of cancer-related genes in human cells.

In addition to causing permanent changes of DNA structures oxidants can exhibit the properties of growth factors and tumor promoters (Zimmerman & Cerutti, 1984; Kozumbo et al., 1985; Cerutti, 1985; Cerutti, 1988; Cerutti & Trump, 1991; Cerutti, 1991). Our work and that of others indicates that they activate signal transduction pathways (Larsson & Cerutti, 1988, 1989) which result in the trascriptional induction of the growth competence-related protooncogenes c-fos, c-jun and c-myc (Crawford et al., 1988; Shibanuma et al., 1988). We have emphasized over the last several years on the elucidation of the molecular mechanism of the induction of c-fos (Cerutti et al., 1990; Amstad et al., 1992 in press). Our results indicate that oxidants and <u>bona fide</u> polypeptide growth factors utilize the same 5'-upstream regulatory elements of the c-fos gene albeit with lower efficiency. Only oxidants, but not polypeptide factors, produce DNA damage and it appears likely that the relatively low potency of oxidant promoters is due to their inherent genotoxicity. Indeed, the overall effect of the exposure of a tissue to oxidants is the sum of its action on several cellular targets and the balance of positive and negative effects on growth and differentiation.

The cellular antioxidant defences are bound to modulate both the amount of the structural damage induced to DNA by oxidant carcinogens as well as their potency as growth promoters or cytotoxins. Low molecular weight antioxidants play a role as well as antioxidant enzymes. Our results with stable transfectants with enhanced complements of one or several major antioxidants enzymes indicate that the balance of the activities of these enzymes is more important for the overall sensitivity of a cell than the level of a single component (Amstad et al., 1991). We are presently asking the fundamental question whether antioxidant enzymes can act as anti-mutators which enhance the stability of the genome and decrease the risk for genetic disease, cancer and possibly aging.

GENOTYPIC MUTATIONS ANALYSIS OF CANCER-RELATED GENES

Somatic mutations participate in the etiology of numerous human pathologies and their detection represents an important goal in molecular epidemiology. Only a minute fraction of cells harbours a particular mutation in the earliest stages of the development of a disease and the mutation only rarely gives rise to a selectable altered phenotype. While the type of mutations which are induced by a particular mutagen can be defined in a model system, the actual mutability of a particular nucleotide sequence is expected to vary substantially for different genetic loci. Factors which affects the mutability are local chromatin structure and sequence context, the transcriptional state of the gene, its replication schedule and the repairability of a mutagen-induced lesions. A case in point is the spectrum of base pair mutations found in the tumor suppressor gene p53 in human tumors. The existence of 5 hypermutable domains in this gene may be dictated by protein functions. However, the type and the exact location of the mutations and the choice of the mutated domain were found to be organ- and tumor-specific (Hollstein et al., 1991). This suggest that different mutagens and factors which modulate mutability are involved in the different tumor progenitor cells. For the identification of endogenous and environmental mutagens that play a causative role in the formation of particular mutations in disease-related target genes "genotypic" mutation analysis is required which allows the quantitation of specific mutated sequences without the need for the ex vivo or in vitro selection or expansion of phenotypically altered cells (Mendelson et al., 1989; Zijlstra et al., 1990; Rossiter et al., 1990).

Most presently available mammalian mutation systems rely on the isolation of small fraction of mutated cells with a selectable mutated phenotype. Only mutations in a few genes can be analysed by this approach and growing cell cultures have to be used for ex vivo or in vitro expansion of mutated cells. Ideally, methods are required for the analysis of mutations in disease-related genes and in sequences which are known or suspected to contain mutational hot-spots. Furthermore, mutational analysis should be possible in non-dividing tissue biopsies. In order to avoid phenotypic selection, rare mutated DNA sequences have to be detected in the presence of an enormous excess of homologous wild-type DNA or they have to be separated by biochemical means. The sensitivity of recent experimental approaches to such "genotypic" mutation analysis (Crawford et al., 1989; Cariello et al., 1990) is limited by backgrounds that originate from the large excess of wild type DNA relative to mutated sequences. We have developed the RFLP/PCR approach which measures mutations in restriction enzyme recognition sites, and greatly reduce this problem (Zijlstra et al., 1990; Felley-Bosco et al., 1991; Sandy et al., 1991; Chiocca et al., 1992). Wild-type recognition sites are cleaved by the corresponding endonuclease allowing the selective amplification of mutated recognition sequences. The mixture of amplified mutated fragments are either directly sequenced or cloned into lambda-gt10 followed by the analysis of plaques by oligonucleotide hybridization.

We are presently applying the RFLP/PCR protocol to the measurement of oxidant-induced mutations in codon 12 of the human c-HA-ras1 gene which represents a hot-spot in human bladder- and thyroid cancers and in codon 248 and 249 of exon VII of the human p53 suppressor gene. This p53 codons are frequently mutated in human colorectal- and hepatocellular carcinoma, respectively.

MECHANISM OF INDUCTION OF PROTOONCOGENE C-FOS BY OXIDANTS

The fact that low doses of oxidants can stimulate rather than inhibit cell growth (Muhlematter et al., 1988; Zimmerman et al., 1984; Shibanuma et al., 1988; Murrell et al., 1990) implies that they are capable of re-programming gene expression and it was not expected to find that they induce the immediate early genes c-fos, c-jun and c-myc (Crawford et al., 1988; Shibanuma et al., 1988). An increase in the amount of Fos protein and heterodimer formation with Jun appears to be required for growth stimulation (see e.g. Mehmet & Rozengurt, 1991) and malignant transformation (Shuermann et al., 1989). While the signal transduction pathways which are activated by oxidants appear to have steps in common with growth- and serum factors they are expected to be unique in other respects.

We have compared the mechanisms of the trascriptional induction of c-fos in mouse epidermal cells JB6 (Clone 30) by an extracellular burst of active oxygen of the type produced by inflammatory phagocytes (released extracellularly by xanthine/xanthine-oxidase) to induction by serum and phorbol ester. All three inducers elhicit a characteristic immediate early response of c-fos which is inhibited by the protein kinase inhibitor H-7 but enhanced by the protein synthesis inhibitor cycloheximide. Experiments with stable transfectants containing fos 5'-upstream regulatory sequences linked to an HSV-tk-CAT reporter construct indicate that the joint DSE-AP-1 motifs exert the most potent enhancer effect in response to active oxygen as well as serum. It is concluded that the different signal transduction pathways used by these inducers converge o the same 5'-regulatory sequences of c-fos (Amstad et al., 1992 in press).

Mobility shift experiments indicate increased binding of Fos and Jun to the fos-AP-1 octanucleotide after treatment with oxidants or serum. Similarly, increased binding of Fos and Jun to the collagenase TRE which is closely related to the fos-AP-1 sequence has been reported for nuclear extracts from phorbol ester (Angel et al., 1987; Auwerx & Sassoni-Corsi, 1991) and UVC-treated cells (Stein et al., 1989). This increase in Fos and Jun was suppressed when cells were treated with oxidants in the presence of inhibitors of ADPR-transferase, protein kinases and protein synthesis. From these results we propose that post-translational polyADP-ribosylation and phosphorylation as yet unidentified protein are required for the _de novo_ synthesis of Fos, Jun and possibly other regulatory proteins. Of course our results do not imply that increased Fos and Jun binding to fos-AP-1 occurs _in vivo_. Although our present work concentrates on c-fos it should be noted that oxidants also induce c-Jun in JB6 (C130 cells) according to nuclear run-off transcription experiments (J. Shah, P. Amstad and P. Cerutti, unpublished). Induction of c-Jun has also been observed in Hela cells in response to H_2O_2 (Devary et al., 1991).

Reports in the literature suggest that newly synthesized Fos and Jun proteins participate in the down-regulation of c-fos (Köning et al., 1989). Indeed our mobility shift experiments indicate the presence of Fos and Jun in the protein complex which interacts with the DSE _in vitro_ since preincubation of nuclear extracts with anti-fos antibodies prevented the binding of proteins to the fos upstream fragment -345 to -295 or caused the formation of supershifted high molecular weight protein complexes. However our mobility shift data reveal no difference between extracts from controls and treated cells. Others inducers of c-fos in several types of cells did not cause a discernible change in protein binding to DSE according to mobility shift experiments and *in vivo* footprinting (Herrera et al., 1989).

Active oxygen induction of c-fos requires the poly-ADP-ribosylation of chromosomal proteins. The inhibitors of ADPR-transferase benzamide and 3-amino-benzamide, suppressed the elongation of the c-fos message and as mentioned above, the *de novo* synthesis of nuclear factors, among them c-Fos and c-Jun, which bind to the fos-AP-1 motif *in vitro* only following stimulation with active oxygen. Only active oxygen, but not serum nor phorbol ester, induces DNA breakage. We propose that poly-ADP-ribosylation is required because it participates in the repair of DNA breaks which interfere with transcription because they alter chromatin conformation. We observed that Fos protein is weakly poly-ADP-ribosylated in response to active oxygen but the functional role of this modification remains unclear (Amstad et al., 1992, in press).

THE ROLE OF THE CELLULAR ANTIOXIDANT DEFENSE IN OXIDANT CARCINOGENESIS

The sensitivity of cells to oxidants is modulated by low molecular weight antioxidants and antioxidant enzymes. The biochemistry of most important enzymes, i.e. superoxide dismutase (SOD), catalase (CAT), GSH-peroxidases (GPx), GSH-reductase and GSH-S-transferase has been studied in detail. However, the physiological role of antioxidant enzymes in situ in the cell is only poorly understood because of complex interactions and interrelationships between the individual components.

The first indication that the cellular antioxidant defense affects the capacity of oxidants to stimulate the growth of epithelial cells was obtained in a study comparing promotable and non-promotable epidermal cells JB6 (Crawford et al., 1989). When we measured the specific activity of Cu,Zn-SOD, CAT, GPx in monolayer culture of JB6 cells, we discovered that promotable clone 41 contained approximately twice the activity of SOD and CAT relative to the non-promotable clone 30, whereas the activity of GPx were comparable. The difference between the two clones is particularly remarkable because the two antioxidant enzymes SOD and CAT are increased coordinately in clone 41. Since the product of the action of SOD is H_2O_2, an increase in its activity is only beneficial to the cell if it is counterbalanced by a sufficient capacity for the destruction of H_2O_2. This is apparently accomplished in the clone 41 by an increase in CAT (Crawford et al., 1989). It should be mentioned that SOD and CAT may mutually protect each other from inactivation by active oxygen.

In order to gain further insight into the role of major antioxidant enzymes in oxidant carcinogenesis we have prepared stable transfectants of JB6 cells with Cu,Zn-SOD and CAT. Since only moderate increase in these enzymes are physiologically meaningful we choosed the following 5 clones for indepth characterization: CAT 4 and CAT 12 with 2.6 fold and 4.2 fold increased CAT activities, respectively, SOD 15 and SOD 3 with 2.3 fold and 3.6 fold increased Cu,Zn-SOD activities, respectively, and SOCAT 3 with a 3 fold higher CAT activity and 1.7 fold higher Cu,Zn-SOD activity than the parent JB6 clone 41. The Cu,Zn-SOD overproducers SOD 15 and SOD 3 were hypersensitive to the formation of DNA single strand breaks, growth retardation and killing by an extracellular burst of superoxide plus H_2O_2 while the CAT overproducers were protected relative to the parent clone JB6 clone 41. The double trasfectant SOCAT 3 was well protected from oxidant damage because of its increased content in CAT which counterbalances the increase in Cu,Zn-SOD (Amstad et al., 1991). We propose that the intracellular balance of the major antioxidant enzymes is more important from the protective effect of antioxidant defense than the level of the single enzyme.

The induction of growth-competence-related genes is a necessary prerequisite for growth stimulation. The immediate early gene c-fos is a prototype in this regard, and we have shown previously that its transcription is induced in JB6 cells by oxidants (Muhlematter et al., 1988; Crawford et al., 1988; Muhlematter et al., 1989). Therefore, we compared the increase in stationary concentrations of c-fos mRNA following antioxidant treatment in the antioxidant gene transfectants. The following order of decreasing inducibility of c-fos was observed: parent clone 41 > CAT 4 > SOD 3 > SOCAT 3 > SOD 15 = CAT 12. The reasons for the decrease in c-fos induction are probably quite different for CAT and Cu,Zn-SOD transfectants. The former are well protected from excessive H_2O_2 toxicity, but at the same time the signal which results in c-fos induction is attenuated. In contrast, increases in Cu,Zn-SOD levels alone augment the intracellular formation of H_2O_2 and toxic effect on components of the signal transduction pathways may predominate.

Acknowledgements.
This work was supported by the Swiss National Science Foundation and The Swiss Association of Cigarette Manufacturers and The Association for International Cancer Research.

REFERENCES

Alexander-Williams, J. (1976) Dis. Colon Rectum 19, 579-581.
Angel, P., Imagawa, M., Chiu, R., Stein, B., Imbra, R., Rahmsdorf, H., Jonat, G., Herrlich, P., and Karin, M. (1987) Cell 49, 729-739.
Amstad, P., Krupitza, G., and Cerutti, P. (1992) Cancer Res. 52, (in press).
Amstad, P., Peskin, A., Shah, G., Mirault, M.E., Moret, R., Zbinden, I., and Cerutti, P. (1991) Biochemistry 30, 9305-9313.
Argyris, T., and Slaga, T. (1981) Cancer Res. 41, 5193-5195.
Auwerx, J., and Sassoni-Corsi, P. (1991) Cell 64, 983-993.
Bradley, M., and Erikson, L. (1981) Biochim. Biophys. Acta 654, 135-141.
Cariello, N., Keohavong, P., Kat, A., and Thilly, W. (1990) Mut. Res. 231, 165-176.
Cerutti, P. (1985) Science 227, 375-381.
Cerutti, P. (1988) Carcinogenesis 9, 514-526.
Cerutti, P., Fridovich, I., and Mc Cord, J. (1988) (eds.) Oxy-Radicals in Molecular Biology and Pathology. Alan R. Liss, New

York.
Cerutti, P., Larsson, R., Krupitza, G. (1990) In: Genetic Mechanisms in Carcinogenesis and Tumor Progression (C.C. Harris and L.A. Liotta, eds.) Wiley-Liss, New York, pp. 69-82.
Cerutti, P., and Trump, B. (1991) Cancer Cells 3, 1-7.
Cerutti, P. (1991) Eur. J. Clin. Invest. 21, 1-5.
Chiocca, S., Sandy, M., and Cerutti, P. (1992) Proc. Natl. Acad. Sci. USA 89, (in press).
Crawford, D., Zbinden, I., Amstad, P., and Cerutti, P. (1988) Oncogene 3, 27-32.
Crawford, D., Amstad, P., Yin Foo, D., and Cerutti, P. (1989) Mol. Carcinogenesis 2, 136-143.
Devary, Y., Gottlieb, R., Lau, L., and Karin, M. (1991) Mol. Cell. Biol. 11, 2804-2811.
Felley-Bosco, E., Pourzand, C., Zijlstra, J., Amstad, P., and Cerutti, P. (1991) Nucl. Acid Res. 19, 2913-2919.
Gey, F. (1987) Am. J. Nutr. 45, 1368-1377.
Harris, R., Key, T., Silcocks, P., Bull, D., and Wald, N. (1991) Nutr. Cancer 15, 63-68.
Herrera, R., Shaw, P., and Nordheim, A. (1989) Nature 340, 68-70.
Hollstein, M., Sidransky, D., Vogelstein, B., and Harris, C. (1991) Science 253, 49-53.
Köenig, H., Ponta, H., Rahmsdorf, U., Bürscher, M., Schönthal, A., Rahmsdorf, H., and Herrlich, P. (1989) EMBO J. 8, 2559-2566.
Kozumbo, W., Trush, M., and Kensler, T. (1985) Chem.-Biol. Interact. 54, 199-207.
Larsson, R., and Cerutti, P. (1988) J. Biol. Chem. 263, 17452-17458.
Mass, M., Kaufman, D., Siegfried, D., and Nesnow, S. (1985) (eds.) Cancer of the Respiratory Tract: Predisposing Factors. Raven Press, New York.
Mehemet, H., and Rozengurt, E. (1991) Brit. Med. Bull. 47, 76-86.
Mendelsohn, M., and Albertini, R. (1989) (eds.) Mutation and Environment. Wiley-Liss, New York, Part C.
Menkes, M., Comstock, G., Vuilleumier, J., Helsing, K., Rider, A., and Brookmeyer, R. (1986) New Engl. J. Med. 315, 1250-1254.
Murrell, G., Francis, M., and Bromley, L. (1990) Biochem. 265, 659-665.
Muehlematter, D., Larsson, R., and Cerutti, P. (1988) Carcinogenesis 9, 239-245.
Muehlematter, D., Ochi, T., and Cerutti, P. (1989) Chem.-Biol. Interact. 71, 339-352.
Rossiter, B., and Caskey, C.T. (1990) J. Biol. Chem. 265, 12753-12756.
Sandy, M., Chiocca, S., and Cerutti, P. (1992) Proc. Natl. Acad. Sci. USA 89, 890-894.
Schuermann, M., Neuberg, M., Hunter, J., Jenuwein, T., Ryseck, T., Bravo, R.P., and Müller, R. (1989) Cell 56, 507-516.
Shibanuma, M., Kuroki, T., and Nose, K. (1988) Oncogene 3, 17-21.
Stein, B., Rahmsdorf, H., Steffen, A., Liftin, M., and Herrlich,

H. (1989) Mol. Cell. Biol. 9, 5169-5181.
Zimmerman, R., and Cerutti, P. (1984) Proc. Natl. Acad. Sci. USA 81, 2085-2087.
Zijlstra, J., Felley-Bosco, E., Amstad, P., and Cerutti, P. (1990) In: Mutagens and Carcinogens in the Diet (Pariza et al., eds.) Wiley-Liss, New York, pp. 187-200.

Free Radicals: From Basic Science to Medicine
G. Poli, E. Albano & M. U. Dianzani (eds.)

THE INVOLVEMENT OF OXY-RADICALS IN CANCER CELL KILLING AND GROWTH

T. Galeotti, S. Borrello, M.E. De Leo and L. Masotti[1]

Institute of General Pathology, Catholic University of Rome and [1]Department of Biological Chemistry, University of Bologna, Italy

SUMMARY: Cancer cell populations appear to respond to an oxy-radical attack in such an aberrant way that the surviving cells may become selectively more resistant to the cytotoxic effect of prooxidant agents. Some features of membranes isolated from cells of rapidly growing tumours, endowed with low levels of antioxidant enzymes, support this hypothesis. Together with a low rate of lipid peroxidation, such cells exhibit ultrastructural, physical and chemical alterations which apparently indicate that oxy-radical induced damage has already occurred *in vivo*. Therefore, if oncogenesis is the result not only of increased cell proliferation but also of an imbalance of the cell survival/death ratio in favour of survival, a role of oxygen radicals in controlling this ratio must be envisaged. Moreover, the recent suggestion that reactive oxygen species can function as second messengers in the activation of signal transduction pathways for both cytotoxicity and proliferation increases the probability of their involvement in the cancer process.

The recent finding that oxidant stress may induce the binding to DNA of transcription factors (NF-*k*B and AP-1) in normal and tumoural human cell lines (Schreck et al., 1991; Devary et al., 1991) has prompted researchers to consider oxygen radicals as second messengers in signal transduction pathways activated by different biological, chemical and physical agents. Indeed, oxygen radicals possess the prerequisite for second-messenger molecules: they are small, diffusible and ubiquitous species. This novel function attributed to oxygen radicals would help to explain the pleiotropic and often contradictory nature of the action of certain substances, such as the cytokines TNF (tumour necrosis factor) and IL-1 (interleukin 1), which are known to generate reactive oxygen species (ROSs). Interestingly, it has been shown recently that both TNF-induced activation of NF-*k*B

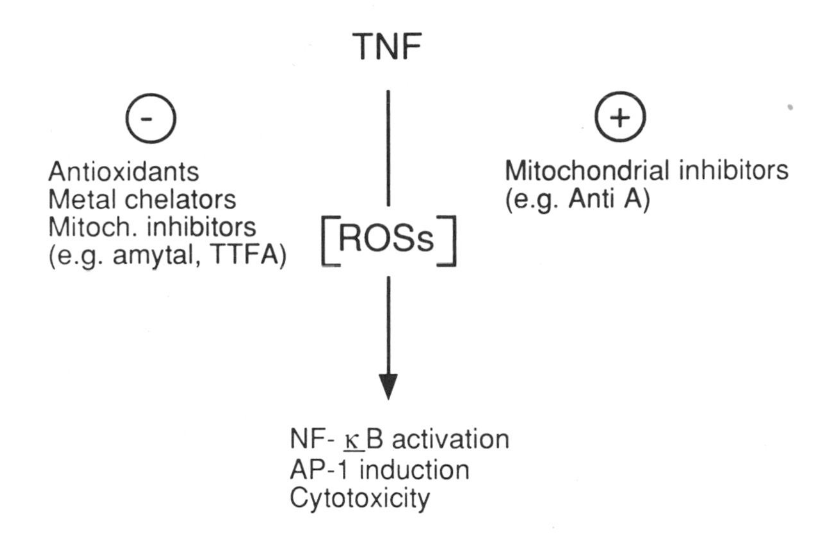

Fig. 1. Reactive oxygen species (ROSs)-mediated activation of transcription factors and cytotoxicity in TNF-sensitive cells. The involvement of oxy-radicals generated in the mitochondrial respiratory chain is indicated.

transcription factor and cytotoxicity in L929 fibrosarcoma cells can be modulated by agents regulating the intracellular level of ROSs (Schulze-Osthoff et al., 1992) (Fig. 1). Thus TNF-dependent mitogenic and cytotoxic signals, although presumably triggered by two distinct ligand-receptor complexes (Tartaglia et al., 1991), might share an (at least partially) common pathway, and the nature of the effect might be defined by the degree of the prooxidant state of the cell. Concentrations of oxy-radicals below the threshold for cytotoxic effect may induce maximal activation of transcription factors and of genes encoding growth-related and defense proteins (Fig. 2).

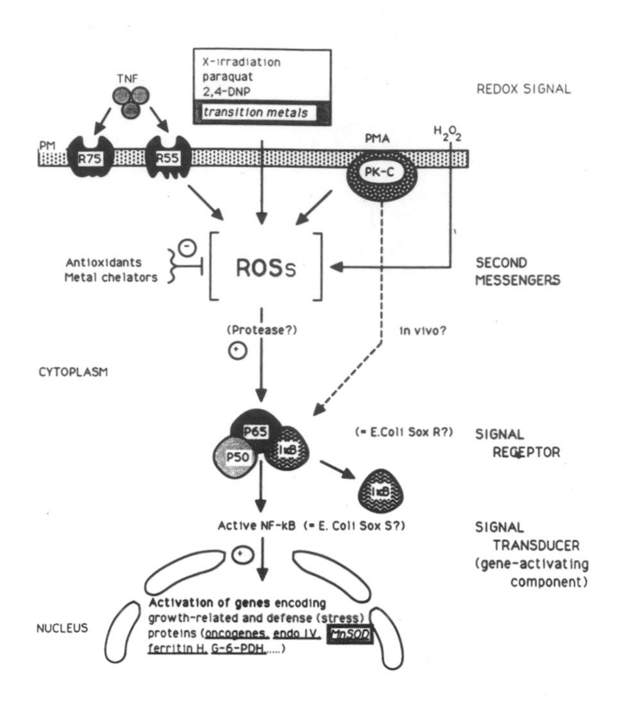

Fig. 2. A model showing the involvement of reactive oxygen species (ROSs) as second messengers in the post-translational activation of the NF-*k*B transcription factor. TNF-R75: growth signal; TNF-R55: cytotoxic signal. Adapted from Schreck et al., 1991.

In view of these considerations, the involvement of oxyradicals in cancer may be considered under two main aspects. First, they may play a pivotal role in tumorigenesis (initiation and growth promotion); second, for the cancer cell, whose antioxidant defenses are often deficient (Galeotti et al., 1989), they may be important determinants in regulating cell number through the active process of *apoptosis* (programmed cell death). These two aspects have many interconnections. Oncogenesis is the result not only of increased cell proliferation (for which many oncogenes are the most responsible factors), but also of

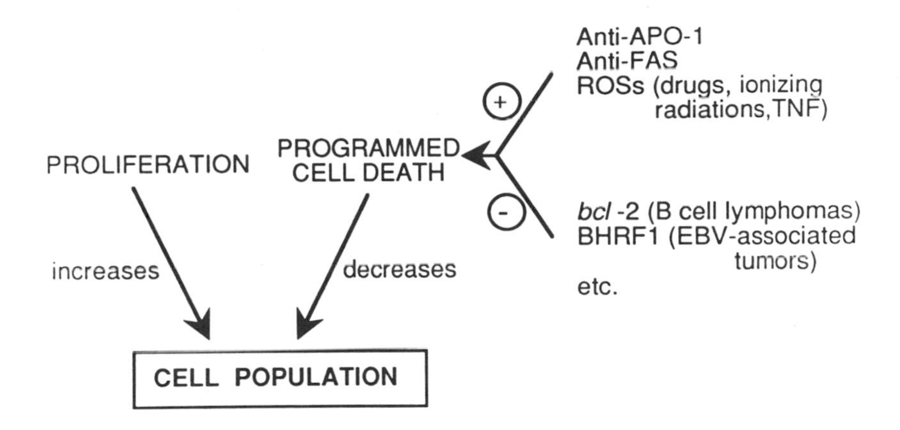

Fig. 3. Control mechanisms regulating cell number in tumour tissues. Adapted from Williams, 1991.

decreased cell death (reviewed by Williams, 1991) (Fig. 3). Cell number in a tumour is thus regulated by the balance between the two sides of this equation.

The monoclonal antibody anti-APO-1 induces *apoptosis* in several human tumours of the B and T line. A similar effect is elicited by anti-Fas antibodies. The Fas antigen is a 36 kd transmembrane receptor with significant homology with the TNF receptor. Another factor which induces *apoptosis* is TNF, whose tumoricidal activity is at present ascribed to the enhancement of mitochondrial generation of O_2 radicals (Wong et al., 1989; Schulze-Osthoff et al., 1992). Certain drugs already employed in cancer therapy, as well as ionizing radiations, eliminate tumour cells by triggering death by *apoptosis* through a mechanism involving overproduction of ROSs in the cell.On the other hand, cell survival may increase by suppressing *apoptosis*, as has been suggested for the oncogene

bcl-2, highly expressed in B cell limphomas, the most common human lymphomas, and for the EBV gene BHRF1.
This novel way of considering oncogenes is indeed attractive: some of them appear to act by inhibiting cell loss rather than by increasing cell proliferation. In conclusion, since experimental evidence is now emerging to suggest that oncogenesis is dependent not only on the rate of cell proliferation but also on the balance between cell survival and death, an important role of oxy-radicals and of their scavenger systems with regard to the regulation of such a balance must not be ignored.
One crucial point concerning oxy-radical dependent cytotoxicity in tumours is related to the vulnerability of the cancer cell and its constituents to prooxidant molecules. Several obser-

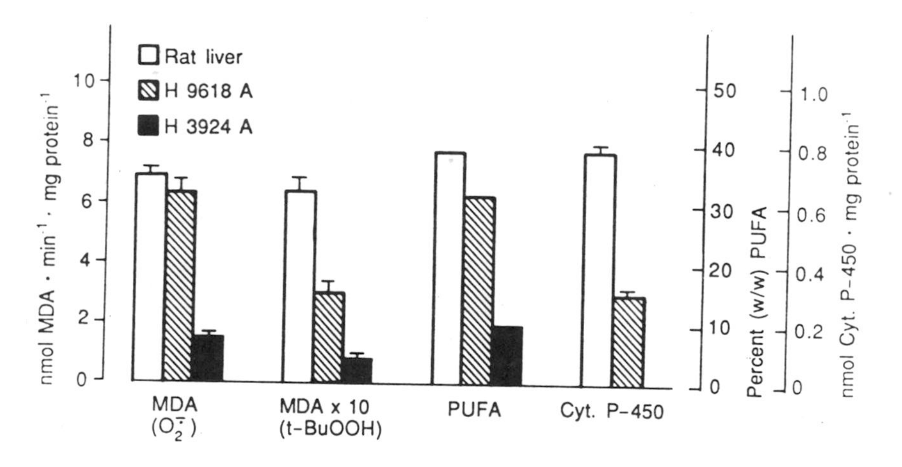

Fig. 4. Superoxide (xanthine + xanthine oxidase)-dependent and hydroperoxide-dependent lipid peroxidation of microsomes from rat liver and Morris hepatomas 9618A (highly differentiated) and 3924A (poorly differentiated), as a function of PUFA and cyt. P-450 content. Data adapted from Borrello et al., 1986.

Table 1
Lipid peroxidation-induced decrease of the fatty acid degree of unsaturation degree (double-bond index) and increase of the molecular order (<P_2>) of 1,6-diphenyl-1,3,5-hexatriene in rat liver and hepatoma 3924A microsomes

Microsomes		LOOH (nmol/mg protein)	MDA (nmol/mg protein)	Double-bond index	<P_2>
Rat liver					
Perox.(min)	0	---	---	102	0.23
	5	88.5	35.8	73	0.42
	10	201.2	68.1	55	0.50
Hepatoma 3924A					
Perox.(min)	0	---	---	32	0.53
	5	41.9	7.1	38	0.53
	10	53.4	13.1	27	0.57

LOOH, lipid hydroperoxides; MDA, malondialdehyde. Data from Masotti et al., 1986.

vations support the existence of such a condition (for review see Galeotti et al., 1989; Sun, 1990). Indeed, intracellular concentration of oxy-radicals may reach cytotoxic levels (reactions with proteins, DNA, lipids and carbohydrates) as a consequence of the unbalanced prooxidant/antioxidant ratio following the decrease of the enzymatic antioxidant defenses (Sies, 1986).

Tumour cells exhibit characteristically low levels of superoxide dismutases (Mn and CuZn dependent), glutathione peroxidases (Se and non-Se) and catalase. There are alterations (structural and functional) of the isolated tumour cell membrane components consistent with oxidative damage. The composition of the lipid bilayer is changed, particularly in terms of a marked decrease in polyunsaturated fatty acid (PUFA) residues, in cyt. P-450 and in the phospholipid/protein ratio, and of an increase of the cholesterol/phospholipid ratio.

Concomitantly, the static and dynamic properties of the membranes change. The intracellular membranes in solid tumors generally show both an increase in order and a decrease in fluidity. Conversely, plasma membranes exhibit only a decrease in fluidity. In systemic tumours, the measurement of such parameters in intact cells show the same trend for plasmalemmas. Mitochondrial, microsomal and plasma membranes isolated from tumour cells have low rates of oxy-radical-dependent lipid peroxidation. Such resistance is correlated to the degree of deviation of the tumour: the less differentiated and faster growing the tumor, the less susceptible it is to reactive oxygen species (Fig. 4).

The array of enzymatic oxy-radical scavengers also reaches minimum values in the most deviated tumours, suggesting a strict correlation between lowered defenses and extensive damage of cellular membranes *in vivo*. Consistently, we have observed changes in fatty acid composition and in physical properties comparable to those of untreated tumour membranes, after subjecting normal membranes to O_2-dependent lipid peroxidation (Table 1).

Finally, mitochondrial inner membrane degenerations, detachment of ribosomes from the endoplasmic reticulum (degranulation) and other alterations of the membranes observed in tumour cells are all events compatible with ROS-induced peroxidative damage under conditions of oxidative stress. It is worth reiterating here that the mitochondria, and sometimes the endoplasmic reticulum, of TNF-treated cells also exhibit morphological abnormalities which can be potentiated or prevented by inhibitors acting as modulators of the intramitochondrial production of O_2 radicals (Matthews, 1983; Schulze-Osthoff, 1992).

Drugs, physical agents and cytokines known to act via oxy-radicals are selectively more dangerous to cancer cells, which tend to counteract their deleterious effects by inducing the expression of defense proteins. The dose of oxy-radicals needed by the cell to activate transcription factors for gene expression (growth-related and defense proteins) is presumably lower than that required for cytotoxicity. Failure of *apoptosis* will then be important not only to increase cell numbers, but also to confer on the escaping cells resistance to natural defenses and therapeutic agents.

Acknowledgements

We wish to thank Dr. Emanuela Bartoccioni for preparing the figures. The work was supported by a grant from *Associazione Italiana per la Ricerca sul Cancro.*

REFERENCES

Borrello, S., Minotti, G., and Galeotti, T. (1986) in Rotilio, G. (ed): Superoxide and Superoxide Dismutase in Chemistry, Biology and Medicine. Amsterdam, Elsevier, pp. 432-434.
Devary, Y, Gottlieb, R.A., Lau, L.F., and Karin, M. (1991) Mol.Cell. Biol. 11, 2804-2811.
Galeotti, T., Borrello, S., and Masotti, L. (1989) in Das, D.K., and Essman, W.B. (eds): Oxygen Radicals: Systemic Events and Disease Processes. Basel, Karger, pp. 129-148.
Masotti, L., Cavatorta, P., Ferrari, M.B., Çasali, E., Arcioni, A., Zannoni, C., Borrello, S., Minotti, G., and Galeotti, T. (1986) FEBS Lett. 198, 301-306.
Matthews, N. (1983) Br. J. Cancer 48, 405-410.
Schreck, R., Rieber, P., and Baeuerle, P.A. (1991) EMBO J. 10, 2247-2258.
Schulze-Osthoff, K., Bakker, A.C., Vanhaesebroeck, B., Beyaert, R., Jacob, W.A., and Fiers, W. (1992) J. Biol. Chem. 267, 5317-5323.
Sies, H. (1986) Angew. Chem. Int. Ed. Engl. 25, 1058-1071.
Sun, Y. (1990) Free Radic. Biol. Med. 8, 583-599.
Tartaglia, L.A., Weber, R.F., Figari, I.S., Reynolds, C., Palladino, Jr., M.A., and Goeddel, D.V. (1991) Proc. Natl. Acad. Sci. USA 88, 9292-9296.
Williams, G.T., (1991) Cell 65, 1097-1098.
Wong, G.H.W., Elwell, J.H., Oberley, L.W., and Goeddel, D.V. (1989) Cell 58, 923-931.

Free Radicals: From Basic Science to Medicine
G. Poli, E. Albano & M. U. Dianzani (eds.)

FREE RADICALS AND ACTIVE STATES OF OXYGEN IN HUMAN CANCER DUE TO ENVIRONMENTAL POLLUTANTS: PUBLIC HEALTH OPTIMISM AND SCIENTIFIC SKEPTICISM

B.D. Goldstein

Environmental and Occupational Health Sciences Institute; Rutgers Univ. and the Univ. of Medicine and Dentistry of New Jersey-Robert Wood Johnson Med School, Piscataway, New Jersey, USA 08855-1179.

SUMMARY:

There is inferential evidence supporting a role for free radicals and active states of oxygen in human cancers caused by many environmental pollutants. This provides grounds for public health optimism in view of the potential for preventing these cancers or arresting their progression. However, there is a need to improve our overall knowledge base about the mechanisms by which environmental pollutants cause human cancer and to develop molecular fingerprints of the in vivo actions of free radicals and oxidative agents on the human genome.

There is an extensive literature on the subject of human carcinogens forming free radicals or active states of oxygen when tested in vitro. There is also substantial research demonstrating in vitro anti-oxidant activity for agents that seem to prevent cancer or slow its progression in humans or laboratory animals. Yet, with the possible exception of radiation, in no case have free radicals definitively been shown to be essential in the causation of a specific cancer by a specific environmental agent.

This review will describe some of the gaps between the concept of a role for free radicals and active states of oxygen in human cancer caused by environmental pollutants and the actual evidence for such a role. It is not intended to be exhaustive. Rather it will focus on certain of the exciting new scientific advances in understanding the mechanism of human cancer which give grounds for public health optimism; and many of the major gaps that require scientific skepticism. For the purposes of this review, the term environment primarily will be used to designate chemical and physical pollutants present in air, water, soil and food; although pertinent examples based on a broader definition will be cited.

A starting point in assessing the role of free radicals and oxidants in human cancer caused by environmental agents is to consider the multiplicity of human cancer types and subtypes; and the multiplicity of pathogenetic mechanisms leading to cancer. It is unlikely that any one pathway, causative agent or mechanistic process accounts for all tumors, or even for all of the cases of a specific tumor. Recent studies demonstrating the multiple steps between normal cells and clinical human cancers, each apparently associated with mutations in specific oncogene or suppressor genes, provides a challenge and an opportunity for research aimed at determining the role of free radicals and active states of oxygen in human cancer (Vogelstein, 1990). Further, as is now evident for mutation of the P53 suppressor gene observed in many, but not all, human cancers, specificity of the type and location of the mutation within the gene suggests that different etiological factors may leave a specific fingerprint allowing understanding of the cause of cancer (Hollstein et al, 1991).

Public health optimism is warranted not only by the possibility that molecular biological approaches will allow pinpointing of specific etiological factors in human cancer. It is also warranted by the potential role of free radicals in human cancer as there is an understanding of the mechanisms by which biological systems protect themselves against such insults and a range of reasonable candidates for antioxidant agents that could act to interrupt any such neoplastic mechanisms. Where there are six specific stages in the development of a cancer there are also six specific targets for the action of free radicals, and as many opportunities for primary prevention or anticarcinogenic action.

THE ROLE OF ENVIRONMENTAL POLLUTANTS IN HUMAN CANCER

There is no question that cancers are caused by environmental pollutants, although there is much argument about the extent to which they are responsible for the totality of human cancers, or for the perhaps 85% of all cancers which originate due to factors in the environment, using the broadest sense of that word. This seems to be mostly an irrelevant argument on three grounds: firstly, whether it is 2% or 25% it is a substantial public health problem given the high risk of cancer in modern society; secondly cancer due to environmental chemical and physical agents represents a correspondingly higher percentage of fully preventible cancers; and, thirdly, and most importantly for present purposes, insight gained through understanding cancer caused by environmental chemical and physical agents may be of major value in understanding cancer causation in general.

One of the most provocative contributions to the literature concerning the extent to which pollutants cause human cancer has been made by Ames and his colleagues (Ames, 1983, 1989; Ames and Gold, 1990). Part of their argument against a major role for external chemical and physical agents in human cancer stems from their important development of the measurement of 8-hydroxy-deoxyguanosine as a marker of oxidative stress and their resultant calculation of a relatively high level of background oxidation of DNA. The relevance of these calculations to the cell specific attack of external chemical and physical agents is not completely clear. There is also some nagging doubt that what is being observed primarily reflects the oxidation of DNA undergoing degradation following normal cell turnover; i.e the high level of oxidized bases in the urine could be a post mortem artifact. Nevertheless, this work is a seminal advance in developing the biological markers of in vivo oxidative stress needed to answer questions about the role of pollutants in human cancer.

THE ROLE OF FREE RADICALS IN HUMAN CANCER

There are basically two types of evidence supporting a role for free radicals and oxidants in environmental cancer: studies showing that known human carcinogens produce such active species in one or more systems, usually in vitro; and evidence that diets high in antioxidants are associated with lower incidence of certain cancers (Clemens, 1991). While the totality of this evidence is strong, perhaps even compelling; for most known human carcinogens the evidence remains circumstantial. This in part reflects the uncertainty caused by the ingenuity of scientists in the field who are able to tease one electron processes or oxidative stress out of almost any chemical, particularly in vitro. As discussed below, there are even two free radical mechanisms that can be ascribed to asbestos, a seemingly inert compound. Epidemiological studies implying an anticancer role for antioxidants are susceptible to alternate explanations; nor has every study found an anticancer action of dietary antioxidants, or shown a link between blood levels of antioxidants and cancer risk (Dorgan and Schatzkin, 1991; Comstock et al, 1992).

The concepts of tumor initiation and promotion have been very useful in providing a framework for understanding the actions of carcinogens. Of note is the increasing evidence of a role for free radicals and oxidizing agents that has developed during the literally decades of steady advances in working out the biochemical process underlying initiation and promotion, subdividing the latter into different stages, and exploring the role of carcinogens and anticarcinogens (Cerutti, 1985; Perchellet and Perchellet, 1989; Goldstein and Witz, 1990). Some of the evidence involves potentially important environmental pollutants. For example, anthrone compounds appear to generate radical intermediates which may account for their tumor promoting activities (DiGiovanni, 1991; DiGiovanni et al, 1988); and benzoyl peroxide, a prototype of environmental organic peroxides which produce free radicals, is a tumor promoter (Rotstein et al, 1987, Taffe et al, 1987).

Molecular biological techniques have led to a recent explosion of information relevant to understanding the stages of human carcinogenesis. An elegant example has been the six stage model of human colon carcinogenesis, many of the stages associated with specific oncogene or suppressor gene alterations (Vogelstein, 1990; Baker et al, 1990; Hollstein et al, 1991). Human colon cancer is a good research target in that it usually progresses through a number of morphologically identifiable stages. It is still too early for confidence that all cancers will have such an orderly progression of genetic changes; in fact, the data from human lung cancers suggest that there are heterogeneities in the multiplicity of genetic defects in the control of growth and differentiation pathways leading to clinical cancers, but some patterns consistent with etiological factors are emerging (Lehman et al, 1991; Vahakangas et al, 1992). Sahu (1990) has reviewed the possible relation between oxygen stress, oncogenes and cancer.

FREE RADICALS IN HUMAN CANCER CAUSED BY ENVIRONMENTAL POLLUTANTS

Broadly speaking, there are three potential roles for free radicals and active states of oxygen in environmental carcinogenesis: the agent itself may be a free radical or active state of oxygen; these active species may serve as the intermediary by which an environmental chemical produces genetic damage; or the metabolism of the chemical through a free radical dependent step may lead to the formation of a carcinogen.

When the first metabolic step of an environmental pollutant constitutes a radical attack on the xenobiotic, not surprisingly a radical product may be formed. Thus, cytochrome P-450 dependent metabolism of carbon tetrachloride leads to the formation of the trichloromethyl radical that in turn is believed to be capable of causing DNA damage (Dianzani and Poli, 1985). On the other hand, the α,ß-unsaturated carbonyl derivative, muconaldehyde, a putative product of hydroxyl radical-dependent ring opening of benzene, is hypothesized to alkylate DNA directly without further formation of radicals or active states of oxygen

(Goldstein et al, 1982). Oxygen free radicals that activate procarcinogens may be formed not only as a result of cytochrome-dependent metabolism but also through the activity of phagocytic cells (Trush and Kensler, 1991; Kozumbo et al, 1992).

In view of the increasing interest in the role of free radicals and active states of oxygen in disease processes it is not surprising that there is a growing literature on these species in cancer caused by environmental carcinogens, including recent studies of the role of reactive species in cancers seen in betel and tobacco chewers (Stich and Anders, 1989), in the renal cancer observed in laboratory animals receiving the food additive potassium bromate (Kurokawa et al, 1990) and in liver cancer due to peroxisome proliferators (Stark, 1991). In addition to the direct action of sunlight, it is also theoretically possible that photochemical sensitizing agents in our general environment, including food, might also provide a potential mechanism for skin cancer. 8-Methoxy psoralen, in therapeutic doses used along with ultraviolet A in the treatment of psoriasis and vitiligo, produces DNA adducts through a mechanism which is believed to involve free radical intermediates (Yang et al, 1989). Psoralens are present in various natural foods.

Rather than detail all of the potential environmental carcinogens that may have free radicals or active states of oxygen in their mechanisms of action, described below are a number of examples chosen in part to demonstrate the ambiguity of our understanding.

<u>1. Asbestos</u>:
Asbestos is undoubtedly a human carcinogen. Like many known human carcinogens it produces a tumor, bronchogenic lung cancer, which is at present indistinguishable from the same cancer caused by other agents. Asbestos also causes mesothelioma, the incidence of which appears to be almost totally due to asbestos.

Of note is that although asbestos is prized as a commercial

substance for being inherently inert, free radicals have been hypothesized to play a major role in its causation of cancer. Two different pathways have been suggested: frustrated phagocytosis leading to the release of radicals from activated phagocytic cells, and direct production of free radicals from iron or perhaps other constituents of asbestos (Walker et al, 1992).

Fiber size is particularly important in the likelihood that asbestos will produce cancer, long and thin fibers being more mesotheliogenic. Frustrated phagocytosis, as the term implies, relates to the attempt of phagocytic cells, particularly alveolar macrophages, to ingest fibers that are larger in length than the phagocyte. This is accompanied by continued release of a variety of products including superoxide anion radical and related species. The phagocyte eventually dies to be replaced by another in a process that continues for the decades that an asbestos fiber may persist in the lung. This continuous production of oxidants fits readily into hypotheses relating activated phagocytes to tumor promotion caused by inflammatory agents (Mossman and Marsh, 1989; Cerutti, 1985; Goldstein and Witz, 1990).

Fiber dimension is not the only consideration in explaining the difference in tumorigenicity among asbestos fibers. Physicochemical properties resulting in the generation of free radicals have been invoked as an explanation for certain asbestos fibers, e.g. erionite, being more tumorigenic in certain system than others. The iron content of asbestos and the surface accessibility of the iron have been particularly considered to be important in generating free radicals (Weitzman and Graceffa, 1984; Walker et al, 1992).

A trial of beta carotene as a chemopreventive agent in individuals with radiographic evidence of pulmonary asbestosis is under way. While a finding of decreased risk of lung cancer among those who receive the beta carotene would be compatible with a role for free radicals, it is instructive to note that such a finding would not be conclusive evidence of a role for free radicals or active states of oxygen in asbestos carcinogenesis.

2. Radon:

An example of molecular biology providing methodology relevant to estimating the role of a free radical producing agent in causing cancer is given by a recent study of uranium miners exposed to radon. While there is no reason to doubt that radon can be a human carcinogen, there has been a debate over the extent radon in homes contributes to lung cancer incidence. A risk estimate prepared by the US Environmental Protection Agency suggesting a substantial annual incidence of lung cancer is caused by radon has depended heavily on extrapolating from radon levels and lung cancer incidence in uranium miners. Abelson (1991) has argued that the risk is greatly overstated, in part because the observed lung cancer incidence primarily reflected cigarette smoking by the miners. However, Vahakangas et al (1992) recently reported that genetic alterations in the P53 gene in lung cancer tissue obtained from 19 uranium miners differs from the pattern that has been observed in lung cancer tissue obtained from cigarette smokers. Note that if cigarette induced lung cancers are due to free radicals, this suggestion of a specificity of genetic damage due to radon as compared to cigarette smoke implies that different free radical processes would have different patterns of genetic damage, a not unexpected result.

3. Benzene:

Benzene is a commonly used bone marrow toxin which undoubtedly produces human acute myelogenous leukemia, and perhaps other bone marrow cancers as well. It is believed that one or more benzene metabolites are primarily responsible for its bone marrow toxicity; that it is predominantly metabolized in the liver, although some bone marrow metabolism does occur; and that despite readily producing circulating chromosomal abnormalities in exposed humans and laboratory animals, it is negative in most short term mutagenesis assays (Goldstein, 1990).

Three different mechanistic roles for free radicals have been suggested: involvement in metabolism to the hematotoxic or leukemogenic metabolite; direct involvement in the hematotoxicity caused by the metabolite; or formation through macrophage involvement in benzene hematotoxicity (Laskin et al, 1989). Benzene metabolism is complex, resulting in a number of reactive intermediates including, at least in vitro, epoxide, peroxides and a dihydrodiol, the latter believed to be central to benzene metabolism to phenol and other hydroxylated intermediates (Snyder et al, 1987). Certain of the metabolic steps are cytochrome P-450 dependent; inhibition of metabolism causing a decrease in hematotoxicity while ingestion of ethanol, an inducer of cytochrome P-4502E1 potentiates hematotoxicity. We have suggested that cytochrome-dependent formation of a highly reactive species results in opening the benzene ring and the formation of trans,trans muconaldehyde (Goldstein et al, 1982). This di-α,β-unsaturated aldehyde has many of the attributes of similar carbonyl compounds derived from lipid peroxidation and has been shown to be hematotoxic in vivo (Witz at al, 1989). Muconaldehyde and its monoalcohol or monoacid derivatives have an important characteristic in the consideration of the mechanism of action of free radicals and active states of oxygen. While highly reactive, their reactivity does not preclude surviving in the circulation for sufficient time to get from the liver to the bone marrow.

A free radical basis for the selectivity of benzene toxicity to the bone marrow has been suggested to be based on the relatively high level of peroxidase in this tissue (Subrahmanyan et al, 1991). Recent findings supporting this hypothesis include the observation that exposure in vitro of HL-60 cells, derived from a human promyelocytic leukemia cell line, to hydroxylated benzene metabolites produces higher levels of 8-hydroxydeoxyguanosine (Subrahmanyan et al, 1991). This and other pathways for the formation of active radicals capable of damaging the bone marrow have been suggested, as has a role for species derived from activated macrophages (Laskin et al, 1989).

4. Ozone:

Ozone is a common environmental pollutant which is itself a reactive form of oxygen but for which a role in human cancer is unproven. Among the active species that can occur in atmospheric pollution, ozone is present in highest concentration and is the most active that can survive long enough to be inhaled deep into the respiratory tract. Singlet oxygen, hydroxyl radicals and other highly energetic species are also formed in photochemical air pollution, but have half-lives too short to be of concern.

Ozone is highly reactive to the lung, producing significant pulmonary edema at concentrations that are no more than an order of magnitude above levels common in the United States, particularly in Southern California. At environmentally relevant levels ozone produces decreases in human lung function and increases in asthma attacks. Yet, despite the undoubted fact that this active form of oxygen has direct effects on the lung, there is no evidence that it is a human carcinogen.

The contradictory nature of the evidence concerning ozone carcinogenesis highlights some of the complexities in considering the role of free radicals and active states of oxygen in cancer. On the one hand, ozone is undoubtedly mutagenic in bacteria; is capable of producing transformation of in vitro cell systems analogous to neoplastic changes (Borek et al, 1986); and, in some but not all studies, increases the incidence of pulmonary tumors in a particularly susceptible mouse strain. There is also some evidence that ozone or nitrogen dioxide exposure in experimental animals may alter the proliferation of pulmonary neuroendocrine cells, the precursor of small cell tumors. On the other hand, despite over three decades of relatively high level exposure, there is no epidemiological evidence as yet indicating an increased risk of lung tumors in the population of Southern California. Further, depending on the dose regimen and sequence, ozone exposure may actually inhibit the production of pulmonary tumors in laboratory animals exposed to known pulmonary carcinogens (Witschi, 1988, 1991; Last et al, 1987). The latter may not be surprising in view of the ability of ozone both to decrease the levels of lung cytochrome P-450 (Goldstein et al, 1975) involved in the activation of certain lung procarcinogens, and to decrease the ability of pulmonary alveolar macrophages to produce superoxide anion radical (Amoruso et al, 1981).

5. Metals and Trace Elements:

Iron is a common dietary ingredient, an essential and widely distributed body constituent present in larger amount than any other metal, and a central feature of many free radical generating schemes considered to be feasible in vivo (Stevens and Kalkwarf, 1990). A role for iron in colon and in hepatic cancer has been suggested (Babbs, 1990; Fargion et al, 1991), and, as described above, the iron content of asbestos has been suggested to be of importance in the carcinogenicity of this seemingly inert fiber.

Both chromium and nickel compounds are among metals classified as pollutants in the workplace or general environment that have been clearly demonstrated to be human carcinogens, causing lung cancer among heavily exposed workers (Coogan et al, 1989; DeFlora et al, 1990; Lees, 1991). In each case there is evidence that responsibility for carcinogenesis is restricted to particular forms; hexavalent chromium, and reduced nickel compounds, particularly the sulfides. A role for free radicals in chromium carcinogenesis has been suggested (Standeven and Wetterhahn, 1991) including the demonstration of a spin adduct consistent with hydroxyl radical following incubation of hexavalent chromium with hydrogen peroxide (Aiyar et al, 1991). While these in vitro studies are of great value, evidence that chromium produces cancer in vivo through a free radical dependent mechanism is lacking, and other mechanisms have been suggested (Snow, 1991). A free radical component to nickel carcinogenesis also has been inferred based on the release of active oxygen species during phagocytosis of nickel particles, but mechanistic studies have mainly focused on the interference with the fidelity of normal DNA replication through selective interaction of nickel ions with heterochromatin leading to alterations of the DNA template or inhibition of DNA polymerase (Coogan et al, 1989, Sunderman, 1989: Costa, 1991).

Other metals or trace elements that could be considered under the heading of environmental pollutants are arsenic, which is clearly a human carcinogen but for which no good animal model exists and mechanistic information is limited; cadmium, a probable human carcinogen for which there is evidence of lipid peroxidation in vivo (Omaye and Tappel, 1975); and copper, which is very active in redox reactions leading to free radicals or oxidants, but for which there is little convincing evidence of human carcinogenicity.

CLOSING THE GAP

Closing the gap between the inference and the proof of roles for free radicals and oxidative states in human cancers caused by environmental agents represents a scientific challenge of major importance to human health. A broad range of disciplines need to be involved. Improved understanding of the various mechanisms by which humans develop cancer is crucial; beginning with the inherited and environmental processes leading to increased cellular susceptibility to the somatic mutation resulting in cancer; continuing through the various subcellular and cellular repair processes; and including all of the steps in progression through the formation of a clinically observable and ultimately metastatic cancer. All of these steps are potentially causable by free radicals, and, thus potentially preventable by appropriate dietary modification or other interventions. Of central importance would be a marked improvement in our basic understanding of how free radicals and active states of oxygen are generated in vivo and interact with the various cellular and subcellular processes involved in carcinogenesis.

Methodology aimed at discovering the in vivo biological

fingerprints of the effects on genetic material of free radicals and active states of oxygen would be particularly of value. Substantial efforts in this area have begun. In addition to studies of urinary 8-hydroxydeoxyguanosine and the work on the P53 gene described above, these include recent work by Loeb et al searching for specific mutagenic markers of oxygen radical effects (Reid and Loeb, 1992; Loeb, 1989; McBride et al, 1991) . Among their studies is a comparison of the genetic changes caused by iron, by phorbol myristic acid and by hydrogen peroxide suggesting specificity to the mutational spectra caused by oxidative stress.

Exploration is also warranted of indirect mechanisms perhaps pertinent to the genotoxicity of environmental oxidants. For example, Preston and Doshi (1990) have suggested that the enzyme-dependent fidelity of DNA replication may be impaired by oxidant damage to the enzyme protein. This would lead to an increase in the background rate of miscopied bases during cell division and hence an increase in mutation rate. The subject of errors in DNA synthesis, including a possible role for active oxygen species, has been reviewed by Loeb and Cheng (1990).

It is rapidly becoming clear that individual cell types often differ markedly in antioxidant defense and DNA repair mechanisms. Further, Breimer (1990), in reviewing the specificity of base mutations in in vitro systems following free radical attack, has pointed out the difficulty in extrapolating among different free radical generating systems and different targets. More information about such specificity in relation to the pathology of free radicals and active states of oxygen could be very helpful in providing an understanding of the potential mechanisms of human carcinogenesis of such environmental agents.

Acknowledgments:

I thank my colleague Dr Gisela Witz for her scholarly contributions, and Domna Hebenstreit and Dan Vierno for their assistance with the manuscript. Supported by NIEHS Center Grant

REFERENCES

Abelson, P.H. (1991) Science 254, 5077.
Aiyar, J., Berkovits, H.J., Floyd, R.A., and Wetterhahn, K.E. (1991) Environ Health Perspec 92, 53-62.
Ames, B.N. (1983) Science 221, 1256-1264.
Ames, B.N. (1989) Free Radic Res Commun 7, 121-128
Ames, B.N. and Gold, L.S. (1990) Science 249, 970-971
Amoruso, M.A., Witz, G. and Goldstein, B.D. (1981) Life Sciences 28, 2215-2221
Babbs, C.F. (1990) Free Radic Biol Med 8, 191-200.
Borek, C., (1991) Free Radic Res Commun 12-13, 745-750.
Breimer, L.H. (1990) Mol Carcinog 3, 188-97.
Cerutti, P.A. (1985) Science 227, 375-381.
Clemens, M.R. (1991) Klin Wochenschr 69, 1123-1134.
Comstock, G.W., Bush, T.L., and Helzlsouer, K. (1992) Am J Epidemiol 135, 115-121.
Coogan, T.P., Latta, D.M., Snow, E.T., Costa, M. (1989) CRC Crit Rev Toxicol 19, 341-384.
Costa, M. (1991) Annual Rev Pharmacol Toxicol 31, 321-37.
DeFlora, S., Bagnasco M., Serra, D., and Zanacchi, P. (1990) Mutat Res 238, 99-172.
Dianzani, M.U., and Poli, G. (1985) in Poli, G., Cheeseman, K.H., Dianzani, M.U. and Slater, T.F. (eds), Free Radicals in Liver Injury, Oxford, IRL Press, 149-158
DiGiovanni, J., Kruszewski, F.H., Coombs, M.M., Bhatt, T.S., and Pezeshk, A. (1988) Carcinogenesis 9, 1437-1443.
DiGiovanni, J. (1991) Prog Exp Tumor Res 33, 192-229.
Dorgan, J.F., and Schatzkin, A. (1991) Hematol Oncol Clin North Amer 5, 43-68.
Fargion, S., Piperno, A., Fracanzani, A.L. and Cappellini, M.D., (1991) Ital J Gastoenterol 23, 584-588.
Goldstein, B.D., Witz, G., Javid, J., Amoruso, M.A., Rossman, T. and Wolder, B. (1982) Biol Reactive Intermediates-II, 331-339
Goldstein, B.D. and Witz, G., (1990) Free Rad Res Commun 11, 3-10
Goldstein, B.D. (1989) Advances in Modern Environmental Toxicology 16, 55-65.
Goldstein, B.D., Solomon, S., Pasternack, B.S. and Bickers, D.R. (1975) Res. Commun. Chem. Pathol. Pharmacol. 10, 759
Hollstein, M., Sidransky, D., Vogelstein, B., and Harris, C.C. (1991) Science 253, 49-53.
Kozumbo, W.J., Agarwal, S. and Koren, H.S. (1992) Toxicol Appl Pharmacol 115, 107-115.
Kurokawa, Y., Maekawa, A., Takahashi, M., and Hayashi, Y. (1990) Environ Health Perspectives 87, 309-335.
Laskin, D.L., MacEachern, L., and Snyder, R. (1989) Environ Health Perspectives 82, 75-79.
Last, J.A., Warren, D.L., Pecquet-Goad, E., and Witschi, H. (1987) J Natl Cancer Inst 78, 149-154.
Lees, P.S. (1991) Environ Health Perspect 92, 93-104.
Lehman, T.A., Bennett, W.P., Metcalf, R.A., Welsh, J.A., Ecker, J., Modall, R.V., Ullrich, S., Romano, J.W., Ettore, A., Testa, J.R., Gerwin, B.I. and Harris, C.C. (1991) Cancer Res

51, 4090-4096.
Loeb, L.A. (1989) Cancer Res 49, 5489-5496.
Loeb, L.A., and Cheng, K.C. (1990) Mutat Res 238, 297-304.
McBride, T.J., Preston, B.D., and Loeb, L.A. (1991) Biochemistry 30, 207-213.
Mossman, B.T. and Marsh, J.P. (1989) Environ Hlth Persp 81, 91-94.
Omaye, S.T. and Tappel, A.L. (1975) Res Commun Chem Pathol Pharmacol 12, 695-711.
Perchellet, J.P. and Perchellet, E.M. (1989) Free Radic Biol Med 7, 377-408.
Preston, B.D. and Doshi, R. (1990) Biological Reactive Intermediates IV, Plenum Press, New York, 193-209.
Reid, T.M., and Loeb, L.A. (1992) Cancer Res 52, 1082-1086.
Rotstein, J.B., O'Connell, J.F., and Slaga, T.J. (1987) in Cerutti, P.A., Nygaard, O.F., and Simic, M.G. (eds) Anticarcinogenesis and Radiation Protection, New York, Plenum Press, 211-219.
Sahu, S.C., (1990) Biomed Environ Sci 3, 183-201.
Snow, E.T. (1991) Environ Health Perspect 92, 75-81.
Snyder, R., Jowa, L., Witz, G., Kalf, G., and Rushmore, T. (1987) Arch Toxicol 60, 61-64
Standeven, A.M., and Wetterhahn, K.E. (1991) Chem Res Toxicol 4, 616-25.
Stark, A.A., (1991) Mutagenesis 6, 241-245.
Stevens, R.G., and Kalkwarf, D.R. (1990) Environ Health Perspect 87, 291-300.
Stich, H.F., and Anders, F. (1989) Mutat Res 214, 47-61.
Subrahmanyan, V.V., Ross, D., Eastmond, D.A., Smith, M.T. (1991) Free Radic Biol Med 11, 495-515.
Sunderman, F.W. Jr. (1989) Scand J Work Environ Health 15, 1-12.
Taffe, B.G., Takahashi, N., Kensler, T.W., and Mason, R.P. (1987) J Bio Chem 262, 12143-12149.
Trush, M.A., and Kensler, T.W. (1991) Free Radic Biol Med 10, 201-209.
Vahakangas, K.H., Samet, J.H., Metcalf, R.A., Welsh, J.A., Bennett, W.P., Lane, D.P. and Harris C.C. (1992) Lancet 339, 576-580.
Vogelstein, B. (1990) Nature 348, 681-684.
Walker, C., Everitt, J., and Barrett, J.C. (1992) Am J Ind Med 21, 253-273.
Weitzman, S.A., and Graceffa, P. (1984) Arch Biochem Biophys 228, 373-376.
Wetterhahn, K.E., Hamilton, J.W., Aiyar, J. Borges, K.M., and Floyd (1989) R. Biol Trace Elem Res 21, 405-11.
Witschi, H. (1988) Toxicology 48, 1-20.
Witschi, H. (1991) Experimental Lung Research 17, 473-483.
Witz, G., Latriano, L. and Goldstein, B.D. (1989) Environ. Health Persp. 82, 19-22
Yang, X.Y., Gasparro, F.P., DeLeo, V.A. and Santella, R.M. (1989) J Investigative Dermatol 92, 59-63

Metabolic Disorders

Free Radicals: From Basic Science to Medicine
G. Poli, E. Albano & M. U. Dianzani (eds.)

HEPATOTOXICITY OF EXPERIMENTAL HEMOCHROMATOSIS

B.R. Bacon, R.S. Britton, R. O'Neill, S.C.Y Li, Y. Kobayashi.

Division of Gastroenterology and Hepatology, Department of Internal Medicine, St. Louis University Medical Center, 3635 Vista Avenue, St. Louis, Missouri 63110, U.S.A.

SUMMARY: In hereditary hemochromatosis, the liver is the major recipient of the excess absorbed iron, and after several years of high tissue iron concentrations, fibrosis and, eventually, cirrhosis develop. Complications of cirrhosis are the most common causes of death in patients with hereditary hemochromatosis. Despite the convincing clinical evidence for the hepatotoxicity of excess iron, the specific pathophysiologic mechanisms responsible for hepatocyte injury and hepatic fibrogenesis in chronic iron overload are poorly understood. In our laboratory, experimental hemochromatosis is achieved by feeding rats a diet supplemented with carbonyl (elemental) iron. The pattern of distribution and the degree of hepatic iron overload are qualitatively and quantitatively comparable to that seen in human hereditary hemochromatosis. We have demonstrated evidence of iron-induced mitochondrial and microsomal lipid peroxidation and a variety of associated organelle functional abnormalities, followed by portal fibrosis, in rats with chronic iron overload. Recent work from our laboratory has shown an increase in the hepatic levels of malondialdehyde (MDA), as well as significantly impaired mitochondrial metabolism of this product of peroxidized lipids, in iron-loaded rats. Additionally, other investigators have reported that MDA causes an increase in both collagen production and in procollagen I gene expression in cultured human fibroblasts. Taken together, these data suggest that aldehydic byproducts resulting from iron-induced lipid peroxidation could be an important initiating factor leading to increased hepatic fibrogenesis in iron overload. In the liver, lipocytes (Ito cells) represent the most likely source of increased collagen production, and with dietary iron overload, hepatocytes are the most likely source of the increase in hepatic MDA. Therefore, aldehydic peroxidation products may serve as a link between hepatocellular lipid peroxidation and subsequent hepatic fibrosis in iron overload, either by directly stimulating lipocyte collagen production or by activating Kupffer cells to release profibrogenic cytokines.

It has recently been shown that approximately 1 in 250 individuals of European descent are homozygous for the hemochromatosis gene (Edwards et al., 1988), thus identifying hereditary hemochromatosis as one of the most common inherited disorders. In both hereditary hemochromatosis and in the various forms of secondary iron overload, there is a pathologic expansion of body iron stores, due mainly to an increase in absorption of dietary iron (Bassett et al., 1984; Nichols and Bacon, 1989). Excess deposition of iron as ferritin and hemosiderin in the parenchymal tissues of several organs (e.g., liver, heart, pancreas) results in cell injury, functional insufficiency and fibrosis (Niederau et al., 1985; Bassett et al., 1986; Bacon and Britton, 1990). Clinical evidence for hepatotoxicity has been provided by studies of patients with hereditary hemochromatosis (Niederau et al., 1985; Bassett et al., 1986), African iron overload (Isaacson et al., 1961), and secondary iron overload due to β-thalassemia (Barry et al., 1974) in which a correlation between hepatic iron concentration and the occurrence of liver damage has been demonstrated or in which therapeutic reduction of hepatic iron by either phlebotomy or chelation therapy has resulted in clinical improvement. Despite this clinical evidence for the hepatotoxicity of excess iron, the specific pathophysiologic mechanisms of hepatocellular injury and fibrosis in chronic iron overload are still poorly understood.

Experimental Iron Overload: In the past, one of the major difficulties in elucidating the cytopathologic mechanisms responsible for the hepatotoxicity of chronic iron overload was the unavailability of a suitable animal model of iron overload. In our laboratory, over the past several years, we have successfully developed and exploited such a model (reviewed in Bacon and Britton, 1990). By feeding rats a diet enriched with 2-3% elemental (carbonyl) iron, we have been able to achieve hepatic nonheme iron concentrations of 50-100 times normal (3-6,000 μg Fe/g liver, wet weight) over relatively short time

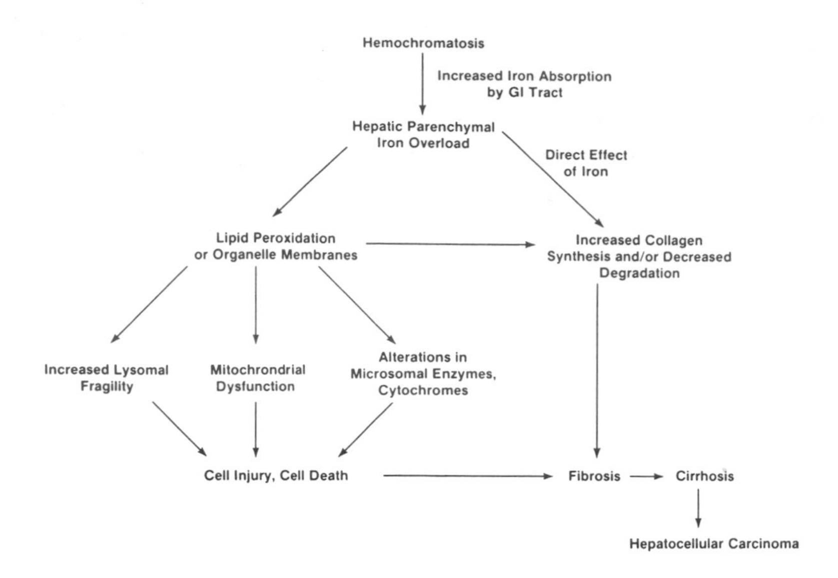

Fig. 1. Proposed pathophysiological mechanisms of liver injury in iron overload (from Bacon and Britton, 1990 with permission).

periods (2-4 months). This experimental model of chronic dietary iron overload produces excessive storage iron deposition predominantly in hepatocytes in a periportal distribution, a pattern identical to that seen in human hereditary hemochromatosis and in African iron overload (Bacon et al., 1983a; Park et al., 1987). We have used this model to investigate mechanisms of liver injury in iron overload (see Figure 1) and have demonstrated direct evidence of hepatic mitochondrial and microsomal lipid peroxidation *in vivo* (Bacon et al., 1983a, 1983b, 1985, 1986, 1989) and an increase in the low molecular weight pool of catalytically active iron (Britton et al., 1990b). Specific membrane-dependent functions of mitochondria (oxidative metabolism, Ca^{2+} sequestration) (Bacon et al., 1985; Britton et al., 1990a, 1991a) and microsomes (cytochrome concentrations, enzyme activities, Ca^{2+} sequestration) (Bacon et al., 1986; Bonkovsky et al., 1987; Britton et al., 1991a) are decreased at hepatic iron concentrations at which mitochondrial and microsomal lipid peroxidation have been observed and are seen at hepatic iron concentrations at which hepatocellular necrosis is seen histologically in the rat (Park et al., 1987) and at which liver disease (fibrosis, cirrhosis) occurs in humans (3-5,000 μg/g) (Bassett et al., 1984). At higher hepatic iron concentrations (> 8,000 μg/g) in rats which were maintained on the iron-supplemented diet for up to 12 months, we have demonstrated evidence of hepatic fibrosis using standard light and electron microscopic and histochemical techniques (Park et al., 1987). This represents the first demonstration of hepatocellular necrosis and hepatic fibrosis in the rat due to chronic iron overload and confirms morphologically the relevance of the biochemical abnormalities seen at lesser degrees of iron overload.

Mitochondrial Lipid Peroxidation and Dysfunction in Chronic Iron Overload: Using the dietary carbonyl iron regimen, hepatic iron concentrations analogous to those seen in typical human hereditary hemochromatosis can be achieved in two to four months (3-6,000 μg Fe/g liver). Hepatic mitochondrial lipid peroxidation was demonstrated *in vivo* by the presence of conjugated dienes in phospholipid extracts of the subcellular fractions (Bacon et al., 1983a, 1983b, 1986; Feldman and Bacon, 1989). An iron concentration threshold was observed for the presence of lipid peroxidation, which for mitochondria was approximately 1,000 to 1,500 μg/g and for microsomes was approximately 3,000 μg/g (Bacon et al., 1983b). Studies were also performed which demonstrated an increase in the low molecular weight pool of catalytically active iron in chronic iron overload (Britton et al., 1990b).

At moderate degrees of chronic hepatic iron overload *in vivo*, significant decreases in the mitochondrial state 3 (ADP-stimulated) respiratory rates and in the respiratory control ratios (RCR) were found for all three substrates studied (glutamate, β-hydroxybutyrate, and succinate) (Bacon et al.,

1985). At hepatic iron concentrations at which there were decreases in oxidative metabolism, there was also evidence of conjugated diene formation, indicative of mitochondrial lipid peroxidation. Appropriate experiments were performed to determine that these abnormalities were not *in vitro* artifacts of preparation. We concluded from these studies that moderate degrees of chronic hepatic iron overload *in vivo* resulted in an inhibitory defect in the mitochondrial electron transport chain as evidenced by a decrease in state 3 respiration and RCR. Subsequent studies showed a reduction in cytochrome oxidase activity (complex IV) (Bacon et al., 1987) and a reduction in hepatic levels of ATP, ADP, and the hepatic energy state (Bacon et al., 1988) in chronic iron overload. Finally, experiments were performed in which a 50% decrease in mitochondrial Ca^{2+} sequestration accompanied by a decrease in RCR was demonstrated in chronic iron overload (Britton et al., 1991a).

Experiments were designed to investigate the effect of iron overload on mitochondrial malondialdehyde (MDA) metabolism. In iron-loaded rats, there was a 2.5-fold increase in hepatic MDA concentration associated with a 47% decrease in mitochondrial RCR. Hepatic mitochondrial and cytosolic fractions were prepared from iron-loaded and control rats and were assayed for aldehyde dehydrogenase (ALDH) activity. There was a 28% decrease in mitochondrial ALDH activity in rats with iron overload, whereas, there were no changes in cytosolic ALDH activity between control and iron-loaded animals (Britton et al., 1990a). The disappearance (metabolism) of MDA added to a suspension of hepatic mitochondria isolated from the iron-loaded rats was also significantly decreased. At each time point after the addition of 6 μM MDA (5, 10, 15 and 30 min), there was a significant reduction in the amount of MDA metabolized by mitochondria from iron-loaded rats compared to controls. Therefore, the increased hepatic MDA levels found in chronic iron overload may result from a combination of increased production, as well as decreased metabolism of MDA, both of which may be due to iron-induced mitochondrial lipid peroxidation (Britton et al., 1990a).

Isolated Iron-loaded Hepatocytes: To examine the relationship between lipid peroxidation and hepatocellular injury in iron overload, we isolated hepatocytes from rats with chronic iron overload and determined the effects of *in vitro* iron chelators and antioxidants on lipid peroxidation, hepatocyte viability and ultrastructure over a 4 hr period (Sharma et al., 1990). Experimental iron overload was achieved by feeding rats a chow diet supplemented with 3.0% carbonyl iron for 8-12 weeks. Rats used as controls were fed a regular chow diet.

Isolated hepatocytes from either control or iron-loaded rats were incubated for 4 hr at 37°C. Cell viability was significantly reduced at 3 and 4 hr in iron-loaded hepatocytes compared to controls, and this decrease in viability was preceded by an increase in iron-dependent lipid peroxidation. *In vitro* iron chelation with either deferoxamine or apotransferrin protected against lipid peroxidation, loss of viability and ultrastructural damage in iron-loaded hepatocytes. In addition, α-tocopherol also protected against lipid peroxidation and loss of viability. The protective effects of iron chelators and an antioxidant support a strong association between iron-dependent lipid peroxidation and hepatocellular injury in iron overload.

DNA Damage: The risk of developing hepatocellular carcinoma (HCC) is increased 200-fold in patients with hereditary hemochromatosis with cirrhosis (Niederau et al., 1985), and it has been postulated that iron-induced DNA damage may be an important initiating event in the development of HCC in these individuals. It has been shown that ionic iron produces DNA damage when incubated *in vitro* with either purified DNA (Loeb et al., 1988) or with isolated rat liver nuclei (Shires, 1982). In addition, the iron chelate, ferric nitrilotriacetate (FeNTA), has been demonstrated to cause transformation of cells. For example, renal adenocarcinoma has been produced in rats after

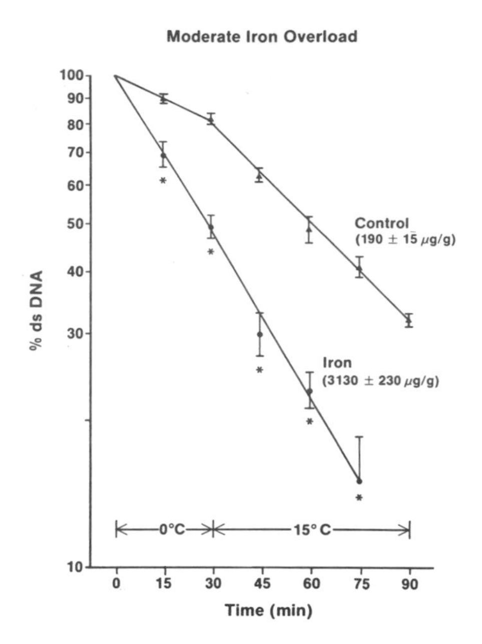

Fig. 2. Unwinding of double-stranded hepatic DNA from rats with chronic iron overload, and controls. The hepatic nonheme iron concentrations in both groups are shown in μg/g wet weight. Samples were incubated at an alkaline pH for 30 min at 0°C and a further 60 min at 15°C. The rate of unwinding was significantly increased after iron overload (mean ± SE, n = 6; * $p < 0.01$).

one year of treatment with FeNTA by intraperitoneal injection (Okada et al., 1983). Furthermore, when cultured rat liver epithelial cells were exposed to a high concentration of FeNTA, some cells showed morphological transformation and induced metastatic carcinomas when injected into newborn rats (Yamada et al., 1990). The authors postulated that oxidative damage to DNA catalyzed by FeNTA may play an important role in the rapid neoplastic transformation of these liver cells.

We have studied the effect of chronic iron overload *in vivo* on hepatic DNA damage (Edling et al., 1990). Hepatic iron overload was produced in rats using dietary carbonyl iron. DNA strand breaks were detected in hepatic DNA using an alkaline unwinding assay (see Figure 2). The rate of unwinding of double-stranded DNA to single-stranded DNA is proportional to the number of strand breaks present. Hepatic DNA from iron-loaded rats with a mean hepatic nonheme iron content of 3,130 μg/g showed a faster rate of unwinding than that from controls, indicating that moderate iron overload increases the number of strand breaks in hepatic DNA. This damage did not occur in rats with mild hepatic iron overload (≈ 600 μg/g), suggesting that the genotoxic effect was dependent on the concentration of iron. The induction of strand breakage in DNA has been associated with both initiation and promotion events in chemically-induced carcinogenesis (Walles and Erixson, 1984; Hartley et al., 1985) and may play a role in the induction of HCC in patients with hereditary hemochromatosis. Additional support for the concept that iron can cause oxidative damage to DNA comes from the recent report that administration of iron-dextran resulted in an increase in the levels of 8-hydroxydeoxyguanosine in hepatic DNA of *Ah*-responsive mice (Faux et al., 1992).

Hepatic Fibrogenesis: It is now well-recognized that complications of cirrhosis are the most common causes of death in patients with hereditary hemochromatosis (Niederau et al., 1985), and the development of increased hepatic fibrosis is iron concentration-dependent (Bassett et al., 1986). Despite these clinical observations, little is known about the mechanisms of hepatic fibrogenesis in chronic iron overload. Early biochemical studies concerning hepatic fibrogenesis in experimental iron overload showed either an increase, no change, or a decrease in prolyl hydroxylase activity in rats or baboons with various types of chronic iron overload (reviewed by Bacon and Britton, 1990).

Morphologic evidence of increased fibrogenesis in experimental iron overload is limited. Using the dietary carbonyl iron model, a detailed analysis of hepatic histology and ultrastructure in rats with chronic iron overload was performed in which rats were examined over a 12-month feeding period using a diet containing 3% carbonyl iron (Park et al., 1987). Within the liver, iron deposition was initially confined to periportal hepatocytes, but subsequently extended to midzonal and centrilobular hepatocytes. At three months, histologic evidence of hepatocellular injury was mild and subtle, with occasional foci of spotty necrosis and ultrastructural subcellular organelle damage. At eight months, portal areas were enlarged with collections of iron-loaded macrophages and increased collagenous tissue. This portal fibrous tissue extended between periportal hepatocytes at sites of maximal iron deposition and around Kupffer cells and macrophages. At 12 months, the periportal fibrosis was more pronounced, and in some animals, early cirrhosis was identified. These serial morphologic studies were the first to demonstrate the production of hepatic necrosis and fibrosis by chronic dietary iron overload and illustrate that this dietary model of iron overload is well-suited to study the mechanisms of hepatic fibrogenesis in iron overload. Pietrangelo et al., (1990) have also observed periportal fibrosis in rats with dietary carbonyl iron overload.

In general, in hepatic fibrosis, levels of collagens I, III and IV are increased, and there is usually an increase in mRNA levels of these proteins, suggesting regulation at a transcriptional level (Rojkind, 1986). Cellular localization of increased collagen production in the liver in pathological conditions suggests that nonparenchymal cells (Friedman et al., 1985; Milani et al., 1989; Weiner et al., 1990; Maher and McGuire, 1990) (particularly activated lipocytes) are the principal site(s) of increased collagen synthesis.

Of the nonparenchymal cells in the liver, lipocytes (also known as Ito cells, fat-storing cells) have the greatest capacity for collagen synthesis in culture (Friedman et al., 1985). Lipocytes are stellate cells that are located in the perisinusoidal space. They are the predominant hepatic storage site for vitamin A and are about as numerous as Kupffer cells (about one-sixth the number of hepatocytes) (Wake, 1980). In inflammation, they acquire fibroblastic features, losing their lipid droplets and elaborating abundant rough endoplasmic reticulum (Maher et al., 1988). Once activated, they demonstrate the ability to synthesize collagens I, III and IV, but particularly type I (Friedman et al., 1985), which predominates in hepatic fibrosis *in vivo* (Rojkind, 1986).

In rats with chronic dietary iron overload, the hepatic levels of type I procollagen mRNA are increased (Irving et al., 1988; Pietrangelo et al., 1990, 1992). Pietrangelo and coworkers (1992) have gone on to demonstrate that, in these animals, the mRNA for type I procollagen is located predominantly in hepatic nonparenchymal cells (using *in situ* hybridization), and more specifically, in lipocytes (using cell separation). Our laboratory has studied collagen and total protein production by lipocytes isolated from iron-loaded rats and controls (Li et al., 1992). Chronic iron overload *in vivo* resulted in an elevated production of collagen and total protein by lipocytes (see Figure 3), indicating that these cells were in an activated state. Therefore, the available data suggest that activation of lipocytes may contribute to hepatic fibrogenesis in experimental iron overload.

The initiating factor(s) leading to increased collagen production in iron overload *in vivo* are unknown. Recent work by Chojkier and coworkers has suggested an important role for aldehydic byproducts of lipid peroxidation as such a factor (Chojkier et al., 1989). MDA, a breakdown product of peroxidized lipids, has been shown to cause an increase in the synthesis of type I collagen and an increase in type I procollagen mRNA levels in cultured human fibroblasts (Chojkier et al., 1989). In these studies, iron-induced peroxidation caused increases in MDA, procollagen mRNA concentration and collagen synthesis; the antioxidant α-tocopherol prevented these sequelae (Chojkier et al., 1989; Houglum et al., 1991). Addition of MDA alone to these cells mimicked the effects of iron-induced peroxidation on collagen gene expression and collagen synthesis (Chojkier et al., 1989). Recent work from our laboratory has shown an increase in hepatic MDA levels in iron-loaded rats (Britton et al., 1990a). Taken together, these

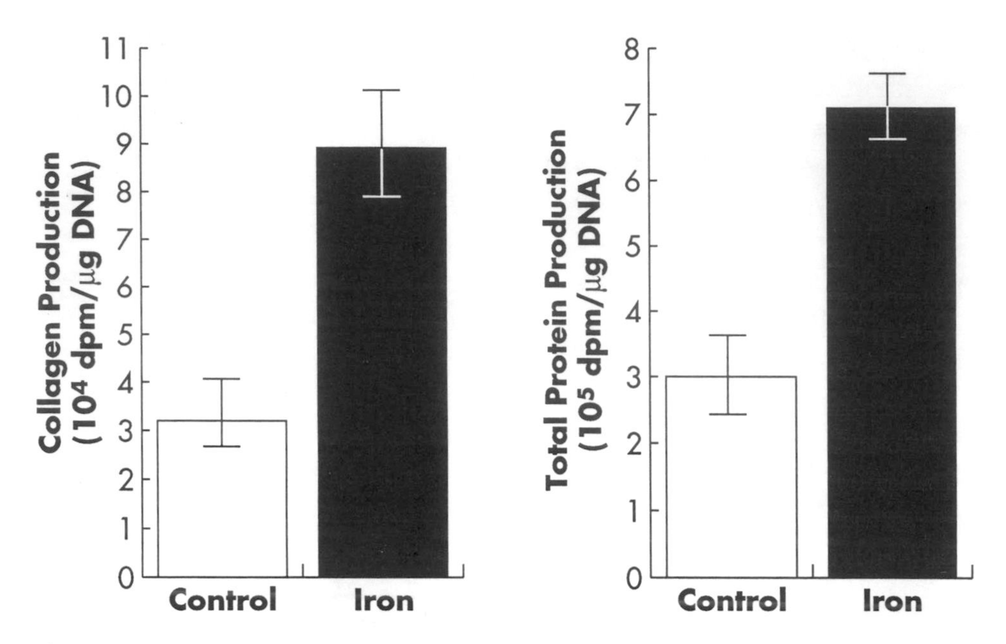

Fig. 3. Collagen and total protein production by lipocytes isolated from rats with chronic iron overload, and controls. Rats were fed a chow diet enriched with 1% carbonyl iron for 15 months. Controls were fed chow diet alone. Lipocytes were cultured on plastic, and after three days, protein production was measured using ^{3}H-proline (24 hr labelling). The collagen component of total protein production was determined using bacterial collagenase. Lipocytes from iron-loaded rats had a marked increase in both collagen and total protein production (mean ± SE, n = 3).

data suggest a working hypothesis concerning the mechanism of fibrogenesis in chronic iron overload, namely that iron-induced lipid peroxidation results in an increase in aldehydic byproducts such as MDA, which in turn could cause lipocytes to increase collagen gene expression and synthesize more collagen.

In addition to the possible direct effect of aldehydic peroxidation products on lipocyte collagen production, there may be interactions between these products, Kupffer cells, and lipocytes. Aldehydic products of lipid peroxidation could act by activating Kupffer cells with release of cytokines and stimulation of collagen synthesis. Thus, aldehydic peroxidation products may serve as a link between hepatocellular lipid peroxidation and subsequent hepatic fibrosis in iron overload, either by directly stimulating lipocyte collagen production or by activating Kupffer cells to release profibrogenic cytokines.

Acknowledgements
The research work of the authors was supported by grants from the U.S. Public Health Service. R.S.B. also received support from the American Cancer Society.

REFERENCES

Bacon, B.R., Tavill, A.S., Brittenham, G.M., Park, C.H., and Recknagel, R.O. (1983a) J. Clin. Invest. 71, 429-439.
Bacon, B.R., Brittenham, G.M., Tavill, A.S., McLaren, C.E., Park, C.H, and Recknagel, R.O. (1983b) Trans. Assoc. Am. Phys. 96, 146-154.
Bacon, B.R., Park, C.H., Brittenham, G.M., O'Neill, R., and Tavill, A.S. (1985) Hepatology 5, 789-797.
Bacon, B.R., Healey, J.F., Brittenham, G.M., Park, C.H., Nunnari, J., Tavill, A.S., and Bonkovsky, H.L. (1986) Gastroenterology 90, 1844-1853.
Bacon, B.R., Dalton, N., and O'Neill, R. (1987) Hepatology 7, 1027.
Bacon, B.R., Britton, R.S., Dalton, N., and O'Neill, R. (1988) Hepatology 8, 1240.
Bacon, B.R., Britton, R.S., and O'Neill, R. (1989) Hepatology 9, 398-404.
Bacon, B.R., and Britton, R.S. (1990) Hepatology 11, 127-137.
Barry, M., Flynn, D.M., Letsky, E.A., and Risdon, R.A. (1974) Br. Med. J. 2, 16-20.
Bassett, M.L., Halliday, J.W., and Powell, L.W. (1984) Semin. Liver Dis. 4, 217-227.
Bassett, M.L., Halliday, J.W., and Powell, L.W. (1986) Hepatology 6, 24-29.
Bonkovsky, H.L., Healey, J.F., Lincoln, B., Bacon, B.R., Bishop, D.F., and Elder, G.H. (1987) Hepatology 7, 1195-1203.
Britton, R.S., O'Neill, R., and Bacon, B.R. (1990a) Hepatology 11, 93-97.
Britton, R.S., Ferrali, M., Magiera, C.J., Recknagel, R.O., and Bacon, B.R. (1990b) Hepatology 11, 1038-1043.
Britton, R.S., O'Neill, R., and Bacon, B.R. (1991a) Gastroenterology 101, 806-811.
Britton, R.S., O'Neill, R., Singh, R., and Bacon, B.R. (1991b) Hepatology 14, 158A.
Chojkier, M., Houglum, K., Solis-Herruzo, J., and Brenner, D.A. (1989) J. Biol. Chem. 264, 16957-16962.
Edling, J.E., Britton, R.S., Grisham, M.B., and Bacon, B.R. (1990) Gastroenterology 98, A585.
Edwards, C.Q., Griffin, L.M., Goldgar, D., Drummond, C., Skolnick, M.H., Kushner, J.P. (1988) N. Eng. J. Med. 318, 1355-1362.
Faux, S.P., Francis, J.E., Smith, A.G., and Chipman, J.K. (1992) Carcinogenesis 13:247-250.

Feldman, E.S., and Bacon, BR. (1989) Hepatology 9, 686-692.
Friedman, S.L., Roll, F.J., Boyles, J., and Bissell, D.M. (1985) Proc. Natl. Acad. Sci. 87, 8681-8685.
Hartley, J.A., Gibson, N.W., Zwelling, L.A., and Yuspa, S.H. (1988) Cancer Res. 45, 4864-4870.
Houglum, K., Brenner, D.A., and Chojkier, M. (1991) J. Clin. Invest. 87, 2230-2235.
Isaacson, C., Seftel, H., Keeley, K.J., and Bothwell, T.H. (1961) J. Lab. Clin. Med. 58, 845-853.
Irving, M.G., Halliday, J.W., and Powell, L.W. (1988) Alcoholism: Clin. Exp. Med. 12, 7-13.
Li, S.C.Y., O'Neill, R., Britton, R.S., Kobayashi, Y., and Bacon, B.R. (1992) Gastroenterology 102, A841.
Loeb, L.A., James, E.A., Waltersdorph, A.M., and Klebanoff, S.J. (1988) Proc. Natl. Acad. Sci. U.S.A. 85, 3918-3922.
Maher, J.J., Bissell, D.M., Friedman, S.L., and Roll, F.J. (1988) J. Clin. Invest. 82, 450-459.
Maher, J.J., and McGuire, R.F. (1990) J. Clin. Invest. 86, 1641-1648.
Milani, S., Herbst, H., Schuppan, D., Hahn, E.G., and Stein, H. (1989) Hepatology 10, 84-92.
Nichols, G.M., and Bacon, B.R. (1989) Amer. J. Gastro. 84, 851-862.
Niederau, C., Fischer, R., Sonnenberg, A., Stremmel, W., Trampisch, H.J., and Strohmeyer, G. (1985) N. Engl. J. Med. 313, 1256-1262.
Okada, S., Hamazaki, S., Ebina, Y., Fujioka, M., and Midorikawa, O. (1983) In Structure and Function of Iron Transport Proteins, Elsevier, New York. pp. 473-478.
Park, C.H., Bacon, B.R., Brittenham, G.M., and Tavill, A.S. (1987) Lab. Invest. 57, 555-563.
Pietrangelo, A., Rocchi, E., Schiaffonati, L., Ventura, E., and Cairo, G. (1990) Hepatology 11, 798-804.
Pietrangelo, A., Gualdi, R., Geerts, A., De Bleser, P., Casalgrandi, G., and Ventura, E. (1992) Gastroenterology 102, A868.
Rojkind, M. (1986) Current Hepatology 6, 111-129.
Sharma, B.K., Bacon, B.R., Britton, R.S., Park, C.H., Magiera, C.J., O'Neill, R., Dalton, N., Smanik, P., and Speroff, T. (1990) Hepatology 12, 31-39.
Shires, T.K. (1982) Biochem. J. 205, 321-329.
Wake, K. (1980) Int. Rev. Cytol. 66, 303-353.
Walles, S.A.S., and Erixson, K. (1984) Carcinogenesis 5, 319-322.
Weiner, F.R., Giambrone, M-A., Czaja, M.J., Shah, A., Annoni, G., Takahashi, S., Eghbali, M., and Zern, M.A. (1990) Hepatology 11, 111-117.
Yamada, M., Awai, M., and Okigaki, T. (1990) Cytotechnology 3, 149-156.

Free Radicals: From Basic Science to Medicine
G. Poli, E. Albano & M. U. Dianzani (eds.)

ROLE OF INDUCIBLE CYTOCHROME P450 IN THE LIVER TOXICITY OF POLYHALOGENATED AROMATIC COMPOUNDS

F. De Matteis, S.J.Dawson, M.E. Fracasso and A.H.Gibbs

MRC Toxicology Unit, Woodmansterne Road, Carshalton, Surrey, U.K.

SUMMARY: Both major inducible forms of liver cytochrome P450 can interact with polyhalogenated biphenyls (PBCs) of the appropriate configuration to produce oxidizing species, as judged by the degradation of bilirubin, a readily oxidizable compound, by isolated microsomes in vitro. Planar PCBs are active with cytochrome P450 IA1, while non-planar di-orthosubstituted compounds are effective with cytochrome P450 IIB1. Evidence for uroporphyrinogen oxidation under similar conditions has also been obtained with 3-methylcholanthrene-induced rat liver microsomes, where the stimulatory effect of a planar biphenyl, though not as great and consistent as with bilirubin, could also be demonstrated. The nature of the oxidizing species produced by the microsomes is still uncertain, but bilirubin undergoing enzymic oxidation shows similar spectral changes as in a chemical oxidation system. We suggest that a PCB of the appropriate steric configuration may first induce the relevant cytochrome in vivo, then become bound to the active site of the induced cytochrome and lead to production of oxidative species. These may then attack not only bilirubin, but also other key targets in the cell, including hexahydroporphyrins (porphyrinogens) and DNA bases.

INTRODUCTION

Following a widespread outbreak of toxic human porphyria in Turkey over 30 years ago, animal experiments demonstrated conclusively that hexachlorobenzene was the causative agent responsible. A marked inhibition of liver uroporphyrinogen decarboxylase was observed in animals treated with hexachlorobenzene and this explained the characteristic feature of the metabolic disorder, i.e. the marked increase in uroporphyrin observed in the liver and urine of poisoned humans and experimental animals (Elder,1978). Several other polyhalogenated aromatic compounds [including polychlorinated and polybrominated biphenyls and the very toxic 2,3,7,8-tetrachlorodibenzo-*p*-dioxin (TCDD)] have since been shown to produce uroporphyria in rodents and chick systems (Marks,1985) and there is also evidence that some of these, for example TCDD, may be similarly active in humans. The liver toxicity of these compounds is not confined to a disorder of porphyrin metabolism, as centrilobular liver necrosis (De Matteis et al., 1961) and, more recently, liver cancer (Smith et al., 1985) have also been described in treated animals.

Two main hypotheses have been put forward to account for the mechanism of liver toxicity of the polyhalogenated chemicals and both of them involve cytochrome P450, of which these componds are powerful inducers. The first proposes that these chemicals may be metabolized by cytochrome P450 to reactive, toxic metabolites, some of which may directly inhibit uroporphyrinogen decarboxylase (Van Ommen et al., 1986), or, by attacking and modifying other important target in the cell, induce other manifestation of liver toxicity. This hypothesis does not readily explain, however, why TCDD, a compound which is very resistant to metabolism, should also be the most active uroporphyria-inducing chemical. The second hypothesis (De Matteis and Stonard, 1977), which has received more support in recent years (Bonkowsky et al., 1987; Smith and De Matteis, 1990) suggests, on the other hand, that these chemicals may act indirectly by binding cytochrome P450 as poor substrates, stimulating reduction of molecular oxygen and increasing production of oxidizing species. These may then be responsible for several manifestations of liver toxicity, by oxidizing key target molecules in the cell. For example, they may be responsible for the induction of uroporphyria by facilitating oxidation of uroporphyrinogen to uroporphyrin (which cannot be metabolized) and also to a long-lived inhibitor of uroporphyrinogen decarboxylase, such as that detected (Cantoni et al., 1984) in porphyric livers. An attractive feature of this oxidative stress mechanism is that it is easier to visualize on this basis the potentiation of the disorder by iron (Smith et al., 1986; De Matteis, 1988), since it is known that iron can interact with (and exagerate) oxidative free radical reactions.

In support of this general oxidative stress mechanism, polyhalogenated chemicals and similar inducers have been shown to activate a microsomal oxidative mechanism in the liver, which results in increased oxidation of uroporphyrinogen (Jacobs et al., 1989) and also of bilirubin (De Matteis et al., 1989), another easily oxidizable compound, in vitro. The microsomal bilirubin-oxidizing system depends on cytochrome P450 IA1 (De Matteis et al., 1991) and requires NADPH and oxygen, but has the unusual feature of being markedly stimulated by addition of a planar polyhalogenated aromatic compound in vitro, while the monooxygenated activity of the induced microsomes is correspondingly inhibited (De Matteis et al., 1989). These findings therefore suggest that a polyhalogenated chemical first acts as an inducer of cytochrome P450 IA1 and then interacts with the induced cytochrome, inhibiting its monooxygenase activity and stimulating production of an oxidizing species.

The purpose of this paper is to summarize recent work from our laboratory in this area, under the following three main headings.

Inducible bilirubin- and porphyrinogen-oxidizing system of liver microsomes.

We have now compared bilirubin and hexahydro-uroporphyrin (uroporphyrinogen) as substrates for the oxidizing system of liver microsomes from β-naphthoflavone-induced chick embryos and 3-methylcholanthrene-induced male rats. With chicken microsomes the rate of oxidation of both substrates was stimulated by planar PCBs (3,3',4,4'-tetrachlorobiphenyl and 3,3'4,4'5,5'-hexachloro-biphenyl), while their non-planar di-orthosubstituted congeners were virtually inactive. Microsomes obtained from induced rats behaved in most cases as described above for chicken microsomes. However in the case of 4 out of a total 14 rats, the basal rate of oxidation of both bilirubin and porphyrinogen (the rate, that is, without addition of a PCB) was significantly higher and the planar PCBs failed to stimulate further the oxidation of the porphyrinogen and were less effective at stimulating bilirubin oxidation. These findings help substantiate the hypothesis that interaction of a polyhalogenated aromatic compound with the induced cytochrome may initiate an oxidative mechanism leading to oxidation of target molecules in the cell. However we cannot yet provide an explanation for the anomalous behaviour of the microsomes obtained form some rats.

Evidence for two Pathways of Iron-catalysed oxidation of Bilirubin. Effect of Desferrioxamine.

The bilirubin oxidizing system of the liver microsomes obtained form rats induced with TCDD and 3-methylcholanthrene can be inhibited by a monoclonal antibody raised against cytochrome P450 IA1 (De Matteis et al., 1991), clearly implicating this isoenzyme, but the nature of the oxidizing species involved has not yet been clarified. Information on the mechanism of the oxidation reaction can be drawn from investigation of the structure of the products of microsomal-dependent oxidation of bilirubin. The acquired knowledge should also be of relevance to the alternative pathway of bilirubin disposal [which operates in the congenitally jaundiced Gunn rat (Kapitulnik and Ostrow, 1978)], since there is evidence that bilirubin degradation by the inducible microsomal system may contribute significantly to this pathway. We have therefore investigated a chemical model system and compared the products of chemical oxidation to those obtained by the alternative pathway with induced microsomes.

When bilirubin is incubated with hydrogen peroxide in presence of iron/EDTA, there is rapid loss of the 450 nm absorbance and this is accompanied by bleaching and complete loss of the chromophore. A similar loss of chromophore has now been documented with induced microsomes *in vitro* and may also be seen during the alternative pathway of bilirubin disposal *in vivo*. The rate of bilirubin disappearance in the chemical oxidation system can be stimulated by adding either reduced glutathione (GSH) or desferrioxamine (DES). However, whereas in presence of GSH, oxidation still proceeds by bleaching, with DES, bleaching is prevented and the chromophore undergoes a time-dependent change from yellow to green and then to red. These colour changes, reminiscent of the Gmelin reaction (Gray, 1953), can be interpreted as resulting from electron (and hydrogen) abstraction to the two electron oxidation derivative of bilirubin (biliverdin), followed by further oxidation of biliverdin to a red pigment.

We conclude that bilirubin can be oxidized by one of two distinct pathways. The first may involve, at least in the first step, electron transfer and is seen when DES is present in the chemical model system. The second leads to loss of chromophore, is seen in the iron-catalyzed chemical system (in absence of DES) and also in the induced microsomal system and may involve oxygen addition and/or fragmentation.

Possible role of Cytochrome P450s of the IIb Group in the liver toxicity of polyhalogenated aromatic compounds.

In previous work (De Matteis et al.,1989 and 1991) we have shown that planar PCBs first induce cytochrome P450 IA1, then interact with the induced cytochrome to produce a bilirubin-oxidizing species. We have now extended these findings to the microsomes from phenobarbitone-induced rats which again respond to challenge by PCBs with increased rate of bilirubin degradation *in vitro.* However non-planar di-orthosubstituted PCBs [like 2,2',4,4'-tetrachlorobiphenyl (2,4-TCB) and 2,2',4,4'5,5,'-hexachlorobiphenyl) are active with phenobarbitone-induced microsomes, in contrast to those active with microsomes induced with TCDD and similar inducers where planarity is required for activity. We have also investigated whether the cytochrome(s) of the IIB group are involved in the accelerated bilirubin degradation seen with phenobarbitone-induced microsomes, by studying the effect of an inhibitory antibody (raised against these isoenzymes) on both their basal and 2,4-TCB-stimulated rates of bilirubin degradation. We find that although both basal and stimulated rates are inhibited by the antibody, the effct on the 2,4-TCB-dependent rate is far greater and resembles in its dose-inhibition curve the inhibitory effect of the antibody on the 7-pentoxy resorufin *O*-depentylase (an enzymatic reaction typical of cytochrome P450 IIB1).

We conclude that both major inducible forms of liver

cytochrome P450 can interact with PCBs of the appropriate configuration to produce oxidizing species. Planar PCBs are active with cytochrome P450 IA1, while non-planar compounds are effective with cytochrome P450 IIB1.We suggest that a PCB of the appropriate steric configuration may first induce the relevant cytochrome *in vivo*, then become bound to the active site of the induced cytochrome and produce oxidative species. These may then attack not only bilirubin, our model target compound, but also other key molecules in the cell, potentially initiating in this way a nunber of toxic reactions. Among these targets are uroporphyrinogen (Jacobs et al., 1989) and DNA bases (Faux et al., 1992), with cellular iron participating, as discussed (De Matteis, 1988; Smith and De Matteis, 1990).

Ackowledgement:

We thank Dr. Alan Paine for the gift of a polyclonal antibody against Cytochrome P450 IIB1.

REFERENCES

Bonkowsky, H.L., Sinclair, P.R., Bement, W.J., Lambrect, R.W., and Sinclair, J.F.(1987) New York Acad. Sci. *514*, 96.
Cantoni, L., Dal Fiume, D., Rizzardini, M., and Ruggieri, R.(1984) Toxicol. Lett. *20*, 211.
De Matteis, F.(1988) Mol. Pharmacol. *33*, 463.
De Matteis, F., and Stonard, M.(1977) Semin. Hematol. *14*, 187.
De Matteis, F., Prior, B.E., and Rimington, C.(1961) Nature (London) *191*, 363.
De Matteis, F.,Trenti,T.,Gibbs,A.H., and Greig, J.B. (1989) Mol. Pharmacol.*35*, 831.
De Matteis, F., Dawson, S.J., Boobis, A.R., and Comoglio, A. (1991) Mol. Pharmacol. *40*, 686.
Faux, S.P., Francis, J.E., Smith, A.G., and Chipman, J.K. (1992) Carcinogenesis, *13*, 247.
Elder, G.H.(1978) Handbook Exp, Pharmacol. *44*, 157.
Gray, C.H. (1953) The Bile Pigments, p.16, Methuen & Co. Ltd, London.
Jacobs, J.M., Sinclair, P.R., Bement, W.J., Lambrecht, R.W., Sinclair, J., and Goldstein, J.A.(1989) Biochem. J. *258*, 247.
Kapitulnik, J., and Ostrow, J.D.(1978) Proc.Nat.Acad.Sci.USA *75*, 682.
Marks, G.S.(1985) CRC Crit. Rev. Toxic. *15*, 151.
Smith, A.G., and De Matteis, F.(1990) Xenobiotica, *20*, 865.

Smith, A.G., Francis, J.E., Dinsdale,D., Manson,M.M., and Cabral, J.R.P.(1985) Carcinogenesis, *6*, 631.
Smith, A.G., Francis, J.E., Kay, S.J.E., Greig, J.B., and Stewart, F.P.(1986) Biochem. J. *238*, 871.
Van Ommen, B., Adang, A.E.P., Brader, L., Posthumus, M.A., Muller, F., and Van Bladeren, P.J.(1986) Biochem. Pharmacol. *35*, 3233.

Free Radicals: From Basic Science to Medicine
G. Poli, E. Albano & M. U. Dianzani (eds.)

CBrCl3-TOXICITY IN ISOLATED RAT HEPATOCYTES: SURVEY ON REASONABLE CYTOTOXIC MECHANISMS

C. Jochmann*, S. Klee*, M. Nürnberger*, F.R. Ungemach*, M. Younes**, T. Grune***, W. Siems***, G. Gerber***

* Institute of Pharmacology and Toxicology, Faculty of Veterinary Medicine, Free University of Berlin, Germany
** WHO, European Centre for Environment and Health, Bilthoven, Netherlands
*** Institute of Biochemistry, Faculty of Medicine (Chariti), Humboldt University Berlin, Germany

SUMMARY: Isolated rat hepatocytes were exposed to 0.5, 1.0 and 1.5 mmol/l of the radical-inducing agent $CBrCl_3$.
Viability as assessed by trypan blue exclusion, LDH release and intracellular K^+ decreased in a dose-dependent manner and correlated well with lipid peroxidation as determined by MDA-formation. Early changes after haloalkane addition were a lower steady state-level of ATP and a marked inhibition of lysosomal neutral red uptake. Rise of cytosolic free Ca^{2+} and β-glucuronidase release by lysosomes followed distinctly later. MTT-reduction, predominantly reflecting mitochondrial dehydrogenase activity, was diminished in a manner which revealed only a minor sensitivity to the radical insult. $CBrCl_3$-induced consumption of GSH was critical for the cytotoxic process only under conditions of inhibited GSSG-reduction or under artificial GSH-depletion.
Thus, the concerted action of different pathobiochemical reactions rather than a single event seemed to ultimately cause the haloalkane-mediated lethal injury in hepatocytes.

The cascade of radical-induced pathobiochemical mechanisms leading to cell death is still controversially discussed (Nicotera et al., 1990; Boobis et al., 1989; Reed et al., 1990; Albano et al., 1989; McGirr et al., 1990; Berger et al., 1986). In order to investigate the consequences of radical challenge to cellular integrity freshly isolated rat hepatocytes were exposed to the haloalkane $CBrCl_3$ which is known to rapidly generate the highly reactive $\cdot CCl_3$-radical at the microsomal electron transport chain (Sipes, 1977; Bini, 1975). This radical is supposed to initiate lipid peroxidation (Ungemach, 1987) and trigger a variety of functional disorders in animals (Recknagel, 1967) as well as in isolated organs and cells (Ungemach, 1987). The relationship between different cellular targets of radical attack and the chain of pathobiochemical reactions is not well understood. The lack of detailed knowledge can mainly be attributed to the fact that this subject was investigated under largely varying experimental conditions.

Therefore, hepatocytes treated with 0.5, 1.0 and 1.5 mmol/l $CBrCl_3$ were kept under strictly defined conditions and were examined with regard to viability, plasma membrane integrity, thiol homeostasis, redox and energy state, lysosomal integrity and cytosolic free Ca_{2+}-concentration. The experimental design allows a more detailed insight into the relationship of $\cdot CCl_3$-radical-induced pathological reactions with regard to temporal and causative coincidence not only at the hepatocellular level but also for the general mechanisms of radical-triggered cytotoxicity.

The methods used in this study were:

Male Wistar rats, 200-250 g body weight and fasted 24 hours prior to the experiments, were used.

Hepatocytes were isolated by a modified collagenase perfusion technique as described by Baur et al.(1975).

Lactate dehydrogenase-activity (LDH) in the incubation medium was determined using a standard test kit (Boehringer, Mannheim). The measurement of intracellular K^+ was performed by flame photometry, of cytosolic Ca^{2+} fluorimetrically with Quin-2-AM as trapping agent (Tsien et al., 1982). Intracellular glutathione (GSH) and protein thiols were determined with Ellman's reagent according to Ellman (1959) and Albano et al.(1985). Malondialdehydeformation (MDA) was measured with the thiobarbituric acid method (Ohkawa, 1979). Lysosomal β-glucuronidase-release into the incubation medium was quantified by the hydrolysis of p-nitrophenol-glucuronide to p-nitrophenol (p-NP) (Fishman, 1974). Intracellular nucleotides (ATP, ADP, AMP, GTP, GDP, hypoxanthine, uric acid, adenine and guanosine) were examined by the HPLC method of Werner et al.(1989).

The neutral red- and thiazolyl (MTT)-test developed for cytotoxicity studies in cell culture systems (Borenfreund et al., 1988) were adapted for hepatocyte suspensions by incubating aliquots of $CBrCl_3$-treated suspensions at the indicated time for a further 20 minute-period with 0.008% of neutral red and extraction of the accumulated dye from the lysosomes of washed cells. MTT-reduction was similarly determined after a 10 minute-incubation period of the cell suspension with 0.5 mg/ml of MTT.
In some experiments hepatocytes were preincubated prior to haloalkane-treatment with 75 µmol/l 1,3-bis(2-chloroethyl)-1-nitrosurea (BCNU) to inhibit GSSG-reductase (Babson et al., 1978) and 0.25 mmol/l phorone (Traber et al., 1992) to deplete the free cytosolic GSH.

As shown in Table I a dose-dependent loss of viability was observed after 60 minutes of incubation with $CBrCl_3$. An almost maximum effect was reached after 30 minutes. Only the highest concentrations induced further damage upon prolonged incubation (data not shown), which was in agreement with earlier investigations (Ungemach, 1987).
Malondialdehyde-formation doubled within 15 minutes of exposure and declined again only at the lowest concentration (Table I). A dose-dependent release of the lysosomal β-glucuronidase was observed (Table I) which showed a marked time-delay at the lower concentrations as compared to 1.5 mmol/l $CBrCl_3$. In the presence of $CBrCl_3$ the mitochondrial reduction of MTT exhibited an almost linear decrease over the whole incubation period without apparent dose-dependence (Table I).
An increase of cytosolic free Ca^{2+} could only be observed at concentrations of at least 1.0 mmol/l $CBrCl_3$. 1.5 mmol/l $CBrCl_3$ induced a biphasic course with an initial moderate Ca^{2+}-peak of short duration followed by a continuous substantial rise, after 15 minutes (Fig.1).
Intracellular GSH-levels decreased by 50% to about 4 mmol/l cellular water independently of the $CBrCl_3$ concentration used (Fig.2).

Table I

$CBrCl_3$-dependent alterations of hepatocellular viability, membrane integrity and MDA-formation after one hour of incubation

mmol/l $CBrCl_3$	0(contr.)	0.5	1.0	1.5
Trypan blue exclusion % viable cells	75 ± 1.0	70 ± 1.2	62 ± 1.4	61 ± 2.1
LDH release [mU/mg wet weight of cells]	47 ± 2.1	68* ± 9.4	79* ± 5.6	102* ± 4.7
Intracellular K^+ [mmol/l cell water]	111 ± 4.6	102 ± 6.2	88 ± 7.5	76* ± 8.4
MDA-formation [nmol/mg wet weight of cells]	0.093 ± 0.0129	0.152 ± 0.0319	0.205* ± 0.0256	0.220* ± 0.0484
β-Glucuronidase release [5mol p-NP/lxmin]	5.49 ± 0.561	6.45 ± 0.395	7.42* ± 0.584	10.62* ± 1.184
MTT-reduction % of control	100	82	86	78

Controls were treated with 1% ethanol serving at the same final concentration as a vehicle for $CBrCl_3$. For further details see above.
All values are mean ± SEM. from at least five experiments.
* significant difference from control ($p<0.05$).

Inhibition of GSSG-reductase by BCNU caused a further loss of GSH at 1.0 and 1.5 mmol/l $CBrCl_3$, but not at 0.5 mmol/l indicating the importance of the reductase for the maintenance of GSH-homeostasis at the higher concentrations of $CBrCl_3$ (Fig.3).
Pretreatment with BCNU was tolerated by control cells, but increased the sensitivity of hepatocytes towards $CBrCl_3$ even at 0.5 mmol/l (Fig.3).
Depletion of GSH by phorone did not impair control cells as assessed by trypan blue exclusion but also increased the susceptibility of hepatocytes to $CBrCl_3$, probably due to the considerable loss of GSH-dependent radical defence. Thus, preincubation

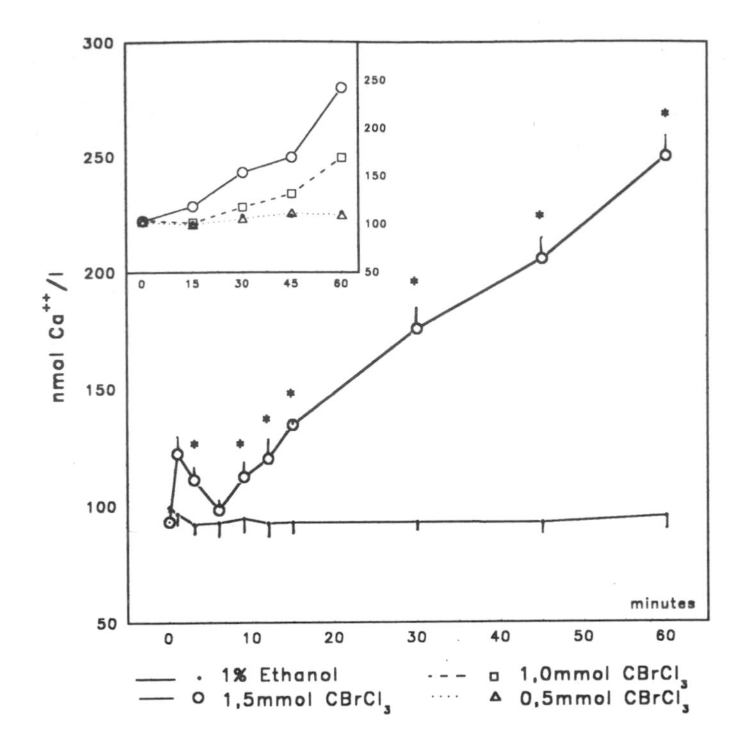

Fig. 1. Variation of cytosolic free calcium of isolated hepatocytes treated with ethanol and $CBrCl_3$.
Isolated rat hepatocytes were incubated at a density of 70 mg/ml in Krebs-Henseleit-buffer with ethanol and $CBrCl_3$ as described above. Values are means ± SEM of 5 experiments. The insert shows the time course of cytosolic Ca^{2+} of one typical experiment for each concentration of $CBrCl_3$.

with phorone resulted in complete cell death after 30 minutes of incubation with concentrations of $CBrCl_3$ above 1 mmol/l (data not shown).

The total amount of free protein thiols was not affected after one hour of $CBrCl_3$-treatment. A decline was observed only after most of the cells had died (data not shown).

The ATP-content of hepatocyte suspensions decreased markedly

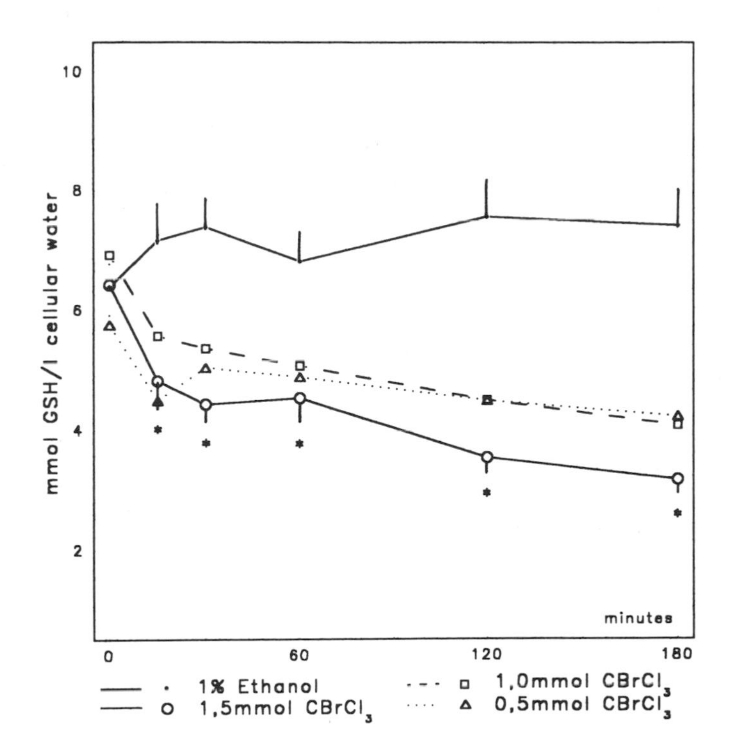

Fig. 2. Variation of intracellular GSH in isolated hepatocytes challenged with ethanol and $CBrCl_3$.
Isolated rat hepatocytes were incubated at a density of 70 mg/ml in Krebs-Henseleit-buffer with ethanol and $CBrCl_3$ as described above. Values are mean ± SEM of at least 5 experiments. *significant difference from control cells.

immediately after addition of $CBrCl_3$. A slight initial decrease was also observed in controls after the addition of 1% ethanol (Fig.4). While the control cells exhibited a rapid, however incomplete recovery, $CBrCl_3$-treated cells maintained a lower steady state level of ATP (about 70% of control).
Further changes of purine concentrations in $CBrCl_3$-treated hepatocytes were a decreased ATP/AMP-ratio up to 60% of controls. Additionally, increased levels of hypoxanthine and uric acid and

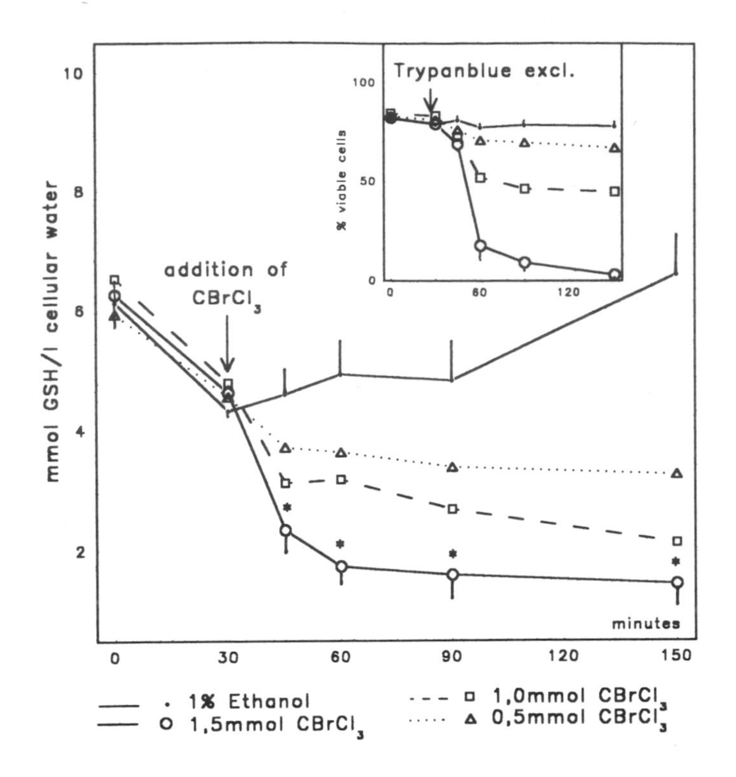

Fig. 3. Effect of BCNU pretreatment on the level of intracellular GSH of isolated hepatocytes treated with ethanol and $CBrCl_3$. Isolated rat hepatocytes were preincubated for 30 minutes with 75 µmol/l BCNU followed by addition of ethanol and $CBrCl_3$ at the indicated concentrations. Values are mean ± SEM of at least 5 experiments. *significant difference from control cells. The insert shows cellular viability under these conditions as determined by trypan blue exclusion.

a reduced guanosine concentration could be detected (data not shown).

The lysosomal accumulation of neutral red (Allison et al., 1964) was rapidly reduced by about 50% at 0.5 mmol/l $CBrCl_3$ and almost completely inhibited at 1.0 mmol/l as compared to the capacity of controls after the same incubation period. Up to three hours of incubation no recovery could be observed (Fig.5).

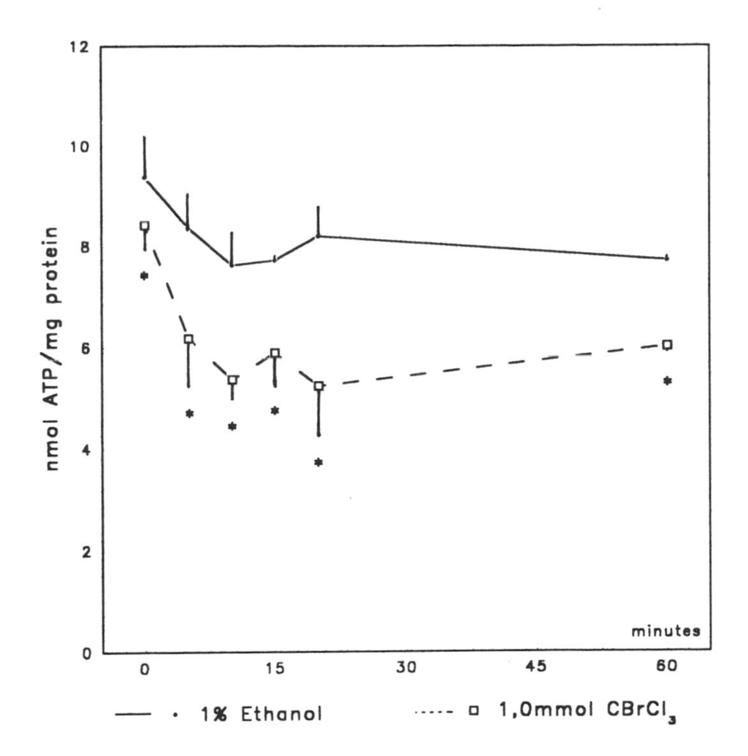

Fig. 4. Variation of intracellular ATP in isolated hepatocytes incubated with ethanol or $CBrCl_3$.
Isolated rat hepatocytes were incubated at a density of 35 mg/ml in the presence of 1% ethanol or 1 mmol/ml $CBrCl_3$. Values are mean ± SEM of at least 5 experiments. *significant difference from control cells.

Radical challenge by the haloalkane $CBrCl_3$ caused a wide spectrum of cellular disorders in isolated rat hepatocytes. Increased permeability of the plasma membrane was reflected by diminished trypan blue exclusion and loss of LDH into the incubation medium (Table I). Corresponding time courses of these parameters with MDA-formation indicate that lipid peroxidation is one of the initial critical events in $CBrCl_3$-toxicity as shown previously by different investigators (Ungemach, 1987).

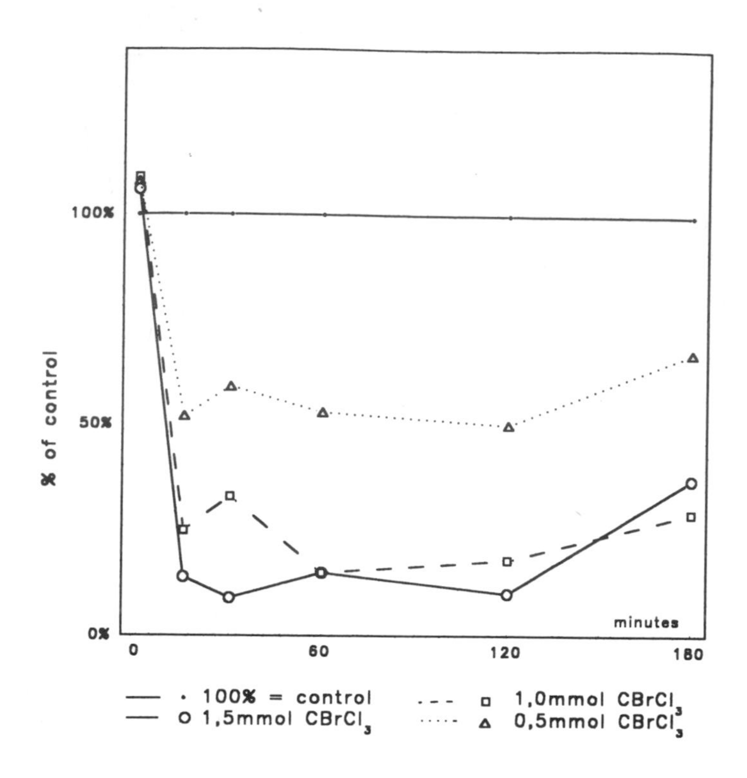

Fig. 5. Neutral red uptake of isolated hepatocytes intoxicated with ethanol or $CBrCl_3$.
Isolated rat hepatocytes were treated with ethanol or $CBrCl_3$. At the indicated time points aliquots of the suspension were taken and incubated for 20 minutes with neutral red to accumulate the dye. The uptake rate of controls after the same incubation interval were expressed as 100%. Values are mean of at least 5 experiments.

The intracellular K^+-concentration is another sensitive parameter of plasma membrane integrity, however, its dependence on the availability of ATP has to be considered. Thus, the decrease of K^+ might also be due to the lower steady state level of ATP (Fig.4) in CBrCl3-treated hepatocytes. This loss of ATP was not quantitatively balanced by an increase in secondary products of

ATP degradation as hypoxanthine and uric acid. Probably the main share of purines lost is degraded to the final end product allantoin which was not detected by the HPLC-method used.
Lysosomal neutral red uptake was much stronger affected as could be expected from the moderate loss of ATP and the slow release of β-glucuronidase. Thus, the indirectly energy-dependent uptake mechanism, which is driven by an ATP consuming proton-gradient (Winckler, 1974) across the lysosomal membrane, reveals a particular sensitivity towards changes of the cellular energy state.
The capacity of MTT-reduction, which is predominantly catalyzed by mitochondrial dehydrogenases, reflects the redox state of the cells (Carmichael et al., 1987). Probably due to the electron-consuming generation of radical species by $CBrCl_3$ cleavage the reduction of the dye was significantly lowered after haloalkane treatment (Table I). However, neither a complete inhibition, as for the lysosomal neutral red uptake, nor a dose-dependence was observed, indicating that alterations of the redox state and subsequently of mitochondrial dehydrogenase activity is not a critical target of radical challenge.
$CBrCl_3$ induced consumption of GSH was not dose-dependent (Fig. 2). As long as the compensating activity of the GSSG-reductase maintained the intracellular GSH at about 4 mmol/l this loss did not interfere with the dose-dependent cytotoxic process. Only after inhibiting GSSG-reductase with BCNU (Fig.3) or depleting GSH with phorone (data not shown) which was both tolerated by control cells an increased sensitivity towards $CBrCl_3$ and a direct correlation of the cytotoxic events with the loss of GSH was obvious (data not shown). Since protein thiols were apparently not affected under any experimental condition their involvement in $CBrCl_3$-induced cell death remains questionable.
The time course of the cytosolic Ca^{2+}-rise did not correlate with the early changes of cellular viability and energy state. The initial Ca^{2+}-peak at 1.5 mmol/l $CBrCl_3$ did not amplify cellular damage as compared to the lower concentrations, which did not induce this early rise. Thus, in hepatocytes disorders of cytoso-

lic Ca^{2+}-levels seem to be more likely a subsequent than a causative event of $CBrCl_3$-toxicity.
From the results obtained the following conclusions can be drawn: The development of $CBrCl_3$-induced cell injury was independent of GSH-consumption as long as cells disposed of sufficient GSH-levels and the capacity of regenerating the oxidized GSSG.
Lysosomes markedly lost the capacity of neutral red accumulation in the early phase of cellular damage whereas their structural integrity was impaired considerably later. Therefore, though they revealed a high sensitivity towards $CBrCl_3$, their causative contribution to the cytotoxic process seems to be unlikely.
A sustained rise of cytosolic Ca^{2+} occured too late as to plausibly explain the early events of cellular injury. Thus, for the experimental system of isolated hepatocytes a central role of Ca^{2+} in radical-induced cell death remains questionable.

Aknowledgements

This study was supported by the Deutsche Forschungsgemeinschaft.

REFERENCES

Albano, E., Rundgren, M., Harvison, P.J., Nelson, S.P., Moldeus, P. (1985) Mol. Pharm. 28, 306-311.
Albano, E., Carini, R., Parola, M., Bellomo, G., Poli, G., Dianzani, M. (1989) Adv. Bioscience 76, Pergamon Press.
Allison, A.C., Young, M.R., (1964) Life Science 3, 1407-1414.
Babson, J.R., Reed, D., (1978) Bioch. Biophys. Res. 83, 754-762.
Baur, H., Kasparek, S., Pfaff, E. (1975) Hoppe-Seyler`s Z. Physiol. Chem. 356, 827-838.
Berger, M.L., Bhatt, H., Combs, B., Estabrook, R.W. (1986) Hepatology 6(1), 36-45.
Bini, A., Vecchi, G., Vivoli, G., Vannini, V., Lessi, C. (1975) Pharm. Res. Com. 7(2) 143-149.
Boobis, A.R., Fawthrop, D.J., Davies, O.S. (1989) TIPS 10, 275-280.
Borenfreund, E., Babick, H., Martin-Alguacil, N. (1988) Toxic. in Vitro 2(1), 1-6.
Carmichael, J., DeGraff, W.G., Gasdor, A.F., Minna, J.D., Mitchell, J.B. (1987) Cancer Res. 47, 936-942.
Ellman, G.L. (1959) Arch. Biochem. Biophys. 82, 70.
Fishman, W.H. (1974) in: Bergmeyer, H.U. (Hrsg.) Methoden der

enzymatischen Analyse, Verlag Chemie, Weinheim, 964-979.
McGirr, L.G., Khan, S., Lauriault, V., O'Brien, P.J. (1990) Xenobiotica 20(9), 933-943.
Nicotera, P., Bellomo, G., Orrenius, S. (1990) Chem. Res. Toxicol. 3, 484-494.
Ohkawa, H. (1979) Anal. Biochem. 95, 351-358.
Recknagel, R. (1967) Pharm. Rev. 19(2), 145-208.
Reed, D. (1990) Chem. Res. Toxicol. 3, 495-502.
Sipes, G.J., Krishna, G., Gillette, J.R. (1977) Life Sciences 20, 1541-1548.
Traber, J., Suter, M., Walter, P., Richter, C. (1992) Biochem. Pharm. 43, 961- 964.
Tsien, R.U., Pozzan, T., Rink, T.J. (1982) J. Cell Biol. 94, 325-334.
Ungemach, F.R. (1987) Chem. Phys. Lip. 45, 171-205.
Werner, A., Schneider, W., Siems, W., Grune, T., Schreiter, C. (1989) Chrom. 11/12, 639-643.
Winckler, J. (1974) Progr. Histochem. Cytochem. 6(3), 1-91.

Free Radicals: From Basic Science to Medicine
G. Poli, E. Albano & M. U. Dianzani (eds.)

EVIDENCE FOR A POSSIBLE ROLE OF LIPID PEROXIDATION IN EXPERIMENTAL LIVER FIBROSIS.

M. Parola*, E. Albano*, G. Leonarduzzi*, R. Muraca*, I. Dianzani#, G. Poli* and M.U. Dianzani*.

*Dip. Medicina e Oncologia Sperimentale and Centro di Immunogenetica del CNR, Università di Torino, #Clinica Pediatrica, Università di Torino.

SUMMARY: Liver fibrosis represents a common stage of several conditions of chronic active liver diseases, leading eventually to cirrhosis. Knowledge of the biology and of the pathobiology of fibrogenetic processes in mammalian liver has advanced significantly in the last decade, and we are beginning to understand the complex molecular network of cytokine-mediated cellular interactions which play a major role in the modulation of extracellular matrix deposition. Evidence, coming from both clinical and experimental studies, indicates that liver oxidative injury and, in particular, lipid peroxidation of polyunsaturated fatty acids of biological membranes, are often associated with liver fibrosis. In our laboratory we have shown that lipid peroxidation is associated with the development of the most commonly used experimental model of liver fibrosis, chronic administration of carbon tetrachloride (CCl_4) to rats. Moreover, in the same experimental model, vitamin E supplementation, a procedure known to prevent lipid peroxidative processes, can afford significant protection against the production of aldehydic end-products of lipid peroxidation, as well as against the development of liver necrosis and collagen deposition. These effects of vitamin E seem also to be related to a decreased expression of transforming growth factor β1 (TGFβ1), the best known pro-fibrogenic cytokine. Finally, *in vitro* studies indicate that lipid peroxidation can modulate collagen synthesis by fibroblasts and, more important, by fat-storing cells. Thus, we suggest the possibility that, because of their activity on some transmembrane signalling systems and other cellular functions, aldehydic end-products of lipid peroxidation might act as molecular mediators in the process of liver fibrosis consequent on oxidative injuries.

Liver fibrosis and oxidative damage. Introductory remarks.

Liver fibrosis is as an extremely complex and dynamic tissue phenomenon, which involves different hepatic cell types and implies the increased synthesis of various types of collagen and of other proteins of the extracellular matrix (Popper, 1987; Rojkind and Greenwel, 1988; Bissel et al., 1990; Gressner, 1991). The changes of fibrotic liver extracellular matrix are likely to be the result of a disproportionate elevation of the subspecies of individual matrix molecules, and of fine changes of the micro-composition of certain extracellular matrix molecules (ECM), as well as of topographic redistribution of ECM in the injured liver (Biagini and Ballardini, 1989; Schuppan, 1990; Gressner, 1991). The latter event, in particular, leads to the well-known morphological aspects of perisinusoidal fibrosis (e.g. subendothelial deposition of connective tissue in the space of Disse) as well as periportal-, bridging-, diffuse or focal fibrosis, which play a major role in determining the adverse consequences of fibrosis. In recent years, it has become increasingly evident that fibrogenesis in the liver is modulated by the release of soluble factors (e.g., cytokines) in the local micro-environment and, possibly, by the matrix itself and by the lobular localization (Bissel et al., 1990; Castilla et al., 1991; Gressner, 1991; Pinzani and Abboud, 1991). In this connection, studies performed in the last decade have provided convincing evidence that non-parenchymal cells, such as Kupffer cells and, in particular, fat-storing cells (also known as Ito's cells or hepatic lipocytes), rather than hepatocytes, play a major role in the fibrotic process (Friedman et al., 1985; Milani et al., 1989, 1990; Bissel et al. 1990; Gressner, 1991; Pinzani and Abboud, 1991).

Much of our knowledge concerning the biology and the pathobiology of the fibrotic process comes directly from the experimental rat model of chronic CCl_4 administration. This model is easy to run, highly reproducible and leads to the classic histological appearance of cirrhosis in 5 to 14 weeks, depending on the route of administration, the dose administered and the frequency of dosing (Ehrinpreis et al., 1980; Miao et al., 1990). Moreover, administration of a small number of doses of the toxin affords a reliable investigation of the early phases of fibrogenesis. The data obtained are in agreement with those resulting from different experimental models, and are also supported by clinical observations (Armendariz-Borunda et al., 1990; Miao et al., 1990).

It is well known that CCl_4-induced liver injury involves the generation of biologically active free radicals, such as trichloromethyl and trichloromethylperoxy radicals, and the development of oxidative damage. In particular, the toxicity of this compound is believed to depend on the two major mechanisms of covalent binding and lipid peroxidation (Slater, 1984; Comporti, 1985; Brattin et al., 1985; Poli et al., 1987). In our laboratory, we have obtained convincing evidence that the peroxidative derangement of polyunsaturated fatty acids of membrane phospholipids is likely to be the main mechanism responsible for acute liver necrosis (Poli et al., 1987, 1990; Biasi et al., 1991).

These considerations led us to investigate the role played by lipid peroxidation in the development of chronic liver injury and fibrosis, by using the experimental model of chronic CCl_4 administration.

Evidence for the involvement of lipid peroxidation in the CCl_4 experimental model of liver fibrosis. The protective role of vitamin E.

At the end of chronic CCl_4 treatment, the livers of all the rats show the classic cirrhotic appearance, consisting of extensive tissue remodelling with the presence of fibrous septa containing a heterogeneous population of non-parenchymal liver cells. As already described by several authors, bands of collagen bridge portal areas or extend from central regions to portal areas. Regenerative nodules without central vein (e.g. pseudolobules), severe and massive cell injury, coagulative necrosis, inflammatory infiltration and fatty metamorphosis are other constant histological features of these livers. This typical histological picture is accompanied by a 5-fold increase in the hepatic collagen content (see Erinphreis et al., 1980; Parola et al., 1992a).

Evidence for the presence of lipid peroxidation comes either from the simple analysis of increased malonaldehyde production by liver homogenates at the end of the cirrhogenetic protocol (Parola et al., 1992c), or from the analysis of carbonyl compounds in cirrhotic liver (Table 1), which shows unequivocally that at the end of the experimental protocol the total carbonyl content of liver samples obtained from CCl_4-treated animals is dramatically increased. In particular, there was an evident 4-fold increase in 4-hydroxy-2,3-alkenals (Table 1) and this included 4-hydroxy-2,3-nonenal (HNE), the most toxic aldehydic end-product of lipid peroxidation known so far (Poli et al., 1987; Esterbauer et al., 1991).

We have also observed that dietary vitamin E supplementation in fact significantly increases the hepatic content of the vitamin. Vitamin E supplementation was able almost completely to abolish the production of 4-hydroxy-alkenals, including HNE (Table 1), without influencing the metabolic activation of the toxic compound (Parola et al., 1992a,c) nor the covalent binding of the CCl_4-derived free radical to cell structures.

Animals fed on a vitamin E supplemented diet and treated chronically with CCl_4 were also characterized by a consistent protection against the increase in serum liver function parameters, such as GOT, GGT, alkaline phosphatase and bilirubin (Table 1). These data confirm those obtained both in the acute *in vivo* intoxication model and in *in vitro* experiments (Poli et al., 1987,; Biasi et al., 1991). Furthermore, no protection is afforded against CCl_4 induced triglyceride accumulation (Table 1), an event mostly associated with covalent binding (Poli et al., 1990).

More interestingly, the protective effect of vitamin E supplementation also includes a significant decrease in liver collagen content, and in the liver-weight/body-weight percentage ratio, a reliable marker of fibrosis (Table 1). These analyses are fully

Table 1. Protective effects exerted by dietary vitamin E supplementation against liver alterations induced in male Wistar rats at the end of 5 weeks of chronic administration of CCl_4.

Parameters	CCl_4	CCl_4 + vit.E
Vitamin E (nmol/gm liver)	30.0 ± 7.5	89.5 ± 15.6 *
GOT (unit/l)	1860 ± 903	440 ± 110 **
Total carbonyls (nmol/mg prot.)	7.1 ± 1.4	4.4 ± 1.3 *
4-hydroxyalkenals (nmol/mg prot.)	1.9 ± 0.7	0.8 ± 0.2 *
liver weight/body weight (%)	7.2 ± 0.8	5.7 ± 0.4 **
collagen (mg/gm liver)	5.7 ± 0.6	2.3 ± 1.0 *
triglycerides (mg/100 gm b.w.)	76.1 ± 24.5	84.6 ± 39.4
mRNA procollagen type I	100	24
mRNA TGFβ1	100	43

procollagen I and TGFβ1 mRNA levels are expressed as percentages of the mean values obtained in the livers of CCl_4-treated rats.
Data refer to 8-12 animals for each group.
* $p < 0.01$; ** $p < 0.05$;

confirmed by conventional histology of liver samples: in vitamin E supplemented animals, we observe a consistent reduction of liver necrosis and inflammation, as well as a more regular liver architecture, in which collagen deposition is greatly reduced, and generally limited to fibrous bands connecting portal areas. Fatty accumulation is, on the contrary, always present in these livers (Parola et al., 1992a).

Taken together, these data suggest the involvement of lipid peroxidation in the pathogenesis of liver fibrosis and cirrhosis induced by CCl_4. Consistent with this, Mourelle et al. (1989) have obtained a high degree of protection against CCl_4-induced oxidative damage, liver necrosis and collagen deposition, by employing the plant-derived antioxidant silymarin. Hepatoprotection related to an inhibition of lipid peroxidation has also been reported by Bursch et al. (1989), who used the prostacyclin derivative iloprost. Interestingly, the latter report shows clearly that iloprost protected against oxidative damage by reducing the amount of histochemically-detectable aldehydic end-products of lipid peroxidation.

Mechanisms by which lipid peroxidation may modulate the fibrotic process.

The pathogenesis of liver fibrosis is extremely complex and Figure 1 outlines a simplified model of cellular and molecular events believed to be involved in the development of fibrotic process (see Rojkind, 1988; Bissel et al., 1990; Gressner, 1991 for comprehensive reviews). In the figure, the points at which lipid peroxidation might be involved in modulating the process are also indicated. From a general point of view, liver necrosis, regardless of the cause of hepatocyte injury and death, represents the driving force in both initiation and progression of hepatic fibrosis. The inflammatory response which follows the necrotic event can play a double role, either in perpetuating liver necrosis through the recruitment of inflammatory cells and the release of toxic oxygen reactive species, or in sustaining the fibrogenic process.

In certain cases, lipid peroxidation may represent the most

important way in which the aetiological agent can provoke acute cell death. In this connection, the protection afforded by vitamin E may be related to the antinecrogenic effect we detected in our experiments (Parola et al., 1992a), and to the reduction in the inflammatory response which usually follows. This simple explanation, however, may not apply to other models, as we will discuss below, and does not seem to be completely adequate even for the CCl_4 model we used.

Molecular biology studies have indeed shown that vitamin E supplementation is not only able to suppress the CCl_4-induced expression of procollagen type I, but also that it strongly decreases the synthesis at both the mRNA and the protein levels of transforming growth factor β1 (TGFβ1) (Table 1) (Parola et al. 1992b). TGFβ1 is the most relevant profibrotic cytokine in the injured liver and is responsible for transforming FSC into myofibroblast-like cells (Bissel et al., 1990; Miao et al., 1990; Armendariz Borunda et al., 1990; Gressner, 1991; Castilla et al., 1991). Moreover, it is of primary interest that vitamin E supplementation alone appears able to down-modulate the basal expression of TGFβ1 in control liver (Parola et al., 1992b), suggesting that lipid peroxidation might be involved not simply in determining cell death but also in the cellular and molecular events which follow necrosis. This concept is emphasised by the knowledge that TGFβ1, rather than acting simply as a pro-fibrotic factor for FSC and myofibroblast-like cells, may also contribute to fibrogenesis in several other ways: a) as chemotactic molecule for inflammatory cells and fibroblasts, b) as a growth inhibitory stimulus for parenchymal cells, c) as a positive stimulus for the synthesis of the tissue inhibitor of metalloproteinases, d) as a negative stimulus for the synthesis of collagenase, and e) as a stimulus for collagen-matrix contraction by fibroblasts (see Gressner, 1991).

On the other hand, through the production of reactive aldehydic end products, lipid peroxidation might be involved in the modulation of the inflammatory reaction, and possibly in that

of the fibrogenic process. Indeed, 4-hydroxy-2,3-alkenals have been demonstrated to exert chemotactic and chemokinetic effects on leukocytes (Curzio, 1986; Esterbauer et al., 1991) and to modulate transmembrane and intracellular signalling via their effect on adenylate cyclase (Paradisi et al., 1985), protein kinase C (Poli et al., 1988) and phospholipase C (Rossi et al., 1990).
Therefore, aldehydic end-products of lipid peroxidation might either serve as amplifying messengers for leukocyte recruitment to the site of injury, or to act on non-parenchymal cells by stimulating Kupffer cells to produce profibrotic cytokines, such as TGFβ1, or by stimulating FSC to produce extracellular matrix proteins. In addition, it has been suggested that human skin fibroblasts are induced to produce collagen by the induction of lipid peroxidation in cell culture (Geesin et al., 1990; Houglum et al., 1991). We have recently found that human FSC increase their basal production of procollagen type I after exposure to the ascorbate/iron pro-oxidant complex; moreover, in the same experiments the synthesis of procollagen type I was completely stopped by pre-treatment of FSC culture with vitamin E or DPPD, which also completely stopped lipid peroxidation (Parola, Pinzani et al., 1992, submitted for publication).

Evidence for the involvement of lipid peroxidation in other experimental models and in clinical conditions.

In the previous paragraph we offered evidence to support a possible role of lipid peroxidation in the development of experimental fibrosis, suggesting, in particular, that aldehydic products might have a multiple role in modulating inflammatory response and ECM production. Lipid peroxidation and, more generally, oxidative damage are known to be associated with clinical conditions and other experimental models, in many of which there is no direct link between the etiological agent, lipid peroxidation and cell necrosis, or only a weak one.

The first example is the liver damage associated with chronic alcohol consumption. Although the exact pathogenesis of alcohol-induced liver injury is still a matter of debate (Lieber, 1990), there are clear indications that free-radical reactions, including lipid peroxidation, are associated with either human alcoholic liver disease or animal models of ethanol-induced liver fibrosis (Nordmann et al., 1992; Albano et al. in this volume). Similar results have been obtained on baboons exposed chronically to ethanol, where signs of lipid peroxidation and of GSH depletion are evident upon the appearance of liver fibrosis (Lieber, 1990). Moreover, there is now clear evidence that microsomes obtained either from experimental animals or from human liver after chronic exposure to alcohol are more susceptible to lipid peroxidation. This effect is related to the induction of a specific ethanol-inducible form of cytochrome P-450 (CYP2E1), and to the production by this cytochrome of oxygen-derived free radicals, as well as of the 1-hydroxyethyl free radical (Albano et al. in this volume).
Recent observations on rats fed by continous intragastric infusion with ethanol and a high-fat diet indicate that the development of liver damage and fibrosis in these animals is associated with a 10-fold rise in the CYP2E1 content and a dramatic stimulation of lipid peroxidation in liver microsomes (French et al. 1992). The animals receiving ethanol also show a significant increase in the serum protein adducts, with both MDA and 4-HNE (French et al. 1992). These results not only confirm that peroxidative reactions consequent on alcohol-mediated CYP2E1 induction take place *in vivo* as well as *in vitro*, but also suggest the oxidative events might be associated with the processes leading to alcohol-induced liver cirrhosis.

The involvement of lipid peroxidation has also recently been proposed to explain the development of liver fibrosis and cirrhosis occurring as a consequence of excess hepatic iron deposition in humans affected by genetic hemochromatosis, as well as the fibrosis found in an experimental model of iron overload (Park et al., 1987; Bacon and Britton, 1990; Bacon et al., 1992 in this volume). Once again, typical markers of the free-radical-mediated process have been unequivocally detected in both the clinical and the experimental conditions, and have been related to the well known pro-oxidant action of free iron in biological systems (see Bacon and Britton, 1990 and Bacon et al., 1992 in this volume, for a more comprehensive review of the topic).

The two examples reported here are of course the more relevant, if one considers the clinical incidence of liver fibrosis and cirrhosis related to alcohol abuse and excess deposition of iron, but other examples also support a possible link between lipid peroxidation and fibrogenesis. Here it is worth mentioning some recent data obtained by Sokol et al. (1990, 1991) in two different rat models of liver fibrosis: bile-duct ligation, which in its terminal stage produces a type of damage that closely resembles human secondary biliary cirrhosis, and chronic copper overload, which has some analogies to human pathologies related to copper-overload disorders. In both models, signs of lipid peroxidation, such as MDA and conjugated diene generation, were found in the mitochondrial fraction. Furthermore, it has been suggested that liver vitamin E modulates both lipid peroxidation and the release of cellular enzymes in the serum. This was particularly evident in vitamin E-deficient rats but, unfortunately, the experiments were limited to early stages of the models, so no information is available in this case on the relevance of lipid peroxidation and antioxidant status on the development of fibrosis.

In conclusion, data provided by experimental models, as well as by clinical observations indicate that lipid peroxidation and, more generally, oxidative damage might be associated with the development of the fibrogenic process. Our laboratory is currently involved in a series of studies devoted to investigating the significance of this association and the precise role played by lipid peroxidation.

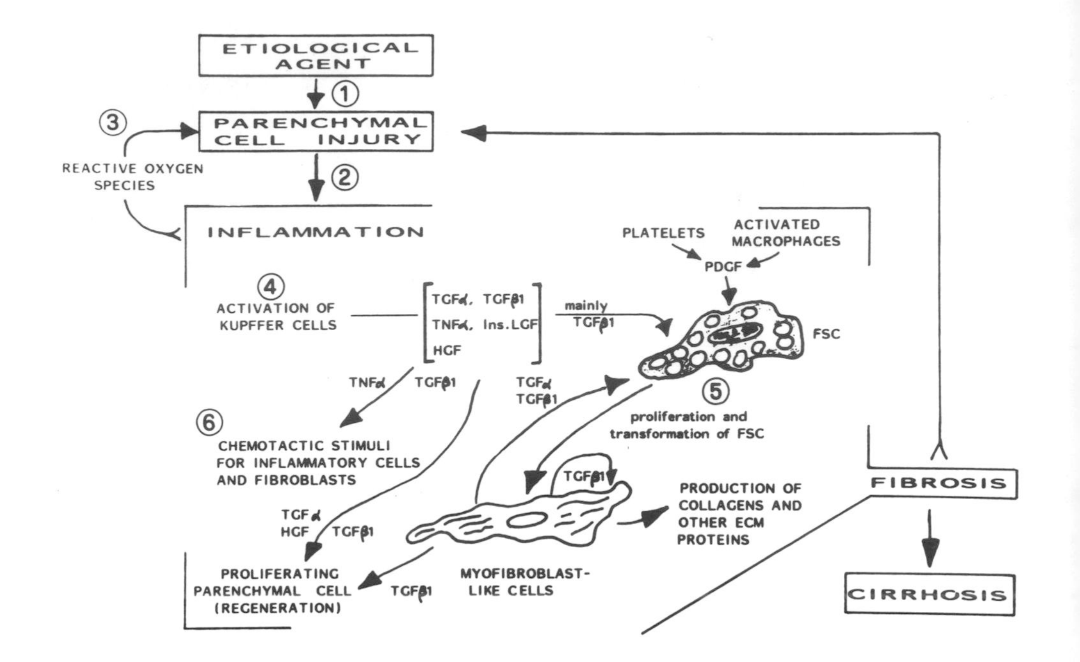

Fig. 1. General model illustrating events involved in the development of liver fibrosis. Liver necrosis is sustained by chronic exposure to the etiological agent and by perpetuation of inflammatory reactions and excess deposition of collagens and other extracellular matrix proteins (ECM) leading eventually to cirrhosis. From a cellular and molecular point of view, relevant events are the activation of: a) Kupffer cells, which are a primary source of cytokines that modulate the process; and b) fat storing cells (FSC), which then proliferate and change their apparent phenotype to myofibroblast-like cells, becoming the primary source of collagens and other ECM proteins as well as an additional source of TGFβ1 able, then, to amplify the process via either paracrine or autocrine loops (see Bissel et al., 1990; Gressner, 1991; Pinzani and Abboud, 1991). Lipid peroxidation may interfere with the process: (1) as a mechanism for agent-induced liver injury; (2) by amplifying inflammatory reactions (chemotactic effects of aldehydes, (6); (3) by mediating liver necrosis due to the release of active oxygen species by inflammatory cells, and by activating (4) Kupffer cells and (5) FSC.
TGFα: transforming growth factor α; TGFβ1: transforming growth factor β1; TNFα: tumor necrosis factor α; Ins.LGF: insulin like growth factor; HGF: hepatocyte growth factor; PDGF: platelet-derived growth factor.

Acknowledgments.
The research presented in this paper was supported by Ministero della Università e della Ricerca Scientifica (MURST) and by Consiglio Nazionale delle Ricerche (CNR), Rome, Italy.

REFERENCES.

Albano, E., Tomasi, A., Goria-Gatti, L. and Dianzani, M.U. (1988) Chem. Biol. Interact. 65, 223-234.
Armendariz-Borunda, J., Seyer, J.M., Kang, A.H. and Raghow, R. (1990) FASEB.J. 4, 215-221.
Bacon, B.R. and Britton, R.S. (1990) Hepatology 11, 127-137.
Biagini, G. and Ballardini, G. (1989) J. Hepatol. 8, 115-124.
Biasi, F., Albano, E., Chiarpotto, E., Corongiu, F.P., Pronzato, M.A., Marinari, U.M., Parola, M., Dianzani, M.U. and Poli, G. (1991) Cell Biochem. Funct. 9, 111-118.
Bissel, D.M., Friedman, S.L., Maher, J.J. and Roll, F.J. (1990) Hepatology 11, 588-498.
Brattin, W.J., Glende, E.A. and Recknagel, R.O. (1985) J. Free Rad. Biol. Med. 1, 27-38.
Bursch, W., Taper, H.S., Somer, M.P., Meyer, S., Putz, B. and Schulte-Hermann, R. (1989) Hepatology 9, 830-838.
Castilla, A., Prieto, J. and Fausto, N. (1991) New Engl. J. Med. 324, 933-940.
Curzio, M. (1986) Biol. Chem. Hoppe-Seyler 367, 321-329.
Ehrinpreis, M.N., Giambrone, M.A., and Rojkind, M. (1980) Bio chim. Biophys. Acta 629, 184-193.
Esterbauer, H., Schaur, R.J. and Zollner, H. (1991) Free Rad. Biol. Med. 11, 81-128.
Friedman, S.L., Roll, F.J., Boyles, J. and Bissel, D.M. (1985) Proc. Natl. Acad. Sci. USA 82, 8681-8685.
French, S.W., Wong, K., Jui, L., Albano, E., Hagbjörk, A.-L. and Ingelman-Sundberg, M. (1992) Exp. Mol. Pathol., (submitted for pubblication).
Geesin, J.C., Gordon, J.S. and Berg, R.A. (1990) Arch. Biochem. Biophys. 278, 350-355.
Gressner, A.M. (1991) Eur. J. Clin. Chem. Clin. Biochem. 29, 293-311.
Houglum, K., Brenner, D.A. and Chojkier, M. (1991) J. Clin. Invest. 87, 2230-2235.
Lieber, C.S. (1990) Pharm. Ther. 46, 1-41.
Nordmann, R., Ribière, C. and Rouach, H. (1992) Free Rad. Biol. Med. 12, 219-240.
Miao, S., Bao-En, W., Annoni, G., Degli Esposti, S., Biempica, L. and Zern, M. (1990) Lab. Invest. 63, 467-475.
Milani, S., Herbst, H., Schuppan, D., Hahn, E.G. and Stein, H. (1989) Hepatology 10, 84-92.
Milani, S., Herbst, H., Schuppan, D., Kim, K.Y., Riecken E.O. and Stein, H. (1990) Gastroenterology 98, 175-184.
Mourelle, M., Muriel, P., Favari, L. and Franco, T. (1989) Fund. Clin. Pharmacol. 3, 183-191.
Park, C.H., Bacon, B.R., Brittenham, G.M. and Tavill, A.S. (1987)

Lab. Invest. 57, 555-563.
Paradisi, L., Panagini, C., Parola, M., Barrera, G. and Dianzani, M.U. (1985) Chem.-Biol. Interact. 53, 209-217.
Parola, G., Leonarduzzi, G., Biasi, F., Albano, E., Biocca, M.E., Poli, G. and Dianzani, M.U. (1992a) Hepatology 16, in press.
Parola, M., Muraca, R., Dianzani, I., Barrera, G., Leonarduzzi, G., Bendinelli P., Piccoletti, R. and Poli, G. (1992b) F.E.B.S. Lett. 308, 267-270.
Parola, M., Leonarduzzi, G., Biasi, F., Albano, E., Biocca, M.E., Poli, G., and Dianznai, M.U. (1992c) In Role of Free Radi cals in Biological Systems (Feher, J., Matkovics, B., Blazo vits, A., Mezes, M., eds.) Akademiai Kiadò, Budapest, in press.
Pinzani, M. and Abboud, H.E. (1991) In: Experimental and Clinical Hepatology (Gentilini, P., Dianzani, M.U., eds.) Elsevier Sci. Publ., Amsterdam, 63-75.
Poli, G., Albano, E. and Dianzani, M.U. (1987) Chem. Phys. Lipids 45, 117-142.
Poli, G., Albano, E., Dianzani, M.U., Melloni, E., Pontremoli, S., Marinari, U.M., Pronzato, M.A. and Cottalasso, D. (1988) Biochem. Biophys. Res. Commun. 153, 591-597.
Poli, G., Cottalasso, D., Pronzato, M.A., Chiarpotto, E., Biasi, F., Corongiu, F.P., Marinari, U.M., Nanni, G. and Dianzani, M.U. (1990) Cell Biochem. Funct. 8, 1-10.
Rojkind, M. and Greenwel, P. (1988) In: The Liver: Biology and Pathobiology (Arias, I.M., Jakoby, W.B., Popper, H., Schachter, D., Shafritz, D.A., eds.), Raven Press, New York, 1269-1285.
Rossi, M.A., Fidale, F., Garramone, A., Esterbauer, H. and Dianzani, M. (1990) Biochem. Pharmacol. 39, 1715-1719.
Schuppan, D. (1990) Sem. Liv. Dis. 10, 1-10.
Slater, T.F. (1984) Biochem. J. 222, 1-15.
Sokol, R.J., Deveraux, M., Mierau, G.W., Hambidge, K.M. and Shikes, R.H. (1990) Gastroenterology 99, 1061-1071.
Sokol, R.J., Deveraux, M. and Khandwala, R.A. (1991) J. Lip. Res. 32, 1349-1357.

Free Radicals: From Basic Science to Medicine
G. Poli, E. Albano & M. U. Dianzani (eds.)

ETHANOL-INDUCIBLE CYTOCHROME P450 2E1. REGULATION, RADICAL FORMATION AND TOXICOLOGICAL IMPORTANCE.

Magnus Ingelman-Sundberg.

Department of Physiological Chemistry, Karolinska institutet, S-104 01 Stockholm, Sweden.

SUMMARY: The microsomal monooxygenase system does produce reactive radicals that might cause cytotoxic effects. These radicals can originate from the oxidase activity, peroxygenase activity, reductase activity or hydroxylase activity of the components, mainly those of cytochrome P450. It appears that the isoform of cytochrome P450 which does produce radicals to the highest extent is ethanol-inducible cytochrome P450 2E1 (CYP2E1). This overview summarizes the results dealing with the importance of P450 for radical production and in particular that of CYP2E1.

THE CYTOCHROME P450 SUPERGENE FAMILY AND ITS EVOLUTION

Cytochrome P450 constitutes a superfamily of today 21 different gene families identified. The total number of cytochromes P450 cloned is over 170. In mammals, having in total ten different gene families, mainly P450s belonging to gene families 1-4 are active in the metabolism of xenobiotics. Of those, gene family 2 is the largest with more than 55 different members (Nebert et al., 1991). Every form of cytochrome P450 has a specific regulation and substrate specificity, although the latter property often overlaps between different isoforms.

All the different cytochromes P450 can be explained by the existence of one single ancestor gene. A broad spectrum of various P450s have subsequently evolved by the occurrence of gene duplications, gene conversions, insertions and mutations (see Gonzalez & Nebert, 1990). These authors have proposed that the diversity in the P450 genes is the result of continuous molecularly driven coevolution of plants producing phytoalexins and animals responding to this by the production of new enzymes to detoxify these chemicals. In favour of this hypothesis is the fact that there is evidence for numerous gene duplications about 400 millions years ago, i. e. at the time when animals became terrestrial.

REACTIONS CATALYZED BY CITOCHROME P450

The major functions of the P450s are as:

Monooxygenases:

$$RH + NADPH + H^+ + O_2 \longrightarrow ROH + NADP^+ + H_2O$$

Reductases:

$$2\ Cl_3\text{-}C\text{-}Cl + NADPH + H^+ \longrightarrow 2\ Cl_3\text{-}C. + 2\ Cl^- + NADP^+$$

Oxidases:

$$NADPH + H^+ + O_2 \longrightarrow NADP^+ + H_2O_2$$

$$2\ NADPH + 2\ H^+ + O_2 \longrightarrow 2\ NADP^+ + 2\ H_2O$$

Peroxidases:

$$RH + XOOH \longrightarrow ROH + XOH$$

By these different catalytic capabilities the different forms of P450 can contribute to radical formation which might be of importance for cytotoxic action of drugs and for lipid peroxidation processes.

CYTOCHROME P450 IN LIPID PEROXIDATION

There is an increasing evidence that cytochrome P450 can participate in cellular lipid peroxidation. There are numerous examples where the amount of P450 in various cell types correlates to the extent of lipid peroxidation in the cells.

P450 in initiation

The most established role for P450 in this step of lipid peroxidation is in reduction of haloalkanes yielding radical species that in turn can initiate lipid peroxidation. This kind of reaction might be true not only for carbon tetrachloride (Johansson & Ingelman-Sundberg, 1985) but also for eg chloroform, trichloroethylene and similar agents. In the ADP-Fe^{3+}-dependent peroxidation system it has been suggested that P450 has an important role in the reduction of the ADP-iron complex yielding species that are able to initiate lipid peroxidation (Sevanian et al., 1990). In addition, it appears that P450 can contribute to initiation of lipid peroxidation by release of superoxide anions that react with non heme iron to form a species being able to initiate the process (Ekström & Ingelman-Sundberg, 1986, 1989).

P450 in propagation

It was early proposed that P450 participated in propagation of lipid peroxidation (Svingen et al., 1979). Results by Weiss & Estabrook (1986) indeed strongly indicate that P450 might cleave peroxides homolytically yielding radicals that might in turn stimulate further lipid peroxidation.

P450 in prevention of lipid peroxidation

The peroxygenase activities of P450 surely contributes to inactivation of lipid peroxidation. In addition, Sevanian et al. (1990) proposed that P450 could inactivate peroxyradicals with the formation of the corresponding hydroperoxide. Furthermore, Vaz and Coon (1987) have shown that P450 can reduce peroxides to yield alkanes in a reaction which is catalyzed by various isozymes at different efficiencies, with CYP2E1 being the most potent one.

P450 and metabolism of products from lipid peroxidation

Ethanol-inducible P450 CYP2E1 has been shown to effectively metabolize the common lipid peroxidation product pentane with a very high affinity (Terelius & Ingelman-Sundberg, 1986). In addition, it has been found that acetaldehyde provides an efficient CYP2E1 substrate (Terelius et al., 1991) and indicates that also other aldehydic products from the lipid peroxidation processes would constitute substrates for P450.

The species of P450 found to be most effective in radical formation and initiation of lipid peroxidation has been ethanol-inducible cytochrome P4502E1.

ETHANOL-INDUCIBLE CYTOCHROME P450, CYP2E1

CYP2E1, general characteristics

Ethanol-inducible cytochrome P450, termed CYP2E1, was first isolated from rabbit liver in 1982 (Koop et al., 1982; Ingelman-Sundberg & Hagbjörk, 1982). It turned out that the enzyme was active in the oxidation of ethanol (Morgan et al., 1982) and thus constituted the major part of the molecular basis for the alcohol-induced microsomal oxidation of ethanol. The corresponding enzyme has subsequently been isolated from rat, human and mouse liver.

CYP2E1 is distributed in the highest concentrations in the liver, followed by the kidneys, whereas significant amonts are also found in most organs investigated, among them the brain. Here, higher amounts are found in the hippocampus, in particular the CA3 region, in substantia nigra and in the striatum (Hansson et al., 1990). The enzyme is localized in neuronal cell bodies, fibers and terminals as well as in some glial cells. Also, a large number of small and large blood vessels, mainly the endothelial cells, throughout the brain contain CYP2E1 (Hansson et al., 1990).

CYP2E1 is not homogeneously distributed within the liver acinus. Clearly, the expression of the enzyme, both constitutively and after induction with e. g. ethanol, is restricted to the centrilobular region of the liver and, in particular, to the three-to four layers of hepatocytes most proximal to the central vein (Ingelman-Sundberg et al., 1988). It can be calculated that the concentration of CYP2E1 in these hepatocytes after induction is as high as about 0.1 mM. The molecular basis for this heterogenous distribution has been shown to be inherent in a regioselective expression of the CYP2E1 gene in the perivenous hepatocytes (Johansson et al., 1990). The transcriptional factors responsible for this are unknown, but our recent data indicate that an identical acinar distribution, i e expression in exactely the same hepatocytes, is evident also for CYP2B1 and CYP3A1 (Bühler et al., 1991).

The regioselective hepatic CYP2E1 expression is of interest because ethanol, acetaminophen, N-nitrosoamines, solvents and other CYP2E1-specific substrates cause a selective destruction of the centrilobular region. It appears probable that the localization of CYP2E1 to a great extent determines the site of their toxic action, perhaps also in the brain.

Substrate specificity of CYP2E1

Today over 80 different chemicals of diverse structures have been found to selectively be metabolized by CYP2E1 (Table I). These include drugs, such as halothane, acetaminophen, organic solvents, nitrosoamines and aliphatic alcohols. It is difficult to find a common structure among these substrates, but it appears that small and hydrophobic compounds in general provide efficient targets for CYP2E1-dependent catalysis. In comparision to specificities exerted by other P450s, it is striking that the rate of metabolism of several different exogenous compounds by CYP2E1 is so much higher than the rate exerted by any other isozyme. If one consider metabolism of solvents, several of those have a very high affinity for CYP2E1. Thus the apparent Km for trichloroethylene, chloroform and dichloromethane are 3, 4 and 10 mM, respectively (Terelius & Ingelman-Sundberg, 1991). This implies that, in relation to the concentrations of organic solvents in blood after moderate human exposure for these chemicals, indeed CYP2E1 has an important role in this kind of solvent metabolism. This relationship is further strengthened by the fact that ethanol intake, as well as treatment with a variety of CYP2E1 inducers, have a pronounced synergistic effect on the hepatotoxicity of these solvents.

Table I

Examples of substrates for CYP2E1

1. Aromatic compounds: Benzene, Phenol, Aniline, Toluene, Acetaminophen, p-Nitrophenol
2. Halogenated alkanes/alkanes: Chloroform, Carbon tetrachloride, Trichloroethylene, Halothane, Pentane
3. Alcohols/ketones/aldehydes: Ethanol, 2-Propanol, 1-Butanol, Acetone, Acetol, Acetaldehyde
4. Nitrosamines: N,N-dimethylnitrosamine, N,N-diethylnitrosamine

CYP2E1 and oxygen radical production

In 1981 it was proposed that the microsomal oxidation of ethanol, as studied with CYP2B4, was carried out to the main part in a Haber-Weiss catalyzed reaction, whereby hydroxyl radicals generated from superoxide anions and hydrogen peroxide oxidized the alcohol (Ingelman-Sundberg & Johansson, 1981). If P450 induced by alcohol would oxidize ethanol by a similar mechanism, the induction of such an enzyme species might by itself be harmful to the cell. Indeed it was found that ethanol-inducible CYP2E1 did generate much more reactive oxygen species than other P450-forms (Ingelman-Sundberg & Hagbjörk, 1982; Ingelman-Sundberg & Johansson, 1984). It turned out that CYP2E1 was a "leaky" enzyme and had the capability to effectively reduce dioxygen (Ingelman-Sundberg & Johansson, 1984, Gorsky & Coon, 1984). The same capacity is utilized by the enzyme in its reductive metabolism of carbon tetrachloride to chloroform and reduction of dioxygen to reactive species (Johansson & Ingelman-Sundberg, 1985). Thus, dioxygen and carbon tetrachloride compete with each other for CYP2E1 electrons (Persson et al., 1990).

It appeared that oxygen radicals generated by CYP2E1 have the capability to initiate membranous lipid peroxidation (Ekström & Ingelman-Sundberg, 1989). Thus, this isozyme turned out to be the most efficient isozyme in the initiation of NADPH dependent lipid peroxidation in reconstituted membranes among five different P450 forms investigated. Furthermore, anti CYP2E1 IgG inhibited microsomal lipid peroxidation dependent on P450, but not lipid peroxidation initiated by the action of NADPH-cytochrome P450 reductase (Ekström & Ingelman-Sundberg, 1989). The rate of NADPH consumption, and formation of either hydrogen peroxide or superoxide anions in 42 different microsomal preparations from variously treated rats, correlated well with the amount of CYP2E1 being present in the membranes.

Our finding about CYP2E1 as such an efficient consumer of dioxygen prompted us to investigate whether the enzyme actually was induced by oxygen treatment of rats. A four-fold induction of CYP2E1 was achieved in both liver and lung microsomes after 60h treatment of rats with 95% oxygen (Tindberg & Ingelman-Sundberg, 1989). The increase of CYP2E1 accounted for 60% of the total increase of P450 in liver microsomes and the CYP2E1 level in these microsomes was about 0.2 nmol/mg. Pretreatment of the rats with acetone caused a significant reduction of the survival time of the rats in the oxygen athmosphere, from 78 hours to about 60 hours.

The conclusion from these studies are that indeed CYP2E1 generates reactive oxygen species that have the capability to initiate microsomal lipid peroxidation, and thus that induction of this enzyme might contribute to oxidative stress in the cells. This might be relevant for the situation in the the liver after ethanol-treatment and perhaps also in other organs, e.g. the brain.

CYP2E1 and liver cirrhosis

Using an intragastric infusion model for ethanol supply to rats, Samuel French and coworkers have developed a model that actually can induce alcohol related liver damage of similar type

Possible relationship between CYP2E1 and hepatic damage

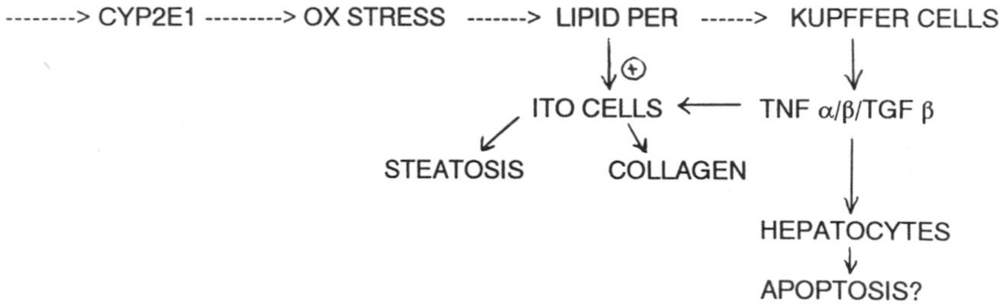

--------> CYP2E1 ---------> OX STRESS -------> LIPID PER ------> KUPFFER CELLS

(+)

ITO CELLS ⟵ TNF α/β/TGF β

STEATOSIS COLLAGEN

HEPATOCYTES

APOPTOSIS?

Fig 1. Hypothetical mechanisms for action of CYP2E1 in development of alcohol-induced liver injury. A central point of action appears to be related to the activation of Ito cells and Kupffer cells by lipid peroxidation products.

as registered in the human (Tsukamoto et al.,1985). Using this TEN (total enteral nutrition)-model they have shown that ethanol-treatment in combination with corn oil gives severe liver damage after about 2 month of treatment. However, replacing corn oil with tallow completely prevents from these sorts of liver damage (Nanji et al., 1989). This effect has been suggested to be attributed to the difference in linolenic acid content between tallow

(0.7%) and corn oil (37%) (Nanji et al.,1989). Examination of the extent of CYP2E1 induction in the livers from such rats revealed a 2-fold higher CYP2E1 content in livers from rats fed corn oil, as compared to tallow, in combination with ethanol in the TEN model (Takahashi et al., 1991). Together with the studies mentioned above, it becomes evident that linolenic acid has a synergistic effect on the ethanol-dependent induction of CYP2E1 and that this fatty acid also has the capability to induce liver damage in combination with the alcohol. The relationship between CYP2E1 generated lipid peroxides and hepatotoxicity might involve activation of Ito cells and Kupffer cells as outlined in Figure 1. Several recent reports, presented at a Symposium about hepatic liver damage at the RSA-meeting in La Jolla June 1992 (Tsukamoto (Ed.), support this scheme.

Regulation of CYP2E1

The expression of CYP2E1 is very much influenced by hormonal and nutritional factors. A summary of the modes of CYP2E1 regulation is given in Figure 2. The CYP2E1 gene is inactive under the prenatal period, but is activated at birth. The mechanism of gene activation is unknown, but this activation is accompanied by demethylation of cytosine residues located within the 5∩ end of the gene (Umeno et al., 1988). The promotor region of CYP2E1 contains a motif specific for hepatocyte nuclear factor i (HNF-1) about 100 bp upstreams transcriptional start and it might be that this transcription factor is important for tissue specific expression of the enzyme (Ueno & Gonzalez, 1990). Furthermore, this part of the promotor region, which is conserved between human and rat, does not bind any nuclear protein in the brain (Ooi & Ingelman-Sundberg, 1990). By contrast, a brain specific motif has been identified 40 bp upstreams that bind brain nuclear proteins, but not any nuclear hepatic protein (Ooi & Ingelman-Sundberg, 1990). These findings can form part of the explanation for the pronounced different modes of regulation of CYP2E1 in brain as compared to liver.

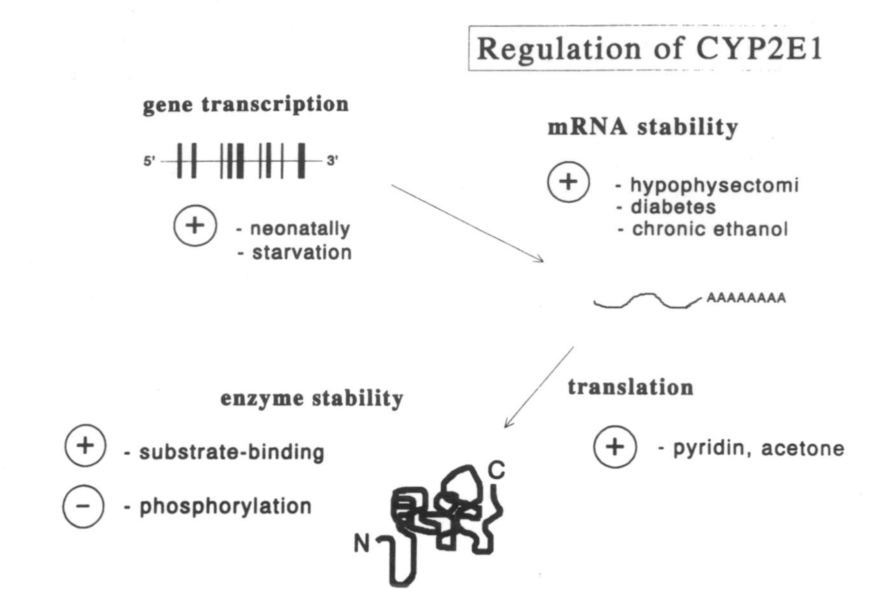

Fig 2. Cellular levels for the regulation of CYP2E1.

Starvation of rats is known to increase the amount of CYP2E1 and of the corresponding mRNA (Hong et al., 1987). Furthermore, starvation does exert a pronounced synergistic effect on the induction of CYP2E1 by eg ethanol, which is manifested in more CYP2E1 mRNA (Johansson et al., 1988). An evaluation about the mechanism behind this starvation effect revealed that the CYP2E1 gene was 3-4-fold more active in nuclei isolated from starved as compared to control rats (Johansson et al., 1990). This difference corresponds to the difference in mRNA between starved and control rats previously registered.

Regulation of CYP2E1 at posttranscriptional levels

A major level for regulation of CYP2E1 appears to be post-translational (Figure 3). CYP2E1-specific substrates have the capability to protect the enzyme from degradation in hepatocytes

(Eliasson et al., 1988). This protection takes place at an efficiency correlating to the binding affinity of the compounds.

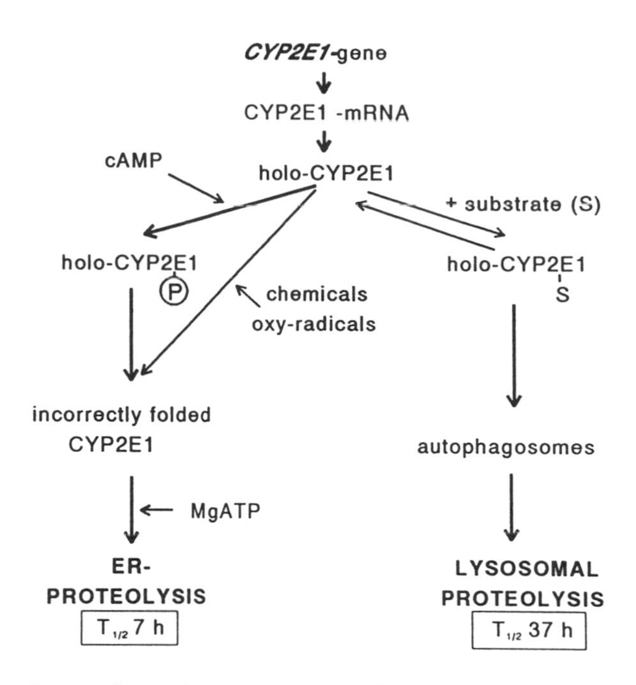

Fig 3. Mechanisms for the posttranslational regulation of CYP2E1 in rat liver. The rate of degradation is regulated by the action of CYP2E1 specific substrates and by cAMP. In the absence of substrate phosphorylation of the enzyme on Ser129 leads to heme loss and rapid degradation in the endoplasmic reticulum. In the presence of substrate, this phosphorylation reaction is prevented and CYP2E1 is degraded at a much slower rate according to the autophagosomal-lysosomal pathway.

The mechanism involves a specific inhibition by the substrates of a cAMP-dependent phosphorylation of CYP2E1 on Ser129 (Eliasson et al., 1990). In the absence of substrate it appears that the heme is lost from the enzyme (Eliasson et al., 1990) and that CYP2E1 is rapidly degraded. This degradation is enhanced by hormones like glucagon and adrenaline which cause elevated cAMP-levels (Johansson et al., 1991). Recent evidence in our laboratory indicate that this fast phase of degradation takes place in the endoplasmic reticulum (Eliasson et al., 1992), whereas the enzyme in the substrate-protected state is degraded at a slower rate via the autophagosomal-lysosomal pathway (Ronis et al., 1991). Thus, CYP2E1 can provide its own receptor for its substrate in the regulation of its rate of degradation.

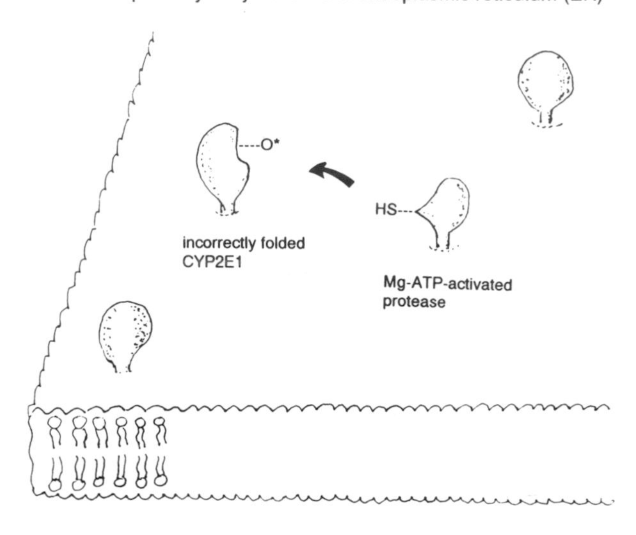

Fig 4. Hypothetical meachnism for the action of Mg-ATP-activated proteolysis of CYP2E1 in the endoplasmic reticulum membranes. The membrane bound protease is integrated with the microsomal monooxygenase system and does only recognize improperly foled variants of the P450 enzymes.

Since the rate of CYP2E1 synthesis continues, the ligand-dependent stabilization mechanism provides a very fast way whereby elevated levels of CYP2E1 occurs, ie a doubling in amount about 2 hours after introduction of the inducer (Ronis et al. 1991).
The fast degradation of CYP2E1 also occurs if the enzyme is modified by other means, eg by reaction with reactive oxy-radicals or reactive intermediates, that appear to change the proper folding of the enzyme, making it an efficient substrate for the ER-bound proteolytic system (Fig 4).

Table II

Characteristics of ethanol-inducible cytochrome P450 2E1

* Produces oxygen radicals that are able to initiate lipid peroxidation

* Metabolically activates paracetamol, organic solvents and N-dimethylnitrosoamine

* Catalyzes oxidation of ethanol and acetaldehyde

* Localized solely in the centrilobular region of the liver acinus where it can reach a very high concentration (0.1 mM)

* Induced by its substrates, by diabetes, by starvation and by hypophysectomy by regulation at both transcriptional, translational and posttranslational levels

* An important level of CYP2E1 regulation is substrate and cAMP-regulated phosphorylation on Ser129 and subsequent proteolytic degradation by a Mg-ATP-dependent proteolytic system in the ER

Genetic polymorphism of CYP2E1

Recent results in our lab indicate that the CYP2E1 gene is polymorphically distributed among Caucasians as evident from RFLP and eg the restriction enzymes Taq I, Rsa I and Dra I (Persson et al., in preparation). The RFLP:s using these three restriction enzymes are only partially linked and the frequencies of the rare alleles are in the range of 0.10 to 0.17. Examination of livers from kidney donors has revealed that the distribution of one of these polymorphic alleles is correlated to decreased rate of radical production and ethanol oxidation by CYP2E1. Future research will be carried out in order to evaluate whether this can form a basis for the differences in individual susceptibility for alcohol-induced liver cirrhosis.

CONCLUSIONS

The properties of CYP2E1 are summarized in Table II. It is

evident that CYP2E1, because of its participation in toxicologically important reactions, appears to be the most studied isoform of cytochrome P450 at present. The stage of CYP2E1 research now appears to be at the level where specific inhibitors should be developed in order to provide us with detailed knowledge about the contribution of CYP2E1 radical mediated reactions to alcohol, drug precarcinogen and solvent toxicity under in vivo conditions.

Acknowledgements
The work in the authors laboratory was supported by grants from the Swedish Alcohol Research Fund and from the Swedish Medical Research Council.

References

Bühler, R., Lindros, K.O., Johansson, Å., Johansson, I. and Ingelman-Sundberg, M. (1991) Eur. J. Biochem. 204, 407-412.
Ekström, G. and Ingelman-Sundberg, M., (1986) Eur J Biochem, 158, 195-201.
Ekström, G. and Ingelman-Sundberg, M, (1989)Biochem Pharmacol,38, 313-1319.
Eliasson, E., Johansson, I. and Ingelman-Sundberg, M., (1988) Biochem. Biophys. Res. Commun. 150, 436-443.
Eliasson, E., Johansson, I. and Ingelman-Sundberg, M., (1990) Proc Natl Acad Sci (USA) 87, 3225-3227.
Gonzalez, F.J. and Nebert, D.W., (1990)Trends Genetics, 6, 182-186.
Gorsky, L.D., Koop, D.R. and Coon, M.J., (1984) J Biol Chem, 259, 6812-6817.
Hansson, T., Tindberg, N., Ingelman-Sundberg, M and Köhler, C., (1990) Neuroscience 34, 451-463.
Hong, J., Pan, J., Gonzalez, F.J., Gelboin, H.V. and Yang, C.S., (1987) Biochem. Biophys. Res. Commun. 142, 1077-1083.
Ingelman-Sundberg, M. and Hagbjörk, A.-L., (1982) Xenobiotica, 12, 673-686.
Ingelman-Sundberg, M., Johansson, I., Penttilä, K.E., Glaumann, H. and Lindros, K.O. (1988) Biochem. Biophys. Res. Commn. 157, 55-60.
Ingelman-Sundberg, M. and Johansson, I., (1981) J. Biol. Chem., 256, 6321-6326.
Ingelman-Sundberg, M. and Johansson, I., (1984) J. Biol. Chem., 259, 6447-6458.
Johansson, I., Lindros, K.O., Eriksson, H. och Ingelman-Sundberg, M., (1990) Biochem. Biophys. Res. Commun. 173, 331-338.
Johansson, I. and Ingelman-Sundberg, M., (1985) FEBS letters, 183, 265-269.
Johansson, I., Eliasson, E. and Ingelman-Sundberg, M., (1991) Biochem. Biophys. Res. Commun. 174, 37-42.

Johansson, I., Ekström, G., Scholte, B., Puzycki, D., Jörnvall, H. and Ingelman-Sundberg, M. (1988) Biochemistry, 27, 1925-1932.
Morgan, E.T., Koop, D.R. and Coon, M.J., (1982) J. Biol. Chem., 257, 13951-13957.
Nanji, A.A., Mendenhall, C.L. and French, S.W., (1989) Alcoholism: Clin Exp Res, 13, 15-19.
Nebert, D.W., Nelson, D.R., Coon, M.J., Estabrook, R.W., Feyereisen, R., Fujii-Kriyama, Y., Gonzales, F.J., Guengrich, F.P., Gunsalus, I.C., Johson, E.F., Loper, J.C., Sato, R., Waterman, N. and Waxman, D.J. (1991) DNA Cell Biol., 10, 1-10.
Ooi, Y. and Ingelman-Sundberg, M., (1990) in Drug Metabolizing Enzymes: Genetics, Regulation and Toxicology (Ingelman-Sundberg, M., Gustfsson, J.-A and Orrenius, S., Eds.) Karolinska Institutet, p.86.
Persson, J.-O., Terelius, Y. and Ingelman-Sundberg, M., (1990) Xenobiotica, 20, 887-900.
Ronis, M.J.J, Johansson, I., Hultenby, K., Lagercrantz, J., Glaumann, H. and Ingelman-Sundberg, M., (1991) Eur. J. Biochem. 198, 383-389.
Sundberg, M., (1991) Eur. J. Biochem. 198, 383-389.
Ryan, D.E., Ramanthan, L., Iida, S., Thomas, P.E., Haniu, M., Shively, J.E., Lieber, C.S. and Levin, W., (1985) J. Biol. Chem, 260, 6385-6393.
Sevanian, A., Nordenbrand, K., Kim, E., Ernster, L. and Hochstein, P. (1990) Free Rad. Biol. Med. 8, 145-152.
Svingen, B.A., Buege, J.A., OnNeal, F.O. and Aust, S.D. (1979) J. Biol Chem 254, 5892-5899.
Takahashi, J., Johansson, I., French, S.W. and Ingelman-Sundberg, M., (1992) Pharmacology and Toxicology, 70, 347-352.
Terelius, Y. and Ingelman-Sundberg, M., Europ. J. Biochem. 161, 303-308.
Terelius, Y. and Ingelman-Sundberg, M., (1992) , submitted.
Terelius, Y., Norsten-Höög, C., Cronholm, T. and Ingelman-Sundberg, M., (1991) Biochem. Biophys. Res. Commun., 179, 689-694.
Tindberg N., and Ingelman-Sundberg, (1989) Biochemistry 28, 4499-4504.
Tsukamoto, H., French, S.W., Benson, N., Delgado, G., Rao, G.A., Larkin, E.C. and Largman, C., (1985) Hepatology, 5, 224-232.
Ueno, T. and Gonzalez, F.J., (1990) Mol Cell Biol, 10, 4495-4505.
Umeno, M., McBridge, O.W., Yang, C.S., Gelboin, H.V. and Gonzalez, F.J., (1988) Biol. Chem. 263, 4956-4962.
Vaz, AA. and Coon. M.J. (1987) Proc Natl Acad Sci (USA) 84, 1172-1176.
Weiss, R.H. and Estabrook, R.W. (1986) Arch Biochem Biophys 251, 336-347.

Free Radicals: From Basic Science to Medicine
G. Poli, E. Albano & M. U. Dianzani (eds.)

FREE RADICAL-INDUCED IMPAIRMENT OF LIVER GLYCOSYLATION PROCESSES IN ETHANOL INTOXICATION

G. Nanni, D. Cottalasso, D. Dapino, C. Domenicotti, M.A. Pronzato, and U.M. Marinari

Institute of General Pathology, University of Genova, Italy

Summary. Ethanol intoxication selectively and precociously affects the intracellular system of glycosylation, maturation and secretion of glycoproteins at the level of the liver endoplasmic reticulum and Golgi apparatus. Both uptake and release of glycoprotein precursors are impaired early on, only 1.5 h after ethanol administration, and they are completely inhibited within 6 h; at this time the level of each glycoprotein oligosaccharide class shows a significant reduction, the galactosyl- and sialyl-transferase activities and the amount of dolichols drop to values significantly lower than the control group. The pathogenesis of such damage has been ascribed to free radical mechanisms activated during ethanol metabolism.

The liver is the most important target of ethanol-induced injury (Comporti, 1978). Several studies have demonstrated that this hepatotoxin is capable of inhibiting the synthesis of both constituent and secretory proteins (Sorrell et al., 1979; Tuma et al., 1981; Baraona et al, 1980).
Our previous investigations indicated that acute ethanol intoxication induces also an impairment of the lipoprotein synthesis and metabolism at the level of the liver Golgi apparatus (Casu et al., 1982; Marinari et al., 1984). It has been established that plasma lipoproteins are also glycoproteins (Lo et al., 1970). In the liver the protein and lipid moieties are first synthesized in the endoplasmic reticulum and then transferred as very low density lipoprotein particles (VLDL) to the Golgi apparatus where their terminal glycosylation occurs (Stein et al., 1967; Wetmore et al., 1974; Dolphin et al., 1977).

Nearly all the steps concerning the terminal glycosylation of proteins occur in the Golgi apparatus where the distal sugars, N-acetylglucosamine, galactose and sialic acid are added via their nucleotide precursors (Hirschberg et al., 1987). A specific multiglycosyltransferase system is required for the synthesis of each type of glycoprotein (Schachter et al., 1980) and most of the glycosyl-transferase enzymes have been recognized to be associated with liver Golgi membranes (Bretz et al., 1980; Fleischer 1981). It has been suggested that UDP-galactose-N-acetylglucosamine galactosyltransferase is the most typical marker enzyme for the Golgi apparatus (Fleischer 1981; Kaplan et al., 1984). The completed glycoproteins are then packaged into the secretory vesicles which fuse with the plasma membrane and discharge the secretory content into the blood stream (Whaley et al., 1979; Hanover et al., 1981; Rothmann, 1981). Unlike the terminal glycosylation, the initial steps of asparagine-linked oligosaccharide synthesis start in the rough endoplasmic reticulum and involve the formation of dolichols which are a family of long chain polyisoprenoid alcohols usually containing 14-24 isoprene units, with the α-isoprene unit being saturated. They exist as free dolichol, dolichyl phosphate, and fatty acyl dolichyl esters (Hemming, 1983; Rip et al., 1985).

Most of the dolichols found in mammalian tissues are present either as free dolichols or as dolichyl esters of fatty acids, which generally account for more than 90% of total dolichol in membranes (Tollbom et al., 1986). In the cell, dolichyl phosphate represents less than 14% of total dolichols and its presence is restricted to the membranes of the endoplasmic reticulum (Rip et al., 1983). A function for such a phosphorylated compound in this membrane area has been established as a glycosyl carrier during the membrane-directed biosynthesis of the N-linked oligosacccharide chains of glycoproteins (Struck et al., 1980; Kornfeld et al., 1985).

In this step of the glycosylation process, dolichyl-phosphate plays an important role in the translocation of activated glycosyl units across the membranes of the rough endoplasmic reticulum; in fact all the sugar residues may cross the endoplasmic reticulum from the cytoplasm to the lumen by means of flip-flop movements of the lipid carrier (Kornfeld et al., 1985; Hirschberg, 1987). The highest concentration of neutral dolichols occurs in the Golgi apparatus (Rip et al., 1983) but little is known about their role in these membranes where they might act either on the terminal glycoprotein processing or on their secretion.

Experiments performed in our laboratory have shown, by monitoring the time sequence of $[^{14}C]$ glucosamine incorporation into the secretory (F1, F2) and formative (F3) fractions of the Golgi apparatus isolated from ethanol-intoxicated female rats (6 g/kg body weight), had demonstrated that both the uptake and the release of this labelled glycoprotein precursor are impaired early on (after 1.5 h of treatment). The glycoprotein release was inhibited after 6 h, especially at the level of the secretory compartments. At the same time, a significant reduction of galactosyltransferase activity was observed. The carbohydrate-protein ratio decreased in all the Golgi fractions. The greatest decrease was observed in the secretory compartments and in their soluble contents where the amounts of hexosamines, neutral sugars and sialic acid dropped to values significantly lower than the control group. The chromatographic separation of individual oligosaccharides showed that the decrease was more evident for the distal sugars (Nanni et al., 1978, 1986; Marinari et al., 1985). The polypeptide profiles of Golgi VLDLs showed that, after ethanol intoxication, an overlapping of two apolipoprotein bands occur (Casu et al., 1982) suggesting that the glycosylation of an apoprotein may have undergone modifications. In fact, carbohydrates are known to affect the electrophoretic mobility of proteins in SDS gels and to characterize polymorphic forms of rat apolipoproteins CIII and apo- ARP (Dolphin et al., 1977).

These data indicate that ethanol may selectively and precociously impair the intracellular glycosylation system at the level of both the endoplasmic reticulum and Golgi apparatus. Several mechanisms could be involved in the pathogenesis of such impairment: ethanol might affect the glycosylation process by inhibiting either the transmembrane movements of sugar nucleotides from the cytoplasm to the Golgi lumen or by reducing the rate of dolichol-mediated synthethic reactions.

In order to investigate these possibilities, the effects of ethanol on the levels and distribution of dolichols in the rat liver endoplasmic reticulum and Golgi apparatus have been studied. HPLC analysis showed that in the microsomal compartment the amount of total dolichols decreases gradually and between 6 and 24 h was significantly lower than that present in microsomes from unintoxicated rats (52% reduction at 12 h). Within the microsomal subfractions, the rough and smooth endoplasmic reticulum is enriched in dolichyl phosphate, which represents 54% and 36% of total dolichols, respectively. In our experiments the levels of dolichyl phosphate represent about 50% of the microsome total dolichols. These values indicate a good correlation between the high enrichment of microsomes in dolichyl-phosphate and its involvement in the biosynthesis of N-linked glycoproteins (Chojnacki et al., 1988). The quantitation of dolichyl phosphate and free dolichol purified from microsomal extracts showed a similar rate of reduction (25%) at 6 h after ethanol administration. Since the concentration of dolichyl phosphate appears to be rate-limiting for the synthesis of glycoproteins (Hanover et al., 1982), our results suggest that ethanol intoxication can impair the functions of microsomal dolichyl-phosphate, which include both the formation of the precursor oligosaccharide (Glc NAc)2-(Man)9-(Glc)3 and the sugar transport from the cytosol to the membranes of the endoplasmic reticulum (Hanover et al., 1982; Kornfeld et al., 1985). Only 3% of dolichyl-phosphate has been found in the Golgi apparatus (Rip et al., 1985) whereas this compartment has the highest content in neutral dolichols, which are strongly enriched with respect to microsomes (Fig.1) (Cottalasso et al., 1990).

In the Golgi apparatus the dolichol amount decreased early with respect to microsomes within 0.5 and 1.5 h and at 6 h the values were 69%, 47% and 50% lower than controls in the secretory frac-

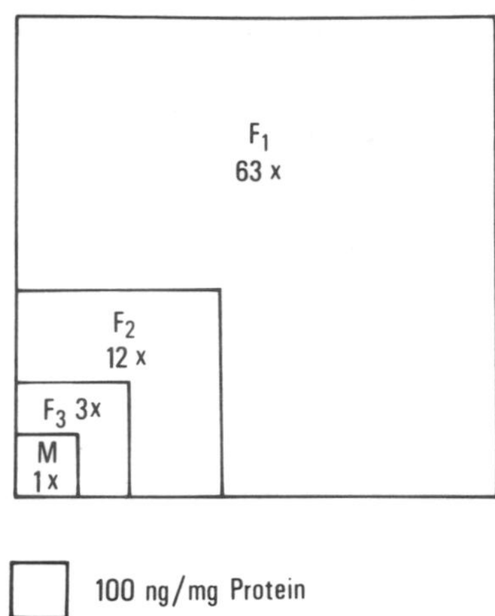

Figure 1. Enrichment of total Dolichol levels in rat liver Golgi apparatus fractions in comparison with microsomes.
F1, secretory side (7544±376); F2, intermediate side (1405±168); F3, formative side (421±45); M, total microsomes (120±15).

tions F1 and F2 and in the formative fraction F3, respectively (Fig.2).

Although the relationship between the Golgi apparatus and dolichols is still not fully understood, the above results suggest that both dolichyl phosphate and neutral dolichols are involved in the ethanol induced impairment of glycoprotein glycosylation and maturation in the endoplasmic reticulum and Golgi apparatus, respectively. These alterations in dolichol content have been correlated to the possible ethanol-induced stimulation of prooxidant events. In fact, several pieces of evidence indicate that during ethanol metabolism the production of free radicals may initiate the process of lipid peroxidation (Albano et al., 1988) which appears the major event involved in the inactivation of glycosylation reactions (Poli et al., 1990). This is suggested both by the susceptibility of Golgi galactosyl- and sialyltransferase activities to the toxic action of aldehydic products of lipid peroxidation (Marinari et al., 1987) and by the

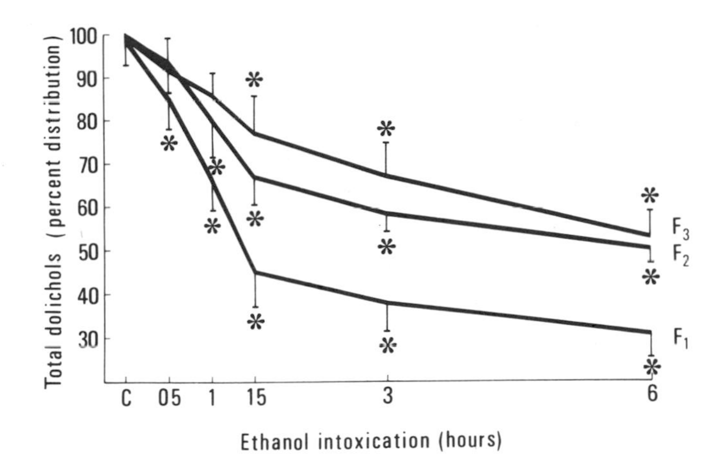

Figure 2. Time course of total dolichol levels in Golgi secretory membranes (F1,F2) and in formative ones (F3) of ethanol-intoxicated rats. The values present the percentage decrease as compared to corresponding controls. Vertical bars denote SD of the means of four to eight experiments. *P<0.01 as compared to control.

marked protection of these activities afforded by rat supplementation with vitamin E (Cottalasso et al., 1984; Poli et al., 1990). Moreover, a good prevention of the ethanol-induced decrease in dolichol levels was observed in animals pretreated with vitamin E (Pronzato et al., 1989) thus indicating that lipid peroxidation plays an important role in the degradation of dolichols. The functional significance of dolichols may be related to the influence exerted on the organization and assembly of cellular and subcellular membranes. In fact, these compounds destabilize lipid membranes and increase the fluidity of phospholipids in bilayers and the permeability properties of membranes (Valtersson et al., 1985; Van Duijn et al., 1986; Monti et al., 1987). These effects exertedby dolichols in membranes may explain their role both in membrane movements between the Golgi region and plasma membrane and in the glycosylation reactions (Hanover et al., 1981; Hirschberg et al., 1987) and may justify their decrease in ethanol-induced liver injury.

On the other hand, the glycoprotein secretory defect observed in ethanol intoxication may be dependent on alterations in the final steps of secretion which occur in other post-Golgi sites (Volentine et al., 1984; Tuma et al., 1986). Thus, ethanol seems able to modify both the early steps of glycoprotein biosynthesis, glycosylation and packaging (pre-Golgi and Golgi-processes) and the microtubular steps of secretion (post-Golgi processes).

Acknowledgements
This work was supported by CNR P.F. FATMA and by MPI 40% and 60% Grants.

REFERENCES

Albano, E., Tomasi, A., Goria Gatti, L. and Dianzani, M.U. (1988) Chem. Biol. Interact. 65, 233-236.
Baraona, E., Pikkarainen, P., Salaspuro, M., Finkelman, F. and Lieber, C.S. (1980) Gastroenterology 79, 104-111.
Bretz, R., Bretz, H. and Palade, G.E. (1980) J. Cell Biol. 84, 87-101.
Casu, A., Cottalasso, D., Pronzato, M.A., Marinari, U.M. and Nanni, G. (1982) Exp. Path. 22, 173-177.
Chojnacki, T. and Dallner, G. (1988) Biochem. J. 251, 1-9.
Comporti, M. (1978) in "Biochemical mechanisms of liver injury", Slater T.F. eds., Academic Press, New York, pp. 449-516.
Cottalasso, D., Pronzato, M.A., Domenicotti, C., Nanni, G. and Marinari, U.M. (1990) in "Chronic liver damage", Dianzani M.U. and Gentilini P. eds.,Elsevier North Holland, pp.39-50.
Cottalasso, D., Pronzato, M.A., Rolla, C., Marinari, U.M., Nanni, G., Chiarpotto, E., Biasi, F., Albano, E., Poli, G. and Dianzani, M.U. (1984) IRCS Med. Sci. 12, 904-905.
Dolphin, P.J. and Rubinstein, D. (1977) Can. J. Biochem. 55, 83-90.
Fleischer, B. (1981) J. Cell Biol. 89, 246-255.
Hanover, J.A. and Lennarz, W.J. (1981) Archs. Biochem. Biophys. 211, 1-19.
Hanover, J.A. and Lennarz, W.J. (1982) J. Biol. Chem. 257, 2787-2794.
Hemming, F.W. (1983) in "Biosynthsis of isoprenoid compounds", Potter J.W.L., Spurgeon S.L. eds., vol.2, John Viley, New York, pp. 305-354.
Hirschberg, C.B. and Snider, M.D. (1987) Annu. Rev. Biochim. 56, 63-87.
Kaplan, F. and Hechtman, P. (1984) Biochem. J. 217, 353-364.
Kornfeld, R. and Kornfeld, S. (1985) Annu. Rev. Biochem. 54, 631-644.
Marinari, U.M., Casu, A., Averame, M.M., Cottalasso, D., Pronzato, M.A. and Nanni, G. (1984) Front. Gastrointest. Res. 8, 24-45.
Marinari, U.M., Pronzato, M.A., Cottalasso, D., Cetta, G., Zanaboni, G. and Nanni, G. (1985) Ital. J. Gastroenterol.

17, 48.
Marinari, U.M., Pronzato, M.A, Cottalasso, D., Rolla, C., Biasi, F., Poli, G., Nanni G. and Dianzani, M.U. (1987) Free Rad. Res. Commun. 3, 319-324.
Monti, J.A. , Christian, S.T. and Schutzbach, J.S. (1987) Biochim. Biophys. Acta 905, 133-142.
Nanni, G., Cottalasso, D., Cetta, G., Pronzato, M.A. and Marinari, U.M. (1986) Front. Gastrointest. Res. 9, 50-69.
Nanni, G., Pronzato, M.A., Averame, M.M., Gambella, G., Cottalasso, D. and Marinari, U.M. (1978) FEBS Lett. 93, 242-246.
Poli, G., Cottalasso, D., Pronzato, M.A., Chiarpotto, E., Biasi, F., Corongiu, F.P., Marinari, U.M., Nanni, G. and Dianzani, M.U. (1990) Cell Biochem. and Funct. 8, 1-10.
Pronzato, M.A., Cottalasso, D., Domenicotti, C., Marinari, U.M. and Nanni, G. (1989) Alcohol and Alcoholism, 24, 384.
Rip, J.W., Chaudary, N. and Carroll, K.K. (1983) Can. J. Biochem. Cell Biol. 61, 1025-1031.
Rip, J.W., Rupar, C.A., Ravi, K., Carroll, K.K. (1985) Prog. Lipid Res. 24, 269-309.
Rothmann, J.E. (1981) Science 213, 1212-1219.
Schachter, H. and Roseman. S. (1980) in "The biochemistry of glycoproteins and proteoglycans", Lennarz ed., Plenum Publishing, New York, pp.85-160.
Sorrell, M.F. and Tuma, D.J. (1979) Clin. Sci. 57, 481-489.
Stein, O. and Stein, Y. (1967) J. Cell Biol. 33, 319-339.
Struck, D.K. and Lennarz, W.J. (1980) in "The biochemistry of glycoproteins and proteoglycans", Lennarz W.J. ed., New York, Plenun Press, pp. 35-73.
Tollbom, O. and Dallner, G. (1986) Br. J. Exp. Pathol. 67, 757-764.
Tuma, D.J., Jennett, R.B. and Sorrell, M.F. (1981) Hepatology 1, 590-598.
Tuma, D.J., Mailliard, M.E., Casey, C.A., Volentine, G.D. and Sorrell, M.F. (1986) Biochim. Biophys. Acta 856, 571-577.
Valtersson, C. Van Duijn , G., Verkleij, A.J., Chojnacki, T., de Kruijff, B. and Dallner, G. (1985) J. Biol. Chem. 260, 2742-2751.
Van Duijn, G., Valtersson, C., Choinacki, T., Verkleij, A.J., Dallner, G. and de Kruijff, B. (1986) Biochim. Biophys. Acta 861, 211-223.
Volentine, G.D., Tuma, D.J. and Sorrell, M.F. (1984) Gastroenterology 86, 225-229.
Wetmore, S. Mahley, R.W., Brown, W.V. and Schachter, H. (1974) Can. J. Biochem. 52, 655-664.
Whaley, W.G. and Dauwalder, M. (1979) Int. Rev. Cytol. 58, 199-245.

Free Radicals: From Basic Science to Medicine
G. Poli, E. Albano & M. U. Dianzani (eds.)

OXIDATIVE DAMAGE AND HUMAN ALCOHOLIC LIVER DISEASES. EXPERIMENTAL AND CLINICAL EVIDENCE

E. Albano, P. Clot, M. Tabone*, S. Aricò*, and M. Ingelman-Sundberg#.

Department of Experimental Medicine and Oncology, University of Torino, Torino, Italy. *Division of Gastroenterology, Mauriziano Hospital, Torino and #Department of Physiological Chemistry, Karolinska Institutet, Stockholm, Sweden.

SUMMARY. An increasing number of studies have shown that hepatic lipid peroxidation and depletion of the liver content of antioxidants is evident in experimental animals exposed to ethanol, suggesting the possible involvement of oxidative damage in the pathogenesis of some of the toxic effects of alcohol. Recent findings demonstrate that hydroxyethyl radicals are generated during ethanol metabolism by the alcohol-inducible form of cytochrome P-450 (CYP2E1) and that human liver microsomes are similary capable of producing ethanol-derived radicals. These intermediates, along with reactive oxygen species, also produced by CYP2E1, can be regarded as possible causes of the stimulation of lipid peroxidation, detectable in liver biopsies and in the blood of patients suffering from alcohol-related liver diseases.

Experimental Evidence of Alcohol-Related Oxidative Damage.

Liver diseases consequent on alcohol abuse represent an important medical and social problem in most industrialized countries. However, in spite of decades of research, the pathogenesis of hepatic injuries in alcoholics is still largely unknown (see Lieber, 1990 for review). One of the aspects of alcohol toxicity which has been investigated in recent years concerns the possible involvement of free-radical-mediated oxidative damage. Several authors have, in fact, reported that acute and chronic alcohol intoxication of rats are associated with the appearance of lipid peroxidation, as measured by conjugated diene absorbance and malonyl dialdehyde (MDA) accumulation in the liver, as well as by the breath exhalation of pentane and ethane (Dianzani, 1985; Albano et al. 1991; Nordmann et al. 1992 for reviews). Furthermore, lipid-derived free radicals have been detected in the liver of rats fed ethanol along with a high fat diet (Reincke et al. 1987), while an increase in the fluorescent adducts between aldehydes derived from lipid peroxidation and plasma proteins is present in rats receiving a similar high fat ethanol-containing diet by intragastric infusion (French et al. 1992). The occurence of oxidative injury during ethanol intoxication is further supported by several reports concerning the decrease of the hepatic content of antioxidants, such a glutathione (GSH) and α-tocopherol (see Videla and Valenzuela 1982; Nordmann et al. 1992 for reviews). These findings are not limited to rodents, since stimulation of lipid peroxidation and GSH depletion have also been observed in baboons treated with alcohols (Lieber, 1990).

Nonetheless, the most compelling evidence in favour of a possible role of free radicals in the pathogenesis of alcohol toxicity has come from the observation that these reactive species can actually be produced as a consequence of exposure to ethanol. Chronic ethanol consumption is, in fact, associated with an increase in the hepatocyte content of a particular form of cytochrome P-450 (CYP2E1), which has an especially high production of both superoxide anion (O_2^-) and hydrogen peroxide (H_2O_2) through the NADPH oxidase activity (Persson et al. 1990). In the presence of iron, the leakage of O_2^- and H_2O_2 might cause the formation of hydroxyl radicals ($OH^\cdot$) and stimulation of lipid peroxidation (Ingelman-Sundberg and Johansson, 1984). Indeed, microsomes obtained from rat chronically exposed to alcohol show an enhanced susceptibility to lipid peroxidation, which can be selectively inhibited by antibodies directed against CYP2E1 (Ekström and Ingelman-Sundberg 1989). Interestingly, the analysis of several microsomal preparations from different human livers has show a direct correlation between CYP2E1 content and the NADPH oxidase activity (Ekström et al. 1989).

Oxygen-derived free radicals are not the only reactive species produced as a result of exposure to alcohol, since, using

Electron Spin Resonance (ESR) spectroscopy in combination with the spin trapping technique, we have demonstrated that a free radical intermediate, identified as 1-hydroxyethyl radical, is produced by rat-liver microsomes incubated in the presence of ethanol and NADPH (Albano et al. 1988). The formation of hydroxy-ethyl radicals is specifically enhanced following induction of CYP2E1 by chronic ethanol-feeding of the rats. The free radical formation, however, is inhibited by antibodies against CYP2E1, indicating that the ethanol-inducible CYP2E1 is responsible for the free radical activation of ethanol (Albano et al. 1991b). Therefore, the capacity of CYP2E1 to generate reactive oxygen species and hydroxyethyl radicals can be regarded as a possible cause for the stimulation of lipid peroxidation consequent on chronic alcohol consumption. Moreover, other possible roles of hydroxyethyl radicals in ethanol toxicity might result from the ability of these species to react with GSH and to alkylate liver proteins (Albano et al. 1988; 1992).

The recent demonstration that hydroxyethyl radicals are produced *in vivo* in deer-mice and rats intoxicated with ethanol (Knecht et al. 1990; Reinke et al. 1991) gives further support the potential importance of free radical intermediates in the hepatic toxicity of alcohol.

Evidence for the Formation of Hydroxyethyl Free Radicals in Human Liver Microsomes.

We have recently observed that free radical intermediates, showing spectral features identical to those ascribed to hydroxyethyl radicals (Albano et al. 1988), can be spin-trapped during ethanol metabolism by human liver microsomes (Fig. 1). The origin of the radical species trapped has been confirmed by using ^{13}C-labelled ethanol, which causes a characteristic change in the ESR spectrum (Fig. 1) and allows its unambiguous identification as due to hydroxyethyl radicals (Albano et al. 1988) As in rat-liver microsomes, free radical generation by human microsomal preparations appears to depend upon the presence of oxygen and

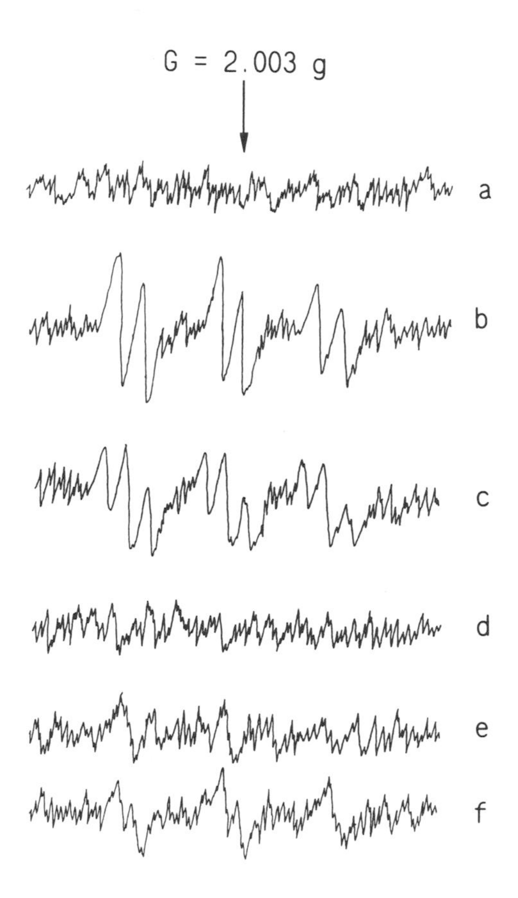

Figure 1: ESR spectrum of hydroxyethyl-4-POBN adduct detected in the chloroform extract of human liver microsomes incubated with ethanol.

Microsomes prepared from human liver fragments were incubated for 30 min at 37° C with an NADPH regenerating system, 25 mM 4-POBN and in the absence (trace a) or in the presence (trace b) of 20 mM ethanol or 20 mM ethanol, enriched with ^{13}C-isotope (trace c). Microsomal suspensions containing ethanol were also incubated without $NADP^+$ (trace d) or under hypoxic conditions (10 min flushing with nitrogen) (trace e), and in the presence of antibodies directed against CYP2E1 (2 mg/mg protein) (trace d). ESR spectra were recorded at room temperature using a Brucker D 200SRC spectrometer with the following instrument settings: microwave power 10 mW; modulation frequency 100 KHz; modulation amplitude 1G; field scan 100 G. The recorder amplification was 10^6 for all the traces.

NADPH, and is strongly decreased by the addition of antibodies against CYP2E1 (Fig.1). Furthermore, experiments employing microsomes obtained from the liver of five kidney donors demonstrate that there is a direct relationship between the levels of CYP2E1 measured in the different preparations and their capacity to form hydroxyethyl radicals (Table I).

TABLE I

Variations in the intensities of the ESR signals due to hydroxyethyl-4-POBN adducts in relation to the content of CYP2E1 in liver microsomes from five different human livers.

	ESR signal intensity (arbitrary units)	CYP2E1 content (pmol/mg proteins)
Liver n° 1	7	<5
Liver n° 2	17	9
Liver n° 3	5	<5
Liver n° 4	29	35
Liver n° 5	25	34

Microsomal suspensions (5 mg protein/ml) were incubated for 30 min at 37°c with a NADPH regenerating system, 25 mM of the spin-trapping agent 4-pyridyl-1-oxide-t-butyl nitrone (4-POBN) and 20 mM ethanol, as previously reported (Albano et al. 1988). CYP2E1 content was measured as described by Ekström et al. (1989).

These observations not only confirm the capacity of human CYP2E1 to catalyze the free radical activation of ethanol, but also suggest a possible increase in the risks connected to the formation of free radicals in relation to the levels of CYP2E1. This might be particularly important in patients who are heavy drinkers, in whom the content of CYP2E1 in the centrilobular areas of the liver is greatly enhanced (Buhler et al. 1991).

In this connection, it is well known that there is an inter-individual difference in the susceptibility to ethanol-induced liver disease. Genetic analysis has evidenced that the *CYP2E1* gene is highly polymorphic in a Caucasian population, and we have found that unlinked Restriction Fragment Length Polymorphisms (RFLPs) can be obtained with the restriction enzymes Dra I, Taq I, Rsa I and Pst I (Persson et al. umpublished results). In order to evaluate whether such genetic polymorphism could be associated with enhanced susceptibility for ethanol-induced liver disease, we investigated the frequency of rare alleles among 33 alcoholics with liver cirrhosis as compared to 156 healthy controls. Interestingly, we found that a particular allele, characterized by a 4.7 kb fragment using Dra I, was much rarer among the alcoholics (allele frequency = 1.5%) than in the control group (allele frequency = 11%). Further analysis will reveal whether this difference is statistically significant and related to functionally different forms of CYP2E1 among these subjects.

Critical Overview of the Clinical Evidence of Alcohol-Related Oxidative Damage.

The findings concerning the formation of free radial species in human liver raise the question of whether oxidative damage, namely stimulation of lipid peroxidation, can be detected in alcoholic patients.

The first report on this topic appeared in 1981, and showed that lipoperoxide levels, measured by the thiobarbituric acid (TBA) method, were higher in the liver and in the serum of heavy drinkers than of non-drinkers (Suematsu et al. 1981). The paper also presented a positive correlation between the lipoperoxide values in the serum and the hepatic leakage of transaminases (Suematsu et al. 1981). Consistently, Shaw and coworkers (1983) reported that lipid extracts of hepatic biopsies from 16 alcoholics with different stages of liver disease presented a significant increase in the conjugated diene content, as compared to those of 8 patients with liver diseases unrelated to alcohol. The same patients also presented a 30% decrease in hepatic GSH levels which was unrelated to their nutritional status (Shaw et al. 1983). These observations were substantially confirmed by Tanner et al. (1986), who showed that elevation of serum lipoperoxides in alcoholics was associated with a decrease in the α-tocopherol content, and by Situnayake et al. (1990), who reported an inverse correlation between hepatic levels of GSH and those of octadeca-(9,11)dienoic acid, a diene isomer of linoleic acid. Patients with alcoholic cirrhosis were also found to exhale more pentane, an end-product of lipid peroxidation (Moscarella et al. 1984; Peters et al. 1986). Moreover, fluorescent adducts between plasma proteins and aldehydes derived from lipid peroxidation were detected in the plasma of alcoholic patients in the absence of biochemical signs of liver damage (Carini et al 1988).

Some authors, however, have expressed doubts about the specificity of ethanol in causing oxidative lesions, since an increase in liver or serum TBA reactive substances was observed in patients suffering from alcoholic as well as non-alcoholic liver disease (Mezes et al. 1986; Mazzanti et al. et al. 1989).

To explain these discrepancies, it should be noted that the analytical methods most often used in the above studies are rather aspecific and susceptible of artifacts (Halliwell and Gutteridge 1990). For instance, much of the evidence concerning the presence of lipid peroxidation during ethanol intoxication has been obtained using a colorimetric method known as thiobarbituric acid (TBA) test, in which the substances forming the chromogen are generated by the decomposition of lipid hydroperoxides catalyzed by traces of transition metals (Halliwell and Gutteridge 1990). This method is often improperly referred to as the estimation of lipoperoxide content of tissues. However, since sugars and amino acids also react with TBA, forming an identical chromogen, it greatly overestimates the effective content of lipid peroxides (Halliwell and Gutteridge 1990). A further problem in applying this test as an index of lipid peroxidation *in vivo* concerns the heterogeneous composition of the samples, and in particular the fact that a decrease in the concentration of antioxidants or, alternatively, an increase in the levels of transition metals, can substantially accelerate the formation of TBA reactive substances during the assay, thus affecting the reliability of the test. (Halliwell and Gutteridge 1990). This point should be taken into account considering that the plasma levels of α-tocopherol are below normal in alcoholics (Tanner et al. 1986; Bell et al. 1992), while ethanol consumption increases the amount of tissue free iron (Nordmann et al. 1992). Caution should also be used in interpreting the results concerning conjugated diene analysis. Indeed, recent research indicates that, in human tissues, the main compound containing conjugated dienes is represented by octadeca-9(cis)11 (trans)-dienoic acid, an isomer of linoleic acid (Dormandy and Wickens 1987) which, in spite of the diene configuration, does not originate from lipid peroxidation (Thompson and Smith, 1985), but probably comes from the diet and from bacterial metabolism in the colon. The lack of specificity of this index is suggested by the increase in the plasma levels of octadeca-9(cis)11(trans)-dienoic acid in alcoholics (Dormandy, 1988) as well as in patients suffering from a number of diseases unrelated to alcohol (Dormandy and Wickens 1987).

Further Studies on the Association Between Oxidative Damages and Alcoholic Liver Injury.

Since analytical procedures used in previous studies of lipid peroxidation in alcoholic liver disease are not above criticism, we reinvestigated the problem, using the plasma lipid hydroperoxides assay proposed by Onishi et al. (1985), and the measure of free malonyl dialdehyde (MDA) in erythrocytes. The former is a colorimetric assay which takes advantage of the reaction of organic hydroperoxides (LPO) with a leuco-derivative of methylen blue (10-methylcarbamoyl-3,7-dimethylamino-10-phenothiazine; MCPD) catalyzed by haemoglobin (Ohishi et al. 1985). The second test is based on a TBA reaction assay, with care being

taken to carry out the assay in a protein and lipid free supernatant, where only free MDA produces the chromogen (Esterbauer and Cheeseman 1990). Both tests are of simple clinical application but, at the same time, reliable and not subject to the interferences mentioned above. As shown in Figure 2, in all 26 cirrhotic patients examined we found a significant increase in both erithrocyte MDA and plasma LPO, whose mean values are about 4-fold higher than the 24 healthy controls.
A good correlation coefficient (r = 0.76) between the results of the two tests in each patient is also evident (Fig. 2).

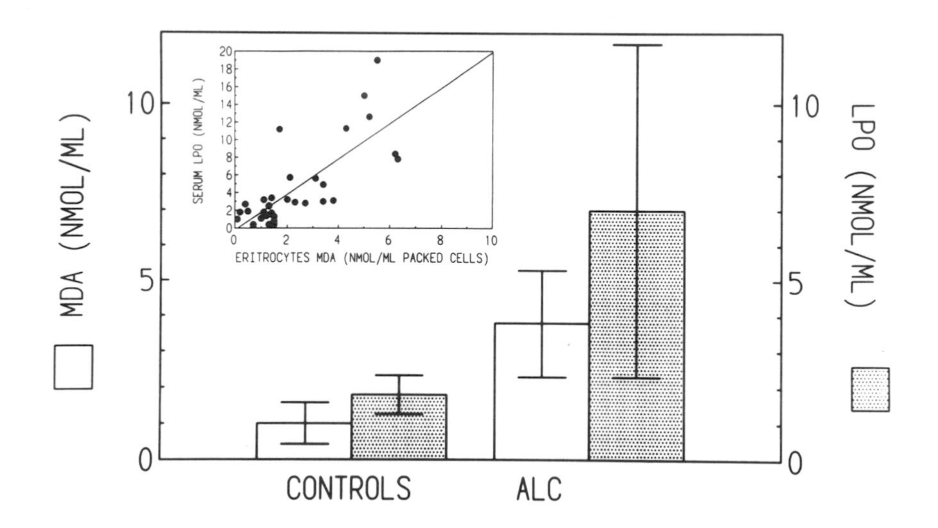

Figure 2: Erythrocyte malonyl dialdehyde (MDA) and plasma lipohydroperoxide (LPO) content in 24 healthy controls and 26 patients suffering from alcoholic cirrhosis (ALC).
The insert shows the correlation between the changes in MDA and LPO values measured in the alcoholic patients.

Interestingly, by grouping these patients according to their estimated daily ethanol intake we observed that the levels of MDA in the erythrocytes, and those of plasma LPO, increased with alcohol consumption (Table II). Moreover, the indices of lipid peroxidation in the two alcoholic groups showed significant differences, not only from the controls, but also between subjects drinking 100-200 g ethanol/day (ALC1) and those consuming more than 200 g ethanol/day (ALC2) (Table II). Conversely, no significant differences are evident between the mean values of MDA and LPO when the patients are grouped according to the clinical stage of their disease (not shown).

TABLE II

Erythrocyte malonyl dialdehyde (MDA), plasma lipid hydroperoxides (LPO) and plasma α-tocopherol content in healthy controls and in alcoholic patients with different daily ethanol intake.

	MDA	LPO	α-tocopherol
Controls	1.1 ± 0.59	1.8 ± 0.5	26.2 ± 4.1
ALC1	2.7 ± 1.46[a]	3.7 ± 1.3[a]	18.0 ± 5.2[b]
ALC2	4.7 ± 1.54[b,d]	13.8 ± 4.8[a,c]	13.6 ± 10.7[b,e]

Patients with alcoholic cirrhosis were grouped according to daily alcohol consumption, estimated through two independent interview-questionaires, as reported by Corrao et al. (1991). The first group, ALC1, included 10 patients with an estimated alcohol intake of 100-200 g ethanol/day, and the second group, ALC2, 16 patients with consumption above 200 g ethanol/day. Healthy controls (n° 24) were all drinking less than 100 g ethanol/day. Free MDA was measured by the TBA reaction in the protein-free supernatant, obtained by treatment with 50% TCA solution of washed erithrocytes as described by Poli et al. (1989). Plasma LPO were estimated by Ohishi et al.'s method (1985) using a kit supplied by Kamiya Biomedical Co (Thousand Oaks, CA, USA). Vitamin E was measured by HPLC by Burton et al.'s method (1985).

a $p < 0.05$ as compared to controls.
b $p < 0.001$ as compared to controls.
c $p < 0.05$ as compared to ALC1.
d $p < 0.005$ as compared to ALC1.
e Not statistically significant from ALC1.

Table II also shows that the plasma levels of α-tocopherol are significantly decreased in both ALC1 and ALC2 groups, while the difference between the two alcoholic groups is not significant (Table II). Since no correlation is evident between the MDA and LPO values and the content of α-tocopherol (not shown), the increase in lipid peroxidation observed in alcoholic patients do not depend upon the lowering of the antioxidant content of the blood.

Thus the results using two different assays confirm that stimulation of lipid peroxidation is actually detectable in the blood of patients with alcoholic cirrhosis, and that it increaes with alcohol consumption. This latter finding suggests that free radical species produced during liver metabolism of ethanol might be responsible for the peroxidative damage.

In conclusion, by inducing a specific form of cytochrome P-450 (CYP2E1), ethanol consumption might lead to the formation of hydroxyethyl radicals and reactive oxygen species, either experimental animals or in humans, which may be responsible for the stimulation of lipid peroxidation in the liver. Indeed, signs of peroxidative reactions have been detected in rat fed with alcohol and confirmed in alcoholic patients. It is not yet clear to what extent oxidative injury might contribute to the pathogenesis of alcohol-related liver lesions. Nonetheless, recent observations indicate that stimulation of lipid peroxidation is associated with the development of hepatocyte damage and liver fibrosis in rats fed intragastrically with a high fat diet containing ethanol (French et al. 1992). This suggests that the increase in free radical formation, resulting from ethanol consumption, might play a role in the pathogenesis of alcohol-mediated liver injury.

Acknowledgements

This research was supported by the Italian Ministry of University and Scientific and Technological Research (Projects: Cirrosi Epatica; Patologia da Radicali Liberi e degli Equilibri Redox), by the Italian National Research Council, by Systembolaget AB and by the Swedish Medical Research Council.

REFERENCES

Albano, E., Tomasi, A., Goria-Gatti, L., Dianzani, M.U. (1988) Chem.-Biol. Interact., 65, 223-234.
Albano, E., Ingelman-Sundberg, M., Tomasi, A. and Poli, G. (1991) In: Alcoholism: A Molecular Perspective. (Palmer, T.N. ed.), Plenum Press New York, pp. 45-56.
Albano, E., Tomasi, A., Goria-Gatti, L., Persson, J.O., Terelius, Y., Goria-Gatti, L., Ingelman-Sundberg, M., Dianzani, M.U. (1991) Biochem. Pharmacol., 41, 1895-1902.
Albano, E., Parola, M., Comoglio, A. (1992) Biochem. Biophys. Res. Commun., Alcohol & Alcoholism, 27 (Suppl. 1), 94.
Bell, H., Biorneboe, A., Eidsvoll, B., Norum, K.R., Raknerud, N., Try, K., Thomassen, Y. and Drevon, C.A. (1992) Alcohol & Alcoholism, 27, 39-46.
Burton, G.W., Webb, A. and Ingold, K.U. (1985) Lipids, 20, 29-39.
Carini, R., Mazzanti, R., Biasi, F., Chiarpotto, E., Marmo, G., Moscarella, S., Gentilini, P., Dianzani, M.U., Poli, G., (1988) In: Alcohol Toxicity and Free Radical Mechanisms (Normann, R., Ribière C., Rouach, H., eds.), Pergamon Press, Oxford, pp. 61-64.
Corrao, G., Aricò, S., Russo, G., Carle, F., Galatola, G., Torchio, P.F., Ruggerini Moiraghi, A. and De La Pierre, M. (1991) Int. J. Epidemiol. 20, 1037-1042.
Dianzani, M.U. (1985) Alcohol & Alcoholism, 20, 161-173.
T.L. Dormandy, (1988) In: Alcohol Toxicity and Free Radical Mechanisms, R. Normann, C. Ribière and H. Rouach, eds., pp. 55-59, Pergamon Press, Oxford.
Dormandy, T.L. and Wickens, D.G. (1987) Chem. Phys. Lipids, 45, 353-364.
Ekström, G., Ingelman-Sundberg, M. (1989) Biochem. Pharmacol. 38, 1313-1318.
Ekström, G., Von Bahr, C., Ingelman-Sundberg M. (1989) Biochem. Pharmacol. 38, 689-693.
Esterbauer, H. and Cheeseman, K.H. (1990) Methods Enzymol. 186, 407-420.
French, S.W., Wong, K., Jui, L., Albano, E., Hagbjörk, A.-L. and Ingelman-Sundberg, M. (1992) Exp. Mol. Pathol., (submitted for publication).
Gorsky, L.D., Koop, D.R., Coon, M.J. (1984) J. Biol. Chem. 259, 6812-6817.
Halliwell, B. and Gutteridge J.M.C. (1990), Methods Enzymol. 186, 1-85.
Ingelman-Sundberg, M., Johansson, I. (1984) J. Biol. Chem. 259, 6447-6458.
Knecht, K.T., Bradfort, B.U., Mason, R.P. and Thurman, G.R. (1990) Mol. Pharmacol. 38, 26-30.
Lieber, C.S. (1990) Pharmac. Ther. 46, 1-41.
Mazzanti, R., Moscarella, S., Bensi, G., Altavilla, E. and Gentilini, P. (1989) Alcohol & Alcoholism, 24, 121-128.
Mézes, M., Par, A., Neèmeth, P. and Javor, T. (1986) Int. J. Clin. Pharm. Res., 6,333-338.

Moscarella, S., Laffi, G., Coletta, D., Arena, U., Cappellini, A.P. Gentilini, P. (1984) In: Frontiers of Gastrointestinal Research, vol. 8, (Gentilini, P. Dianzani, M.U. eds.), Krager Verlag, Basel, pp.208-216.
Nordmann, R., Ribière, C., Rouach, H. (1992) Free Rad. Biol. Med. 12, 219-240.
Ohishi, N. Ohkawa, H., Miike, A., Tatano, T., and Yagi, K. (1985) Biochem.Int. 10, 205-211.
Persson, J.O., Terelius, Y., Ingelman-Sundberg, M. (1990) Xenobiotica, 20, 887-900.
Peters, T.J., O'Connell, M.J., Venkatesan S., Ward, R.J. (1986) In: Free Radicals Cell Damage and Disease (Rice-Evans, C. ed.), Richelieu Press, London, pp. 99-110,.
Poli, G., Biasi, F., Chiarpotto, E., Dianzani, M.U., De Luca, A., and Esterbauer, H. (1989) Free Rad. Biol. Med. 6, 167-170.
Reinke, L.A., Lai, E.K., DuBose, C.M., Mc Cay, P.B. (1987) Proc. Natl. Acad. Sci. 84, 9223-9227.
Reinke, L.A., Kotake, Y., Mc Cay, P.B., Janzen E.G. (1991) Free Rad. Biol. Med. 11, 31-39.
Shaw, S., Rubin, K.P., Lieber, C.S. (1983) Dig. Dis. Sci. 28, 585-589.
Situnayake, R.D., Crump, B.J., Thurnham, D.I., Davies, J.A., Gearty, J., Davis, M. (1990) Gut, 31, 1311-1317.
Suematzu, T., Matsumura,T., Sato, N., Miyamoto, T., Ooka, T., Kamada, T. Abe, H. (1981) Alchol. Clin. Exp. Res., 5, 427-430.
Tanner, A.R., Bantock, I., Hinks, L., Lloyd, B., Turner, N.R., Wright, R. (1986) Dig. Dis. Sci. 31, 1307-1312.
Thompson, S. and Smith, M.T. (1985), Chem.-Biol.Interact. 55, 357-364.
Videla, L.A. and Valenzuela, A. (1982) Life Sci. 31, 2395-2407.

Free Radicals: From Basic Science to Medicine
G. Poli, E. Albano & M. U. Dianzani (eds.)

OXIDISED LOW DENSITY LIPOPROTEINS

Catherine Rice-Evans

Free Radical Research Group, Division of Biochemistry, United Medical & Dental Schools of Guy's and St Thomas's Hospitals, St Thomas Street, London SE1 9RT

SUMMARY: The evidence for the oxidative modification of low density lipoprotein (LDL) and its subsequent recognition by the scavenger receptors on macrophages in vitro is clear. The presence of oxidatively modified LDL in atherosclerotic lesions is well-recognised. However, the mechanism whereby LDL becomes oxidised in vivo is yet to be clarified. New case control studies and cross-sectional random surveys in several European countries suggest a role for dietary antioxidants in protecting LDL from oxidation.

In this review of the series of presentations on Oxidised Low Density Lipoproteins, several experts have contributed novel data on:

i) potential modes of oxidation of LDL in vivo;
ii) evidence for erythrocyte-mediated LDL oxidation;
iii) the significance of hydroperoxide levels;
iv) the importance of plasma and LDL antioxidant status, their concerted action and the relevance of minor, e.g. ubiquinol-10, as well as major components;
v) the interpretation of the lag phase and its relationship with oxidisability of LDL.

INTRODUCTION

There is now overwhelming evidence that, following damage by some mechanism to the endothelium, the oxidation of low density lipoproteins (LDL) in the sub-endothelial space is one of the earliest events in atherosclerosis. Endothelial damage is proposed to be followed by increased permeability and release of chemotactic factors which attract monocytes from the circulation to the site of damage (Willerson & Buja, 1980). These develop into macrophages in the vessel wall. Scavenger receptors on the macrophages recognise oxidatively modified LDL molecules and convert them to cholesterol-laden foam cells. Products of lipid peroxidation such as lysophosphatidyl choline, 4-hydroxy nonenal, may act as chemotactic factors for blood monocytes, encouraging their recruitment into the lesioned area (Steinbrecher, 1990). It has been hypothesised that secretion of superoxide radicals, hydrogen peroxide and hydrolytic enzymes by activated monocytes and macrophages injure neighbouring endothelial cells; in addition, factors released by macrophages stimulate the proliferation of smooth muscle cells, which may break through the elastic lamina and form a mass of cells which eventually will form the main mass of the atherosclerotic plaque (Ross et al, 1974).

Evidence from cellular studies in vitro initially showed how oxidative processes could play a central role in the pathological changes involved in the genesis of athero-sclerosis. LDL can be oxidatively modified in culture by a range of cell types including endothelial cells (Henriksen et al, 1981), arterial smooth muscle cells (Henriksen et al, 1983) as well as macrophages (Parthasarathy et al, 1986; Rankin & Leake, 1987, Leake & Rankin, 1990) and is subsequently taken up by the scavenger receptors on target macrophages . (Human fibroblasts are capable of oxidatively modifying LDL and predisposing it to uptake by the macrophage scavenger receptors). The relative contribution of the different cell types may achieve different levels of importance at the various stages in the development of the lesion. Normal arterial wall contains endothelial cells and smooth muscle cells, whereas atherosclerotic lesions may also contain macrophages and T-lymphocytes. Thus when the atherosclerotic lesion develops, what is responsible for the initiation of LDL oxidation? Stimulation of endothelial cells, smooth muscle cells or macrophages may induce the secretion of components capable of promoting mechanisms of initiation of LDL oxidation.

WHAT IS THE MECHANISM BY WHICH LDL BECOMES OXIDISED IN VIVO - IS A RADICAL SPECIES NECESSARY?

The major question arising is what is the mechanism by which LDL becomes oxidsed in vivo. Are specific initiating radical species essential or is the propagation of peroxidation subsequent to lipoxygenase-mediated hydroperoxide formation the major likely priming event? If the former, what is the proabable initiating agent, where is it located and what activates it? If the latter, are the lipoxygenases macrophage-derived or from other cell sources?

Cell-induced modification of LDL in vitro has been demonstrated to be mediated by free radicals. All the cell types mentioned have been shown to release superoxide radicals, albeit by different mechanisms and at different rates. Thus, addition of superoxide dismutase has an inhibitory effect on the oxidative modification, although the response varies according to cell type, implicating superoxide radical in the mechanism of cell-mediated modification (Morel et al, 1984; Heinecke et al, 1986). The significance of superoxide radical in the initiation, but not in the propagation, of LDL oxidation in cultures of monocytes/macrophages is indicated by experiments showing inhibition of the oxidative modification by superoxide dismutase only if the antioxidant is added within a few hours of the initiation of the incubation. However, an antioxidant such as butylated hydroxytoluene, a lipid chain breaking antioxidant as well as a hydroxyl radical scavenger, is effective in inhibiting the oxidative modification as late as 11 hr after the onset of incubation. It is important to note that small amounts of iron in the medium are an absolute requirement for oxidation of LDL by cultured macrophages (Leake & Rankin, 1990). Lipoxygenase inhibitors are effective in inhibiting the oxidative modifications mediated by endothelial cells and macrophages, lipoxygenase inhibitors also showing antioxidant properties (Parthasarathy et al, 1988; Rankin et al, 1991). Studies from the laboratory of Parthasarathy et al suggest that endothelial cells can initiate the oxidation of LDL through a superoxide-independent pathway that involves lipoxygenase and this pathway may predominate in endothelial cells. Chisolm's group has reported that monocyte-mediated oxidation of LDL involves monocyte lipoxygenase products which induce release of superoxide radical from the monocytes

(McNally et al. 1990).

However, the superoxide released from these cells and hydrogen peroxide generated therefrom are not very reactive per se. Their reactivity may, in principle, be amplified in the presence of available delocalised haem proteins or metal ions to generate more reactive species.

$$O_2^{\cdot -} \dashrightarrow H_2O_2 \begin{cases} \rightarrow {}^{\cdot}OH \text{ hydroxyl radical} \\ \rightarrow {}^{\cdot}X\text{-}[Fe^{IV}{=}O] \text{ or } HX\text{-}[Fe^{IV}{=}O] \text{ ferryl species} \end{cases}$$

The ferryl haem protein species is more selective and may, perhaps, be more relevant in vivo than the hydroxyl radical. Delocalised haem proteins have been detected in several locations in vivo; for example, haemoglobin release and microbleeding in the eye causes retinal damage (Doly et al, 1986), in the brain (Panter et al, 1985), at sites of inflammation in the rheumatoid joint (Yoshino et al, 1985); release of myoglobin has been observed in excessive exercise, in kidney disorders and immediately after an acute myocardial infarction (Drexel et al, 1983). Thus in atherogenesis, the trapping of released haem proteins from ruptured erythrocytes in the artery wall during the early stages in the oxidising locality of activated macrophages may create a scenario for initiating or propagating species. This hypothesis presupposes the presence of available haemoproteins in the sub-endothelial space. It should also be noted, however, that haemoglobin has been observed to occur freely in older atherosclerotic plaques where haemorrhaging occurs but, in this case, this is a later event. Iron may be released from certain haemoproteins when exposed in a local region to a relative excess of hydrogen peroxide arising continually from superoxide, perhaps released from inflammatory or other superoxide-producing cells. Any released iron may, in an appropriate environment, exert catalytic effects in the generation of other highly reactive toxic initiating species. Delocalised haem proteins may also be significant, in the appropriate location, for converting hydroperoxides generated through a lipoxygenase-dependent pathway into alkoxyl and peroxyl radical species, which are capable of initiating oxidation damage in LDL, in the same way as haem proteins and transition metal complexes can propagate oxidation damage in LDL.

In the presence of pre-formed lipid hydroperoxides, induced by enzymic pathways (lipoxygenase-mediated) or non-enzymic pathways (radical mediated), propagation of peroxidation can be effected in the vicinity of haem-containing and iron-containing species, generating alkoxyl and peroxyl radicals which can amplify the damage by initiating further rounds of lipid peroxidation (Labeque & Marnett, 1988; O'Brien, 1969).

$$LOOH + Fe^{III}\text{-complex} \longrightarrow LOO^{\cdot} + Fe^{II}\text{-complex}$$

$$LOOH + Fe^{III}\text{-complex} \longrightarrow LO^{\cdot} + [Fe^{IV}{=}O]\text{-complex}$$

$LOOH + Fe^{II}$ -complex ----> $LO^{\cdot}$ + Fe^{III}-complex

$LO^{\cdot}$ + LH ----> LOH + $L^{\cdot}$

$L^{\cdot}$ + O_2 ----> $LOO^{\cdot}$

$LOO^{\cdot}$ + LH ----> LOOH + $L^{\cdot}$

Thus the alkoxyl radical formed is susceptible to interaction with polyunsaturated fatty acid chains, effectively re-initiating further damage, or interaction with a chain-breaking antioxidant such as α-tocopherol or probucol, forming the hydroxy fatty acyl derivative, LOH, terminating the interaction for this species. The fate of lipid peroxyl radical passing through the same sequence of events will be a lipid hydroperoxide formation, which can then re-enter the same propagative cycle catalysed by haem proteins or transition metal complexes, leading to further lipid peroxidation and oxidative modification of the LDL but this does not explain the nature and origins of the initiating species.

NEW INSIGHTS INTO CELLS CAPABLE OF PARTICIPATING IN THE OXIDATIVE MECHANISM: ERYTHROCYTE-MEDIATED LDL OXIDATION

Paganga in the laboratory of Rice-Evans (University of London) has explored the potential for haem proteins to mediate the oxidative modification of low density lipoproteins. Previous studies from this group have demonstrated that ferryl myoglobin radicals and ruptured cardiac myocytes, which generate ferryl myoglobin species on activation (Turner et al.

1991), oxidatively modify LDL (Dee et al, 1991; Rice-Evans & Bruckdorfer, 1992; Green et al, 1992). The interaction of ruptured erythrocytes and myocytes with LDL induces oxidative damage to the LDL as detected by alterations in electrophoretic mobility and the peroxidation of the polyunsaturated fatty acyl chains. Difference spectroscopy reveals that the amplification of the oxidative process to the LDL by the haem proteins is apparently dependent on the transition of the oxidation state of the haemoglobin in the erythrocyte lysate from the oxy ($HX\text{-}Fe^{II}\text{-}O_2$) to the ferryl ($HX\text{-}Fe^{IV}{=}O$) state via the deoxy form (Paganga et al. 1992). The timescale of this haem conversion is related to the antioxidant status of the LDL and that of the erythrocyte lysate. The incorporation of lipid-soluble antioxidants such as tocopherol, butylated hydroxytoluene at specific time points during the LDL-erythrocyte interaction prolongs the lag phase to oxidation, eliminates the oxy to ferryl conversion of the haemoglobin and delays the oxidative modification of the LDL. Ascorbate also prevents the erythrocyte-induced oxidation of the LDL in vitro, presumably by exerting its dual role in sparing the tocopherol in the LDL, as well as enhancing the antioxidant status of the erythrocyte lysate (Andrews et al. 1992).

The findings here suggest that, after an initial slow phase corresponding to the antioxidant capacity of the LDL, hydroperoxides can interact with haemoglobin in a similar manner to hydrogen peroxide, forming ferryl haemoglobin, which is then rapidly reduced to mixtures consisting mainly of oxy- and met-forms, possibly by the synproportionation reaction (Guilivi & Davies, 1990). Incorporation of the chain-breaking antioxidant arrests the rapid transition from the oxy to the ferryl form and inhibits LDL oxidation, supporting the idea that it is the interaction between lipid hydroperoxides and oxyhaemoglobin which is essential for the haemoglobin-mediated modification to LDL to take place. Enhancement of the antioxidant status of the LDL increases the resistance of LDL to oxidation and to oxidative damage induced by erythrocyte lysate; thus the antioxidant capacity of the LDL is a controlling factor in the oxidation of oxyhaemoglobin to more reactive, damaging forms.

Balla et al (1991) have recently reported the destruction of haem and the release of iron on interaction of LDL with haemin/ hydrogen peroxide mixtures (ten- or twenty-fold molar excess of peroxide). Our studies clearly show that, during interaction between LDL and erythrocyte lysate, LDL oxidation and oxidative activation of the oxyhaemoglobin occur with no requirement for exogenous oxidants, involving no haem destruction nor iron release during the timescale studied. It has been shown that probucol or BHT lower the frequency of occurrence of atherosclerotic lesions in animals (Bjorkhem et al, 1991; Carew et al, 1979). Our examination of the incorporation of the lipid-soluble antioxidant butylated hydroxytoluene into the LDL/haem protein systems reveals a delay in the onset of the oxidative conversion of the oxyhaemoglobin, the abolition of the transition to the ferryl and met state and the inhibition of LDL oxidation. Thus haem proteins leaking from ruptured cells may be capable of enhancing the oxidation of LDL which has penetrated the endothelium of the coronary vessels. This may occur by haem protein-mediated decomposition of preformed peroxides in LDL which has already been minimally oxidised by contact with neighbouring cells or the enzymatic activity of lipoxygenases.

ANTIOXIDANT STATUS OF LDL AS A DETERMINING FACTOR IN ITS OXIDISABILITY

The extensive studies of Esterbauer et al have demonstrated the relative importance of the endogenous antioxidants within the LDL molecule in protecting it from oxidative modification. The antioxidant content of plasma low density lipoproteins is summarised in Esterbauer et al (1989) and Rice-Evans & Bruckdorfer (1992).

Puhl et al in the laboratory of Esterbauer (University of Graz) have used the lag phase measurement at 234nm of the development of conjugated dienes on copper-stimulated LDL oxidation to define the oxidation resistance (OR) of different LDL samples. During the lag phase the antioxidants in LDL (vitamin E, carotenoids, ubiquinol-10) are consumed in a distinct sequence with α-tocopherol as the first followed by γ-tocopherol, thereafter the carotenoids cryptoxanthine, lycopene and finally ß-carotene. α-Tocopherol is the most prominent antioxidant of LDL (6.4±1.8 mol/mol LDL) whereas the concentration of the other γ-tocopherol, ß-carotene, lycopene, cryptoxanthine, zeaxanthine, luteine, phytofluene is only 1/10 to 1/300 of α-tocopherol. Since the tocopherols reside in the outer layer of the LDL molecule, protecting the monolayer of phospholipids and the carotenoids are in the inner core protecting the cholesterol, and the progression of oxidation is likely to occur from the aqueous interface inwards, it seems reasonable to assign to α-tocopherol the rank of the "front-line" antioxidant. In vivo, the LDL will also interact with the plasma water-soluble antioxidants in the circulation, not in the artery wall, as mentioned above.

In a screening study with 78 subjects the lag phase varied from 34 to 114 min. Interestingly, only a weak correlation was found between the α-tocopherol content and the lag phase ($r=0.2, p<0.01, n=78$). Increasing the α-tocopherol content of individual LDL samples in vitro or by oral supplementation led always to a proportional increase of OR, according to the equation $y=kx+a$. The slope k is the efficacy of vitamin E and the intercept a represents a vitamin E independent parameter. Strong, individual variations were observed for k and a (0.7 to 17 for k and .68 to 108 min for a), which probably explains that the vitamin E content alone is not predictive for the OR of an individual LDL.

Other antioxidants not contained in LDL (probucol, α- and gamma tocotrienol, trolox C, ascorbate, glutathione) were also tested in vitro. All of them prolonged the lag phase, with probucol as the most efficient one. In contrast to ascorbate, trolox C and glutathione, probucol did not spare vitamin E in

LDL. Indeed, studies of the efficacy of probucol in vivo in animal models of atherosclerosis have demonstrated protection from LDL oxidation and a decrease in the rate of the progression of the disease. The protective effects are independent of the anti-atherogenic properties of the drug through its cholesterol-lowering properties.

Electron spin resonance spectroscopy (ESR) from Kalyanaraman's laboratory (University of Wisconsin) have shown that the α-tocopheroxyl radical is a primary free radical formed during oxidation of LDL. In some oxidation systems, a secondary free radical, presumably derived from the other endogenous antioxidants in LDL, also appeared with time. ESR studies have also demonstrated that both α-tocopheroxyl and probucol phenoxyl radicals are formed during oxidation of LDL supplemented with probucol. Ascorbate supplementation has been shown to inhibit significantly the oxidation such that lipophilic antioxidant-derived radicals undergo recycling in its presence, with the concomitant formation of ascorbyl radicals. The ESR results are consistent with the finding that ascorbate supplementation inhibits degradation of oxidised LDL mediated by macrophages.

The role of ubiquinol-10 in the inhibition of the early stages of lipoprotein lipid oxidation has recently been defined by Stocker et al (Heart Research Institute, Sydney). Using ultra-sensitive HPLC assays for lipid hydroperoxides (LOOH), these workers have previously concluded that (i) ubiquinol-10 ($CoQ_{10}H_2$) is associated with LDL, and (ii) under conditions of constant rate of initiation the oxidation chain length in LDL was low (0.2-0.4) as long as $CoQ_{10}H_2$ was present but increased 25-fold upon its consumption even though 80-90% of α-tocopherol and carotenoids were still present (Stocker et al, 1988). This has now been applied to in vivo consideration by dietary supplementation of human volunteers with ubiquinone-10 (CoQ_{10}). This resulted in increased concentration of $CoQ_{10}H_2$ within circulating LDL and such supplemented LDL was found to be more resistant towards the initiation of lipid oxidation, to an extent that was proportional to the initial conceentration of $CoQ_{10}H_2$ in the LDL.

In sharp contrast to the situation with the LDL particle, the peroxyl radical-mediated oxidation chain length of extracted LDL lipids in homogeneous systems was very low (ca 0.03) and not significantly influenced by the addition of physiological ($CoQ_{10}H_2$). Also in contrast to LDL, oxidation of isolated high-density lipoprotein resulted in LOOH formation without delay and at a linear rate throughout the incubation. Competition experiments carried out in fresh plasma showed that HDL lipids were oxidised before those in LDL. This was likely due to the absence of $CoQ_{10}H_2$ and α-tocopherol from most freshly isolated HDL particles. Indeed, in plasma these antioxidants were preferentially located in LDL and this was reflected in an uneven distribution of plasma LOOH: 85% of the detectable plasma cholesterylester hydroperoxide were carried in HDL particles and only 15% in LDL. Linked with high concentration of cholesteryl ester hydroperoxide was (i) the presence of phospholipid hydroperoxides in HDL, and (ii) a low $CoQ_{10}H_2$:CoQ_{10} ratio. Neither $CoQ_{10}H_2$ nor α-tocopherol alone correlated with plasma LOOH. The results suggest an important role for $CoQ_{10}H_2$ in preventing the early stages of lipoprotein lipid oxidation.

NEW INTERPRETATIONS FROM COMPARISON OF INITIATORS

Noguchi, in the laboratories of Niki (University of Tokyo) and Shimasaki (Tokyo University School of Medicine) has studied the oxidative modification of LDL induced by copper, water-soluble radical initiator, 2,2'-azobis(2-amidinopropane) dihydro- chloride (AAPH) or lipid-soluble radical initiator, 2,2'-azobis(2,4-dimethylvaleronitrile) (AMVN). Oxygen consumption and the changes in vitamin E, lipid hydroperoxides, thiobarbituric acid-reactive substances, relative electrophoretic mobility and aggregation and fragmentation of apolipoprotein B were measured. These initiators all induced the free radical-mediated, chain oxidation of LDL and gave phosphatidylcholine hydroperoxide and cholesteryl ester hydroperoxide as major products. When compared at the same extent of oxidation AMVN gave cholesterol ester hydroperoxide in the highest yield, suggesting that AMVN initiates lipid peroxidation in the core part of LDL. Furthermore, the rates of accumulation of hydroperoxides were enhanced after depletion of vitamin E in the oxidations induced by copper or AAPH but those induced by AMVN were less dependent on the presence or absence of vitamin E, implying that vitamin E does not suppress the oxidation taking place in the core efficiently. Lipid peroxidation also increased in all oxidation systems and copper was more effective than both azo compounds. A good correlation was observed between the formation of thiobarbituric acid-reactive substances and change in electrophoretic mobility, or fragmentation of apolipoprotein B in all oxidation systems, although the manner observed in the oxidations induced by AMVN was different from those induced by copper or AAPH. These results suggest that the oxidative modification of LDL is dependent on the type of chain initiation, above all the presence or absence of metal ions and site of radical formation.

In the laboratories of Darley-Usmar (Wellcome Research Laboratories, Kent) and Wilson (University of Essex), Hogg has used kinetic simulation software to model a proposed mechanism of copper-dependent peroxidation based on both the production of conjugated dienes and the uptake of oxygen shown to follow synchronous time courses consisting of a slow 'lag' period followed by a fast phase. From these simulation studies, the following conclusions are drawn: (i) the length of the lag period is dependent upon the concentration of endogenous peroxide, the presence of antioxidant and the concentration of copper; (ii) in the absence of antioxidants the lag period is proportional to the log of the concentration of peroxide when copper concentration is constant and in excess over peroxide; (iii) the observed inhibition of LDL oxidation by antioxidants is strongly dependent on the content of endogenous peroxide and on the point during the oxidation at which the addition of antioxidant is made.

Acknowledgements

The work in the author's laboratory in the University of London is generously supported by the British Heart Foundation, the British Technology Group, St Thomas' Endowments Trust and Bioxytech, Paris.

REFERENCES

Andrews B, Rice-Evans C, Paganga G (1992). Submitted
Balla G, Jacob HS, Eaton JW, Belcher JD, Vercellotti GM (1991) Arteriosclerosis & Thrombosis 11, 1700-1711
Berghund I, Henriksson P (1991) Arteriosclerosis & Thrombosis 11, 15-22
Bjorkhem J, Henriksson-Freyschuss A, Breuer O, Diczfalusy U, Berghund I, Henriksson P (1991) Arteriosclerosis & Thrombosis 11, 15-22
Carew TE, Schwenke DC, Steinberg O (1987) Proc Natl Acad Sci 84, 7725-7729
Dee G, Rice-Evans C, Obeyesekera S, Meraji S, Jacobs M & Bruckdorfer KR (1991) FEBS Lett 294, 38-42
Doly M, Bonhomme B & Vennat JC (1986) Opthalmic Res 18, 21-27
Drexel H, Durozak E, Kirchmair W, Milz M, Puschendorf B & Dienstl F (1983) Am Heart J 105, 641-651
Esterbauer H, Rothender M, Stregl G, Waeg G, Ashby A, Sattler W & Jurgens G (1989) Fat Sci Technol 91, 316-324
Esterbauer H, Streigl G, Pul H & Rothender M (1989) Free Rad Res Comm 6, 67-75
Giulivi C, Davies K (1990) J Biol Chem 265, 19342-19460
Heinecke JW, Rosen H & Chait A (1984) J Clin Invest 74, 1890-1894
Henriksen T, Mahoney EM & Steinberg D (1981) Proc Natl Acad Sci USA 78, 6499-6503
Henriksen T, Mahoney EM & Steinberg D (1983) Arteriosclerosis 3, 149-159
Labeque R & Marnett L (1977) Biochemistry 27, 7060-7070
Leake D & Rankin SM (1990) Biochem J 270, 741-748
McNally AK, Chisolm GM, Morel DW, Cathcart MK (1990) J Immunol 145, 254-259
Morel DW, DiCorleto PE & Chisholm GM (1984) Arteriosclerosis 4, 357-364
O'Brien PJ (1969) Can J Biochem 47, 485-492
Paganga G, Rice-Evans C, Rule R, Leake D (1992) FEBS Lett. In the press
Panter SS, Sadrzadeh SM, Hallaway PE, Haines JL, Anderson VE,

Eaton JW (1985) J Exp Med 161, 748-754
Parthasarathy S, Putz DJ, Boyd D, Joy L, Steinberg D (1986) Arteriosclerosis 26, 505-510
Parthasarathy S, Willard E, Steinberg D (1988) Proc Nat Acad Sci 86, 1046-1050
Rankin SM, Leake D (1987) Biochem Soc Trans 15, 485-486
Rankin SM, Parthasarathy S, Steinberg D (1991) J Lipid Res 32, 449-456
Rice-Evans C, Bruckdorfer KR (1992) Molecular Aspects of Medicine 13, 1-111
Ross R, Glomset J & Kariya B (1974) Proc Natl Acad Sci USA 71, 1207-1210
Steinberg D, Parthasarathy S, Carew TE, Khoo JC & Witztum JL (1989) New Engl J Med 320, 915-924
Steinbrecher UP, Zang H, Lougheed M (1990) Free Radical Biol Med 9, 155-168
Stocker R (1988) Proc Natl Acad Sci 88, 1646
Turner JJO, Rice-Evans C, Davies MJ & Newman ESR (1990) Biochem Soc Trans 18, 1056-1059
Turner JJO, Rice-Evans C, Davies MJ & Newman ESR (1991) Biochem J 277, 833-837
Willerson WT & Buja LM (1980) Am J Med 69, 903-914
Yoshino S, Blake DR, Hewitt S, Morris C, Bacon PA (1985) Ann Rheum Dis 44, 485-490

Free Radicals: From Basic Science to Medicine
G. Poli, E. Albano & M. U. Dianzani (eds.)

PROTEIN PEROXIDES: FORMATION BY SUPEROXIDE-GENERATING SYSTEMS AND DURING OXIDATION OF LOW DENSITY LIPOPROTEIN

A.V. Babiy, S. Gebicki and J.M. Gebicki

School of Biological Sciences, Macquarie University, Sydney, Australia

SUMMARY: BSA incubated in solution in presence of several superoxide producing systems acquired peroxide groups which were detected by an iodometric assay. Activated neutrophils, xanthine oxidase acting on xanthine in presence of Fe(III), NADH oxidase with NADH and Fe(III) or Cu(II), and the combination of ascorbate/Fe(II)/EDTA were all effective. The role of Fenton reaction in the oxidation was shown by its inhibition with SOD, catalase, or absence of metal. Apoprotein B isolated from human LDL oxidised by gamma radiation or by incubation with Cu(II) was also peroxidised. The extent of peroxidation was linear with time, showing that the process was not affected by the normal antioxidants present in the LDL particles.

Actions of free radicals on proteins have been extensively studied in recent years. Most of the investigations were carried out in presence of oxygen, using reactive oxygen derivatives such as the hydroxyl, superoxide, peroxyl and hydroperoxyl free radicals, or higher oxidation states of iron. In general, the consequences of exposure of a range of proteins to these agents led to crosslinking and scission of polypeptide chains, destruction of amino acids, loss of biological function and increased susceptibility to proteolysis (Okada et al., 1960; Schuessler & Herget, 1980; Levine et al., 1981; Schuessler & Schilling, 1984; Wolff & Dean, 1986; Stadtman, 1986; Davies et al., 1987; Amici et al., 1989; Pacifici & Davies, 1990; Lissi & Clavero, 1990). These findings acquired special significance from discoveries of similar alterations in proteins found in living organisms, especially those subjected to deteriorative or diseased conditions (Oliver et al., 1987a, 1987b).

We have recently reported that BSA and lysozyme exposed to

free radicals acquire two types of covalently bound reactive moieties: a reducing variety, tentatively identified as DOPA, and an oxidising one, identified as hydroperoxide (Simpson et al., 1992). The free radical responsible for their formation was the HO·; $O_2^{\cdot -}$ was ineffective. The significance of these observations lies in the role the reactive groups may play in a living system exposed to a flux of free radicals. Since proteins constitute the largest mass component of living organisms after the unreactive water, the probability of reactions between proteins and any reactive oxygen species released must be high. This was demonstrated in studies which estimated the total peroxyl radical trapping potential of albumins in plasma at 10 - 50% (Wayner et al., 1987) and in reports that protein sulphydryl groups in plasma challenged with peroxyl radicals disappeared at a similar rate to its principal antioxidant, ascorbate (Frei et al., 1989). Such findings led to the suggestion that oxyradical interaction with proteins such as albumins is beneficial, because the affected proteins act as sacrificial antioxidants, are damaged and then quickly removed from circulation (Halliwell, 1988). Our results showed that the damaged proteins may themselves be potentially dangerous. The formation of relatively stable oxidising and reducing groups on the molecules attacked by free radicals effectively traps some of their chemical potential, extending their lifetime and radius of operation. Although considerably less reactive than the original free radicals, the altered proteins can oxidise essential antioxidants such as ascorbate and GSH, or reduce metals, allowing them to take part in radical-generating processes (Simpson et al., 1992). This, in turn, is likely to compromise the ability of the organism to withstand further oxidant challenge.

In our previous studies, the protein-bound reactive moieties were produced by radicals generated by gamma radiation. Now we report that other, more physiologically plausible systems, can also peroxidise BSA. We also show that in the process of oxidation of LDL, which produces the atherogenic form of the lipoprotein, the apo B is peroxidised in parallel with the lipids.

MATERIALS AND METHODS

All solvents and chemicals were of highest obtainable purity. Glass distilled water was further purified by passage through a filtering system (Millipore-Waters, Sydney). BSA, xanthine oxidase, diaphorase, catalase, phorbol myristate acetate and Nonidet P-40 were purchased from Boehringer (Mannheim) or Sigma (St. Louis) and their activities checked by standard assays. Neutrophils were isolated from fresh human blood, as in Winn et al. (1991). LDL was also prepared from blood by density gradient centrifugation (Babiy et al., 1990). Protein and LDL lipid peroxide assays were carried out by iodometric methods (Babiy et al., 1990; Simpson et al., 1992). Protein concentrations were measured by 280 nm absorbance or by the Lowry technique (Lowry et al., 1951). Low molecular weight substances were removed from proteins by Sephadex G-50 M (Pharmacia, Uppsala) gel filtration or by dialysis. Isolation of delipidised apo B from the LDL utilised four methods: extraction by methanol:chloroform (6:5) (Socorro & Camejo, 1979), by ether:ethanol (3:1) (Cardin et al., 1982) or by Nonidet P-40-DEAE Sepharose ion exchange chromatography followed by elution with 1 M NaCl (Socorro & Camejo, 1979) or 6 M guanidinium HCl (Shireman et al., 1977). Residual phospholipid and cholesterol were assayed by Sigma kit assays. Irradiations were performed in a cobalt-60 source in solutions saturated with air at a dose rate of about 50 Gy/min (1 Gy = 1 J/kg) measured by Fricke dosimetry (O'Donnell & Sangster, 1970).

RESULTS

The results summarised in Table I show that all of the tested systems capable of generating reactive oxygen species led to the formation of BSA peroxides. In these experiments, the concentration of BSA was 10 mg/ml. In the first system, the protein was incubated with 15 mM ascorbate, 2.5 mM Fe(II) and

5 mM EDTA for 30 min, treated with 260 units/ml of catalase and passed through Sephadex column before assay for peroxides. BSA incubated with the system NADH (2 mM), NADH oxidase (2.5×10^{-4} units) in presence of 75 μM Fe(III) or Cu(II) was treated similarly. In the other enzyme system, the BSA was equilibrated for 30 min with 150 μM Fe(III) and then incubated at 37° for 30 min with 500 μM xanthine and 2×10^{-3} units of the oxidase per ml. After catalase treatment and dialysis, peroxides were assayed. Neutrophils ($1 - 2 \times 10^6$ cells) in Dulbecco's single strength buffered saline medium with Ca(II) or Mg(II) were added to the BSA and stimulated with 0.1 μg/ml of phorbol myristate acetate (PMA). After incubation at 37° for 45 - 60 min with shaking, and treatment with catalase, the protein peroxides were measured.

TABLE I: BSA PEROXIDE FORMATION BY SUPEROXIDE-GENERATING SYSTEMS

Oxidation System		BSA Peroxide (μM)
Ascorbate/Fe(II)/EDTA		6
NADH/NADH oxidase/Fe(III) or Cu(II)		11
Xanthine/xanthine oxidase/Fe(III)		15
	+ SOD	6
	+ catalase	0
	- Fe(III)	5
Neutrophils stimulated by PMA		5-10

Of the four methods used to extract the LDL protein, the detergent-ion exchange was the most satisfactory (Table II).

In these experiments the LDL was oxidised by 30 min gamma irradiation before extraction. This led in every case to a drastic decrease in the recovery of the apoprotein which can most likely be attributable to direct damage to the protein or formation of crosslinks between protein and lipid. There is considerable experimental evidence for either possibility. Proteins exposed to oxygen free radicals can form high molecular weight aggregates, and the interaction between proteins and lipids can also lead to a range of modifications which may include formation of covalent lipid - protein bonds.

TABLE II: ISOLATION OF SOLUBLE APO B FROM NATIVE AND FREE RADICAL OXIDISED LDL

Extraction Method	% Recovery of Soluble Protein		Lipid contamination (% of original in LDL)			
			Phospholipid		Cholesterol	
	Native LDL	Oxidised LDL	Native LDL	Ox. LDL	Native LDL	Ox. LDL
Methanol : Chloroform	56	14	18	18	0.9	0.5
Nonidet/DEAE Sepharose	50-90	27	ND	ND	ND	ND
Ether:ethanol	30-42	17-23	ND	ND	0.5	0.3
Ether:ethanol/ guanidinium-HCl	61	-	29	-	0.5	-

ND - not detected (less than 1%)

For these reasons, shorter irradiations were employed in subsequent tests (Fig. 1). Irradiation of LDL (1 mg/ml) in buffered solutions followed by measurements of total and apo B peroxides showed that the protein was peroxidised at a linear rate. In these experiments, the apo B was isolated by Nonidet chromatography, followed by NaCl elution, and the maximum radiation dose was 750 Gy (Fig. 2). In contrast, the rate of formation of lipid peroxides showed the well-documented lag, followed by a rapid increase. Qualitatively similar results were obtained when the LDL was incubated with 20 μM Cu(II) (Fig. 3).

In Fig. 3 the sigmoidal shape of the lipid and total peroxidation response was based on many experiments documenting the existence of a lag period (Esterbauer et al., 1990).

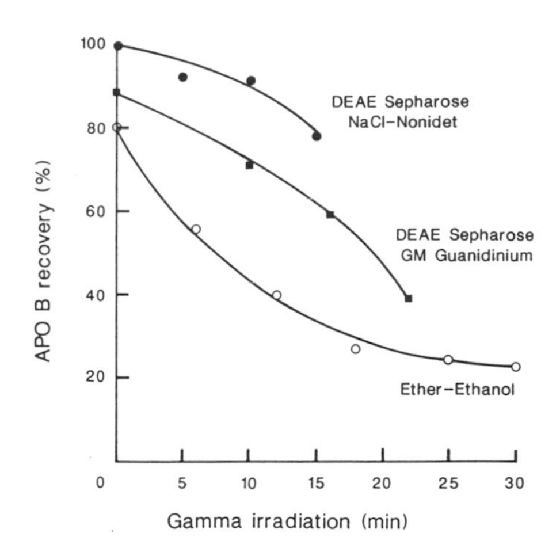

Fig. 1. Recovery of apoprotein B from LDL irradiated with gamma rays at 50 Gy/min. Results of three different extraction methods are shown.

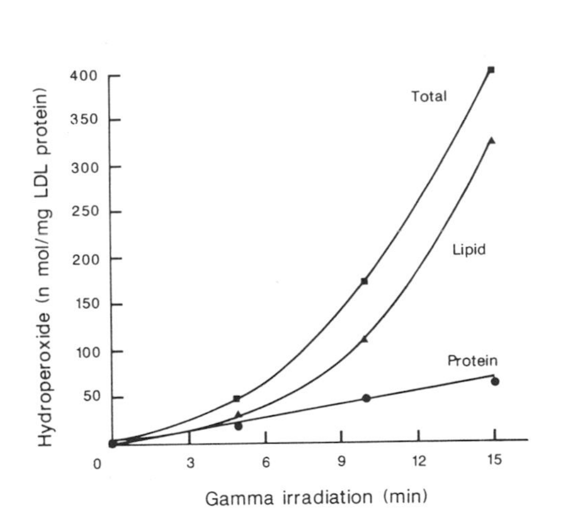

Fig. 2. Formation of lipid and protein hydroperoxides in LDL irradiated with gamma rays at 50 Gy/min.

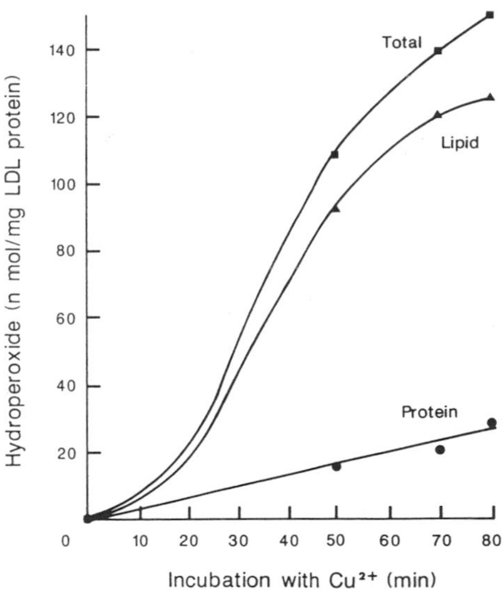

Fig. 3. Formation of lipid and protein hydroperoxides in LDL incubated with 20 μM Cu(II) at 37°.

DISCUSSION

Formation of BSA peroxides during exposure of the protein to a variety of conditions sometimes grouped as "mixed function oxidation systems" (Oliver et al., 1987a) confirms the possibility of their formation in complex biological systems. While the peroxide yields were low compared to results obtained with irradiation, decomposition of AAPH (a peroxyl radical generating compound) or the Fenton system (Gebicki & Gebicki, submitted), they were significant (Table I). Since the conditions employed led to generation of $O_2^{\cdot -}$, which cannot peroxidise BSA directly, peroxide formation must have been initiated in metal-catalysed reactions, probably directly at the protein surface. The operation of such processes is confirmed by the partial reduction in peroxide yields by SOD and omission of Fe(III) and by their complete elimination by catalase. One reason for the fairly low peroxide yields observed may be the presence or formation of reducing agents in all the systems used - ascorbate, Fe(II) or Cu(I). All of these reduce protein peroxides, the metals rapidly and ascorbate slowly.

The results shown in Figs. 1 and 2 are the first demonstration that apo B peroxidation accompanies the formation of lipid peroxides in conditions used in routine preparation of oxidised LDL. This form of the lipoprotein has been studied widely, because it is presently the best candidate for a primary substance responsible for atherogenesis (Gebicki et al., 1991). Whether apo B peroxidation itself plays a role in the recognition of the modified LDL by the macrophage scavenger receptor is unknown. However, peroxidation constitutes a chemical modification of the protein, giving it the capacity to cause further reactions. It also appears to be a modification of the LDL not affected by the presence of the lipid soluble antioxidants, as shown by the immediate linear formation of protein peroxides (Figs. 1 & 2). Since several studies have shown lack of correlation between the content of the most abundant of these antioxidants, vitamin E, and the resistance of LDL to oxidation (Babiy et al., 1990; Jessup et al., 1990;

Esterbauer et al., 1991) other factors must be involved in determining the antioxidant potential of LDL. Susceptibility of the apo B to peroxidation may be one such factor.

Acknowledgements
This work was supported by Australian Research Council and Macquarie University Research grants.

REFERENCES

Amici, A., Levine, R.L., Tsai, L. and Stadtman, E.R. (1989) J. Biol. Chem. 264, 3341-3346.
Babiy, A.V., Gebicki, J.M. and Sullivan, D.R. (1990) Atherosclerosis 81, 175-182.
Cardin, A.D., Witt, K.R., Barnhart, C.L. and Jackson, R.L. (1982) Biochemistry 21, 4503-4511.
Davies, K.J.A., Delsignore, M.E. and Lin, S.W. (1987) J. Bil. Chem. 262, 9902-9907.
Esterbauer, H., Dieber-Rotheneder, M., Striegl, G. and Waeg, G. (1991) Am. J. Clin. Nutr. 53, 314S-321S.
Esterbauer, H., Dieber-Rotheneder, M., Waeg, G., Striegl, G. and Jurgens, G. (1990) Chem. Res. Toxicol. 3, 77-92.
Frei, B., England, L. and Ames, B.N. (1989) Proc. Natl. Acad. Sci. USA 86, 6377-6381.
Gebicki, J.M., Jurgens, G. and Esterbauer, H. (1991) in Oxidative Stress: Oxidants and Antioxidants (Sies, H., ed.) pp 371-397, Academic Press.
Halliwell, B. (1988) Biochem. Pharmacol. 37, 569-571.
Jessup, W., Rankin, S.M., DeWhalley, C.V., Hoult, J.R.S., Scott, J. and Leake, D.S. (1990) Biochem. J. 265, 399-405.
Levine, R.L., Oliver, C.N., Fulks, R.M. and Stadtman, E.R.(1981) Proc. Natl. Acad. Sci. USA, 78, 2120-2124.
Lissi, E.A. and Clavero, N. (1990) Free Rad. Res. Comms. 10, 177-184.
Lowry, O.H., Rosebrough, N.J., Farr, A.L. and Randall, R.J. (1951) J. Biol. Chem. 193, 265-275.
O'Donnell, J.H. and Sangster, D.F. (1970) Principles of Radiation Chemistry, Arnold, London.
Okada, S., Kraunz, R. and Gassner, E. (1960) Radiat. Res. 13, 607-612.
Oliver, C.N., Levine, R.L. and Stadtman, E.R. (1987a) J. Amer. Geriatric Sci. 35, 947-956.
Oliver, C.N., Ahn, B.W., Moerman, E.J., Goldstein, S. and Stadtman, E.R. (1987b) J. Biol. Chem. 262, 5488-5491.
Pacifici, R.E. and Davies, K.J.A. (1990) in Methods in Enzymology (Packer, L. and Glazer, A.N., eds.) Vol. 186, pp 485-502, Academic Press, San Diego.
Schuessler, H. and Herget, A. (1980) Int. J. Radiat. Biol. 37, 71-80.
Schuessler, H. and Schilling, K. (1984) Int. J. Radiat. Biol. 45, 267-281.
Socorro, L. and Camejo, G. (1979) J. Lipid Res. 20, 631-638.

Shireman, R., Kilgore, L.L. and Fisher, W.R.(1977) Proc. Natl. Acad. Sci. USA 74, 5150-5154.
Simpson, J., Narita, S., Gieseg, S., Gebicki, S., Gebicki, J.M. and Dean, R.T. (1992) Biochem. J. 282, 621-624.
Stadtman, E.R. (1986) Trends Biochem. Sci. 11, 11-12.
Wayner, D.D.M., Burton, G.W., Ingold, K.U., Barclay, L.R.C. and Locke, S.J. (1987) Biochim. Biophys. Acta 924, 408-419.
Winn, J.S., Guille, J., Gebicki, J.M. and Day, R.O. (1991) Biochem. Pharmacol. 41, 31-36.
Wolff, S.P. and Dean, R.T. (1986) Biochem. J. 234, 399-403.

Free Radicals: From Basic Science to Medicine
G. Poli, E. Albano & M. U. Dianzani (eds.)

AN IN VITRO APPROACH TO THE STUDY OF INFLAMMATORY REACTIONS IN ATHEROSCLEROSIS.

A.Andalibi, S.Hama, S.Imes, R.Gates, M.Navab, and A.M.Fogelman

UCLA School of Medicine, Los Angeles, CA, USA, 90024-167917

SUMMARY: Mild oxidation of LDL by prolonged storage or by mild tratment with iron produced a modified form of LDL that was still recognized by the LDL receptor. This mildly modified LDL (MM-LDL) contained oxidized lipid(s) that caused artery wall cells to express genes whose protein products could account for the cellular events seen in the developing fatty streak, namely: monocyte adherence (Berliner et al., 1990), monocyte migration by monocyte chemotactic protein MCP-1 (Cushing et al., 1990), and monocyte differentiation by macrophage-colony stimulating factor M-CSF (Andalibi et al. Molecular Biology of Atherosclerosis, 1990). We have used a multilayer coculture of human aortic wall endothelial and smooth muscle cells and studied the modification of LDL in this system and the resulting effects on the artery wall cells.

Modification of LDL by products of lipid peroxidation was proposed to be a pre-requisite to the development of fatty streaks in 1980 (Fogelman et al., 1980). Subsequently, Steinberg and colleagues demonstrated that LDL could be oxidatively modified by endothelial cells in culture if the medium was devoid of serum and contained sufficient iron or copper ions (Henriksen et al., 1992). A decade of research followed in a number of laboratories (Morel et al., 1984; Steinbrecher et al., 1984; Heineke et al., 1984) that focused on the production of higly modified (oxidized) LDL that was no longer recognized by the LDL receptor but was recognized by the scavenger receptor (Steinbrecher et al., 1987) or the ox-LDL receptor (Sparrow et al., 1989) and which was cytotoxic (Hessler et al., 1979). In 1988 the in vivo presence of the products of lipid peroxidation in lesions and their association with apo B was reported (Haberland et al., 1988). This was later confirmed and extended by a number of laboratories providing evidence that lipid peroxidation was involved in fatty streak formation (Palinski et al., 1989). Subsequently it was reported that mild oxidation of LDL by prolonged storage or by mild treatment with iron produced a modified form of LDL that was still recognized by the LDL receptor. This mildly modified LDL (MM-LDL) contained oxidized lipid(s) that caused artery wall cells to express genes whose protein products could account for the cellular events seen in the developing fatty streak, namely: monocyte adherence (Berliner et al., 1990), monocyte migration by monocyte chemotactic protein MCP-1 (Cushing et al., 1990), and monocyte differentiation by macrophage-colony stimulating factor M-CSF (Rayavashisth, 1990).

Subsequent to the reports from this group on the induction of MCP-1 and M-CSF by MM-LDL in endothelial cells, others (Yla-Herttuala et al., 1991; Nelken et al., 1991) have confirmed and extended the importance of MCP-1 demonstrating the presence of its mRNA in lesions of both animals and humans. High expression seen in areas of high monocyte-macrophage density has been demonstrated.

The in vitro prepared MM-LDL in our laboratories was biologically active in vivo (Liao et al., 1991). However, the failure to oxidatively modify LDL in the presence of even small amounts of serum was troublesome. In an interview in Science (Marx, 1987) Dan Steinberg discussed the oxidized LDL hypothesis and was quoted as saying "Blood serum protects against the damage"..."This is the weak spot of the hypothesis". He went on to indicate that the hypothesis might still be correct, however, if the intact arterial lining can prevent the penetration of the protective serum components (Marx, 1987). In 1991 we reported that microenvironments can be created by artery wall cells in a multilayer coculture that exclude aqueous antioxidants and permit the formation of MM-LDL even in the presence of serum. Moreover, we demonstrated that HDL, and specifically HDL_2, inhibits the formation of MM-LDL from native LDL (Navab et al., 1991a).

In these studies, incubation of native LDL with serum containing cocultures of human aortic wall cells for 24-48 hours resulted in a marked induction of monocyte binding to target endothelial cells and their subsequent transmigration and localization in the subendothelial space of the cocultures. The increase in monocyte migration was most likely and largely due to the increased levels of MCP-1 since it was completely blocked by an antibody to MCP-1 and it was associated with increased levels of MCP-1 protein. Inclusion of HDL along with LDL inhibited the observed induction of monocyte transmigration as did the pretreatment of the cocultures with antioxidants. The effect of HDL and antioxidants appeared to be confined to the events that occured during the initial steps of interaction with LDL and the artery wall cells, since HDL or antioxidants did not prevent the induction of monocyte transmigration brought about by LDL that had previously been incubated with and presumably modified by the cocultures, and subsequently incubated with fresh sets of untreated cocultures. Incubation of LDL cell or smooth muscle cells alone did not result in increased monocyte transmigration. Therefore, it seems likely that LDL is modified in the microenvironment formed by the extracellular matrix components produced by the interaction of endothelial cell and smooth muscle cells (Navab et al., 1988; Navab et al., 1991b; Merrilees & Scott, 1981). The serum components that potently suppress cell-dependent modification of LDL (Cathcart et al., 1985) are apparently excluded from this space. The coculture-modified LDL then induces MCP-1 production by endothelial cell and smooth muscle cells which presumably results in the establishment of a chemotactic gradient across the endothelial monolayer in the coculture. This is schematically shown in Figure 1.

Previous studies of the modification of LDL by endothelial cell (Morel et al., 1984; Steinbrecker et al., 1984), by smooth muscle cells (Morel et al., 1984; Steinbrecker et al., 1984; Heineke et al., 1984), and by macrophages (Cathcart et al., 1985) in culture all used serum-free medium and obtained highly modified LDL (Steinberg et al., 1989). In addition, the culture medium used in these studies (Ham's F10) contained eight times

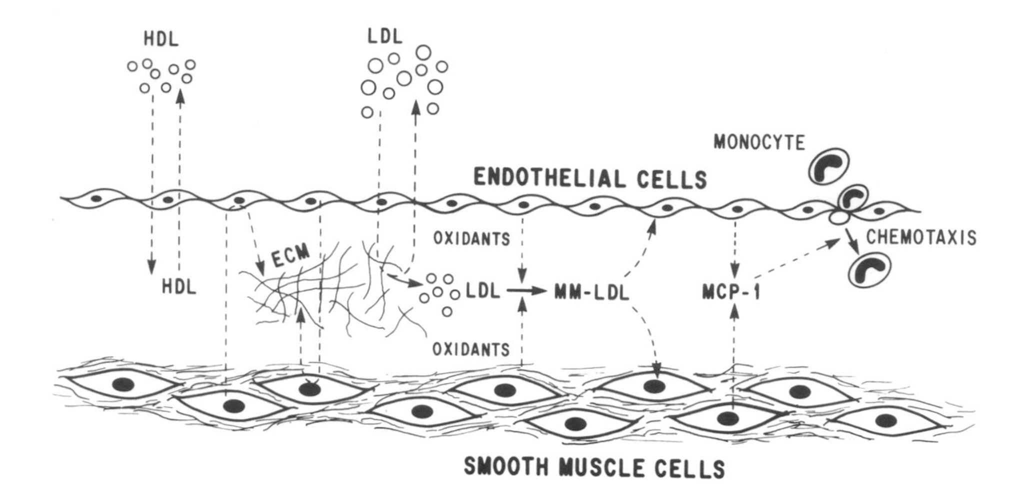

Fig. 1. A schematic demonstration of the proposed interactions of artery wall cells in coculture.

higher levels of iron (compared to medium 199 used in our studies) which is known to catalyze the oxidation of LDL (Steinbrecker et al., 1984; Heineke et al., 1984). The mild degree of modification produced in the coculture system is presumably a result of the action of prooxidants in the microenvironment largely sequestered from the effect of antioxidants present in the serum in the serum in the coculture. We observed that the LDL and the artery wall cells had to be in close contact to modify the LDL so that it would stimulate monocyte transmigration. Similarly, the protective effect of HDL was eliminated if HDL was separated from the cells and LDL by a filter that was impermeable to HDL.

The reisolated cell-modified LDL produced in the serum containing cocultures was examined for the changes associated with highly oxidized LDL (Cathcart et al., 1985). The LDL reisolated from the cocultures had a buoyant density similar to that of native-LDL (d=1.019-1.063), was not toxic to human aortic endothelial cell or smooth muscle cells, or human monocytes, and was not chemotactic for monocytes by itself. Moreover, the LDL reisolated from cocultures had the same electrophoretic mobility on agarose gel, the same content of conjugated dienes, the same fluorescence emission at 430 nm when excited at 360 nm, and the same rate of uptake and degradation by human monocytes-macrophages as did native-LDL.

The mechanism of LDL modification in the artery wall is not known. The modification migth result from the release of superoxide anions from the artery wall cells, the action of membrane bound enzymes on LDL, and/or the transfer of cellular lipid peroxides to LDL (Steinberg et al., 1989; Wiztum & Steimberg, 1989). The loss of antioxidants such as alpha-tocopherol and the peroxidation of polyunsaturated fatty acids in the LDL lipids appears to be the initiation step in the modification of LDL (Esterbauer et al., 1987). Morel and colleagues (Morel et al., 1984) and Steinbrecher et al. (Steinbrecher et al., 1984) have demonstrated that cellular modification of LDL in the absence of serum was completely inhibited by alpha-tocopherol or butylated hydroxytoluene. Copper or iron at sufficiently high concentration (3-5 µM) in the absence of serum are capable of producing oxidatively modified LDL (Steinbrecher et al., 1984). This has prompted the suggestion that the major contributions of the cells in LDL modification is to enhance the oxidative environment (Steinbrecher et al., 1989). Since in previous studies the addition of serum inhibited the LDL modification, it was suggested that in vivo, the process must occur extravascularly in microenvironments protected from naturally occurring antioxidants (Steinbrecher et al., 1989). The coculture system used in the present studies appears to provide such a microenvironment. The present observation that HDL at levels as low as 50 µg/ml was able to almost completely prevent the LDL-induced effects demonstrates the high protective capacity of HDL in these LDL-cell interactions. Hessler and his colleagues (Hessler et al., 1979) originally demonstrated that HDL protected endothelial cell in culture from the cytotoxic effects of oxidized LDL . These authors attributed the HDL effect to its protein-phospholipid components (Hessler et al., 1979).

Van Hinsbergh and coworkers (van Hinsbergh, 1986) subsequently demonstrated that HDL prevented the production of highly modified LDL by endothelial cells. Parthasarathy and colleagues (Parthasarathy et al., 1990a) reported that inclusion of HDL in serum-less cultures of endothelial cells containing LDL had a profound inhibitory effect on the subsequent degradation of the incubated and highly modified LDL by macrophages (Parthasarathy et al., 1990b). Preincubation of the cells with antioxidants before the incubation with LDL, markedly blocked LDL modification. This suggests that the artery wall cells in culture were capable of storing sufficient quantities of antioxidants to prevent the subsequent release of active oxygen species or to inhibit the production of oxidized cellular lipids which have been proposed to be potential contributors to the initiation of lipid oxidation in LDL (Steinberg et al., 1989). There was a significant variability in the effect of LDL preparations from various donors in these experiments. It is possible that the variation in the antioxidant content of different LDL preparations or qualitative and quantitative variations in the fatty acid composition of the different LDL preparations were at least partially responsible for the observed differences as suggested by others (Parthasarathy et al., 1990b, Lenz et al., 1990).

Possible components of the microenvironment that permit LDL to be oxidized in a serum containing coculture include the matrix, the cell membranes, and the lipoproteins themselves.
The determinants of the matrix that might be important include collagen, elastin, fibronectin, laminin, glycosaminoglycans, and

proteoglycans. Examination of autopsy specimen and those from experimental atherosclerosis in laboratory animals has demonstrated quantitative and qualitative alterations of matrix components including collagen and fibronectin.
Cell-cell interactions may be important in generating a microenvironment that excludes aqueous antioxidants. We have reported that in the coculture system (Navab et al., 1991b) a close approximation of the smooth muscle cells to the endothelial cells resulted in increased levels of fibronectin and collagen.

A third component of the microenvironment is LDL itself. As reported previously LDL becomes entrapped in the three dimensional matrix cage work of the subendothelial space within two hours of injection into a rabbit. Here, the lipids in LDL appear to coalesce into large lipid vesicles (Nievelstein et al., 1991). these vesicles have been seen in rabbits fed cholesterol, in WHHL rabbits (126) and similar lipid vesicles have been reported in human lesions (Frank et al., 1989; Guyton et al., 1988; Guyton et al., 1988). The formation of such large lipid vescicles in close proximity to cell membranes would create a very hydrophobic microenvironment that would favour the exclusion of aqueous antioxidants potentiating LDL modification.

The ability of coculture cells to generate MM-LDL from native-LDL was markedly inhibited by the pretreatment of the coculture with antioxidants , such as probucol, α-tocopherol, or β-carotene (Navab et al., 1991a). This finding is consistent with the La Jolla group's hypothesis that the transfer of cellular membrane lipid peroxides to LDL is an essential first step in initiating LDL lipid oxidation (Witztum et al., 1991).

We and others have noted variability in LDL from different donors to under go modification to MM-LDL. This variability may in part be due to difference in the antioxidant contents of these LDL preparations. It has been demonstrated that alteration of antioxidant (Stein, 1991; Sattler et al., 1991; Dieber-Rotheneder et al., 1991; Esterbauer et al., 1991) and monounsaturated fatty acid content (Lenz et al., 1990) affect the ability of LDL to be oxidized by metal ions or cells in serumless medium. However, these studies simply determined TBARS or conjugated diene content and conversion to highly oxidized LDL that recognized by the scavenger receptor. We have found that one can not predict the biologic activity characteristic of MM-LDL in an LDL preparation based on its TBARS or conjugated diene content. All preparations of MM-LDL which are biologically active contain increased TBARS and conjugated dienes. However, the reverse is not true, increased TBARS and conjugated dienes are not predictive of biologic activity (Liao et al., 1991). This means that there are specific oxidized lipids that mediate the biologic activity of MM-LDL.

A number of studies have demonstrated that distinct subclasses of LDL isolated by ultracentrifugation display metabolic and biochemical diversity. La Belle and Krauss (1990) have shown that a decline in carbohydrate content of a dense LDL subclass is associated with increased risk of myocardial infarction. Others have shown that the dense LDL subclass is more susceptible to copper oxidation (de Graaf et al., 1991a; 1991b). However, as noted above, susceptibility to the formation of conjugated dienes is not predictive of the ability of native-LDL to be converted to MM-LDL by the cells of the artery wall.

We have reported that HDL_2 but not HDL_3 prevented the conversion of native-LDL into MM-LDL in the coculture system. Podet et al. (1991) reported that apoE containing HDL_2 almost completely inhibited LDL binding to elastin while HDL_3 which did not contain apoE was relatively ineffective. One possibility to explain our findings is that apoE containing HDL_2 is able to bind to the matrix components in the cocultures. Thus, HDL_2 is able to enter and remain for sufficient time in the microenvironment where LDL is converted to MML-LDL such that the HDL_2 can act as a sink into which lipid peroxides can partition preventing their conversion into biologically active lipid oxides in LDL. The corollary of this hypothesis is that HDL_3 which lacks apoE is not able to enter or remain in the microenvironment to the extent that HDL_2 does.

We have hypothesized that there are natural mechanisms for regulating the inflammatory reaction induced by MM-LDL and mechanisms for preventing the formation of MM-LDL. In studies done with desferoxamine Mn a superoxide oxide dismutase mimic we demonstrated that the incubation of rabbit aortic endothelial cells with this component resulted in a significant reduction in the induction of both M-CSF and MM-LDL. This result then suggests that oxygen free radicals or their by-products may be responsible for the induction in artery wall cells of certain genes involved in the early stages of atherogenesis.

We recently observed that modification of LDL, and the resulting monocyte transmigration in cocultures of human aortic wall cells was inhibited by a new antinflammatory compound, a leucine derivative termed Leumedin (Burch et al., 1991). In these studies, incubation of multilayer cocultures of human aortic endothelial cells and smooth muscle cells with LDL, in the presence of human serum resulted in a marked increase in the transmigration of monocytes. Addition of a new antiinflammatory agent, n-fluorenyl-methoxy-carbonyl-leucine, leumedin, with LDL, inhibited the monocyte migration. Addition of leumedin, 24 h after the addition of LDL, did not prevent the modification of LDL and the resulting monocyte migration. Pretreatment of the cocultures with leumedin prior to the addition of LDL, or pretreatment of LDL with leumedin, inhibited the subsequent monocyte transmigration. LDL that was preincubated with leumedin and refloated by ultracentrifugation did not induce monocyte transmigration in the cocultures, whereas LDL incubated with aspirin and then refloated, induced monocyte transmigration to a degree similar to the control LDL. Incubation of human serum with radiolabeled leumedin and subsequent isolation of the lipoproteins demonstrated that leumedin has associated with lipoproteins to a degree that was 35-fold more than it had associated with the d>1.21 g/ml fraction. We have concluded that: 1) Unlike aspirin, leumedin readily associates with LDL and inhibits LDL modification in cocultures of human aortic cells and 2) The LDL-leumedin complex is highly stable.

Another line of research in our laboratories has focused on HDL obtained from the sera of individuals during the acute phase. Serum amyloid A (SAA) is a family of proteins some of which appear in plasma as acute phase reactants following a variety of stimuli including surgery, myocardial infarction, infection, and even in more chronic illnesses such as arthritis (Whitehead et al., 1992, Strachan et al., 1989).
These proteins are the precursors to amyloid protein A which in turn is the precursor to amyloid fibrils in secondary amyloidosis (Hebert & Gervais, 1990). SAA are associated with lipoproteins and are found mainly in the HDL and VLDL density classes. In the croton oil rabbit model of inflammation up to 80% of proteins in HDL were replaced by SAA 72 hours after injection of the inflammatory stimulus. The SAA-enriched particles became denser but larger, had slower electrophoretic mobility and were depleted of apoAI, cholesterol, triglyceride, and phospholipid. SAA also increased in VLDL while apo E decreased (Cabana et al., 1989). After myocardial infarction as much as 38% of the total apoproteins in VLDL and HDL were found to be SAA (Feussner et al., 1991).

We have recently observed that unlike normal HDL, acute-phase HDL does not prevent, but amplifies the modification of LDL and the resulting monocyte transmigration in cocultures of human aortic wall cells. In these studies, incubation of LDL with multilayer cocultures of human aortic endothelial cells and smooth muscle cells in the presence of human serum, resulted in mild oxidative modification of LDL. Subsequent addition of human monocytes to the endothelial side of cocultures, after LDL modification, resulted in a significant increase in transmigration of monocytes into the subendothelium of the cocultures. Inclusion of HDL along with LDL in cocultures, prevented the LDL modification and monocyte migration. Normal HDL or HDL containing the major acute-phase reactant, serum amyloid A (SAA-HDL) isolated from the sera of patients with inflammatory conditions including arthritis, or from individuals following cardiac surgery, when incubated with cocultures, *without* the addition of LDL, did not induce increased monocyte transmigration. Inclusion of SAA-HDL together with LDL, however, resulted in a marked increase in monocyte transmigration. From these studies we have concluded that: 1)Substitution of SAA for apolipoprotein A in HDL during the acute-phase reaction, renders HDL incapable of preventing LDL modification by cocultures of human aortic wall cells, and 2)SAA-HDL exerts a synergistic effect on LDL modification and amplifies the resulting monocyte transmigration in cocultures.

On hypothesizing the mechanisms involved in the interaction of SAA-HDL or the cells, SAA-HDL could alter the microenviroment of artery wall by stimulating collagenase production. Brinckerhoff et al. (1989) demonstrated that SAA induces collagen synthesis. Another possibility is that SAA-HDL carries lipid peroxides that by themselves are not sufficient to induce an inflammatory response but when transferred to LDL provide additional seeding to that donated from the cells and hence favors MM-LDL formation. In a similar way the lipids in SAA-HDL may facilitate the transfer and retention of cellular lipid peroxides to LDL. Yet another possibility is that SAA-HDL contains enzymes that facilitate the conversion of LDL to MM-LDL or conversely lacks enzymes that are necessary to inhibit this conversion. Based on the work of Prescott and colleagues PAF acetylhydrolase may play a role in this regard. It has been shown that apo-SAA preferentially binds with neutrophil membranes and HDL has been shown to activate protein kinase C in endothelial cells. Consequently, Nel et al. (1988) demonstrated that the only apoprotein in HDL that was phosphorylated by protein kinase C was apo-SAA. Perhaps, SAA-HDL activates protein kinase C and this in turn stimulates cellular lipid peroxidation that enhances the seeding of LDL.

Based on findings from our group and those of other investigators we have proposed the following scheme in Figure 2 for the sequence of events in the development of foam cells in the artery wall:

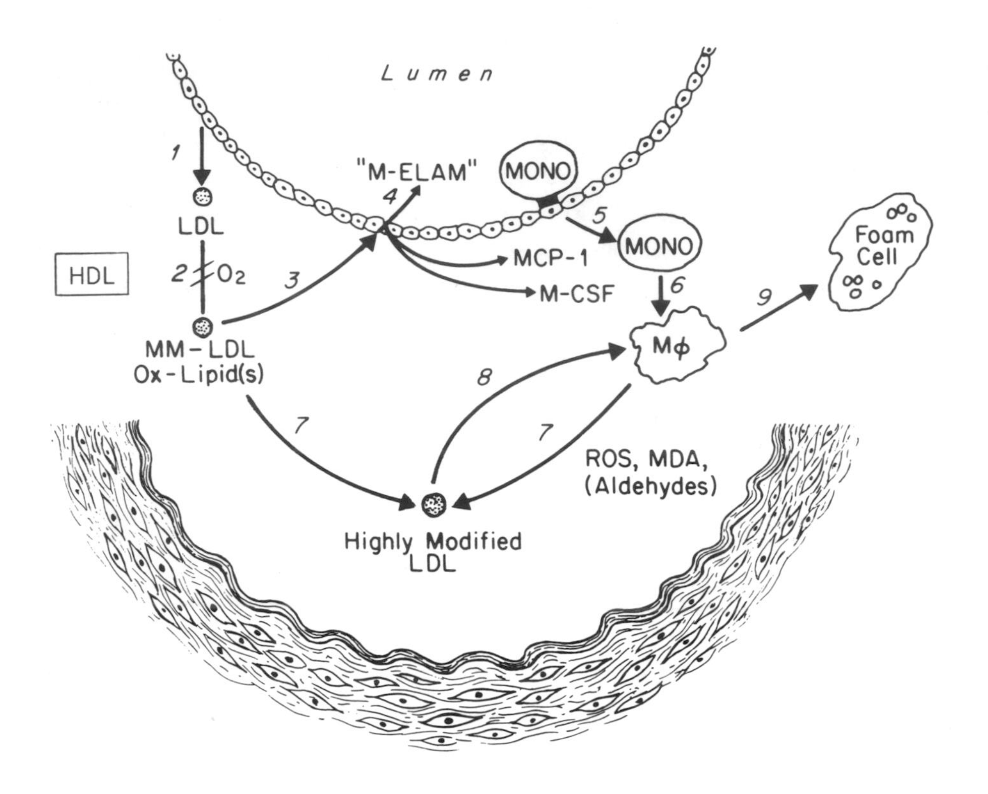

Fig. 2. A proposed model for foam cell formation in the arthery wall.

LDL becomes trapped in microenviroments in the extracellular matrix of the subendothelial space sequestered from plasma anti-oxidants. Reactive oxygen species and/or oxidized cellular lipids can be transferred to LDL and initiate the propagation of LDL lipid peroxidation. Since at the early stages of atherogenesis the subendothelial space is largely acellular and does not contain a significant number of monocyte-macrophages releasing high levels of prooxidants, the resulting LDL is only minimally oxidized. This minimally modified LDL (MM-LDL) can then induce the overlaying endothelium to express an adhesion molecule(s) for monocytes, secrete MCP-1, and M-CSF. These molecular events in turn induce monocyte binding, monocyte migration into the subendothelial space, and monocyte differentiation into macrophages. The macrophages could subsequently release reactive oxygen species and aldehydes further modifying the MM-LDL into a highly modified form which is then recognized and taken up by the macrophage scavenger and/or oxidized-LDL receptor, resulting in foam cell formation. If HDL (presumably HDL_2) is present in sufficient concentrations the formation of biologically active MM-LDL is prevented and the inflammatory reaction may be blocked (Navab et al., 1991a).

ACKNOWLEDGEMENTS

This work was supported in part by the U.S. Public Health Services grants HL 30568, IT 32 HL 07412, and RR 865; by the Laubisch, Rachel Israel Berro; and M. K. Grey Funds.

REFERENCES

Berliner, J.A., Territo, M.C., Sevanian, A., Ramin, S., Kim, J.A., Bamshad, B., Esterson, M., and Fogelman, A.M. (1990) J. Clin. Invest. 82:1260-1266.
Brinckerhoff, C.E., Mitchell, T.I., Karmilowicz, M.J., Kluve-Beckerman, B., and Benson, M.D. (1989) Science 243, 655-657.
Burch, R.M., Weitzberg, M., Blok, N., Muhlhauser, R., Martin, D., Farmer, S.G., Bator, J.M., Connor, J.R., Ko, C., Kuhn, W. et al. (1991) Proc. Natl. Acad. Sci. USA 88, 355-359.
Cabana, V.G., Siegel, J.N., and Sabesin, S.M. (1989) J. Lipid Res. 30, 39-49.
Cathcart, M.K., Morel, D.W., and Chisolm, G.M. (1985) J. Leukocyte Biol. 38, 341-350.
Cushing, S.D., Berliner, J.A., Valente, A.J., Territo, M.C., Navab M., Parhamai, F., Gerrity, R., Schwartz, C.J., and Fogelman, A.M. (1990) Proc. Natl. Acad. Sci. 87, 5134-5138.
de Graaf, J., Hak-Lemmers, H.L., Hectors, M.P., Demacker, P.N., Hendriks, J.C., Stalenhoef, A.F. (1991a) Arterioscler. Thromb. 11, 298-306.
de Graaf, J., Hendriks, J.C., Demacker, P.N., and Stalenhoef, A.F. (1991b) Circulation 84, 484-489.
Dieber-Rotheneder, M., Puhl, H., Waeg, G., Striegl, G., and Esterbauer, H. (1991) J. Lipid Res. 32, 1325-1332.

Esterbauer, H., Jurgens, G., Quehenberger, O., and Koller, E. (1987) J. Lipid Res. 28, 495-509.
Esterbauer, H., Dieber-Rotheneder, M., Striegl, G., and Waeg, G. (1991) Am. J. Clin. Nutr. 53, 314-321.
Feussner, G., Schuster, M., and Ziegler, R. (1991) 12, 283-286.
Fogelman, A.M., Seager, J., Hokom, M., Child, J.S., and Edwards, P.A. (1980) Proc Natl Acad Sci USA 77, 2214-2218.
Frank, J.S., and Fogelman, A.M. (1989) J. Lipid Res. 30, 967-978.
Guyton, J.R., and Klemp, K.F. (1988) J. Histochem. cytochem. 36, 1319-1328.
Guyton, J.R., and Klemp, K.F. (1989) Am. J. Pathol. 134, 705-717.
Haberland, M.E., Fong, D., and Cheng, H. (1988). Science 241, 215-218.
Heinecke, J.W., Rosen, H., and Chait, A. (1984) J. Clin. Invest. 74, 1890-1894.
Henricksen, T., Mahoney, E.M., and Steinberg, D. (1981) Proc. Natl. Acad. Sci. USA 78, 6499-6503.
Hebert, L., and Gervais, F. (1990) Scand. J. Immunol. 31, 167-173.
Hessler, J.R., Robertson, Jr. A.L., and Chisolm, G.M. (1979) Arteriosclerosis 32, 213-229.
La Belle, M., and Krauss, R.M. (1990) J. Lipid Res. 31, 1577-1588.
Lenz, M.L., Hughes, H., Mitchell, J.R., Via, D.P., Guyton, J.R., Taylor, A.A., Gotto, A.M., Jr., and Smith, C.V. (1990) J. Lipid Res. 31, 1043-1050.
Liao, F., Berliner, J.A., Mehrabian, M., Navab, M., Demer, L.L., Lusis, A., and Fogelman, A.M. (1991) J. Clin. Invest. 87, 2253-2257.
Marx, J, (1987) Science 235, 529-531.
Merrilees, M.J., and Scott, L. (1981) Atherosclerosis 39, 147-161.
Morel, D.W., DiCorleto, P.E., and Chisolm, G.M. (1984) Arteriosclerosis 4, 357-364.
Navab, M., Hough, G.P., Stevenson, L.W., Drinkwater, D.C., Laks, H., and Fogelman, A.M. (1988) J. Clin. Invest. 82, 1853-1863.
Navab, M., Imes, S.S., Hough, G.P., Hama, S.Y., Ross, L.A., Bork, R.W., Valente, A.J., Berliner, J.A., Drinkwater, D.C., Laks, H., and Fogelman, A.M. (1991a) J. Clin. Invest. 88, 2039-2046.
Navab, M., Liao, F., Hough, G.P., Rossi, L.A., Van Lenten, B.J., Rajavashisth, T.B., Lusis, A.J., Laks, H., Drinkwater, D.C., and Fogelman, A.M. (1991b) J. Clin. Invest. 87, 1763-1772.
Nel, A.E., De Beer, M.C., Shephard, E.G., Strachan, A.F., Vandenplas, M.I., and De Beer F (1988) Biochem. J. 255, 29-34.
Nelken, N.A., Coughllin, S.R., Gordon, S.R., Gordon, D., and Wilcox, J.N. (1991) J. Clin. Invest. 88, 1121-1127.
Nievelstein, P.F., Gogelman, A.M., Mottino, G., Frank, J.S. (1991) Arterioscler. Thromb. 11, 1795-1805.
Palinski, W., Rosenfeld, M.E., Yla-Herttuala, S., Gurtner, G.C., Socher, S.S., Butler, S.W., Parthasarathy, S., Carew, T.E.,

Steinberg, D., and Witztum, J.L. (1989) Proc. Natl. Acad. Sci. 86, 1372-1380.
Parthasarathy, S., Barnett, J., and Fong, L.G. (1990a) Biochim. Biophys. Acta 1044, 275-283.
Parthasarathy, S., Khoo, J.C., Miller, E., Barnett, J., Wtiztum, J.L., Rajavashisth, T.B., Andalibi, A., Territo, M.C., Berliner, J.A., Navab, M., Fogelman, A.M., and Lusis, A.J. (1990) Nature 344, 254-257.
Podet, E.J., Shaffer, D.R., Gianturco, S.H., Bradley, W.A., Yang, C.Y., and Guyton, J.R. (1991) Arterioscler. Thromb. 11. 116-122.
Rajavashisth, TB, Andalibi, A., Territo, M.C., Berliner, JA, Navab, M., Fogelman, A.M., and Steinberg, D. (1990) Nature 344, 254-257.
Sattler, W., Puhl, H., Hayn, M., Kostner, G.M., and Esterbauer, H. (1991) Anal. Biochem. 198, 184-190.
Sparrow, C.P., Parthasarathy, S., and Steinberg, D.(1989) J Biol. Chem. 264, 2599-2604.
Stein, Y. (1991) Am. J. Clin. Nutr. 53, 899-907.
Steinberg, D. (1990) Proc. Natl. Acad. Sci. USA 87: 3894-3898.
Steinberg, D., Parthasarathy, S., Carew, T.E., Khoo, J.C., and Witztum, J.L., (1989) N. Engl. J. Med. 320, 915-924.
Steinbrecher, U., Parthasarathy, S., Leake, D.S., Witztum, J.L., Steinberg, D. (1984) Proc. Natl. Acad. Sci. USA 83, 3883-3887.
Steinbrecher, U. (1987) J. Biol. Chem. 262, 3603-3608 1987.

Strachan, A.F., Brandt, W.F., Woo, P., van der Westhuyzen, D.R.,
Coetzee, G.A., de Beer, M.C., Shephard, E.G., and de Beer, F.C. (1989) J. Biol. Chem. 264, 18368-18373.
van Hinsbergh, V.W., Scheffer, M., Havekes, L., and Kemper, H.J. (1986) Biochim. Biophys. Acta 878: 49-64.
Yla-Herttuala, S., Lipton, B.A., Rosenfeld, M.E., Sarkioja, T., Yoshimora T, Leonard, E.J., Witztum, J.L., and Steinberg, D. (1991) Proc. Natl. Acad. Sci. USA 88: 5252-5256.
Whitehead, A.S., de Beer, M.C., Steel, D.M., Rits, M., Lelias, J.M., Lane, W.S., de Beer FC (1992) J. Biol. Chem. 267, 3862-3867.
Witztum, J.L., and Steinberg, D. (1991) J. Clin. Invest. 88, 1785-1792.

Antioxidants

Free Radicals: From Basic Science to Medicine
G. Poli, E. Albano & M. U. Dianzani (eds.)

ANTIOXIDANT DEFENSES IN EUKARIOTIC CELLS: AN OVERVIEW

Etsuo Niki

Research Center for Advanced Science and Technology, The University of Tokyo, 4-6-1 Komaba, Meguro-ku, Tokyo 153, Japan

SUMMARY: Our bodies are protected from free radical-induced oxidative damage by various antioxidants with different functions which constitute a defense system either independently, cooperatively, or even synergistically.

The aerobic organisms are protected against oxidative damage induced by free radicals by far more potent defense systems than those in foods, oils and plastics. The functions and actions of the antioxidant defense have been studied extensively both *in vitro* and *in vivo*. The epidemiological studies imply the involvement of free radical-mediated oxidative damage in various diseases and also the protective role of various antioxidants. The therapeutic applications of natural and synthetic antioxidants have been explored.

In the last decade, there have been a substantial progress in both fundamental and practical aspects, but there still remain a number of questions and problems that should be addressed and clarified. The antioxidants acting in the defense systems are classified into 3 categories from their functions (Fig.1). The fourth line of defense which is not shown in Fig. 1 is the adaptation (see text).

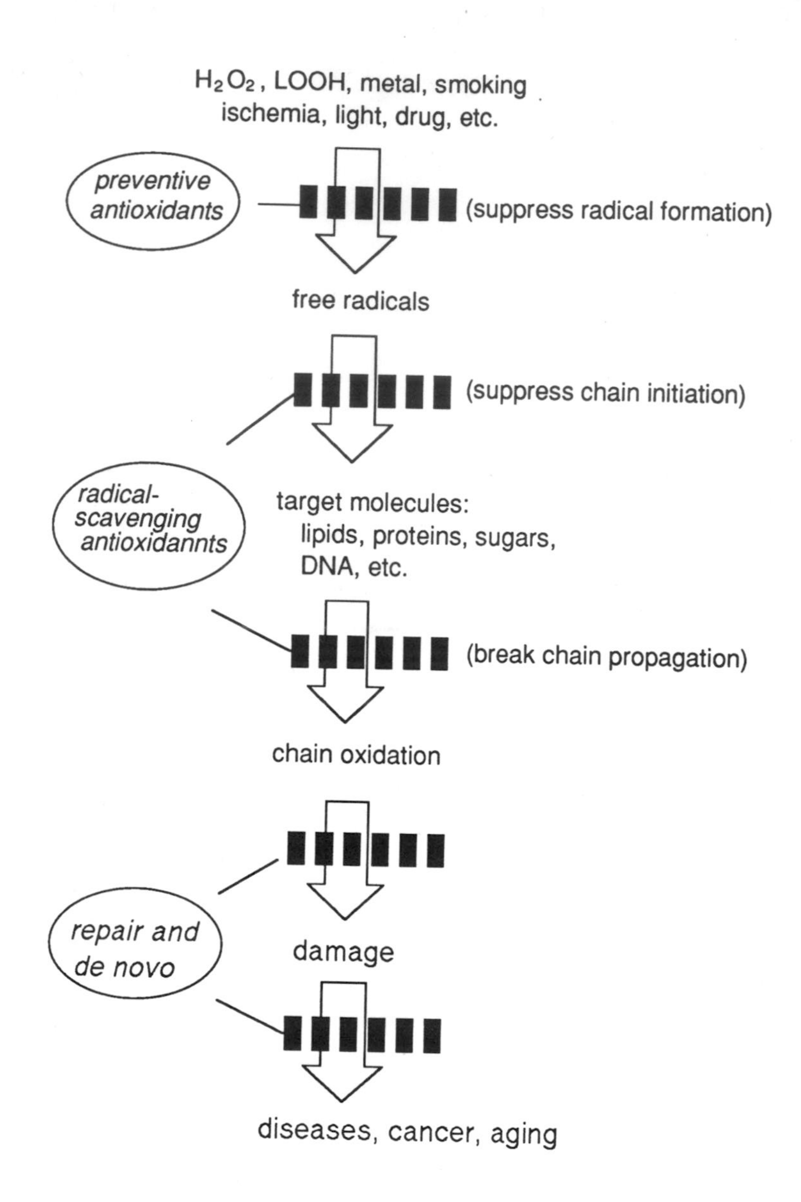

Fig. 1 Antioxidants in the defense system against free radical-induced oxidative damage.

Table I Decomposition hydroperoxides and hydrogen peroxide by enzymes

Enzymes	Function
Glutathione peroxidase	Reduction of fatty acid hydroperoxides and hydrogen peroxide $LOOH + 2GSH \longrightarrow LOH + H_2O + GSSG$ $H_2O_2 + 2GSH \longrightarrow 2H_2O + GSSG$
Glutathione-S-Transferase	Reduction of fatty acid hydroperoxide
Phospholipid hydroperoxide glutathione peroxidase	Reduction of phospholipid hydroperoxide $PLOOH + 2GSH \longrightarrow PLOH + H_2O + GSSG$
Peroxidase	Reduction of fatty acid hydroperoxide and hydrogen peroxide $H_2O_2 + AH_2 \longrightarrow 2H_2O + A$
Catalase	Reduction of hydrogen peroxide $2H_2O_2 \longrightarrow 2H_2O + O_2$

PREVENTIVE ANTIOXIDANTS

The first line of defense is the preventive antioxidants, which suppress the formation of free radicals. Although the precise mechanism and site of radical formation *in vivo* are not well elucidated yet, the metal-induced decompositions of hydroperoxides and hydrogen peroxide must be one of the important sources. To suppress such reactions, some antioxidants reduce hydroperoxides and hydrogen peroxide beforehand to alcohols and water respectively without generation of free radicals and some proteins sequester metal ions.

Table I summarizes the enzymes which reduce hydroperoxides and hydrogen peroxide. Glutathione peroxidase, glutathione-

S-transferase, phospholipid hydroperoxide glutathione peroxidase (PHGPX) and peroxidase are known to decompose lipid hydroperoxides to corresponding alcohols. PHGPX is unique in that it can reduce hydroperoxides of phospholipids integrated into biomembranes (Ursini et al. 1985). Its molecualr structure has been also studied (Flohe et al. 1992). Glutathione peroxidase, peroxidase, and catalase reduce hydrogen peroxide to water.

The transition metal ions such as iron and copper are also accepted to play a vital role in the formation of free radials *in vivo* (Halliwell et al., 1984). Transferrin, ferritin, and lactoferrin sequester iron, while ceruloplasmin sequesters copper. Wilson's disease, an inherited metabolic defect characterized by low concentrations of ceruloplasmin in the blood, is accepted to be induced by copper-stimulated free radical oxidations. The chelating agents such as penicillamine which promote copper excretion are used for treatment of Wilson's disease.

Carotenoids such as β-carotene and lycopene are also accepted as important antioxidants. Singlet oxygen oxidizes unsaturated lipids rapidly to give lipid hydroperoxides which are important precursors of alkoxyl and peroxyl radicals. Carotenoids quench singlet oxygen quite rapidly and inhibit the hydroperoxide formation.

Superoxide is one of the active oxygen species which is being watched with the keenest interest. Although its chemical reactivity is not high, its direct attack on target molecule is not ruled out. More importantly, it may reduce sequestered iron to give active iron (II) or it may be protonated to yield hydroperoxyl radical which acts as a real active species. It may also attack carbon-halogen bonds to give active radicals. In any event, superoxide dismutase (SOD) catalyzes the disproportionation of superoxide to hydrogen peroxide. The application of SOD, especially for ischemia-reperfusion and organ transplantation, has received much attention.

RADICAL-SCAVENGING ANTIOXIDANTS

Although the generation of free radicals *in vivo* is minimized by the above-mentioned preventive antioxidants, small amounts of free radicals may be formed *in vivo* or taken into body exogenously. The second line of defense is the antioxidants which scavenge the active radicals to suppress chain initiation and/or break the chain propagation reactions. Various endogenous radical-scavenging antioxidants are known: some are hydrophilic and others are lipophilic. Vitamin C, uric acid, bilirubin, albumin and thiols are hydrophilic, radical-scavenging antioxidants, while vitamin E and ubiquinol are lipophilic radical-scavenging antioxidants. The function of carotenoids as a radical-scavenging antioxidant has been also reported (Burton et al., 1984) in addition to the quenching of singlet oxygen.

The antioxidant activities of radical-scavenging antioxidants are determined not only by their inherent chemical reactivities and concentrations but also by the location and mobility of antioxidants at the micro-environment. Vitamin E is accepted as the most potent radical-scavenging lipophilic antioxidant, but it has been shown that its mobility, especially vertical mobility, in the membranes is much reduced and hence the efficiency for radical scavenging is reduced in the membranes (Takahashi, et al., 1989).

Other important factors which determine the overall antioxidant activities are the fate of the radical derived from the antioxidant and the site of radical formation. In general, the radical-scavenging antioxidant donates hydrogen to active radicals and new radical is formed from the antioxidant. This antioxidant-derived radical may undergo various reactions. Figure 2 shows the fate of vitamin E radical as an example. It may scavenge another radical to give non-radical product, react with reducing agent to regenerate vitamin E, or, under some circumstances, it may react with lipids or lipid hydroperoxide to give active radicals, which initiate another chain reaction. The relative importance of these reactions, and hence the overall antioxidant activity, depends on various conditions such as concentrations of radicals and reducing agents.

The action of ubiquinol, a reduced form of coenzyme Q, in the membranes and low density lipoprotein has been the subject of extensive studies (Beyer, 1990).

The site of free radical formation is also important in determining the antioxidant activity. The hydrophilic antioxidants such as vitamin C effectively scavenge aqueous radicals but they can not scavenge radicals within the membranes and can not act as a chain-breaking antioxidant. Even a small fraction of radicals formed initially in the aqueous phase but penetrated into the membrane induce the chain reaction and amplify the damage. The lipophilic antioxidants are responsible for scavenging such lipophilic, chain-carrying radicals within the membranes.

REPAIR AND *DE NOVO* ANTIOXIDANTS

The third line of defense is the repair and *de novo* antioxidants. Various phospholipases repair the oxidatively damaged phospholipids and their activities are accepted to be stimulated by oxidation. Among others, phospholipase A_2 has

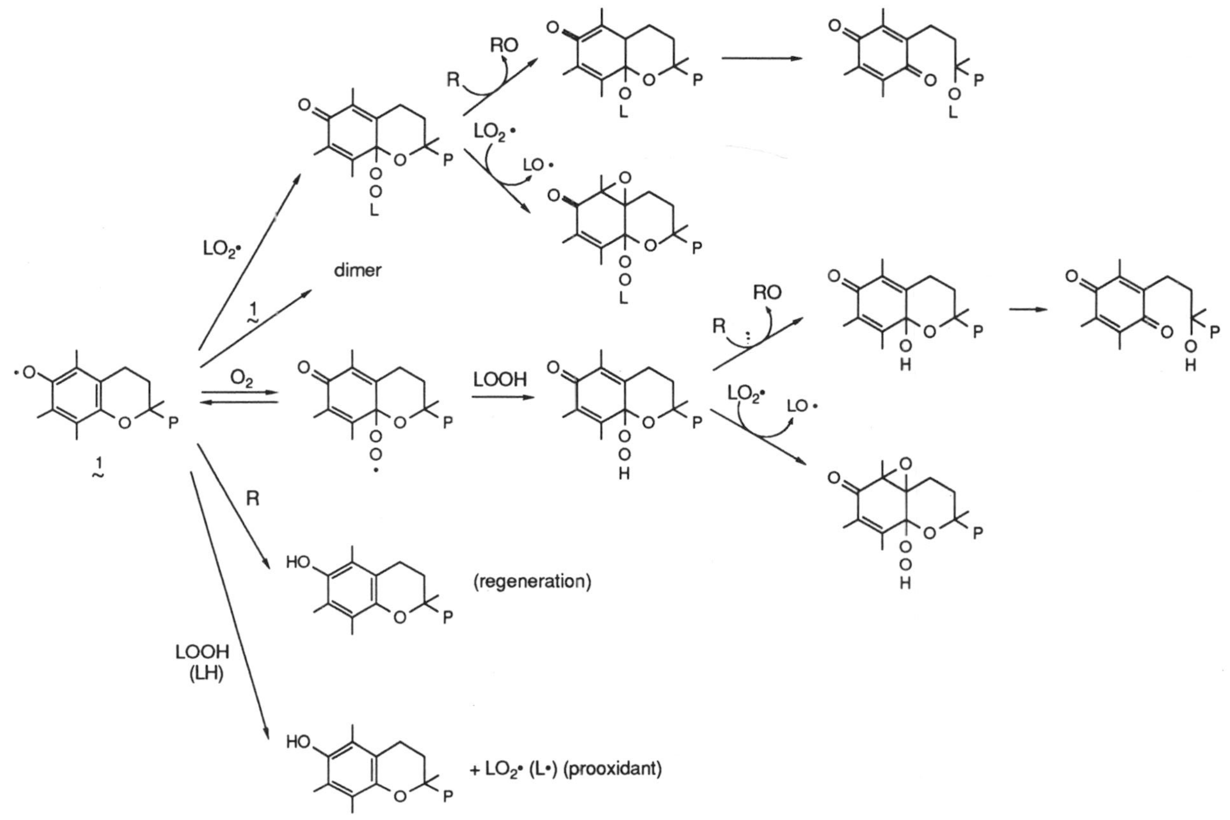

Fig. 2 Fate of vitamin E radical. R: reductant; P: phytyl side chain

received much attention since this enzyme is assumed to selectively cleave peroxidized lipids from the membranes, thus preventing the accumulation of toxic products (Sevanian, 1985). Upon action of phospholipase A_2 to the peroxidized phospholipid, lysophospholipid and free fatty acid hydroperoxide are formed. The released fatty acid hydroperoxide is reduced by peroxidases to corresponding alcohol and detoxicated. The lysophospholipid is reacylated by a fatty-acyl-coenzyme A to complete the repair.

The proteolytic enzymes, proteinases, proteases, and peptidases, present in the cytosol and in the mitochondria of mammalian cells recognize, degrade and remove oxidatively-modified proteins and prevent the accumulation of oxidized proteins. (Davies, 1988). The oxidatively-modified proteins are, in general, more susceptible to proteolytic degradation than are native proteins. The exception is the cross-linked proteins which exhibit decreased susceptibility to proteolytic digestion. Interestingly, the aging process has been reported to involve alterations in proteolytic capacity as well as primary and secondary defense capabilities.

The DNA repair systems also play an important role in the total defense system against oxidative damage. Various kinds of enzymes such as glycosylases and nucleases which repair the damaged DNA are known (Demple et al., 1991). A glycosylase enzyme removes a damaged base by cleaving the base-sugar bond to leave an apurinic/apyrimidinic (AP) site. The AP site is recognized by AP endonuclease. The damaged part of the strand is removed, new DNA synthesis fills the gap and finally a DNA ligase enzyme joins the newly synthesized DNA to the rest of the strand.

ADAPTATION MECHANISM

There is another important function called adaptation where the signal for the production and reactions of free radicals induces formation and transport of the appropriate antioxidant to the right site. In fact, it has been known that the formation of superoxide dismutase and catalase is induced by oxidative stress.

REFERENCES

Beyer, R. E. (1990) Free Rad. Biol. Med. 8, 545-565.

Burton, G. W. and Ingold, K. U. (1984) Science 224, 569-573.

Davies, K. J. A. (1988) in Cellular Antioxidant Defense Mechanisms, vol. II (Chow, C. K. ed.) CRC Press, Boca Raton, pp. 25-67.

Demple, B. and Levin, J. D. (1991) in H. Sies ed. Oxidative Stress: Oxidants and Antioxidants, Academic Press, London, pp. 119-154.

Flohe, L., Brigelius-Flohe, R., Maiorino, M., Roveri, A., Renmkens, J., Schuckelt, R., Strazburger, W., Ursini, F., and Wolf, B. (1992) in Biological Free Radical Oxidations and Antioxidants (Ursini, F and Cadenas, E. eds.) CLEUP, University Pub., Padova, pp. 123-125.

Halliwell, B. and Gutteridge, J. M. C. (1984) Biochem. J. 219, 14.

Sevanian, A. and Kim, E. (1985) J. Free Rad. Biol. Med., 1, 263-271.

Takahashi, M., Tsuchiya, J., and Niki, E. (1989) J. Am. Chem. Soc. 111, 6350-6353.

Ursini, F., Maiorino, M., and Gregolin, C. (1985) Biochim. Biophys. Acta, 839, 62-70.

Free Radicals: From Basic Science to Medicine
G. Poli, E. Albano & M. U. Dianzani (eds.)

ELECTRON PARAMAGNETIC RESONANCE STUDIES OF FLAVONOID COMPOUNDS.

Wolf Bors, Werner Heller*, Christa Michel and Kurt Stettmaier

Institut für Strahlenbiologie and *Institut für Biochemische Pflanzenpathologie, GSF-Forschungszentrum für Umwelt und Gesundheit, 8042 Neuherberg, Germany

SUMMARY

Flavonoids, while functioning as antioxidants, necessarily produce radical intermediates after scavenging initially formed oxidizing radicals. These predominantly pulse-radiolytic investigations have now been extended to include EPR spectroscopy as a powerful tool to obtain structural information of radicals. In the present work we used horseradish peroxidase/hydrogen peroxide-catalysed oxidation at pH 9 to obtain coupling constants of radical structures of six out of eight flavanones, flavones and flavonols. The EPR data were quite similar to those reported for the same compounds after autoxidation in strongly alkaline solutions containing an excess of DMSO (Kuhnle et al., 1969). Oxidation of the B-ring catechol structure can be assumed to be the predominant reaction also under our conditions. In additional experiments, the cations Mg^{2+} and Zn^{2+}, which have been shown to act as 'spin stabilizing' agents with other *o*-semiquinones, rather destabilized the radicals during the EPR observation period. We rationalize this by analogous complex formation, which in the case of flavonoids would rather lead to an <u>enhanced</u> decay of the intermediary *o*-semiquinones via nucleophilic rupture of the C_2-$C_{1'}$ bond.

INTRODUCTION

The antioxidative potential of flavonoids as widespread plant polyphenols is increasingly being recognized. However, except for pulse-radiolytic data (Bors & Saran, 1987, Bors et al., 1990, 1992), little information exists on the mechanisms of the formation and decay of the respective radicals. Aside from kinetic spectroscopy in pulse radiolysis, EPR spectroscopy is the best method to resolve structures of intermediary radical species. Only few attempts towards a structural assignment of the observed radical signals have so far been performed using EPR spectroscopy (Dixon et al., 1975, Jensen & Pedersen, 1983, Kuhnle et al., 1969). Jensen & Pedersen (1983) limited their studies to the intermediary radicals formed during the oxidative degradation of flavans such as (+)-catechin, while Dixon et al. (1975) investigated a few flavonoids, but mostly coumarins, in their EPR studies of naturally occurring hydroxypyrones. Only the early studies of Kuhnle et al. (1969) can be considered relevant, even though the flavonoid radicals were produced in strongly alkaline solutions containing 80% DMSO.

Hodnick et al. (1988) first described spin stabilization with Mg^{2+} and Zn^{2+} salts of *o*-semiquinones formed from strongly autoxidizable flavonoids. These authors as well as Canada et al. (1990) were mainly interested in the formation of oxygen radicals such as hydroxyl (·OH) or superoxide anion radicals ($O_2^{-}\cdot$) during redox cycling of these flavonoids. Both groups trapped these radicals with 5,5-dimethyl-1-pyrroline-N-oxide (DMPO). Interestingly, the formation of DMPO spin adducts of ·OH or $O_2^{-}\cdot$, and the competitive inhibition by several flavonoids was used repeatedly to demonstrate the antioxidative potential of these polyphenols (Husain et al., 1987, Ueno et al., 1984a,b, Yoshikawa et al., 1990). Husain et al. (1987) used HPLC with electrochemical detection to determine the DMPO spin adducts.

In the present study, we re-examined EPR spectra of radicals of a group of structurally related flavonoids, formed *in situ* by autoxidation in slightly alkaline aqueous solutions as well as by horseradish peroxidase/hydrogen peroxide (HRP/H_2O_2, Ohnishi et al., 1969). Complementary experiments were performed, adding acetate salts of Mg^{2+} and Zn^{2+} for spin stabilization studies (Kalyanaraman, 1990, Kalyanaraman & Sealy, 1982, Kalyanaraman et al., 1986,1987), in order to gain additional information on the structures and reactivities of flavonoid aroxyl radicals. In particular, we wanted to investigate,

(i) whether radicals formed with HRP/H_2O_2 are different from those generated under autoxidizing conditions, in particular since those of Kuhnle et al. (1969) were rather unphysiological;

(ii) whether potential radical sites at the 3/5-hydroxy plus 4-oxo structure could be spin-stabilized:

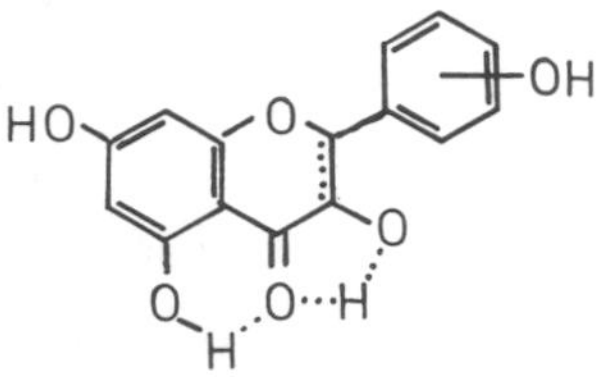

Compounds of four different structural groups of flavonoids, flavanones and flavones and their respective 3-hydroxy derivatives, shall be presented.

Scheme I

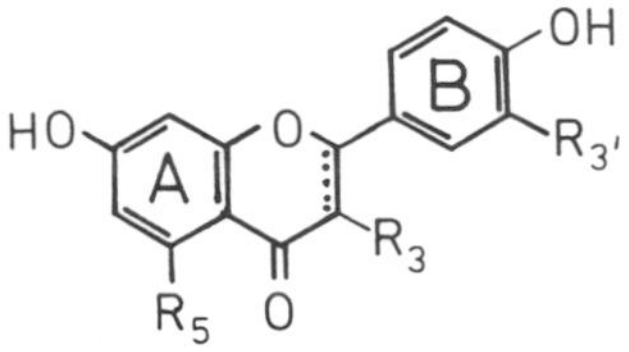

flavanones: 2,3-bond saturated

flavones: 2,3-double bond, no 3-OH group

flavonols: 2,3-double bond and 3-OH group

#1	eriodictyol	5,7,3',4'-tetrahydroxyflavanone
#2	dihydrokaempferol	3,5,7,4'-tetrahydroxyflavanone
#3	dihydrofisetin	3,7,3',4'-tetrahydroxyflavanone
#4	dihydroquercetin	3,5,7,3',4'-pentahydroxyflavanone
#5	luteolin	5,7,3',4'-tetrahydroxyflavone
#6	kaempferol	3,5,7,4'-tetrahydroxyflavonol
#7	fisetin	3,7,3',4'-tetrahydroxyflavonol
#8	quercetin	3,5,7,3',4'-pentahydroxyflavonol

MATERIALS AND METHODS

Most of the flavonoids were from Roth of chromatographically pure quality, dihydroquercetin was from Serva, kaempferol and quercetin as well as magnesium(II)acetate [$Mg(Ac)_2$] and zinc(II)acetate [$Zn(Ac)_2$] from Fluka. Dihydrokaempferol was isolated from heartwood of Larix europaea. Horseradish peroxidase type VI-A from Sigma was used throughout the experiments. Hydrogen peroxide (as Perhydrol) was from Merck.

Substrate concentrations were: flavonoids, 0.1 mM; $Mg(Ac)_2$ and $Zn(Ac)_2$, 0.1 M; H_2O_2, 1 mM; HRP, 47 nM; pH was adjusted with NaOH. The reaction was started by the addition of H_2O_2.

The EPR experiments were performed using the following settings: modulation amplitude, 0.4 G, sweep, 14 G, scan time, 84 sec; in some cases scans at higher resolutions were also performed.

RESULTS

In preliminary experiments, we investigated oxidation of some flavonoids by Fenton chemistry (Fe^{2+}/chelates + H_2O_2), by photolysis of H_2O_2 and by thermolysis of the water-soluble 2,2'-azo-bis(2-amidinopropane) hydrochloride. These systems were eventually discontinued in favor of autoxidation at pH 9.0 and HRP/H_2O_2-catalyzed oxidation in N_2-saturated aqueous solutions. Comparable results were obtained under the latter two conditions, with a higher radical level in the HRP/H_2O_2 system as shown in **Fig. 1** for eriodictyol and dihydrofisetin.

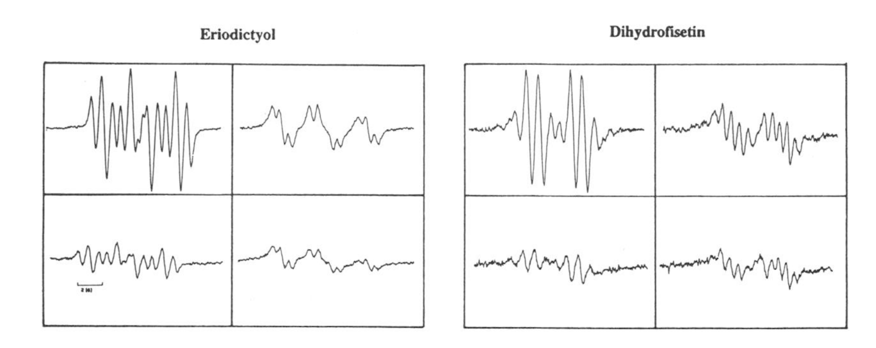

Figure 1: EPR spectra of eriodictyol and dihydrofisetin after reaction with HRP/H_2O_2 or autoxidation at pH 9.
(left panel: without additions, right panel: with $Mg(Ac)_2$; upper line series: HRP, lower line series: autoxidation; ordinate: amplitude in <u>identical</u> arbitrary units, except that the scale for eriodictyol is twice that of dihydrofisetin)

The cations Mg^{2+} or Zn^{2+} are supposed to interact with *o*-semiquinones, causing a 'spin stabilization' of the respective EPR spectra (Kalyanaraman, 1990). Changes in the EPR spectra with these salts can therefore be considered evidence of *o*-semiquinone formation. However, for the two compounds studied which contain a catechol structure in the B-ring (#3,4), 'spin stabilization' is a misnomer. In contrast to the example shown with DOPA (Kalyanaraman, 1990), both the Mg^{2+} and Zn^{2+} salts diminish the EPR signal over a 10 min period, whereas with HRP/H_2O_2 alone the signal decays much more slowly. **Fig. 2** shows these results for dihydroquercetin, which were very similar to those of dihydrofisetin.

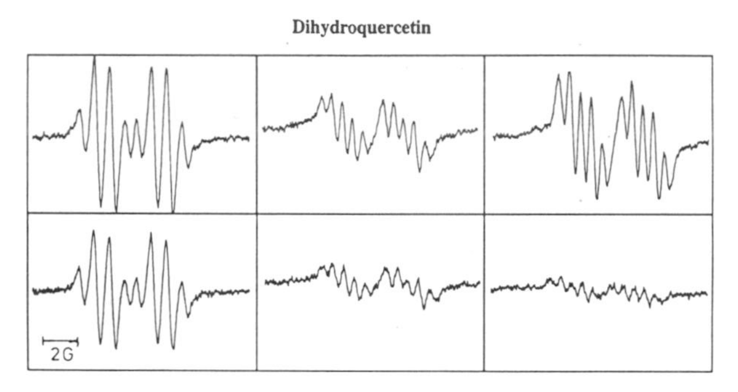

Figure 2: Effect of 0.1M $Mg(Ac)_2$ or 0.1M $Zn(Ac)_2$ on the stability of the EPR signal of dihydroquercetin after reaction with HRP/H_2O_2 at pH 9.
(left panels: HRP/H_2O_2 alone, center panels: with $Mg(Ac)_2$, right panels: with $Zn(Ac)_2$); top: EPR spectra recorded after 1 min, bottom: after 10 min; ordinate: amplitude in identical arbitrary units)

HRP/H_2O_2 at pH 7.5 consistently gave reasonable radical yields, which were further enhanced by raising the pH-value to 9.0. The structurally most interesting set of flavanones (#1-4) and their 2,3-unsaturated counterparts (#5-8) were studied both at pH 7.5-7.8 and at pH 9.0, only the latter results were used for evaluating radical structures. The respective EPR spectra are compiled in **Fig. 3**. The first column is complementary to Table I (see below), but is more illustrative for those unfamiliar with EPR spectroscopy.

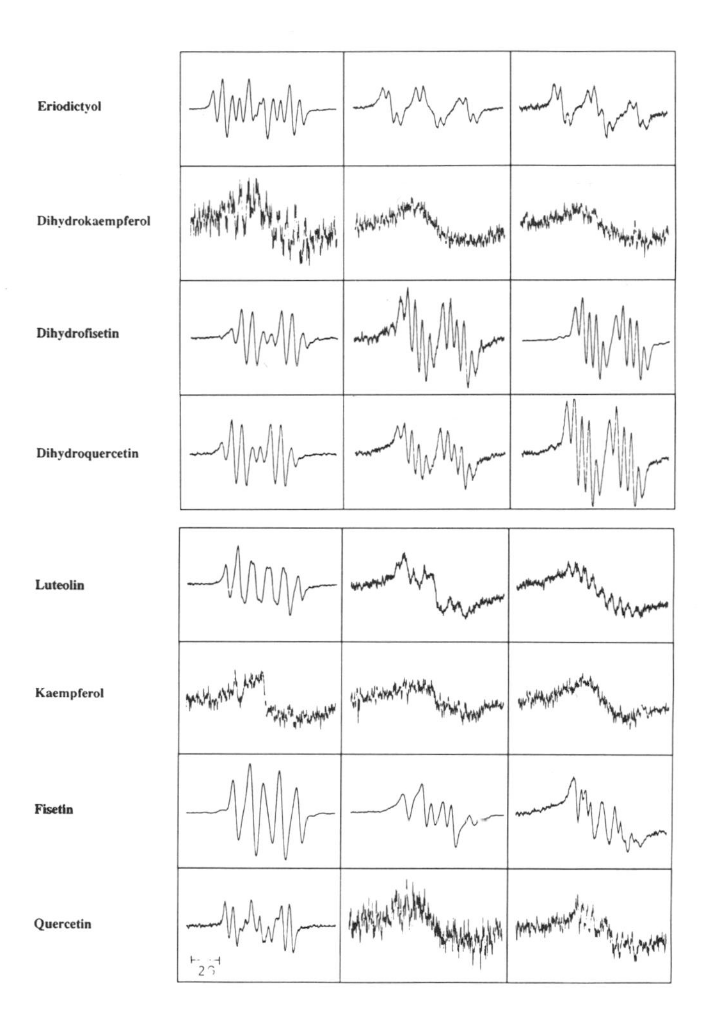

Figure 3: EPR spectra of flavones, flavonols and their 2,3-dihydro analogs after reaction with HRP/H_2O_2 at pH 9.
(left panels: HRP/H_2O_2 alone, center panels: with $Mg(Ac)_2$, right panels: with $Zn(Ac)_2$; ordinate: amplitude in <u>different</u> arbitrary units)

The strong signals of all compounds are particularly evident, except for kaempferol and its 2,3-dihydro derivative, which gave only weak spectra. Fisetin and quercetin showed time-dependent spectral changes, most strongly in the presence of $Mg(Ac)_2$, the depicted spectra were taken over the first 3 min. This is in contrast to **Fig. 2**, where the EPR signal is only diminished but not altered over a 10 min period.

Table I

Coupling Constants of Flavonoid Radicals

#	name	a_2	a_3	$a_{2'}$	$a_{5'}$	$a_{6'}$
1	eriodictyol	2.3	-	0.8	0.8	3.5
3	dihydrofisetin	0.9	-	0.9	0.9	3.4
4	dihydroquercetin	0.9	-	0.9	0.9	3.4
5	luteolin	-	1.40	1.0	1.0	2.8
7	fisetin	-	-	1.4	1.4	2.4
8	quercetin	-	-	1.25	0.7	3.25

Results from oxidation by HRP/H_2O_2 at pH 9; subscripts of the coupling constants denote the hydrogen atoms at the respective C-atoms - the assignments of $a_{2'}$ and $a_{5'}$ are interchangeable (after Kuhnle et al, 1969).

If we compare the coupling constants for the radicals derived from these substances, as calculated by Kuhnle et al. (1969), it is evident that under the strongly alkaline conditions and in the presence of 80% DMSO, which were used to enhance the stability of the radicals, different coupling constants result. Yet the values which we obtained under rather more physiological conditions are not too dissimilar, suggesting that the same radicals are observed.

The EPR spectra in the presence of Zn^{2+} could only be evaluated for the three flavanones (#1,3,4), the respective flavones (#5) and flavonols (#7,8) showing asymetries and temporal instabilities indicative of further reactions. For each of the three flavanones an increase in the coupling constants (i.e. spin densities) at C_2 and $C_{6'}$ corresponds to a decrease of the values assigned to $C_{2'}$ and $C_{5'}$, in effect corroborating complex formation of Zn^{2+} with the B-ring *o*-semiquinone (data not shown).

DISCUSSION

As already shown by Pardini's group (Hodnick et al., 1986, 1988a,b) and by Canada et al. (1990), a number of flavonoid aglycones readily autoxidize even in neutral solutions. In line with other semiquinone compounds, the redox cycling occurring under these conditions results in the formation of $O_2^{-\cdot}$ (and eventually $\cdot OH$) radicals. This may cause further oxidative degradation of the flavonoids - especially rupture of the heterocyclic ring (cf. the EPR observation of a *p*-benzo-semiquinoid structure with kaempferol by Kuhnle et al., 1969, and in our own preliminary studies, or identification of fragmentation products of kaempferol (Takahama, 1987)). In contrast, Nishinaga et al. (1979) report that only flavonols lacking a 7-hydroxy group - which are practically non-existent in nature (Harborne, 1988) - undergo heterocyclic ring cleavage.

The use of HRP/H_2O_2 in N_2-saturated aqueous solution as a radical-generating system (Ohnishi, 1969) and comparing the results to autoxidation in slightly alkaline aqueous solution was chosen as a method closely approaching physiological conditions, e.g. to those in apoplasts where both H_2O_2 and peroxidases may coexist (Castillo & Greppin, 1986) and where oxygen radical formation may take place after ozone exposure or during apoplast 'respiratory burst' in plant-pathogen interaction (Rosemann et al., 1991, Sandermann et al., 1989).

The HRP reaction leads also to degradation products after heterocyclic ring cleavage, as reported for kaempferol (Miller & Schreier, 1985), quercetin (Schreier & Miller, 1985) and its 3-

glycoside rutin (Takahama, 1986) - possibly with concomitant B-ring *o*-quinone formation where possible. The conditions were sufficiently mild, however, to observe intermediary aroxyl radicals in *in situ* EPR experiments. Since the autoxidation and HRP/H_2O_2-catalysed oxidation, both at pH 9, led to similar radical spectra for most of the investigated flavonoids, the results might allow a comparison with the pulse-radiolytic data, in which case the observed transient spectra represent aroxyl radicals generated by univalent oxidation of individual but <u>unspecified</u> phenolic positions (Bors et al., 1990, 1992).

We expected to obtain more structural information of the aroxyl radicals from the EPR spectra in addition to the transient spectra and kinetic data resulting from pulse radiolysis experiments (Bors & Saran, 1987). The most strikingly different results between the two techniques shall be highlighted:

(i) while the three flavanones with B-ring catechol structure (#1,3,4) gave nearly identical transient spectra in pulse radiolysis representing the *o*-semiquinones (Bors & Saran, 1987), the EPR spectrum of eriodictyol lacking a 3-hydroxy group differs quite strongly from the 3-hydroxy-flavanones (#3,4). This is also reflected in the 2.5-fold difference of the coupling constant for the C-2 hydrogen atom (see Table I), yet the radical identified represents the analogous *o*-semiquinone in the B-ring.

(ii) the flavonols kaempferol and quercetin show almost identical transient spectra in pulse radiolysis, yet the EPR spectra of kaempferol are weak and look different to those of quercetin. Considering the fact that the kaempferol aroxyl radical decays faster by three orders of magnitude than the quercetin aroxyl radical (Bors et al., 1990), it seems reasonable to propose that it is merely this kinetic instability which, due to the absence of a catechol structure in the B-ring, cannot be ameliorated by Mg^{2+} or Zn^{2+}, is the cause of the weak spectra. Furthermore, <u>all</u> flavonoids with catechol structures in the B-ring were shown to form relatively more stable radicals than those compounds lacking this moiety (Bors et al., 1990).

Thus, if one compares time-resolved absorption spectra from pulse radiolysis with EPR spectra, one has to keep in mind that the first technique basically reflects kinetic processes. EPR spectroscopy, in contrast, is a feature of the thermodynamic equilibria occurring in these reactions, i.e. it reflects only stationary phases.

The first query of the introductory presentation concerned the similarity (or even identity) of the radicals produced with HRP/H_2O_2 and during the autoxidation conditions as employed by Kuhnle et al. (1969). In our own autoxidation experiments at pH 9, spectral comparisons did not suggest gross dissimilarities, except for higher radical intensities with HRP/H_2O_2 (Fig. 1). Therefore the coupling constants in Table I may be compared with those obtained by Kuhnle et al. (1969) for the same substances. The hydrogen splittings are basically similar aside from smaller coupling constants for the indistinguishable C-2' and C-5' hydrogen atoms and occasional minor deviations. It thus seems possible to record these radicals after generation by an enzymatic reaction under near physiological conditions, and we are presently carrying out experiments with HRP/H_2O_2 at lower pH values.

We already confirmed in preliminary experiments, that dissociated phenolate groups, present at higher pH, are easier oxidized by electron transfer (c.f. oxidation by azide radicals as reported by Alfassi & Schuler, 1985). In the case of flavonoids, the sequence of hydroxy group dissociation has generally been assumed to be 7-OH >> 4'-OH > 3-OH (Mabry et al., 1970, Slabbert, 1977).

Yet, as shown in the scheme, the 7-hydroxy group is an unlikely radical site, as it interacts either with *m*-hydroxy groups at C-5 or via the 4-oxo structure. Indeed, such a radical structure, which should be especially apparent with eriodictyol and luteolin, was not observed. To the contrary, we can confirm the general statement of Kuhnle et al. (1969), that the predominant attack occurs at the B-ring.

The apparent effects of the spin stabilizing cations with some of the flavonoids were different from previous suggestions (Kalyanaraman, 1990). Complex formation with *o*-semiquinone radical sites is considered the prime mechanism of these cations. Evidently such complex formation does <u>not</u> occur with radical structures involving the 3- and/or 5-hydroxy groups in combination with a 4-oxo function as we originally posited. While the EPR spectra of eriodictyol on one hand, and dihydrofisetin and dihydroquercetin on the other hand are quite distinct, considering the coupling constants they all represent the *o*-semiquinone structure in the B-ring. This is further corroborated by the evaluation of the EPR spectra in the presence of Zn^{2+} for these three compounds, which showed an enhancement of the spin densities at C_2 and $C_{6'}$ and a corresponding decrease of the $C_{2'}/C_{5'}$ coupling constants (the fact that in the absence of a 3-hydroxy group the C_2 coupling as in eriodictyol, is considerably higher than in its presence, is in accordance with a negative inductive effect of this group).

Since we thus confirm Zn^{2+} complexation with the *o*-semiquinone in the B-ring, we can now explain the <u>destabilizing</u> effect of this cation (as well as of Mg^{2+}) on the EPR spectra (**Fig. 2**). Taking into account the various fragmentation products found for kaempferol (Miller & Schreier, 1985, Takahama, 1987) and quercetin (Schreier & Miller, 1985), we suggest that these *o*-semiquinone intermediates are degraded further. Complex formation leads to higher spin densities at C_2 and $C_{6'}$ which facilitates nucleophilic attack at $C_{1'}$ with subsequent loss of the B-ring. In the presence of the cations the decay of the *o*-semiquinones would thus be enhanced which is exactly what we observed. The alternative fate which we observed in pulse radiolysis in the absence of reactants is second-order decay via dismutation (Bors et al., 1990), an inherently slow process at low radical concentrations.

We have thus far not been able to evaluate the EPR spectra of the 2,3-unsaturated analogs (#5,7,8) in the presence of Zn^{2+}. As seen in Fig. 3, there are considerable changes from the uncomplexed spectra as well as asymetries (#7), evidence for higher instabilities of these radicals. With the interpretation of the pulse-radiolytic transient spectra that the 2,3-double bond enables electron delocalization over all three ring systems (Bors et al., 1990), it can be expected that determination of spin densities are more complex than for the saturated compounds.

CONCLUSIONS

Using HRP/H_2O_2 at pH 9 (and occasionally at pH 7.5), we could demonstrate that under these nearly physiological conditions basically the same radicals are formed as under autoxidizing conditions at pH 9 and at higher pH in the presence of 80% of DMSO (Kuhnle et al., 1969). Evaluation of the coupling constants confirms the early observations of Kuhnle et al. (1969), who found preferential oxidation in the B-ring. The coupling constants of the three flavanones eriodictyol, dihydrofisetin, and dihydroquercetin in the presence of Zn^{2+} furthermore corroborated the complexation with the B-ring *o*-semiquinone. The diminished radical stability under these conditions is explained by the lability of these semiquinones towards nucleophilic attack at $C_{1'}$, a reaction which is favored in the presence of the cation. These results are also significant for the interpretation of the role of flavonoids and possibly other phenolic constituents in antioxidative processes *in vivo*.

REFERENCES

Alfassi, Z.B. and Schuler, R.H. (1985) J. Phys. Chem. 89, 3359-63.
Bors, W. and Saran, M. (1987) Free Rad. Res. Comm. 2, 289-294
Bors, W., Heller, W., Michel, C., and Saran, M. (1990) In: Oxygen Radicals in Biological Systems. Part B: Oxygen Radicals and Antioxidants. Meth. Enzymol. (Ed.: Packer L, Glazer AN; Academic Press, New York), 186, 343-354
Bors, W., Heller., W, Michel, C. and Saran, M. (1992) in: Free Radicals in Liver Injury (G. Csomos, J. Feher, eds., Springer, Berlin), 77-95.
Canada, A.T., Giannella, E., Nguyen, T.D., and Mason, R.P. (1990) Free Rad. Biol. Med. 9, 441-449
Castillo, F.J., and Greppin, H. (1986) Physiol. Plant. 68, 201-208
Dixon, W.T., Moghimi, M., and Murphy, D. (1975) JCS, Perkin II, 101-103
Harborne, J.B. (1988) The Flavonoids. Advances in Research since 1980. (Chapman & Hall, London)
Hodnick, W.F., Kung, F.S., Roettger, W.J., Bohmont, C.W., and Pardini, R.S. (1986) Biochem. Pharmacol. 35, 2345-57
Hodnick, W.F., Milosavljevic, E.B., Nelson, J.H., and Pardini, R.S. (1988a) Biochem. Pharmacol. 37, 2607-11
Hodnick, W.F., Kalyanaraman, B., Pritsos, C.A., and Pardini, R.S. (1988b) In: Oxygen Radicals in Biology and Medicine Basic Life Sciences (Ed.: Simic MG, Taylor KA, Ward JF, von Sonntag C,; Plenum Press, New York), 49, 149-152
Husain, S.R., Cillard, J., and Cillard, P. (1987) Phytochem. 26, 2489-91
Jensen, O.N. and Pedersen, J.A. (1983) Tetrahedron 39, 1609-15
Kalyanaraman, B. (1990) In: Oxygen Radicals in Biological Systems. Part B: Oxygen Radicals and Antioxidants. (Ed.: Packer L, Glazer AN; Academic Press, New York), Meth. Enzymol. 186, 333-342
Kalyanaraman, B. and Sealy, R.C. (1982) Biochem. Biophys. Res. Comm. 106, 1119-25
Kalyanaraman, B., Hintz, P., and Sealy, R.C. (1986) Feder. Proc. 45, 2477-84
Kalyanaraman, B., Premovic, P.I., and Sealy, R.C. (1987) J. Biol. Chem. 262, 11080-87
Kuhnle, J.A., Windle, J.J., and Waiss, A.C. (1969) J. Chem. Soc. B, 613-616
Mabry, T.J., Markham, K.R., and Thomas, M.B. (1970) eds.: The Systematic Identification of Flavonoids. Part 2. Springer, Berlin.
Miller, E. and Schreier, P. (1985) Food Chem. 17, 143-154
Nishinaga, A., Tojo, T., Tomita, H., and Matsuura, T. (1979) JCS, Perkin I, 2511-16
Ohnishi, T., Yamazaki, H., Iyanagi, T., Nakamura, T., and Yamazaki, I. (1969) Biochim. Biophys. Acta 172, 357-369
Rosemann, D., Heller, W., and Sandermann, H. jr. (1991) Plant Physiol. 97, 1280-86

Sandermann, H. jr., Heller, W., and Langebartels, C. (1989) in: Proc. Int. Congress on 'Forest Decline Research: State of Knowledge and Perspectives'. (Ed.: Ulrich, B, KfK Karlsruhe), 517-525
Schreier, P. and Miller, E. (1985) Food Chem. 18, 301-317
Slabbert, N.P. (1977) Tetrahedron 33, 821-824.
Takahama, U. (1986) Biochim. Biophys. Acta 882, 445-451
Takahama, U. (1987) Plant Cell Physiol. 28, 953-957
Ueno, I., Kohno, M., Yoshihira, K., and Hirono, I. (1984a) J. Pharm. Dyn. 7, 563-569
Ueno, I., Kohno, M., Haraikawa, K., and Hirono, I. (1984b) J. Pharm. Dyn. 7, 798-803
Yoshikawa, T., Naito, Y., Oyamada, H., Ueda, S., Tanigawa, T., Takemura, T., Sugino, S., and Kondo, M. (1990) In: Antioxidants in Therapy and Preventive Medicine; Adv. Exp. Med. Biol. (Ed.: Emerit I, Packer L, Auclair C,; Plenum Press, New York), 264, 171-174

Free Radicals: From Basic Science to Medicine
G. Poli, E. Albano & M. U. Dianzani (eds.)

ANTIOXIDANT MECHANISMS OF VITAMIN E AND β-CAROTENE

Graham W. Burton[1], Lise Hughes[1], David. O. Foster[2], Ewa Pietrzak[1,3]
Mark A. Goss-Sampson[4] and David P.R. Muller[4]

[1]Steacie Institute for Molecular Sciences and [2]Institute for Biological Sciences, National Research Council of Canada, Ottawa, Ontario, Canada K1A 0R6; [3]Department of Biochemistry, University of Ottawa, Ottawa, Ontario, Canada; [4]Division of Biochemistry, Institute of Child Health, WC1N 1EH, London, U.K.

SUMMARY: Vitamin E is an excellent trap for peroxyl radicals (ROO•) and it is the major lipid soluble antioxidant present in mammalian cells. It therefore occupies a unique position in the arsenal of natural antioxidants providing protection against various diseases. Product studies of the reaction of α-tocopherol with peroxyl radicals suggest that the existence of an α-tocopherol regeneration mechanism is essential for maintaining the antioxidant viability of the vitamin. Evidence now exists that vitamin C may regenerate vitamin E in some tissues. Studies carried out with deuterium-labeled α-tocopherol have confirmed that turnover of vitamin E is very slow in neural tissue, the tissue most susceptible to the effects of a deficiency of vitamin E in humans.

There is much interest in β-carotene because it appears to have anticancer activity. It has been assumed, but not shown, that the anticancer activity is related to β-carotene's antioxidant properties. Possibly, the superior ability of β-carotene to quench singlet oxygen is important but there is no evidence that singlet oxygen is generated in significant quantities in any organs except those exposed to light (i.e., skin and eye). Alternatively, β-carotene may trap peroxyl radicals, but in vitro studies show that it can do this only at low partial pressures of oxygen. This observation is difficult to reconcile with the epidemiological evidence indicating that the strongest association between dietary intake of β-carotene and reduced incidence of cancer occurs for the tissue exposed to the highest partial pressures of oxygen, i.e., the lung.

Our understanding of how eukariotic cells are protected against free-radical mediated damage is far from complete. As a first step, in vitro studies are important in providing direction to understanding the far more complex situation that exists in vivo. Further progress in our understanding depends upon an ability to realistically elaborate models which, for example, reflect the microstructure of the in vivo environment by incorporating some of the physical structural features of biological membranes or lipid particles. In vitro model systems of enhanced complexity increase the opportunity of encountering, exploring and understanding potentially significant antioxidant interactions of medical relevance.

VITAMIN E: It is well-established that α-tocopherol, biologically the most active form of vitamin E (Diplock 1985; Farrell 1988), is an excellent chain-breaking antioxidant in homogeneous organic solvents (Burton et al., 1981; Burton et al., 1985). Also, it has been demonstrated that vitamin E is the major lipid-soluble antioxidant present in mammalian cells and blood (Burton et al., 1982; Cheeseman et al., 1984; Cheeseman et al., 1986; Ingold et al., 1987b; Cheeseman et al., 1988). However, there now are indications that other lipid-soluble substances may be of importance in modulating the effects of lipid peroxidation in vivo. For example, it has been discovered recently that ubiquinol-10 in low density lipoproteins (LDL) is better at protecting LDL lipids against peroxidation under certain circumstances than is the substantially larger amount of α-tocopherol that is normally present (Stocker et al., 1991).

α-Tocopherol functions as an antioxidant by rapidly transferring its phenolic hydrogen atom to a lipid peroxyl radical, resulting in the formation of two molecules that are relatively unreactive towards polyunsaturated lipid, i.e., a lipid hydroperoxide and the α-tocopheroxyl radical (Fig. 1). Ideally, the rate of the reaction of a peroxyl radical with the antioxidant is so much greater than that of the competing reaction with another polyunsaturated fatty acid moiety that a small amount of α-tocopherol is sufficient to protect a large amount of polyunsaturated lipid.

The fate of the tocopheroxyl radical in vivo is not known. In fact, rather little is known about the metabolism of α-tocopherol. Although two closely related metabolites (Simon et al., 1956a; Simon et al., 1956b) have been found in the urine of rats (Weber et al., 1963), rabbits (Simon et al., 1956b) and humans (Simon et al., 1956a; Schmandke 1965; Schmandke et al., 1968), most of the α-tocopherol that

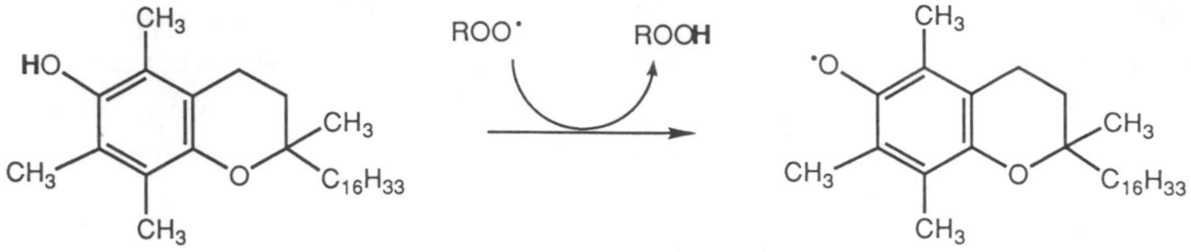

ROO• ROOH

α-Tocopherol α-Tocopheroxyl Radical

is excreted from the rat, for example, appears to be secreted from the liver via the bile into the small intestine, mostly in some unknown, metabolized form(s) (Gallo-Torres 1980; Bjørneboe et al., 1986; Bjørneboe et al., 1987). Study of the bile α-tocopherol metabolites may eventually permit determination of the relative amounts of α-tocopherol that are lost from the body by free radical-mediated and homeostatic processes, respectively.

Even product studies of the in vitro reaction of α-tocopherol with peroxyl radicals (in polar organic solvents or in aqueous phospholipid liposomes) have only recently been reported (Matsuo et al., 1989; Yamauchi et al., 1989; Liebler et al., 1990; Liebler et al., 1991). The results unexpectedly revealed the formation of several epoxide products (Fig. 2). The first-formed tocopherone intermediates are depicted in Fig. 2 as arising from attack of peroxyl radicals either at the *para* (path (a)) or *ortho* (path (b) or (c)) carbons of the α-tocopheroxyl radical. Subsequent reaction of the tocopherone intermediates with water, either during the reaction or work-up, would give the tocopherol quinone products depicted in Fig. 2.

Reaction at the *para* carbon of the tocopheroxyl radical (path (a)) gives an 8a-(alkyldioxy)tocopherone adduct, representing a net consumption of 2 peroxyl radicals, starting from α-tocopherol (i.e., the overall stoichiometry is 2). This is consistent with the net stoichiometry found earlier for reaction in non-polar organic solvents (Burton et al., 1981). The existence of epoxide products, however, carries an implication of the operation of a non-antioxidant pathway. The most likely mechanism for formation of epoxides is by addition of a peroxyl radical to one of either of the *ortho* carbons (path (b) or

Peroxyls add to *ortho* or *para* carbons

8a-(alkyldioxy)tocopherone

Epoxytocopherone

Epoxytocopherone

(c)), followed by decomposition of the adduct to an epoxy carbon-centered radical which is accompanied by the release of an alkoxyl radical (Fig. 3). The epoxy carbon-centered radical intermediate then reacts rapidly with ambient, dissolved molecular oxygen to give a peroxyl radical. This radical and the previously released alkoxyl radical will each abstract hydrogen atoms from other polyunsaturated fatty acid moieties (RH), thereby continuing rather than stopping the chain reaction.

If the decomposition of the ortho peroxyl radical adducts to epoxytocopherone intermediates occurs rapidly during the course of the reaction, rather than artefactually during the subsequent work up and analysis, then reaction through the *ortho* carbons via paths (b) and (c) results in no net trapping of peroxyl radicals. That is, α-tocopherol behaves as a chain-carrying radical. (Although the epoxide products are obtained even in the presence of unreacted α-tocopherol, indicating that the rate of reaction of a peroxyl radical with the α-tocopheroxyl radical is at least as fast as with α-tocopherol itself, this result does not rule out the build-up of ortho peroxyl adducts that can decompose rapidly during work-up.) Therefore, the overall ability of α-tocopherol to effectively trap peroxyl radicals may very

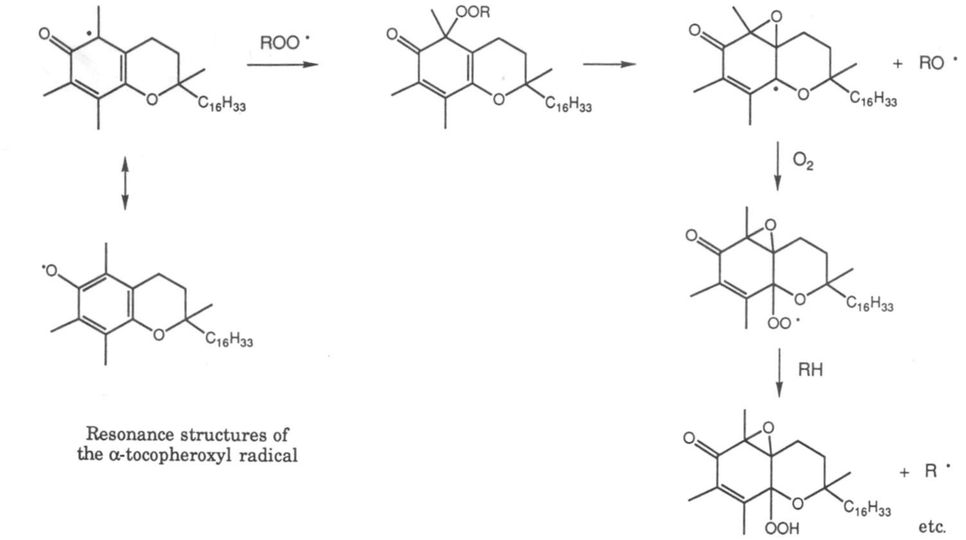

Resonance structures of the α-tocopheroxyl radical

Fig. 3. Peroxyl addition through path (b)

well depend upon the relative extents to which reaction proceeds through path (a) versus paths (b) and (c).

Under this scenario, the recent findings published by Liebler's group (Liebler et al., 1991) imply that the effectiveness of α-tocopherol is significantly compromised in aqueous liposomal environments. In their model reactions the total epoxide yield increased from 29% for reaction with α-tocopherol in pure acetonitrile to 64% for reaction in aqueous liposomes, corresponding to net peroxyl trapping stoichiometries of $2 \times (1 - 0.29) = 1.4$ and $2 \times (1 - 0.64) = 0.7$, respectively. (The calculations are based on stoichiometries of 2 for path (a) and 0 for paths (b) or (c), respectively). The very low overall stoichiometry implied by the high yield of epoxides in the liposomal environment indicates that reaction of the α-tocopheroxyl radical with another peroxyl radical would undermine the effect of the preceding radical-trapping reaction of α-tocopherol. Therefore, under these circumstances, the need for a mechanism of regeneration of α-tocopherol from the tocopheroxyl radical becomes imperative if α-tocopherol is to be a viable antioxidant.

There is extensive evidence of a sparing or regeneration effect of vitamin C upon vitamin E in in vitro systems (Packer et al., 1979;

Doba et al., 1985; Niki et al., 1985; Frei et al., 1988). There also is some evidence for the existence of enzymatic mechanisms of regeneration (Gibson et al., 1985; Maguire et al., 1989; Packer et al., 1989; Scholich et al., 1989). Evidence of an in vivo interaction between vitamins C and E, however, has been much more difficult to obtain. In our own work, we have looked for an effect upon levels and turnover of vitamin E using dietary intake to modulate tissue vitamin C levels in the guinea pig, which, like man, is incapable of endogenous synthesis of vitamin C.

In our first study we found no effect of dietary intake of vitamin C upon vitamin E in the guinea pigs, even when the dietary intake of vitamin E was insufficient to maintain tissue levels over a period of 8 weeks (Burton et al., 1990c). Recently, however, we have found that when guinea pigs are placed on scorbutic levels of vitamin C (10 mg/kg diet), the adrenal gland, lung and heart show statistically significant decreases in vitamin E concentrations, while levels of vitamin E in other tissues are unaffected (E. Pietrzak, D. Foster, G. Burton, unpublished results). This is illustrated in Fig. 4 which shows that both vitamins C and E declined in the adrenal gland ($p < 0.01$ for a non zero slope for regression of vitamin E vs. time), whereas the vitamin E level did not decline in the brain, despite a very marked drop in the level of vitamin C. (It is important to note that in our earlier study which used higher intakes of vitamin C, the same level of dietary vitamin E (35 mg/kg *RRR*-α-tocopheryl acetate) was sufficient to maintain constant levels of vitamin E in all tissues (Burton et al., 1990c)).

Recently, Igarashi and co-workers published results which show that dietary vitamin C can influence levels of tissue vitamin E in a mutant strain of male Wistar rat unable to synthesize vitamin C (Igarashi et al., 1991). Statistically significant increases were observed in the plasma, red cells, heart, liver, kidney, spleen and lung of rats maintained on 50 mg vitamin E/kg diet and fed 600 mg vitamin C/kg diet for 6 weeks compared to rats fed 300 mg/kg for the same period of time (Table 1). The largest effects were seen in plasma and red cells, perhaps reflecting a protective effect of vitamin C upon vitamin E during absorption from the small intestine. In rats fed

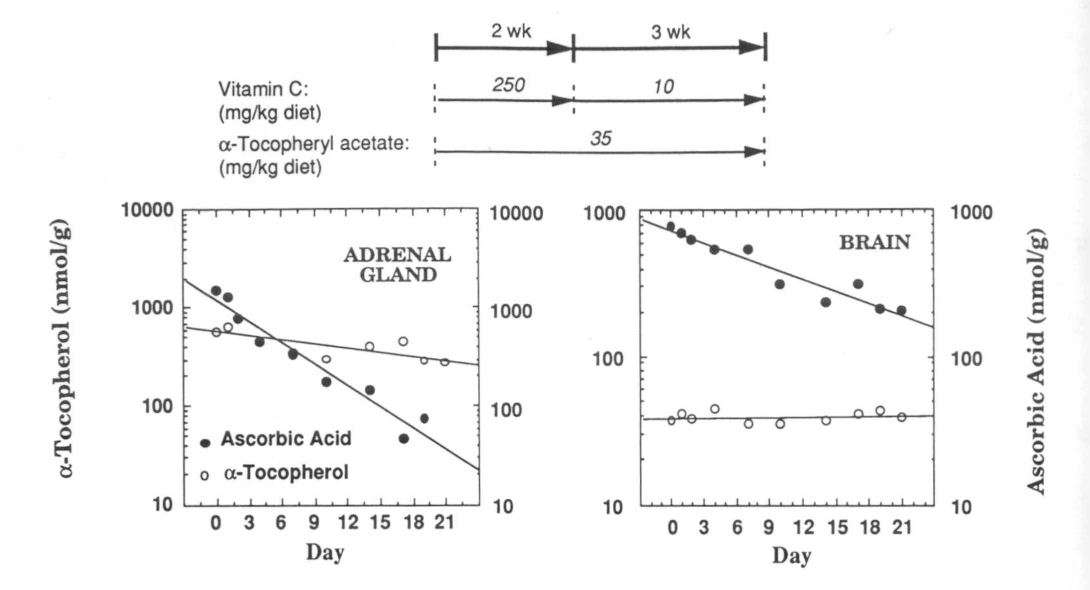

Fig. 4. Concentrations of α-tocopherol and ascorbic acid in adrenal gland and brain of guinea pigs at selected times during a 3 week period on a diet containing a normal concentration of vitamin E and a scorbutic concentration of vitamin C. The animals were maintained beforehand for 2 weeks on a diet containing the same concentration of vitamin E and an adequate concentration of vitamin C.

a diet containing only 1 mg/kg of vitamin E, the higher dietary level of vitamin C produced significantly higher levels of vitamin E in heart, liver, spleen and lung (Table 1). However, the levels of vitamin E in plasma and red cells were lower in the animals fed the high level of vitamin C, which may result from a pro-oxidant effect of vitamin C upon vitamin E in the small intestine or the blood.

Whole animal experiments like these suffer from the drawback that even if an interaction is observed it is difficult, if not impossible, to interpret the nature of the interaction. However, it appears that a tissue-dependent effect does exist under at least some conditions of physiological stress, consistent with the expectation that a regeneration mechanism likely exists for the maintenance of adequate protection of tissues by vitamin E.

As part of our effort to better understand the protection of tissues by vitamin E, we have embarked over the past several years on a comprehensive study of the dynamics of α-tocopherol in vivo (Ingold et al., 1987a; Burton et al., 1988; Traber et al., 1988; Burton

Table 1. Effect of vitamin C upon levels of vitamin E in blood and tissues of mutant male Wistar maintained on diets containing low (1 mg/kg diet) and normal (50 mg/kg diet) concentrations of *all-rac*-α-tocopheryl acetate. The rats in each vitamin E group were subdivided into two equal groups that were fed one of two levels of vitamin C (300 or 600 mg/kg diet) for 6 weeks. Results are expressed as ratios obtained by dividing the tissue vitamin E concentration in animals fed 600 mg/kg vitamin C by the corresponding value in rats fed 300 mg/kg vitamin C (Igarashi et al., 1991).

	α -Toc (600) / α-Toc (300)	
	Low Vitamin E	Normal Vitamin E
Plasma	0.75	1.45
Red Cells	0.81	1.36
Heart	1.11	1.10
Liver	1.15	1.20
Kidney	1.04[a]	1.17
Spleen	1.16	1.06
Lung	1.10	1.16

[a] Difference not significant.

et al., 1990a; Burton et al., 1990b; Burton et al., 1990c; Traber et al., 1990a; Traber et al., 1990b; Traber et al., 1990c). Recently we have used our deuterium labeled vitamin E/gas chromatography-mass spectrometry technique to compare the dynamics of vitamin E in neural tissues of normal and vitamin E deficient rats (M. Goss-Sampson, D. Muller and G. Burton, unpublished work).

Rats maintained on a vitamin E-free diet for 16 weeks (phase 1) were fed a diet containing trideuterium-labeled *RRR*-α-tocopheryl acetate (d_3-*RRR*-α-TAc; 36 mg/kg diet) for 12 weeks (phase 2), after which time the animals were returned to the vitamin E-free diet for another 16 weeks (phase 3). Animals were sacrificed at selected time intervals during phases 2 and 3 and tissues were analyzed for unlabeled (d_0-α-T) and labeled (d_3-α-T) α-tocopherol.

Fig. 5 illustrates the behavior of the neural and non-neural tissues. During phase 2 both types of tissues showed steady accumulation of the labeled α-tocopherol, as is displayed in the d_3/d_0 vs. time plots. However, after the switch back to the vitamin E-free diet in phase 3, a dichotomy developed in the behavior of the neural and non-neural tissues. The d_3/d_0 ratio declined in the non-neural tissues whereas the ratio continued to increase in all neural tissues. Evidently, the animals responded to a lack of vitamin E in their diet

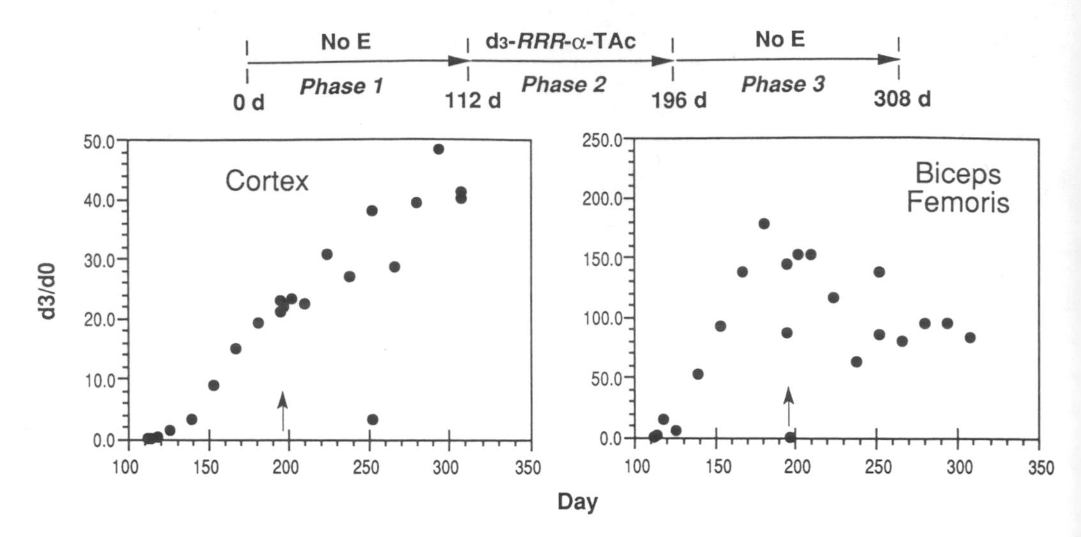

Fig. 5. Ratios of d_3- to d_0-α-tocopherol (d_3/d_0) in cortex and biceps femoris of rats sacrificed at selected time intervals during a 12 week period on a diet containing 36 mg/kg d_3-*RRR*-α-tocopheryl acetate (phase 2) and during a 16 week period on a vitamin E-free diet (phase 3). The animals were maintained beforehand for 16 weeks on a vitamin E-free diet (phase 1).

by redistributing vitamin E from non-neural to neural tissues. It therefore appears that there is a mechanism for the conservation of neural vitamin E when dietary sources are low. This behavior is consistent with the very slow movement of vitamin E in and out of neural tissues (Ingold et al., 1987a; Burton et al., 1990c) and with the importance of vitamin E for the maintenance of neurological health (Muller et al., 1982; Harding et al., 1985; Goss-Sampson et al., 1988). The blood-brain barrier possibly plays a role in this phenomenon.

β-CAROTENE: β-Carotene currently attracts widespread attention because it appears to have anticancer activity. It is widely assumed that underlying the purported anticancer activity is an ability for β-carotene to act an antioxidant. Presumably, the proposed antioxidant effect derives from β-carotene's reputation as an excellent quencher of singlet oxygen in plants and algae. However, it is doubtful that singlet oxygen is a significant threat to eukariotic cells not normally exposed to UV and visible light.

As an alternative, we have suggested that β-carotene may act as an unusual kind of chain-breaking antioxidant (Burton et al., 1984;

Burton 1989). However, although we have shown that β-carotene does indeed have modest inhibiting capabilities in lipid peroxidations conducted in vitro, the inhibitory effect is obtained only at low partial pressures of oxygen (e.g., 15 Torr). The basis of the antioxidant effect is the ability of β-carotene to intercept a peroxyl radical to form a strongly delocalized carbon-centered radical that reacts reversibly with oxygen. The reversibility of the reaction with oxygen and the relative stability of the carbon-centered radical means that at lower oxygen tension there is proportionately more of the much less reactive carbon-centered radical. Consequently, the radical chain reaction can terminate by reaction with another carbon-centered radical or by cross-reaction with a peroxyl radical and this can occur more rapidly than the normal termination between two peroxyl radicals. However, the enhancement of inhibition of peroxidation at low oxygen partial pressure is difficult to reconcile with the epidemiological evidence that β-carotene is most effective in lung, the tissue exposed to the highest partial pressure of oxygen.

Most, if not all, published results reporting an antioxidant effect for β-carotene reflect the extraordinary high reactivity of β-carotene towards peroxyl radicals. This criterion alone, however, is not sufficient to establish antioxidant activity. Obviously, if a more reactive substance is added to a peroxidizing system it will be consumed first and it will appear to confer a "sparing" effect on less reactive substances. By this criterion, arachidonic acid would be an "antioxidant" for linoleic acid! It is important, therefore, not to confuse a temporary sparing effect with an antioxidant effect. A true antioxidant will protect polyunsaturated lipid by substantially decreasing the length of the chain reaction, thereby decreasing the rate of oxygen uptake.

β-Carotene is a linear polyene that is very prone to autoxidation. In fact, β-carotene may be a pro-oxidant! (There are anecdotal stories of β-carotene undergoing spontaneous combustion upon warming up to room temperature after long-term storage at low temperatures.)

In the past year, three publications appeared describing some of the oxidation products that are formed during the autoxidation of β-carotene in organic solvents (Handelman et al., 1991; Kennedy et al., 1991; Mordi et al., 1991). Of the myriad products that are formed, retinal (vitamin A) is produced in small amount. From the foregoing, it appears, therefore, that there is a need to critically re-examine the mode of action of β-carotene in vivo.

Acknowledgments

The authors wish to acknowledge the support of the Association for International Cancer Research, the National Foundation for Cancer Research, the Leverhulme Trust, NATO, the Natural Source Vitamin E Association, Eastman Chemicals, Eisai Ltd., Henkel Inc., and Hoffmann-La Roche (Basel).

REFERENCES

Bjørneboe, A., Bjørneboe, G.-E. A., Bodd, E., Hagen, B. F., Kveseth, N. and Drevon, C. A. (1986) Biochim. Biophys. Acta 889, 310-315.
Bjørneboe, A., Bjørneboe, G.-E. A. and Drevon, C. A. (1987) Biochim. Biophys. Acta 921, 175-181.
Burton, G. W. and Ingold, K. U. (1981) J. Am. Chem. Soc. 103, 6472-6477.
Burton, G. W., Joyce, A. and Ingold, K. U. (1982) Lancet ii, 327.
Burton, G. W. and Ingold, K. U. (1984) Science 224, 569-573.
Burton, G. W., Doba, T., Gabe, E. J., Hughes, L., Lee, F. L., Prasad, L. and Ingold, K. U. (1985) J. Am. Chem. Soc. 107, 7053-7065.
Burton, G. W., Ingold, K. U., Foster, D. O., Cheng, S. C., Webb, A., Hughes, L. and Lusztyk, E. (1988) Lipids 23, 834-840.
Burton, G. W. (1989) J. Nutr. 119, 109-111.
Burton, G. W., Ingold, K. U., Cheeseman, K. H. and Slater, T. F. (1990a) Free Radical Res. Comm. 11, 99-107.
Burton, G. W. and Traber, M. G. (1990b) Ann. Rev. Nutr. 10, 357-382.
Burton, G. W., Wronska, U., Stone, L., Foster, D. O. and Ingold, K. U. (1990c) Lipids 25, 199-210.
Cheeseman, K. H., Burton, G. W., Ingold, K. U. and Slater, T. F. (1984) Toxicol. Pathol 12, 235-239.
Cheeseman, K. H., Collins, M., Proudfoot, K., Slater, T. F., Burton, G. W., Webb, A. C. and Ingold, K. U. (1986) Biochem. J. 235, 507-514.
Cheeseman, K. H., Emery, S., Maddix, S. P., Slater, T. F., Burton, G. W. and Ingold, K. U. (1988) Biochem. J. 250, 247-252.
Diplock, A. T. (1985) Fat-soluble vitamins. Lancaster, Pennsylvania, Technomic Publishing Co. 154-224.
Doba, T., Burton, G. W. and Ingold, K. U. (1985) Biochim. Biophys. Acta 835, 298-303.
Farrell, P. M. (1988) Modern Nutrition in Health and Disease. Philadelphia, Lea & Febiger 340-354.
Frei, B., Stocker, R. and Ames, B. N. (1988) Proc. Nat'l. Acad. Sci. USA 85, 9748-9752.
Gallo-Torres, H. E. (1980) Vitamin E: A Comprehensive Treatise. New York, Marcel Dekker, Inc. 193-267.
Gibson, D. D., Hawrylko, J. and McCay, P. B. (1985) Lipids 20, 704-711.
Goss-Sampson, M. A., MacEvilly, C. J. and Muller, D. P. R. (1988) J. Neurol. Sci. 87, 25-35.

Handelman, G. J., van Kuijk, F. J. G. M., Chatterjee, A. and Krinsky, N. I. (1991) Free Radical Biology & Medicine 10, 427-437.
Harding, A. E., Matthews, S., Jones, S., Ellis, C. J. K., Booth, I. W. and Muller, D. P. R. (1985) N. Engl. J. Med. 313, 32-35.
Igarashi, O., Yonekawa, Y. and Fujiyama-Fujihara, Y. (1991) J. Nutr. Sci. Vitaminol. 37, 359-369.
Ingold, K. U., Burton, G. W., Foster, D. O., Hughes, L., Lindsay, D. A. and Webb, A. (1987a) Lipids 22, 163-172.
Ingold, K. U., Webb, A. C., Witter, D., Burton, G. W., Metcalfe, T. A. and Muller, D. P. R. (1987b) Arch. Biochem. Biophys. 259, 224-225.
Kennedy, T. A. and Liebler, D. C. (1991) Chem. Res. Tox. In press.
Liebler, D. C., Baker, P. F. and Kaysen, K. L. (1990) J. Am. Chem. Soc. 112, 6995-7000.
Liebler, D. C., Kaysen, K. L. and Burr, J. A. (1991) Chem. Res. Toxicol. 4, 89-93.
Maguire, J. J., Wilson, D. S. and Packer, L. (1989) J. Biol. Chem. 264, 21462-21465.
Matsuo, M., Matsumoto, S., Iitaka, Y. and Niki, E. (1989) J. Am. Chem. Soc. 111, 7179-7185.
Mordi, R. C., Walton, J. C., Burton, G. W., Hughes, L., Ingold, K. U. and Lindsay, D. A. (1991) Tet. Lett. 32, 4203-4206.
Muller, D. P. R. and Lloyd, J. K. (1982) Ann. N. Y. Acad. Sci. 393, 133-144.
Niki, E., Kawakami, A., Yamamoto, Y. and Kamiya, Y. (1985) Bull. Chem. Soc. Jpn. 58, 1971-1975.
Packer, J. E., Slater, T. F. and Willson, R. L. (1979) Nature 278, 737-738.
Packer, L., Maguire, J. J., Mehlhorn, R. J., Serbinova, E. and Kagan, V. E. (1989) Biochem. Biophys. Res. Comm. 159, 229-235.
Schmandke, H. (1965) Internat. Z. Vitaminforsch. 35, 321-327.
Schmandke, H. and Schmidt, G. (1968) Internat. Z. Vitaminforsch. 38, 75-78.
Scholich, H., Murphy, M. E. and Sies, H. (1989) Biochim. Biophys. Acta 1001, 256-261.
Simon, E. J., Eisengart, A., Sundheim, L. and Milhorat, A. T. (1956a) J. Biol. Chem. 221, 807-817.
Simon, E. J., Gross, C. S. and Milhorat, A. T. (1956b) J. Biol. Chem. 221, 797-805.
Stocker, R., Bowry, V. W. and Frei, B. (1991) Proc. Natl. Acad. Sci. USA 88, 1646-1650.
Traber, M. G., Ingold, K. U., Burton, G. W. and Kayden, H. J. (1988) Lipids 23, 791-797.
Traber, M. G., Burton, G. W., Ingold, K. U. and Kayden, H. J. (1990a) J. Lipid Res. 31, 675-685.
Traber, M. G., Rudel, L. L., Burton, G. W., Hughes, L., Ingold, K. U. and Kayden, H. J. (1990b) J. Lipid Res. 31, 687-694.
Traber, M. G., Sokol, R. J., Burton, G. W., Ingold, K. U., Papas, A. M., Huffaker, J. E. and Kayden, H. J. (1990c) J. Clin. Invest. 85, 397-407.
Weber, F. and Wiss, O. (1963) Helv. Physiol. Acta 21, 131-141.
Yamauchi, R., Matsui, T., Satake, Y., Kato, K. and Ueno, Y. (1989) Lipids 24, 204-209.

Free Radicals: From Basic Science to Medicine
G. Poli, E. Albano & M. U. Dianzani (eds.)

A NUCLEAR POOL OF GLUTATHIONE IN HEPATOCYTES

Giorgio Bellomo, Mariapia Vairetti[1], Giusy Palladini, Francesca Mirabelli, Lucianna Stivala[2] and Plinio Richelmi[1]

Dipartimento di Medicina Interna e Terapia Medica, Clinica Medica I, [1]Instituto di Farmacologia Medica II and [2]Istituto di Patologia Generale, University of Pavia, 27100 PAVIA, Italy

SUMMARY: Using fluorescent probes to selectively label glutathione (GSH) in living, cultured rat hepatocytes, and image analysis, we detected a compartmentation of GSH in the nucleus. A nuclear GSH concentration of 15 mM, and a cytosolic GSH concentration of 4.5 mM, with a concentration gradient of approximately 3, was present in these cells. This gradient was maintained by ATP-requiring process(es) and a GSH-stimulated ATPase activity was discovered in isolated rat liver nuclei. A critical role of nuclear GSH as cofactor of peroxidase(s), transferase(s) and thiol-disulfide oxidoreductase(s) can be envisaged in protecting DNA and other nuclear structures from oxidative and chemical injury as well as in the physiological control of chromatin conformation.

The tripeptide glutathione (γ-glutamyl-cysteinyl-glycine) is involved in many important cellular functions, ranging from the control of physical-chemical properties of cellular proteins and peptides (and thus modulating the activity of many enzymes) to the detoxification of xenobiotics and free radicals (Meister and Anderson, 1983; Orrenius et al., 1983; Tanigouchi et al., 1989; Orrenius and Bellomo, 1989). It also protects cells against the toxic effects of oxygen, by reacting directly and enzymatically with reactive oxygen intermediates, and less directly, by maintaining other compounds which have antioxidant activity, such as ascorbate and α-tochopherol, in reduced form (Meister, 1992). The antioxidant and reducing activity is exhibited by the reduced form of glutathione (GSH), whose concentration over the oxidized forms (glutathione disulphide, GSSG and glutathione mixed disulphides, R-S-SG) is relatively high. The intracellular level of glutathione in most mammalian cells is in the millimolar range, and is kept so by continuous synthesis (Griffith and Meister, 1979).

Conventional cell fractionation studies have provided evidence for the compartmentation of glutathione in subcellular organelles. Approximately 10 - 20 % of total cellular glutathione in rat liver is sequestered in the mitochondrial matrix (Wahllander et al, 1979; Meredith and Reed, 1982). The size of this pool depends on cytosolic GSH synthesis and the active transport of GSH into mitochondria via a recently-described multicomponent system (Martensson et al., 1990). Intramitochondrial GSH apparently plays an essential role in protecting mitochondrial structures against oxidative damage by oxygen reactive species produced during the physiological activity of the respiratory chain, as well as during the metabolism of toxic compounds. In addition to mitochondria, however, conventional techniques have shown no other organelles to contain specific pools of glutathione.

Despite the known functions of GSH in DNA synthesis and protection from oxidative DNA damage, little is known about the nuclear localization of GSH or the factors regulating the nuclear GSH level. Using fractionation and centrifugation techniques in non-aqueous media to prevent redistribution of GSH, Tirmenstein and Reed (1988a) measured nuclear glutathione concentration in rat kidney, and found values similar to those in the cytosol. However, the inhibition of GSH synthesis by buthionine-L-sulphoximine (BSO) caused a preferential GSH depletion in the nucleus, indicating that this pool of glutathione was specifically regulated. Other fractionation techniques (such as selective permeabilization of various cell organelles with different detergents) have provided equivocal results. Britten et al. (1991), however, reported that the sensitivity of cancer cell lines to chemotherapeutic agents was inversely correlated to the intranuclear glutathione content. Although not conclusive, these investigations, taken together, have suggested that a nuclear pool of glutathione may exist in intact cells.

Fluorescent probes to investigate thiol distribution in single cells

Recent advances in image-analysis technology, together with the development of additional, non-toxic, fluorescent indicators which can be used in intact cells, have greatly facilitated the study of various aspects of cell physiology (DiGiuseppi et al., 1985). Among the available indicators are derivatives of bimanes, which are themselves non-fluorescent, but which develop a strong blue fluorescence upon binding to thiol groups (Kosower et al., 1979). The major advantages of these compounds are the high reaction rate of the labeling agent at neutral pH, the strong preference for thiols, the photostability of the labeled products and the small size of the bimane moiety. Several derivatives of bimane are commercially available, including monobromo-bimane (BmBr), bisbromo-bimane (BmbBr) and monobromo-trimethyl-ammonium-bimane (BmqBr). While the first two compounds are neutral and easily penetrate intact living cells, BmqBr labels extracellular thiols as well as dead and permeabilized cells. These properties have been employed to label extracellular and intracellular thiols selectively in living cells (Mirabelli et al, 1992).
None of the above-mentioned bimane derivatives exhibits any kind of selectivity toward specific thiols. Interestingly, the reaction rate of monochlorobimane (BmCl) with GSH or other thiols is relatively low, whereas the rate of formation of the GSH adduct increases markedly in the presence of glutathione-S-transferases (Shrieve et al., 1988). Because of this property, BmCl has been proposed as a fluorescent label for GSH in living cells. However, the heterogeneity of GSH-transferase isoforms among various species causes the quantitative specificity of BmCl for the substrate (GSH) to differ. For example, the Km for the reaction catalyzed by rat liver GSH transferases is approximately 6 mM, while in some human cell lines it ranges from 200 to 350 mM (Cook et al., 1991).

One of the major advantages of using fluorescent probes to label glutathione is the possibility of mapping the intracellular distribution of the tripeptide spatially, using the same procedure and the same technological facilities (such as image analysis) that have been employed to investigate the intracellular distribution of ionized Ca^{2+} and pH, using the fluorescent probes Fura-2 and BCECF (Tsien and Harootunian, 1990). However, unlike the fluorescent probes for Ca^{2+} and pH, monochlorobimane forms an adduct with GSH, and the fluorescence which develops in fact represents the intra-cellular distribution of BmCl-GSH adduct. For this reason, one must be extremely careful in interpreting image analysis data of BmCl-labeled cells, and several control experiments must be done. In the following sections we will describe some of our recent work performed on isolated rat hepatocytes labelled with monochlorobimane, and dealing with the nuclear localization of glutathione in these cells.

Demonstration of a nuclear compartmentation of glutathione in hepatocytes

The incubation of 24 hour cultured rat hepatocytes with 80 mM BmCl for 90 seconds results in the development of a strong blue fluorescence (excitation 393 nm, emission through a 445 nm dichroic mirror with a 470 nm barrier filter). Image analysis of BmCl-labeled hepatocytes revealed an uneven distribution of the fluorescence characterized by a highly fluorescent region near the centre of the cell (Bellomo et al., 1992). This kind of distribution occurred in the large majority of the cells investigated (over 95 %), and was independent of culture conditions (it was detected in hepatocytes 1 hour after isolation as well as 72 hours after isolation and culturing).
When BmCl-labeled hepatocytes were permeabilized with low concentrations of digitonin (which selectively permeabilizes the plasma

membrane), more than 98 % of the fluorescence was released. Since this permeabilization treatment does not release GSH sequestered in mitochondria, it can be assumed that, in the experimental conditions employed, the mitochondrial pool of GSH is not labeled by BmCl. After digitonin treatment, the subsequent labeling of the permeabilized cells with the DNA dye Hoechst 33258 revealed the location of the nucleus. Comparison of BmCl and Hoechst 33258 fluorescence images indicated a perfect identity between the regions exhibiting the highest BmCl fluorescence and those labeled with Hoechst 33258 in all cells examined (Fig. 1). A series of control experiments have revealed that the enhanced nuclear fluorescence developed after BmCl labeling was due to neither (1) changes in cell thickness in the nuclear region, nor (2) selective compartmentation of glutathione transferases in the nucleus, nor (3) translocation of the GSH-BmCl adduct from the cytosol into the nucleus.

Taken together, these findings demonstrate that high concentrations of glutathione are present in the nucleus of rat hepatocytes. Assuming that BmCl fluorescence in labeled cells underestimate the total GSH concentrations (the fraction of GSH sequestered in mitochondria) by 15-20 %, and that, in hepatocytes, the cytosolic and nuclear compartments occupy a volume of 54 % and 6 % respectively (Alberts et al., 1983), one can calculate a nuclear GSH concentration of approximately 15 mM, and a cytosolic GSH level of approximately 5 mM. Thus a GSH concentration gradient of 3 exists between the cytosol and the nuclear matrix.

<u>Putative mechanisms responsible for GSH compartmentation in the nucleus</u>

BmCl labeling of 24 hour cultured rat hepatocytes pre-incubated with mitochondrial uncouplers or ATP-depleting agents failed to reveal any significant compartmentation of fluorescence (and thus of GSH) in the nucleus (Bellomo et al., 1992). This finding suggests that an active (ATP-dependent) process is required to maintain the nuclear-cytosolic GSH concentration gradient.

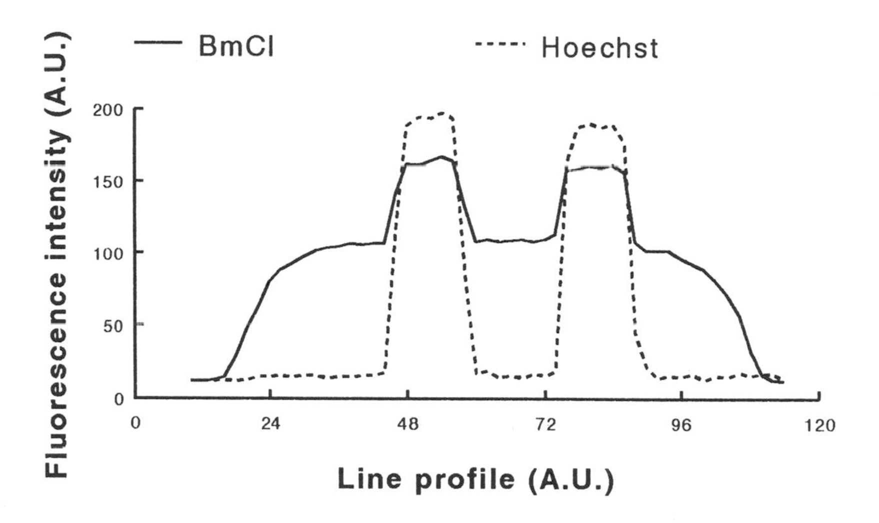

Fig. 1: Intracellular location of GSH in hepatocytes. Cultured rat hepatocytes were incubated with BmCl to label intracellular GSH. The image of intracellular fluorescence was taken before perfusion of the same cells with 10 mg/ml digitonin. Hoechst 33258 was added, and the fluorescent image showing location of the nucleus in the cell previously analysed was taken. Both images were then processed, and the BmCl (solid line) and Hoechst 33258 (dashed line) fluorescence distribution along a line-profile was quantified. The reported data refer to a binucleated cell. For experimental details see Bellomo et al, 1992.

This evidence prompted us to investigate the existence of an ATP-driven glutathione accumulation in a highly purified preparation of rat liver nuclei. However, any attempt to demonstrate glutathione transport in isolated nuclei failed, probably because of a

continuous leakage of nuclear GSH into the extranuclear medium through the nuclear pore, or through leaking sites in the nuclear envelope. Interestingly, freshly-isolated nuclei do not contain measurable amounts of free glutathione (Bellomo G., unpublished results), thus indicating that the simple homogenation and centrifugation procedures in acqueous media caused the release of most of the nuclear GSH. One can additionally speculate that cytosolic factors or cytoskeletal elements which are lost during fractionation procedures could be essential to maintain the nuclear glutathione translocator operative, and experiments in this direction are in progress.
Several reports indicate that, in some instances, glutathione translocation across plasma membrane (Nicotera et al., 1985), and across mitochondrial inner membrane (Martensson et al., 1990), requires ATP, and that, in most cases, a glutathione-stimulated ATPase is involved. To test the possibility that a GSH-stimulated ATPase activity could also be involved in the trasport of GSH across the nuclear membrane, the existence of a GSH-dependent ATP hydrolysis was investigated in isolated nuclei.
A significant stimulation of ATP hydrolysis was detected in rat liver nuclei incubated with GSH at concentrations above 3 mM. The kinetic parameters of this ATPase activity are reported in Table I, and are in close agreement with the measured concentrations of GSH and ATP found in the cytosol. This ATPase activity was exclusively detected in the nuclear preparation, and was specific for GSH, since neither constituent amino acids nor the oxidized form of glutathione (GSSG) exhibited any appreciable effects (Fig. 2). In addition, the thiol reductant dithiothreitol was not as powerful as GSH in favouring ATP hydrolysis, thus indicating that the simple reduction of -SH groups in a Mg2+-dependent ATPase molecule was not the mechanism responsible for GSH-stimulated ATP hydrolysis in liver nuclei. Furthermore, with the exception of vanadate, none of the available inhibitors of the various known ATPases caused a significant decrease in the GSH-dependent ATPase activity (Vairetti M, Palladini G, Mirabelli G

and Bellomo G,, manuscript in preparation). Taken together, these findings suggest that a GSH-stimulated ATPase could be linked to active GSH transport in the nuclear matrix and responsible for the nuclear-cytoplasmic concentration gradient observed.

Table I

Kinetic parameters of GSH-stimulated ATP hydrolysis in isolated liver nuclei

Km for GSH (mM)	7.01 1 0.4
Km for ATP (mM)	98.2 1 14
Vmax (nmol Pi/min/mg protein)	25.3 1 2.2
pH optimum	6.7 10.25

Experiments were performed with isolated rat liver nuclei

Physiological and toxicological relevance of nuclear glutathione

The presence of high concentrations of glutathione in the nucleus of hepatocytes would be important for the several GSH-dependent enzymes associated with the nuclear structures. For example, Bennet et al. (1986) reported that a 30 kDa DNA-binding protein isolated from rat cell nuclei exhibited chemical and immunological properties of glutathione-S-transferase Yb subunits. Furthermore, Tan et al. (1988) purified and characterized glutathione transferases from rat liver nuclei. The specificity of isoforms of glutathione transferases for nuclear location was pointed out by Bennet and Yeoman (1987) who demonstrated that microinjected glutathione-S-transferase Yb sub-units translocated to the cell nucleus. In addition, reports from Tirmestein and Reed (1988b, 1989) have demonstrated that lipid peroxidation may occur in isolated liver nuclei, and it can be efficiently prevented by GSH-dependent enzyme systems. Finally, a nuclear and perinuclear localization of the phospholipid hydroperoxide glutathione peroxidase has recently been discovered (Ursini F, personal communication).

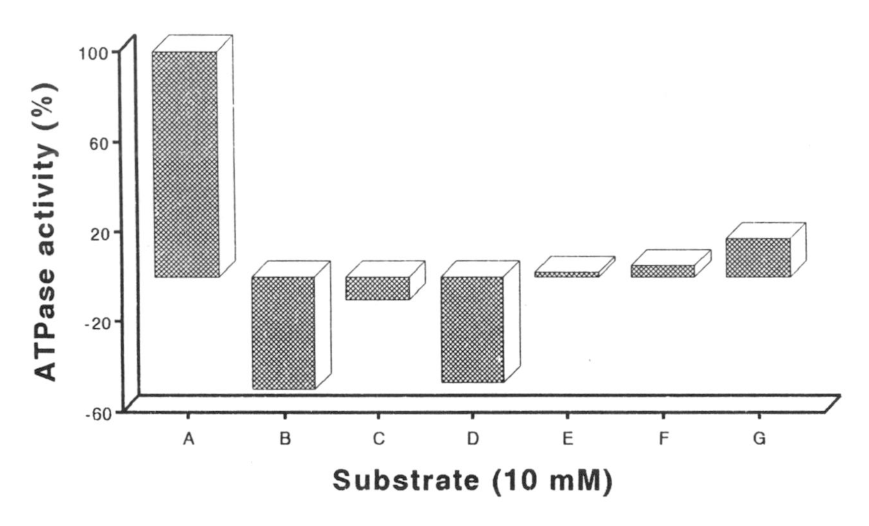

Fig.2: Effects of various substrates on ATPase activity in isolated nuclei. Isolated rat liver nuclei (0.1 mg protein/ml) were incubated in MOPS/KCl/MgCl2 buffer, pH 7, with 1 mM ATP and 10 mM of various substrates (A: GSH; B: GSSG; C: cysteine; D: cystine; E: glycine; F: glutamic acid; G: dithiothreitol) at 37°C. After 45 min, the reaction was stopped by adding perchloric acid, and the amount of phosphate generated by ATP hydrolysis was measured. GSH-stimulated ATP hydrolysis was taken as 100 % ATPase activity.

It is possible to postulate a concerted operation between nuclear GSH, the mechanisms responsible for GSH accumulation in the nuclear matrix and the cytosolic and nuclear glutathione-S-transferases and peroxidases, playing a critical role in the protection of DNA and other nuclear structures against chemical and oxidative damage (see Fig. 3 for a general scheme).

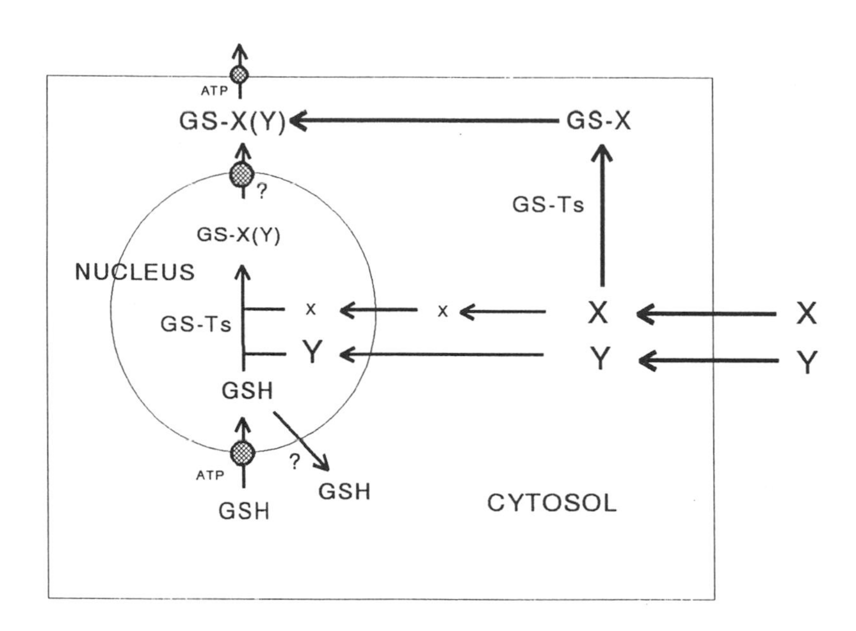

Fig. 3: Schematic representation of the concerted operation between nuclear GSH, the mechanisms responsible for GSH accumulation in the nuclear matrix, and the cytosolic and nuclear glutathione-S-transferases. X, Y: xenobiotics; GS-Ts: glutathione-S-transferases; GS-X(Y): glutathione conjugate with the xenobiotic X or Y.

In addition to its participation in various detoxication reactions, GSH is critically involved in many physiological processes, including the regulation of nuclear matrix organization (Dijkwel and Wenik, 1986). The nuclear matrix, in fact, undergoes some structural changes on treatment with thiol reagents (Stuurman et al., 1992) and it has been hypothesized that physiological fluctuations of the thiol-disulphide balance in nuclear matrix and nuclear matrix proteins might stabilize or disgregate the chromatin organization. Furthermore, the maintenance in the reduced state of the cysteine residues essential for the zinc-finger structural motif of DNA-binding proteins and transcription factors (Klug and Rhodes, 1987) would require high nuclear GSH concentration. These considerations increase the importance of this tripeptide in the reactions taking place within the nucleus, and encourage further investigation on the biochemical mechanisms involved in regulating this compartmentation process.

Acknowledgements

The work described in this paper was supported by grants from Ministero dell'Università e della Ricerca Scientifica e Tecnologica and from Boehringer Mannheim Italia.

References

Alberts, B., Bray, D., Lewis, J., Raff, M., Roberts, K., and Watson, J.D. editors (1983) Molecular Biology of the Cell, Garland New York, p 321.
Bellomo, G., Vairetti, M., Stivala, L., Mirabelli, F., Richelmi, P., and Orrenius, S. (1992) Proc. Natl. Acad. Sci. USA 89, 4412-4416.
Bennet, C.F., Spector, D.L., and Yeoman, L.C. (1986) J.Cell Biol. 102, 600-609.
Bennet, C.F., and Yeoman, L.C. (1987) Biochem. J. 247, 109-112
Britten, R.A., Green, J.A., Broughton, C., Browning, P.G.W., White, R., and Warenius, H.M. (1991) Biochem. Pharmacol. 41, 647-649.
Cook, J.A., Iype, S.N., and Mitchell, J.B. (1991) Cancer Res. 51, 1606-1612.
DiGiuseppi, J., Iman, R., Ishihara, A., Jakobson, K., and Herman, B. (1985) Biotechniques 3, 349-403.
Dijkwel, P.A., and Wenick, P.W. (1986) J. Cell Sci. 84, 53-67.
Griffith, O.W., and Meister, A. (1979) J. Biol. Chem. 76, 5606-5610.
Klug, A., and Rhodes, D. (1987) Trends Biochem. Sci. 12, 464-469.
Kosower, N., Kosower, E.M., Newton, G.L., and Ranney, H.M. (1979) Proc. Natl. Acad. Sci. USA 76, 3382-3386.
Martensson, J., Lai, J.C.K., and Meister, A. (1990) Proc. Natl. Acad. Sci. USA 82, 4668-4672.

Meister, A., and Anderson, M.E. (1983) Annu. Rev. Biochem. 52, 711-760.
Meister, A. (1992) in "Biological Free Radical Oxidations and Antioxidants" (Ursini, F. and Cadenas, E. eds.) CLUEP, Padova, pp. 89-93.
Meredith, M.J., and Reed, D.J. (1982) J. Biol. Chem. 257, 3747-3753.
Mirabelli, F., Vairetti, M., and Bellomo, G. (1992) Appl. Fluoresc. Technol. 4, 9-13.
Nicotera, P., Baldi, C., Swensson, S.A., Larsson, R., Bellomo, G., and Orrenius, S. (1985) FEBS Lett. 187, 121-124.
Orrenius, S., Jewell, S.A., Bellomo, G., Thor, H., Jones, D.P., and Smith, M.T. (1983) in "Functions of Glutathione: Biochemical, Physiological, Toxicological and Clinical Aspects" (Larsson, A. et al., eds.) Raven Press, New York, pp. 261-271.
Orrenius, S., and Bellomo, G. (1989) in "Glutathione: Chemical, Biochemical and Medical Aspects" Part B, (Dolphin, D. et al. eds) Wiley, New York, pp. 383-409.
Shrieve, D.C., Bump, E.A., and Rice, G.C. (1988) J. Biol. Chem. 263, 14107-14114.
Stuurman, N., Floore, A., Colen, A., DeJong, L., and van Driel, R. (1992) Exp. Cell Res. 200, 285-294.
Tan, K.H., Meyer, D.J., Gillies, N., and Ketterer, B. (1988) Biochem. J. 254, 841-845.
Tanigouchi, N., Higashi, T., Sakamoto, Y., and Meister, A. (editors) (1989) Glutathione Centennial: Molecular Properties and Clinical Implications, Academic Press, New York.
Tirmenstein, M.A., and Reed, D.J. (1988a) Biochem. Biophys. Res. Commun. 155, 956-961.
Tirmestein, M.A., and Reed, D.J. (1988b) Arch. Biochem. Biophys. 261, 1-11.
Tirmestein, M.A., and Reed, D.J. (1989) Biochim. Biophys. Acta 995, 174-180.
Tsien, R.Y., and Harootunian, A.T. (1990) Cell Calcium 11, 93-110.
Wahllander, A., Soboll, S., and Sies, H. (1979) FEBS Lett. 97, 138-140.

Free Radicals: From Basic Science to Medicine
G. Poli, E. Albano & M. U. Dianzani (eds.)

PHOSPHOLIPID HYDROPEROXIDE GLUTATHIONE PEROXIDASE IS THE MAJOR SELENOPEROXIDASE IN NUCLEI AND MITOCHONDRIA OF RAT TESTIS

Matilde Maiorino, Antonella Roveri and Fulvio Ursini*.

Department of Biological Chemistry, Via Trieste, 75, 35100 Padova, Italy.*Department of Chemistry, Via Cotonificio 108, 33100 Udine, Italy.

SUMMARY: This paper reports the identification of the selenoenzyme Phospholipid Hydroperoxide Glutathione Peroxidase (PHGPX) in rat testis. PHGPX specific activity in the testis was the highest so far measured in mammalian tissues, and the activity distribution pattern was just the opposite of that found in the liver. In fact, while liver PHGPX is essentially a soluble cytosolic protein, in the testis, the highest specific activity was detected in the nuclear and mitochondrial membranes. The identity of PHGPX with the previously-described 15-20 Kd selenoprotein (Wallace et al., 1983), involved in the maturation of spermatozoa, is proposed. The observation that PHGPX is the preponderant selenoprotein in rat testis throws some light on the complex relationship between selenium and spermatogenesis, and suggests that the effects of selenium deficiency may reasonably be produced by the absence of this enzyme.

Mammalian selenocysteine-containing proteins

Selenium, previously known only as a dangerous poison (Franke, 1934), was subsequently revealed to be also an essential trace element (Schwarz, 1961). The biological role of selenium depends on its incorporation into some proteins as selenocysteine. To date, five selenocysteine-containing proteins, which have been shown to be different gene products, have been characterized in mammals (Stadtman, 1991): glutathione peroxidase (GPX) and its plasma variant, phospholipid hydroperoxide glutathione peroxidase (PHGPX), selenoprotein P, and tetraiodothyronine 5'-deiodinase. Of these, cellular GPX and PHGPX appear to be widely distributed in mammalian tissues, whereas tetraiodothyronine 5'-deiodinase is reported to be located mainly in the thyroid, kidney and liver, and plasma GPX and selenoprotein P have only been reported in plasma (Stadtman, 1991).

Studies on selenium deficiency and repletion have shown that a hierarchy exists in the supply of this element to different organs. Particularly, it has been reported that brain, endocrine and reproductive tissues have priority in selenium supply over other tissues, and that selenoproteins other than GPX are synthesized first (Behne et al., 1988).

Selenium and spermatogenesis

The role of selenium in maintaining normal spermatogenesis in rodents has been demonstrated by several investigators (McCoy et al., 1969; Sprinker et al., 1971, Wu et al., 1973, Wallace et al., 1983), and it has been shown that rats, fed a selenium-deficient diet, produced sperm with impaired motility (McCoy et al., 1969, Wu et al., 1969). Transmission electron micrographs through epididymal spermatozoa from those animals showed morphological damage in the portion of sperm tail which houses mitochondria i.e. the so-called "midpiece". The mitochondria appeared irregularly shaped and arranged around the sperm tail, and varied in size, shape and location (Wallace et al., 1984).
Data from Behne (Behne et al., 1986), showed that testis selenium content rose six-fold during organ maturation, despite only a doubling of GPX specific activity.
On the basis of data obtained using tracer 75Se during spermatogenesis, it has been reported that most of the label was incorporated in a single protein, 15-20 kDa M.W., present in the outer membrane of the sperm mitochondria (Calvin et al., 1978, McConnel et al., 1979, Pallini et al., 1979). This protein, which was referred to as the selenoprotein of sperm mitochondrial capsulae (MCS), was also identified in testicular extracts. However, positive evidence for the presence of selenocysteine in this protein was never subsequently obtained. In 1990, the cDNA clones encoding MCS were sequenced, and the TGA codon was not found (Kleene et al., 1990). Since TGA directs a co-translational incorporation of selenium into selenocysteine, these data indicate that either: i) selenium is not present as selenocystein,

but is post-translationally linked to MCS; or ii) another protein with a M.W. similar to MCS actually accounts for the radioactivity band in sodium dodecyl sulphate polyacrylamide gel electrophoresis (SDS PAGE) of sperm mitochondrial ghosts.
This paper reports the identification of PHGPX as a major rat testis protein, suggesting that this enzyme actually accounts for the selenium moiety previously attributed to MCS.

Phospholipid hydroperoxide glutathione peroxidase

PHGPX was first purified as a peroxidation-inhibiting protein in 1982 (Ursini et al., 1982), following the discovery of anti-peroxidant activity in liver cytosol (McCay et al., 1982, Burk, 1983). The antioxidant effect was attributed to the unique capability of the enzyme to reduce lipid hydroperoxides in membranes, i.e phospholipid and cholesterol hydroperoxides, at the expense of thiols (Daolio et al., 1983, Thomas et al., 1990).
In 1985, after the identification of selenium in the active site (Ursini et al. 1985), PHGPX was characterized as a selenium-containing peroxidase different from GPX. In the following years, distinction between the two selenoenzymes was also made on the basis of biochemical parameters (Maiorino et al., 1986, Thomas et al., 1990) and protein and cDNA sequence analysis (Schuckelt et al., 1991).
PHGPX molecular weight is 18 KD on SDS PAGE and gel permeation column chromatography. The enzyme has been purified to homogeneity from pig liver cytosol (Ursini et al., 1982), heart (Maiorino et al., 1982) and brain (Ursini et al., 1984), and it has been identified in many other tissues: rat kidney, heart, lung, muscle, and brain (Ursini et al., 1983, Zhang et al., 1989); and cells: human tumour cell lines MCF7, K562, HL60, HepG2 (Maiorino et al., 1991).
The evidence that the selenium moiety of PHGPX is involved in the catalytic cycle was supported by inactivation kinetics in the presence of iodoacetate and thiols (Ursini et al., 1985). More

recently, the presence of selenocysteine in PHGPX aminoacid analysis, together with the discovery of TGA codon in the open reading frame of cDNA clones, has conclusively demonstrated the presence of selenium in the enzyme as selenocysteine (Schuckelt et al., 1991). Sequence analysis has also indicated that PHGPX is a protein homologous, but poorly related, to GPX.

PHGPX in testis

In adult rat testis homogenate, the ratio between GPX and PHGPX specific activity is below 1, indicating that in this organ, but not in liver, heart or HL 60 human tumour cell lines, PHGPX is the major selenium-containing peroxidase (Table I). While a variable percentage of PHGPX activity (10-30%) is associated with membranes in other organs (unpublished results), 80% of the total activity is membrane-bound in testis. Table II reports a comparison of PHGPX specific activity distribution pattern in liver and testis. PHGPX specific activity is five times higher in testis nuclei and mitochondria than in cytosol,

Table I

Ratio between GPX and PHGPX specific activities in homogenates.

	Liver	Heart	HL60 cells	Testis
GPX/PHGPX	550	500	40	0.65

Table II

Subcellular distribution of PHGPX in rat liver and testis.

Fraction	Liver	Testis
	nmoles/min/mg protein	
Nuclei	5.9 ± 0.6	102 ± 9
Mitochondria	2.3 ± 0.2	94 ± 9
Microsomes	2.5 ± 0.3	2 ± 0.3
Cytosol	8.9 ± 0.6	18 ± 3

while an inverse distribution pattern was observed in liver. It is worth noting that the distribution pattern found in testis was peculiar, and that the specific activiy is by far the highest observed.

PHGPX, but not GPX, specific activity increases during maturation in testis (Roveri et al., 1992), in strict parallel with the increase in selenium uptake (Behne et al., 1986).

Hypophysectomy before puberty prevented PHGPX rise, while in adult animals it led to a progressive decline of activity, which was partially restored by gonadotropin treatment. GPX activity was not affected by either treatment (Roveri et al., 1992).

PHGPX was immunolocalized in rat testis (Roveri et al., 1992) supporting the distribution of the specific activity. Immunostaining also showed a different distribution pattern during maturation of spermatozoa. In fact, a finely granular brown end-product localized throughout the cytoplasma was seen in spermatogonia, but this stain was confined to the peripheral regions of the cytoplasm, the nuclear membrane and mitochondria, in maturing primary spermatocytes.

Testis cytosolic and membrane-bound PHGPX were found to be cross-reactive with the antiserum raised against pig heart cytosol PHGPX (Roveri et al., 1992), and cyanogen bromide fingerprint analysis showed a similar fragmentation pattern for the cytosolic and membrane-bound testis enzyme (unpublished results). At present, the possible modification of PHGPX addressing the enzyme toward the membrane or the soluble compartment is still unknown.

Conclusions

The M.W. of PHGPX, which is a major selenoprotein in testis, is in the range reported for MCS which, on the other hand, does not contain selenocysteine. It is therefore reasonable to identify PHGPX with the selenoprotein previously detected in the testis.

The remarkable abundance of PHGPX in rat testis, and the gonadotropin dependency, deserve some comment concerning the enzyme's possible physiological role. The generally acknowledged function of selenium peroxidases is antioxidant protection. In fact, reducing hydroperoxides, GPX and PHGPX cooperate to decrease the steady-state concentration of hydroperoxides in the aqueous and in the membrane, environments, respectively. In this context, a reasonable role for PHGPX in testis could be envisaged in the protection of genetic material from free radicals generated by decomposition of lipophylic hydroperoxides. However, the peculiar localization of PHGPX in membranes during maturation of spermatozoa, as well as the gonadotropin dependency of its expression, seems to deal not only with a "non specific" oxidative stress, but also with a more specific role played by PHGPX reaction substrates and/or products, during differentiation and maturation. This hypothesis could also see a possible role of selenium in differentiation, as suggested by the recent discovery that the conversion of thyroxine to the active thyroid hormone is mediated by tetraiodothyronine 5'-deiodinase, a seleocysteine-containing enzyme (Berry et al., 1991).

REFERENCES

Behne, D., Duk, M., and Elger, W. (1986) J. Nutr. 116, 1442-1447.
Behne, D., Hilmert, H., Scheid, S., Gessner, H., and Elger, W. (1988) Biochim. Biophys. Acta 966, 12-21.
Berry, M. J., Banu, L., and Larsen, P. R. (1991) Nature 349, 438-440.
Burk, R. F. (1983) Biochim . Biophys. Acta 757, 21-28.
Calvin H. I. (1978) J. Exp. Zool. 204, 445-452.
Daolio, S., Traldi, P., Ursini, F., Maiorino, M., and Gregolin, C. (1983) Biomed. Mass. Spectrosc. 10, 499-504.
Franke, K. W. (1934) J. Nutr. 8, 597-608.
Kleene, K. C., Smith, J., Bozorgzadeh, A., Harris, M., Hahn, L., Karimpour, I., and Gerstel J. (1990) Develop. Biol. 137, 395-402.
Maiorino, M., Ursini, F., Leonelli, M., Finato, N., and Gregolin, C. (1982) Biochem. Intern. 5, 575-583.
Maiorino, M., Roveri, A., Gregolin, C., and Ursini, F. (1986) Arch. Biochem. Biophys. 251, 600-605.
Maiorino, M., Chu, F. F., Ursini, F., Davies, K. J. A., Doroshow, J. H., and Esworthy, R. S. (1991) J. Biol. Chem. 266, 7728-7732.
McCay, P. B., and Gibson, D. D. (1982) Fed. Proc. 49, 911-914
McConnell, K. P., Burton, R. M., Kute, T. and Higgins, P. J. (1979) Biochim. Biophys. Acta 588, 113-119.
McCoy, K. E., and Weswig, P. H. (1969) J. Nutr. 98, 383-389.

Pallini, V., and Bacci E. (1979) J. Submicrosc. Cytol, 11, 165-170.
Roveri, A., Casasco, A., Maiorino, M., Dalan, P., Calligaro, A. and Ursini, F. J. Biol. Chem. (1992) 267, 6142-6146.
Schuckelt, R., Brigelius-Floh, R., Maiorino, M., Roveri, A., Reumkens, J., Straaburger, W., Ursini, F., Wolf, B., and Floh, L. (1991) Free Rad.Res. Commun. 14, 343-361.
Schwarz, K. (1961) Fed. Proc., Fed . Am. Soc. Exp.Biol. 20, 666-673.
Sprinker, L. H., Harr, J. R., Newberne, P. M., Whanger, P. D., and Weswig, P. H. (1971) Nutr. Rep. Int. 4, 335-340.
Stadtman, T. C., (1991), J. Biol Chem. 266, 16257-16260.
Thomas, J. P., Maiorino, M., Ursini, F., and Girotti, A. W. (1990) J. Biol. Chem. 265, 454-461.
Ursini, F., Maiorino, M., Valente, M., Ferri, L., and Gregolin, C. (1982) Biochim. Biophys. Acta 710, 197-211.
Ursini, F., Maiorino, M., and Gregolin, C. (1983) in "Oxy radicals and their Scavenger Systems" (R. Greenwald and G. Cohen Eds.) Elsevier- North Holland New York vol.II, pp. 224-230.
Ursini, F., Maiorino, M., Bonaldo, L., and Gregolin, C. (1984) in "Oxygen radicals in chemistry and biology" (Bors, W., Saran, D., and Tait, D. Eds.) De-Gruyter, Berlin, NewYork. pp. 713-718.
Ursini, F., Maiorino, M., and Gregolin, C. (1985) Biochim. Biophys. Acta 839, 62-70.
Wallace, E., Calvin, H. I., and Cooper, G. W. (1983) Gamete Res., 4, 377-387.
Wallace, E., Calvin, H. I., Ploetz, K., and Cooper G. W. (1984). in: Selenium in Biology and Medicine (Combs G. F., Spallhoz, J. E., Levander O. A., Oldfield J. E. Eds), Van Nostrand Reinhold Company, New York, pp.181-196.
Wu, A. S. H., Oldfield, J. E., Muth, O. H., Whanger, P. D., and Weswig, P. H. (1969) Proc. West. Sec. Am. Soc. Anim. Sci. 20, 85-89.
Wu, A. S. H., Oldfield, J. E., Shull, L. R., and Cheeke, P.R. (1979) Biol. Reprod. 20, 793-798
Zhang, L., Maiorino, M., Roveri, A., and Ursini, F. (1989) Biochim. Biophys. Acta 1006, 140-143.

Free Radicals: From Basic Science to Medicine
G. Poli, E. Albano & M. U. Dianzani (eds.)

MEDICAL APPLICATIONS OF ANTIOXIDANTS: AN UPDATE OF CURRENT PROBLEMS

Helmut Sies

Institut für Physiologische Chemie I, Heinrich-Heine-Universität, Düsseldorf, Moorenstrasse 5, W-4000-Dusseldorf, Germany.

SUMMARY: Oxidative stress, defined as an imbalance in the prooxidant-antioxidant equilibrium in favour of the prooxidants, has been invoked by many investigators to play a role in clinical medicine. This would a rational basis for antioxidant therapy. Based on the variety of biological antioxidants, pharmacological intervention centered on both enzymatic and nonenzymatic compounds. The former is exemplified by superoxide dismutase or enzyme mimics such as ebselen, the latter by antioxidant vitamins and micronutrients (vitamin E, vitamin C, etc.) and other low molecular weight antioxidant compounds, e.g. glutathione (GSH), n-acetyl cysteine and lipoate.

However, appropriate assessing of oxidative stress is difficult clinically, due to methodological problems and to the multifactorial nature of many disease states, precluding a straightforward assignment of the causative or associative role of oxidative stress in medically important situations. Thus, at present a rational basis for antioxidant therapy has yet to be identified for many diseases addressed in the literature.

There is an ample literature on the association of reactive oxygen species with disease states, not to be presented here in detail. The idea is that the disease process is characterized by either an increased production of prooxidants or by a weakening of antioxidant defense, thus leading to a disbalance in the prooxidant/antioxidant steady state, termed 'oxidative stress' (Sies, 1985, 1986). Thus, early on the concept emerged that antioxidant treatment could be beneficial in medicine. This seemed logical also in view of the continuous demand of tissues and body fluids for antioxidants both enzymatic and nonenzymatic.

Antioxidant protection can be grouped into three categories, prevention, interception, and repair. For each of these three entities there are interesting examples. There are multiple strategies in biological antioxidant defences (Sies, 1991), and specific organs required specific ones, e.g. in ischemia-reperfusion (Omar et al, 1991), in the lens (Spector, 1991), the skin (Fuchs and Packer, 1991), in sepsis and shock (Wendel et al, 1991) and in multiple organ failure (Redl et al, 1991), just to name a few. Strategies are different on the time scale as well: short-term and long-term, acute diseases and chronic diseases, requiring selective responses. Thus, the problem is more complex as may have been anticipated. This is also due to the fact that the basal antioxidant potential in an organ can be subject to change in the transition from physiological to pathophysiological states and, furthermore, that there are adaptive response that occur over the time-course of a disease. In short, the problems for better defining the role in clinical medicine arise from the fact that the disease process is multifactorial, that there are overlapping antioxidant functions and that there is a pathophysiology of regulation.

From the point of view of basic science, the first problem is, however, an appropriate demonstration of the role of reactive oxygen species in a disease process in the first place. This has been difficult to achieve, and various methods have been applied (see Hageman et al, 1992, and Packer and Glazer, 1990). Our own recent addition to this field has been the evaluation of the tocopheryl quinone/tocopherol ratio as an indicator (Murphy et al, 1992). In a surgical setting, i.e. in aortic crossclamping, the tocopheryl quinone/tocopherol ratio in plasma obtained from the pulmonary artery was determined. It was found that during ischemia there was a doubling of this ratio returning towards the initial state upon reperfusion. Such a parameter could be helpful in monitoring time-courses in disease state, since some confounding factors are canceled out in the ratio.

Most single parameters that have been evaluated have their pitfalls. In particular there is a vast literature on merits and on the problems associated with the use of the thiobarbituric acid-reactive substances (TBARS) in clinical medicine not to be discussed here in detail. It is likely that multiple parameters need to be determined to obtain sufficient information on what is called the oxidative stress status (OSS) (Pryor, 1991).

ENYMES AND ENZYME MIMICS

The therapeutic principles that have been assessed include the use of enzyme, purified from biological sources or obtained by recombinant technology. A major focus of interest was and is superoxide dismutase (SOD), as discussed by Bulkley (this volume). One recent observation of experimental nature should be mentioned: Nakazone et al (1991) were able to show that targeted SOD, obtained by protein engineering (Inoue et al, 1991), led to a return towards normal levels of blood pressure in hypertensive rats.

Regarding enzyme mimics, just to mention one: glutathione peroxidase activity is exhibited by a selenoorganic compound ebselen (Müller et al, 1984). This compound exhibits antiinflammatory properties, and there may be clinically useful applications.

NONENZYMATIC ANTIOXIDANTS: NATURAL AND SYNTHETIC

The role of the natural antioxidants has been examined extensively in recent years, and the clinical aspects of vitamin E in particular are presented by Mino et al (this volume). Our own recent compilation on the antioxidant properties of vitamin E, vitamin C and β-carotene and other carotenoids (Sies et al, 1992) gives the present status. In short, these natural antioxidants protect basically by interception, notably of peroxyl radicals and of singlet oxygen. Ascorbate (vitamin C) has additional roles in enzymatic reactions, but this is becoming an area of interest for the other antioxidants as well. There is a growing body of knowledge on biochemical functions of these antioxidants and of other micronutrients. It is possible that non-antioxidant functions may be decisive in some aspects, e.g. in chemoprevention.

Clinically, there is the use of nonenzymatic antioxidants in several countries. Not always are the fundamental aspects and the rationale fully understood. Thioctic acid (lipoate) is employed in the treatment of diabetic neuropathy, for example. However, the complex pathogenesis of this disease state makes it a challenge to delineate whether and where a strictly antioxidant function is useful in treatment. A back-up function for the regeneration of tocopherol from the tocopheroxyl radical has been described (Scholich et al, 1989).

Similar remarks could be made for N-acetyl cysteine, glutathione (GSH), and other biological reductants. This topic is beyond merely an antioxidant one, treated comprehensively by Meister (this volume).

Acknowledments:

Work from our own group was supported by National Foundation for Cancer Research, Bethesda, by the Ministerium für Forschung und Technologie, Bonn, and by the Jung-Stiftung für Wissenschaft und Forschung, Hasbur.

REFERENCES

Fuchs, J., Packer, L. (1991) in: Oxidative Stress: Oxidants and Antioxidants (Sies, H., ed.) pp. 559-583, Academic Press, London.

Hageman, J.J., Bast, A. and Vermeulen, N.P.E. (1992) Chem.-Biol. Interactions, 82, 243-293.

Inoue, M., Watanabe, N., Utsumi, T. and Sasaki, J. (1991) Free Rad. Res. Comms., 12-13, 391-399.

Müller, A., Cadenas, E., Graf, P. and Sies, H. (1984) Biochem. Pharmacol. 33, 3235-3239.

Murphy, M.E., Kolvenbach, R., Aleksis, M., Hamsen, R. and Sies, H. (1992) Free Rad. Biol. Med. 13, 95-100.

Nakazono, K., Watanabe, N., Matsuno, K., Sasaki, J., Sato, T. and Inoue, M.(1991) Proc. Natl. Acad. Sci. 88, 10045-10048.

Omar, B., McCord, J. and Downey, J. (1991) in: Oxidative Stress: Oxidants and Antioxidants (Sies, H., ed.) pp. 493-527, Academic Press, London.

Packer, L. and Glaser, A.N. (1990) Academic Press, San Diego, London, 186.

Pryor, W.A. and Godber, S.S. (1991) Free Rad. Biol. Med. 10, 177-

Redl, H., Gasser, H., Hallström, S., Paul, E., Bahrami, S., Schlag, G. and Spragg, R. (1991) in Oxidative Stress: Oxidants and Antioxidants (Sies, H., ed.) pp. 595-616, Academic Press, London.
Scholich, H., Murphy, M.E. and Sies, H. (1989) Biochim. Biophys. Acta 1001, 256-261.
Sies, H. (1985) in: Oxidative Stress (Sies, H., ed.) pp. 1-8, Academic Press, London. Sies, H. (1986) Angew. Chem. Int. Ed. Engl. 25, 1058-1071.
Sies, H. (1991) Amer. J. Med. 91 (Suppl. 3C), 31S-38S.
Sies, H. (1992) "Beyond Deficiency: New Views on the Functions and Health Benefits of Vitamins" (Machlin, L.J. and Sauberlich, H.E., eds.) Vol. XXX, 7-20, New Yory Academy of Sciences, New York.
Spector, A. (1991) in: Oxidative Stress: Oxidants and Antioxidants (Sies, H., ed.) pp. 529-558, Academic Press, London.
Wendel, A., Niehörster, M. and Tiegs, G. (1991) in: Oxidative Stress: Oxidants and Antioxidants (Sies, H., ed.) pp. 585-593, Academic Press, London.

Free Radicals: From Basic Science to Medicine
G. Poli, E. Albano & M. U. Dianzani (eds.)

FREE RADICALS AND ANTIOXIDANTS IN MUSCULAR AND NEUROLOGICAL DISEASES AND DISORDERS

Abraham Z. Reznick[1] and Lester Packer[2]

[1] Department of Morphological Sciences - Technion, Faculty of Medicine, Efron Street, POB 9649, Haifa 31096, Israel.

[2] Department of Molecular and Cell Biology, 251 Life Science Addition, University of California, Berkeley, California 94720 USA

MUSCLE STUDIES:

INTRODUCTION:

Muscle is a very dynamic specialized tissue which serves several functions in humans and animals. Beside providing the main apparatus for body movement (skeletal muscles) it is also the main source of energy reserves. It also provides passive and active protection to our body and it is essential in some vital processes of life such as breathing (diaphragm and intercoastal muscles), blood flow (cardiac muscle), and digestion (intestinal smooth muscles).

Muscle metabolism and energy utilization has been under active research in the last few decades. Indeed aerobic energy consumption was shown to be the main source of ATP production for muscle contraction. Thus, as research continued it became apparent that oxygen is involved directly in muscle cellular activities.

In the 1980's several observations were made showing that some of the oxygen participating in muscle respiration ends up as free radicals (such as $O_2 \bullet^-$ and $\bullet OH^-$) and thus muscle damage could be attributed to these reactive oxygen species (ROS) increasing under pathological conditions. The following discussion will describe several conditions in which free radicals and antioxidants have been shown to be involved in muscle damage.

1) Free Radical Involvement in Muscle Damage Due to Exercise:

The main findings concerning oxidative damage observed in several exercise studies are summarized in Table I.

Table I: Oxidative Damage Indicators Due To Exercise In Animals and Human Studies

Type of system or tissue	Methodology and findings	Reference
Rat muscles	ESR -free radicals * ↑	Davies et al. (1982) Jackson et al. (1985)
Rat liver	ESR-free radicals ↑	Davies et al. (1982) Alessio et al. (1988)
Human breath	Pentane expiration ↑	Dillard et al. (1978) Simon-Schnass & Pabst(1988)
Mouse muscles	Lipid peroxidation ↓	Salminen & Vihko (1983)
Mouse skeletal muscles	SOD activity ↑	Steinhagen-Thiessen E. et al. (1982)
Mouse cardiac muscles	SOD activity ↑	Reznick et al. (1982)
Rat muscle	SOD activity ↔	Ji et al. (1988)
Rat muscle	Protein carbonyls ↑	Witt et al. (1992)

*Legend: ↑ increase, ↓ decrease, ↔ no change

Early studies by Dillard et al (1978) showed that pentane expiration as a measure of lipid peroxidation was increased in human breath due to exercise. In the early 1980's Davies et al (1982) and Jackson et al (1985) showed, using Electron Spin Resonance (ESR), that one can pick up increased signals of free radicals in exercising muscles. Concomitantly with the increase of oxidative stress, Vitamin E, as the major lipid soluble antioxidant was shown to be consumed much faster in exercising muscles (Aikawa K. M. et al., 1984). The increased comsumption of Vitamin E in endurance exercise was probably compensated by increased levels of antioxidative enzymes such as superoxide dismutase, catalase and glutathione peroxidase and reductase (Quintanila A and Packer L. 1983; Reznick et al. 1982)

The role of nutrition in execise and possible mechanisms of free radical generation during exercise were recently outlined by Packer and Singh (1992) and Singh (1992). Among the various ways which exercise may trigger an increase of free radical production, is the increase of $O_2^{\bullet -}$ generation in the mitochondrial electron transport chain, or the increase of metal catalyzed free radical production due to mechanical and morphological damage to muscles. In such cases free iron, which is released from myoglobin and other iron binding proteins, will accelerate free radical generation. Finally, in any type of exercise the levels of catecholamines such as adrenaline are increased many fold, and some of these can generate free radicals through a process of autooxidation (Singh V. H. 1992). Therefore, the observation that certain types of exercise such as strenuous or endurance training may lead to muscle oxidative damage, may indicate that under such conditions muscle antioxidant defense mechanisms are not sufficient to cope with the sudden imposition of oxidative stress.

2) Ischemia Reperfusion Injury Due to Oxidative Damage in Cardiac and Skeletal Muscles:

The evidence for the involvement of oxygen free radicals in myocardial damage following ischemia - reperfusion (I/R) is now quite overwhelming (Guamieri et al 1980, Bolli et al 1989, Zweier et al. 1988). The evidence accumulated can be divided in two categories: I. Direct evidence in which superoxide and/or of hydroxyl radicals were shown to be produced following the reperfusion stage by ESR measurements. II. Indirect evidence, in which exogenous antioxidant enzymes, free radical scavengers and chelators of metal ions, can partially protect the heart against ischemia - reperfusion injury. A recent study from our laboratory has shown that L-Propionyl carnitine is a very potent cardioprotector against I/R damage, acting as a possible iron chelator thus preventing formation of ROS as well as an efficient source of metabolic energy for cardiac cells (Reznick et al., 1992).

In addition to study of I/R involvement in cardiac damage, similar studies were conducted in skeletal muscles. Using canine gracilis muscles, Walker et al (1987) showed that extensive skeletal necrosis may occur after prolonged ischemia-reperfusion of hind leg muscles. Using reduced O_2 delivery and free radical scavengers, these authors were able to reduce muscle necrosis from 78% in control muscles, to 53% in muscles treated with antioxidants and reduced O_2 delivery. Thus showing indirectly that the oxygen free radicals are involved in skeletal I/R necrosis. Other studies by Korthuis et al (1985) on microvascular and parachymal injury induced by I/R, showed that O_2 derived free radicals increased canine muscle vascular permeability considerably . However, studies on ischemia-reperfusion of skeletal muscles by the same authors showed that under such conditions, there was attenuation of postischemic microvascular injury (Korthuis et al 1989).

Recent studies by McCutchan et al., (1990) showed that H_2O_2 derived from xanthine oxidase may contribute to reperfusion injury of ischemic skeletal muscles. Using inhibitors of xanthine oxidase and catalase, these authors could show an increase of muscle function after I/R in muscles treated with these inhibitors. Indeed, under such conditions H_2O_2 production was almost halted and could hardly be measured.

These and other studies indicate that also in skeletal muscles there is active involvement of Reactive Oxygen Species in I/R associated damage which can be reduced to a great extent by applying various antioxidants.

3. Oxidative Damage in Muscle Immobilization and in Muscle Crush Syndrome:

Hind leg muscle immobilization in aging animals causes a sharp decline (30-40%) of muscle weight after four weeks of immobilization. This is accompanied by morphological as well as physiological and biochemical changes (Carmeli et al 1992).

Studies on metal-catalyzed oxidation of proteins as measured by the protein carbonyl assay showed about a 400% increase of protein oxidation after immobilization. In normal untreated old muscles the concentration of protein carbonyls is about 2 nmoles/mg protein. This value increased to over 8 nmoles/mg protein under conditions of immobilization. When animals were treated with growth hormone the values were reduced to about 4 nmoles/mg protein.

Capillary blood volume as measured by photoplethysmography showed a sharp decline in immobilized animals. Thus an inverse correlation of capillary blood volume with carbonyl values ($r= -0.825$ $p< 0.001$) was found. These results indicate that oxidative damage to proteins is quite severe under immobilization conditions, and can be attenuated by the administration of growth hormone as an anabolic hormone.

A recent review article by Odeh (1991) dealt with the question of the role of ROS in rhabdomyolysis, better known as muscle crush syndrome. Since traumatic rhabdomyolysis is a classic case of ischemia due to continuous pressure on the limb, the author argues that muscle damage due to crush should involve ROS. When muscles are relieved of the pressure, a reperfusion-like condition will occur in which free-radical associated damage would take place. Recently, work performed in our laboratory has shown that in a model system of crush syndrome in rats, there was a significant increase of oxidation of proteins which was partially blocked by antioxidants such as mannitol (Reznick, A.Z., unpublished data). Mannitol has been shown to reduce the various crush-associated damage parameters in muscles.

4. Involvement of Free Radicals in Muscle Dystrophies:

Omaye and Tappel (1974) were the first to suggest that oxidative stress plays a role in the genesis of muscular dystrophy. In addition, there have been consistent findings that antioxidant enzymes as well as products of lipid peroxidation are considerably elevated in the muscles of dystrophic chickens (Elbrink et al., 1987; Mizuno, Y., 1983; Murphy and Kehrer, 1986). The above observations may reflect a higher degree of oxidative stress in muscle of dystrophic animals and humans.

Recent studies of Murphy and Kehrer (1989) showed that in muscular dystrophy of genetically affected chickens there was a higher level of protein carbonyls in muscles like pectoralis major but not in the soleus. In addition there was considerable elevation of glutathione, glutathione disulfide, protein-glutathione mixed disulfides, but lower contents of free protein thiol groups in pectoralis major of genetically dystrophic chickens. Other muscles such as soleus and other tissues such as liver and heart were not affected, substantiating the fact that it is the pectoralis major muscle which is most affected by the dystrophic disease in chickens. The oxidative stress in muscles demonstrated by changes in thiol groups and protein oxidation, precede the macrophage infiltration and cellular necrosis that was noted in this disease.

The above discussion may indicate that oxidative stress may be one of the primary events that takes place in dystrophic muscles and subsequently leads to altered protein metabolism and muscle atrophy.

NEUROLOGICAL AND BRAIN DISORDERS INVOLVING FREE RADICALS:

INTRODUCTION:

Halliwell and Gutteridge (1985) were among the first to discuss the potential role of ROS in the brain and neural tissue. The brain has a relatively high lipid content, major requirements for oxidative energy metabolism and relatively low levels of antioxidants enzymes. The brain contains large amounts of unsaturated fatty acids and catecholamines which are substrates susceptible to ROS attack. Moreover, the localization of major antioxidant defense systems in glial cells rather than in neurons may cause the nerve cells to be more susceptible to oxidants present in the brain (Bondy 1992).

Lebel and Bondy (1992) enumerated several factors that can possibly contribute to the formation of ROS in the brain and nervous tissue. Among the factors, they discuss the possibility of iron release in brain damage, which will accelerate the generation of ROS. Other factors such as lowering the pH due to excessive glycolitic activity may impair oxidative ATP production and contribute to the appearance of prooxidants. In addition, elevation of calcium can activate phospholipase activity which will lead to arachidonic acid that is readily oxidizable, thus increasing free radical productions (Lebel and Bondy 1992).

In another review Jesberger and Richardson (1991) have discussed a long list of neurological diseases and disorders in which free radical involvement was implicated. The various diseases are outlined in Table II:

Table II: Brain & neurological diseases and disorders which involve free radicals

1. Parkinson's disease	13. Progeria
2. Alzheimer's disease	14. Werner's syndrome
3. Stroke	15. Cocaine syndrome
4. Alcoholism	16. Vitamin E deficiency
5. Epileptic seizures	17. AIDS (HIV infection)
6. Head trauma (brain)	18. Oxygen inhalation therapy
7. Retinal damage	19. Shock
8. Spinal cord damage	20. Brain edema
9. Demyelination (multiple sclerosis)	21. Tardive dyskinesia
10. Schizophrenia	22. Inflammatory diseases
11. Dementia	23. Shock-brain damage by hyperoxia and hyperbaric oxygen
12. Down's syndrome	

The following discussion will cover only part of this long list, emphasizing mainly these maladies in which more extensive and convincing data have been accumulated concerning the involvement of free radicals

1. Brain Ischemia - Reperfusion Studies:

The mongolian gerbil does not have a complete circle of Willis in the base of the brain. Therefore, there is not an anastomosis between the major arteries leading to brain such as two verterbral and two internal carotid arteries. Thus, it is possible to ligate one side of the carotid interna and by doing so to have a good experimental model for ischemia of the brain.

Using the gerbil model for ischemia reperfusion studies, Floyd 1990 has shown that under conditions of brain ischemia there was an increase of hydroxyl radical generation shown by spin-trapping technique and increase of salicylate oxidation. In addition protein oxidation as measured by protein carbonyls was also increased. Using spin trapping agents (which presumably scavenge free radicals) it was possible to improve the survival rate of animals after the ischemia-reperfusion stage (Floyd 1990). In another study (Carney et al 1991) neurological functions which were impaired after the ischemia-reperfusion stage, could be improved by treating the animals with high concentration of a spin trap (PBN). Other parameters such as protein oxidation and inactivation of glutamic synthetase were also improved following treatment with the spin trapping agent. After removing the spin trapping agent the increase of protein carbonyls and glutamine synthetase inactivation reappeared. This observation strongly indicated that these changes were caused by generation of free radical species which react with the spin trap. Behavioral changes were also observed, mainly in the old gerbils compared to the young ones after ischemia reperfusion treatment. This may imply that brains of old animals may be more susceptible to oxidative damage compared to young animals.

Other studies also demonstrate that free radicals and lipid peroxidation were involved in neuronal damage occuring during ischemia of the brain and the spinal cord (Halat et al 1989).

2. Epilepsy and Seizure Disorders:

The involvement of iron-associated free radical formation in epilepsy has long been postulated (Jesberger and Richardson 1991). Thus release of iron in experimental models of epilepsy has been implicated in membrane lipid peroxidation (Willmore et al 1982). Other studies showed antioxidant enzyme systems were able to protect brain tissue from lipid peroxidative damage during epileptic activity in animal models (Singh and Pathak 1990). Studies by Mori, A. et al (1990), Liu et al (1991), and Kabuto et al (1992) on iron-induced epileptiform model in the rat brain has also shown that following ferric chloride injection into rat cerebral cortex, an increased production of $\bullet O_2^-$ and $\bullet OH^-$ radicals was observed. The accelerated generation of ROS led to increased production of gaunidine compounds in the brain which may in turn lead to epileptogenicity. Treatment of these animals with antioxidants inhibited the formation of malondialdehyde and reduced the epileptic symptoms in these animals (Mori et al 1990).

The possible involvement of free radicals in seizure mechanisms was discussed by Hiramatsu et al (1986). Studies by Armstead et al (1989), showed that seizure activity can elevate reactive oxygen species in the brain, especially superoxide. In another study ROS were shown to inactivate glutamine synthetase (GS) activity in the brain, thus permitting abnormal accumulation of the excitatory transmitter glutamic acid. This oxidative inactivation of GS may be one of the mechanisms by which ROS can induce seizure activity (Oliver et al., 1990).

3. Parkinson's Disease:

Parkinson's disease is a neurological disorder in which the Dopamine neurotransmitter of the substentia nigra in the brainstem is oxidized rapidly by the enzyme monoamine oxidase-B (MAO-B). It can also be oxidized by autooxidation in the presence of trace amounts of metals, thus leading to depletion of dopamine and malfunctioning of the dopaminergic neurons (Bondy 1989). Several lines of evidence support the notion that the etiology of Parkinson's disease involves oxidative stress and free radical production. Dexter et al. (1989) found an increased iron content in the substantia nigra of Parkinson's brain. Saggu et al. (1989) measured higher levels of superoxide dismutase activity within the substantia nigra of Parkinson's brain, thus implying an induced response to oxidative stress.

The discovery of the neurotoxin 1-methyl-4-phenyl, 1,2,3,6 tetra hydropyridine (MPTP) and its induction of Parkinson's like syndrome (Kopin and Schoenberg 1988), has supported the concept that Parkinson's disease may involve some environmental neurotoxins. It is the oxidation of MPTP by MAO-B to form 1-methyl-4-phenyl pyridine ions (MPP^+) which leads to selective accumulation of MPP^+ in dopaminergic neurons. In these neurons MPP^+ has been shown to interfere with mitochondrial respiration and ATP production through the excessive generation of hydrogen peroxide and several other free radicals. Thus, in the last few years the usage of deprenyl (a MAO-B inhibitor) along with provision of Vitamin E have been attempted in order to slow down the degeneration of the dopaminergic neurons in Parkinson's disease (Shoulson 1989, Cadet et al 1989). The mechanism of vitmain E recycling by other antioxidants (e.g. ascorbate and thiols) indicates that other antioxidants may also have a role in the treatment of this disease (Packer, 1992).

CONCLUSIONS:

It is still an open question whether the involvement of ROS and oxidative stress in neurological diseases is the prime cause for the observed pathologies. However, the evidence that has been accumulating in the last several years, indicated that in many degenerative aging diseases of the central nervous system it is possible to postulate with confidence the active role of oxidative stress contributing to degeneration and destruction of nervous tissue.

REFERENCES

Aikawa K.M., Quintanilha, A.T., De Lumer B.O., Brooks G.A., and Packer L. (1984) Biosci. Rep. 4:253-257.
Alessio H.M., Goldfarb, H.H. and Cutler B.G. Am. J. Physiol. (1988) 255 c874-c877.
Armstead, W.M., Mirro, R., Leffler, C.W., Busija, D.W. (1989) J. Cerebral Blood Flow and Met. 9:175-179.
Bolli R., Jeroudi M.O., Patel B.C., DuBose C.M., Lai E.K., Roberts R., and McCay P.B. (1989) Proc. Natl. Acad. Sci, USA 116:4695-4699.
Bondy, S.C. (1992) Neurotoxicology, 13:87-100.
Cadet, J.L., Katz, M., Jackson-Lewis, V., and Fahn, S. (1989) Brain Research 476:10-15.
Carmeli E., Livne E., Hochberg Z., Lichtenstein I., Kestenboim C., Silberman M., and Reznick A. Z. (1992) Submitted for publication.
Carney, J.M., Starke-Reed, P.E., Oliver, C.N., Landum, R.W., Cheng, M.S., Wu, J.F., Floyd, R.A. (1991) Proc. Natl. Acad. Sci. 88:3633-3636.
Davies K.J A., Quintanilha A.T., Brooks G.A., and Packer L. (1982) Biochem Biophys, Res Commen. 107:1198-1205.
Dexter, D.T., Wells, F.R., Lees, A.J., Agid, F., Agid, Y., Jenner, P. and Marsden, C.D. (1989) J. Neurochem. 52:1830-1836.

Dillard C.J., Iitov R.E., Savib W.M., Dumelin E.E and Tappel A.L. (1978) J. Appl. Physiol. 45:927-932.
Elbrink J., Malhorta S.K., and Hunter E.G. (1987) Medical Hypo. 23:131-136.
Floyd, R. (1990) FASEB J. 4:2587-2597.
Guarnieri C., Flamigni F., and Caldarera C.M. (1980) J. Mol. Cell. Cardiol. 1:797-808.
Halat, G., Chavko, K., Lukacova, N., Kluchova, D., and Marsala, J. (1989) Neurochemical Research. 14:1089-1097.
Halliwell, B., Gutteridge, J.M.C. (1985) Trends in Neurosci. 8:22-26.
Hiramatsu, M., Edamatsu, R., Kohno, M., and Mori, A.(1986) Jpn. J. Psychiat. Neurol. 40:349-353.
Jackson M.J., Edwards R.H.T., and Symons M.C.R. (1985) Biochem. Biophys. Acta 847:185-190.
Jesberger, J.A. and Richardson, J.S. (1991) Intern. J. of Neuroscience. 57:1-17.
Ji L.L., Stratman F.W., and Lardy H.A. (1988) Arch Biochem. Biophys. 263:150-160 .
Kabuto, H., Yokoi, I., and Mori, A. (1992) Neurochem. Res. 17:585-590.
Kopin, I.J., and Schoenberg, D.G. (1988) The Mount Sinai Journal of Medicine. 55:43-49.
Korthuis R.S., Smith J.K., and Carden D.L. (1989) Am. J. Physiol. 256:H315-H320.
Liu, M., Liu, J. Okada, S., and Mori, A. (1991) Med. Sci. Res. 19:747-749.
Lebel, C.P. and Bondy, S.C. (1992) Progress in Neurobiology. 38:601-609.
McCutchan H.J., Schwappach J.R., Enquist E.G., Walden D.L., Terada L.S., Reiss O.K., Left J.A., and Repine J.E. (1990) Am. J. Physiol. H1415-H1419.
Mizumo Y. (1985) J. Neurol. Sci. 68:47-60.
Mori, A., Hiramatsu, M., Yokoi, I., Edamatsu, R. (1990) Pav. J. Biol. Sci. 25:54-62.
Murphy M.E., and Kehrer, J.P. (1986) Biochem. Biophys. Res Comm. 134:550-556.
Murphy M.E., and Kehrer, J.P. (1989) Biochem. J. 260:359-364.
Odeh, M. (1991) New Eng. J. of Medicine, 324:1417-1422.
Oliver, C.N., Starke-Reed, P.E., Stadtman, E.R., Lin, G.J., Correy, J.M., and Floyd, R.A. (1990) Proc. Nat. Acad. Sci. 87:5144-5147.
Omaye, S.T., and Tappel A.L. (1974) Life Sci. 15:137-145.
Packer, L. (1992) Proc. Soc. Exp. Biol. Med. 200:271-276.
Packer, L., and Singh, V.N. (1992) J. of Nutrition 122:758-801.
Quintanilha, A. and Packer L. (1983) Ciba Found. Symp. 101:56-61.
Reznick, A.Z., Kagan, V.E., Ramsey, R., Tsuchiya M., Khwaja S., Serbinova E.A., and Packer L. (1992) Arch. Biochem. Biophys. 296(2):349-401.
Reznick, A.Z., Steinhagen-Thiessen E., and Gershen D. (1982) Biochem. Med. 28:347-352.

Saggu, H., Cooksey, J., Dexter, D., Wells, F.R., Lees, A., Jenner, P., and Mardsen, C.D. (1989) J. Neurochem. 53:692-697.
Salminen A. and Vihko V. (1983) Acta Physiol. Scand. 117:109-113.
Simon- Schnass I., and Pabst H. (1987) Int. J. Vitamin Nutr. Res. 58:49-54.
Shoulson, I. (1989) Acta Neurol. Scand. 126:171-175.
Singh, R. and Patnak, D.N. (1990) Epilepsia. 31:15-26.
Singh V.N. (1992) J. of Nutr. 122:760-765.
Walker P. M., Lindsay T.F., Labbe R., Mickle P.A., and Romaschin A.D. (1987) J. Vasc. Surg. 5:68-75.
Willmore, L.J., and Rubin, J.J. (1982) Brain Research. 246:113-119.
Witt E.H., Reznick A.Z., Viguie C.A., Starke-Reed P., and Packer L. (1992) J. Nutr. 122:766-773.
Zweier L. (1988) J. Biol. Chem. 263:1353-1357.

Free Radicals: From Basic Science to Medicine
G. Poli, E. Albano & M. U. Dianzani (eds.)

PHARMACEUTICAL INTERVENTION FOR THE PREVENTION OF POST-ISCHEMIC REPERFUSION INJURY

Toshihiko Mayumi, M.D.,

Henry J. Schiller, M.D.,

Gregory B. Bulkley, M.D.

Department of Surgery
The Johns Hopkins Medical Institutions
Blalock 685
The Johns Hopkins Hospital
600 North Wolfe Street
Baltimore, MD, 21205, U.S.A.
(410) 955-8500
FAX (410) 955 0834

Supported by National Institutes of Health grant No. DK31764.

SUMMARY

In many organs a substantial proportion of the injury sustained as a consequence of ischemia is actually caused by a cascade of toxic oxygen metabolites, triggered by the generation of the superoxide by xanthine oxidase at the time of reperfusion. This mechanism, first elucidated in the feline small intestine, has also been found to be operative in the heart, lung, stomach, liver, pancreas, kidney, skin, skeletal muscle, central nervous system and other organs. Its ubiquity is based upon the ubiquity of endothelial cell xanthine oxidase in the microvasculature. Pre-clinical studies in animal models of human disease have demonstrated a variable contribution of this reperfusion injury mechanism to the total injury sustained consequent to ischemia. While ill-conceived and poorly designed clinical trials, which have failed to take this into account, and therefore have included heterogeneous and unstratifiable patient populations, have been predictably unenlightening, a few well-designed clinical trials, particularly in renal transplantation and in elective cardioplegia for cardiopulmonary bypass have confirmed the efficacy of free radical ablation for the prevention of reperfusion injury in man.

THE FUNDAMENTAL MECHANISM OF REPERFUSION INJURY

The rescue of an organ from an episode of ischemia necessarily requires the resumption of blood flow. However the reperfusion of an ischemic organ can also result in further injury. This reperfusion injury can be defined as the damage that occurs to an organ during reflow after an episode of ischemia. It must be distinguished from the injury induced during the ischemic period, although an ischemic episode is necessary to generate the conditions required to sustain injury at reperfusion. One hallmark of reperfusion injury is that it may be ameliorated by interventions initiated after the onset of ischemia, but before reperfusion. In some clinical situations, injuries that have traditionally been attributed to ischemia and considered a *fait accompli* following ischemia, may actually be preventable by therapy initiated at reperfusion.

Toxic oxygen metabolites involved in the evolution of this postischemic reperfusion injury are illustrative of the importance of free radicals to human disease processes. In the schema first proposed by Granger and his colleagues from their experiments in the feline small intestine (Granger 1981, Parks 1982), ischemia results in the accumulation of the purine metabolites of ATP, as well as in the proteolytic conversion of the enzyme xanthine oxidoreductase from the dehydrogenase (D form) to the free radical-generating xanthine oxidase (O form) [Figure 1]. Oxygen, the only substrate absent during ischemia, is reintroduced into the system in excess at reperfusion, resulting in the explosive generation of the superoxide free radical which initiates a cascade of secondary oxidant generation, ultimately resulting in tissue injury. In many organs, especially the small intestine, neutrophils, which generate superoxide (not with xanthine oxidase but with a membrane-associated NADPH oxidase) as well as a large number of non-radical toxic mediators, including elastase, collagenase, and other proteases (Weiss 1989), are also essential to this injury.

The activation of xanthine oxidase from xanthine dehydrogenase during ischemia, with the subsequent generation of superoxide at reperfusion, appears to be the fundamental initiator of free radical generation at reperfusion. It has become clear that specific interruption at a number of disparate points in the distal oxidant cascade is equally effective (Grisham 1986); thus scavenging of hydrogen peroxide with catalase (Forman 1981, Granger 1986), or scavenging of hydroxyl radicals with dimethyl

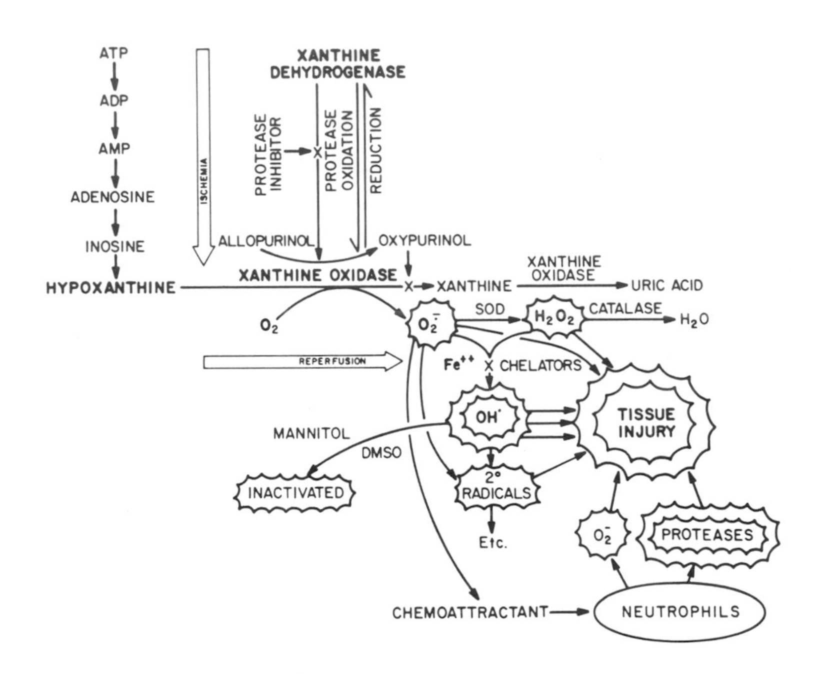

Figure 1. Biochemical Mechanism of Reperfusion Injury. During ischemia, the breakdown of high energy phosphate compounds (ATP) results in the accumulation of the purine metabolites hypoxanthine and xanthine. At the same time, xanthine dehydrogenase is converted to xanthine oxidase. At reperfusion, oxygen is reintroduced, suddenly and in excess, driving the rapid oxidation of purines, producing urate and the superoxide radical as a byproduct. This superoxide can then secondarily generate the highly toxic hydroxyl radical via an iron-catalyzed reaction. In addition to causing injury themselves, these radicals may also lead to the adherence, the accumulation, and the activation of neutrophils. In many organs, such as intestine, neutrophil-generated radicals and proteases contribute substantially to tissue injury. Selected free radical inhibitors are also listed where they have been shown to interrupt this cascade. (Modified from Granger 1981. From: Sussman,M.S., Schiller,H.J., Buchman,G.B., Bulkley,G.B. (1990) Mechanisms of organ injury by toxic oxygen metabolites. In *Multiple Organ System Failure* (Fry,D.E., ed.), Mosby-Yearbook Inc., St.Louis, pp 143-165, with permission)

sulfoxide (DMSO) or mannitol (Parks 1983), or preventing formation of these hydroxyl (or perferryl) (Koppenol 1985, Minotti 1987) radicals by chelating iron with deferoxamine (Aust 1985) or transferrin (Hernandez 1987a) is as beneficial as scavenging superoxide with superoxide dismutase (SOD) (Granger 1981), or blocking its generation from xanthine oxidase with allopurinol (Parks 1982, Granger 1986).

THE ENDOTHELIAL CELL TRIGGER MECHANISM

While the initial studies had been conducted in the (feline) small intestine precisely because it was known to contain large quantities of xanthine oxidoreductase, the levels of this enzyme in tissue homogenates of different organs vary greatly (Parks 1986). This called into question the universal application of the above mechanism to some organs, such as the human heart (Eddy 1987) where it could have the greatest therapeutic potential. However, subsequent studies had suggested that this was indeed an operant mechanism in a large number of organs, including the stomach (Itoh 1985), intestine (Granger 1981, Parks 1982, Haglund 1988, Ozasa 1990), liver (Atalla 1985, Adkinson 1986), pancreas (Sanfey 1985, Sanfey 1986, Salim 1991), lung (Heffner 1989), kidney (Paller 1984, Koyama 1985), skeletal muscle (Korthius 1985, Perler 1988), skin (Manson 1983, Im 1985), heart (Shlafer 1982, Hearse 1986, Naslund 1990, Rashid 1991, Tabayashi 1991) and central nervous system (Davis 1987, Liu 1989), some of which did not appear to contain measurable levels of xanthine oxidase in tissue homogenates.

For the resolution of this apparent contradiction, microvascular endothelial cells have received much attention, as not only an initial target for oxidant injury at reperfusion, but also as the source of xanthine oxidase-generated superoxide. We have found significant concentrations of xanthine dehydrogenase, which was rapidly converted to xanthine oxidase, in cultured rat pulmomary artery endothelial cells subject to relatively short periods of anoxia (Ratych 1985). Moreover, anoxia followed by reoxygeneration (to mimic ischemia/reperfusion) in this system resulted in endothelial cell lysis, which was prevented by either allopurinol or SOD and catalase administered following anoxia, coincident with reoxygenation.

These studies suggest that the entire xanthine oxidase-based free radical-generating system is present and operative, even to the extent that it produces endothelial lysis, within the endothelial cell itself, in the absence of neutrophils or parenchymal cells. We have proposed this **endothelial cell trigger mechanism** as the ubiquitous initiator of free radical-mediated reperfusion injury [Figure 2].

Immunohistochemistry has subsequently confirmed this microvascular

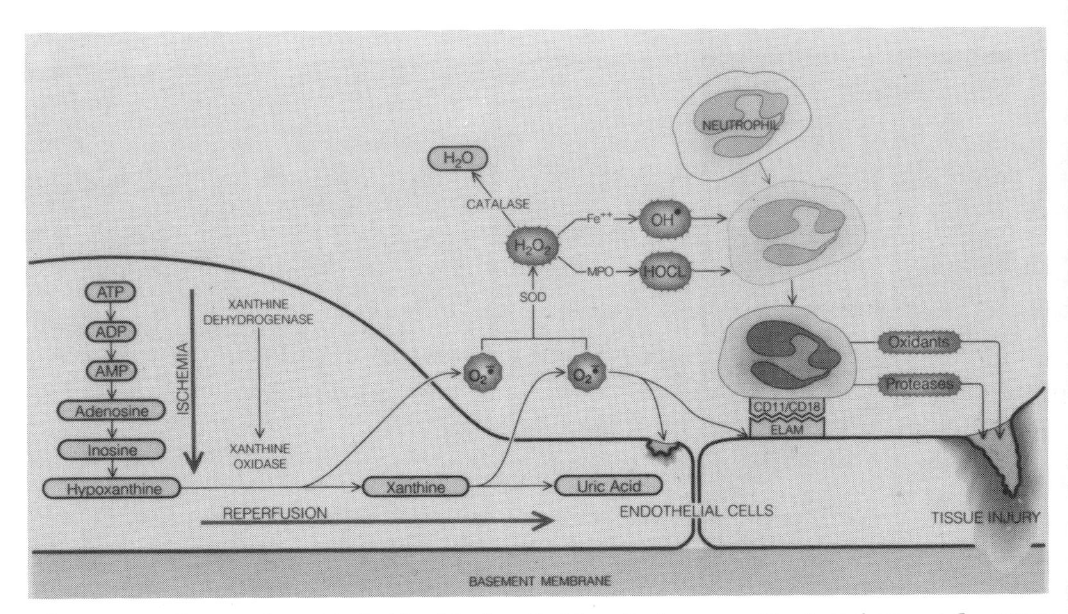

Figure 2. The Endothelial Cell Trigger Mechanism of Free Radical-Mediated Reperfusion Injury. The entire xanthine oxidase-based free radical-generating system is present in the endothelial cell itself. In some systems, these oxidants alone can produce substantial microvascular injury and consequent parenchymal organ injury. In other organs, however, such as feline intestine, these xanthine oxidase-derived oxidants act primarily by stimulating the adhesion of circulatimg neutrophils to endothelial cell surface, probably by the up-regulation of neutrophil adhesion glycoproteins (integrins) and of the endothelial leukocyte adhesion molecules (ELAMs) to which they bind. This process not only arrests the circulating neutrophils but also activates them to produce neutrophil-derived oxidants. In addition, these adherent neutrophils secrete highly toxic proteases, such as elastase. These not only damage tissue directly but also inactivate available enzymatic antioxidants, such as superoxide dismutase and catalse. This serves to amplify the toxic effects of the amebient levels of oxidants, especially within the microenviroment between the endothelium and the adherent neutrophil. This latter component constitutes the neutrophil amplifier (From: Reilly 1991, with permission).

localization for xanthine oxidase by the unequivocal staining of the microvascular endothelium of a number of organs, including the human heart (Bruder 1983, Jarasch 1986, Vickers 1990a, Vickers 1990b) where xanthine oxidase had not previously been detectable in tissue homogenates (Eddy 1987). Moreover, histochemical staining of these microvascular endothelia based on the allopurinol-inhibitable reduction of nitro blue tetrazolium (NBT) confirms the enzymatic activity of this microvascular xanthine oxidase (Miyachi 1991).

This endothelial cell trigger mechanism based upon the apparently ubiquitous distribution of microvascular endothelial cell xanthine oxidase, could therefore explain the role of xanthine oxidase as an initiator of reperfusion injury in many organs. While in some organs this primary focal endothelial cell injury may itself be sufficient to result in microvascular thrombosis and ultimate loss of organ function (Marzella 1988), in other organs, including the intestine (Suzuki 1990), heart (Romson1983, Werns 1988) and lung (Pillai 1990), this injury may act only as an initial trigger to attraction, accumulation and activation of neutrophils, which subsequently mediate the major portion of the injury. This mechanism of **neutrophil amplification** has been most carefully studied by Granger and his colleagues in the cat small intestine (Suzuki 1990). Here, neutrophil ablation with antineutrophil serum, or blockade of the neutrophil CD11/CD18 adhesion glycoprotein with a specific monoclonal antibody, not only blocks neutrophil accumulation within the tissue but also prevents the microvascular and consequent epithelial injury (Hernandez 1987b, Inauen 1990, Kvietys 1990). (Indeed, xanthine oxidase inhibition with allopurinol also blocks neutrophil accumulation during and after ischemia by blocking the trigger mechanism.) The microenviroment created by the adherence of the neutrophil to the endothelial cell plasma membrane and the high concentrations of O_2^- (from both endothelial cells and neutrophils) at this site are sufficient to inactivate circulating antioxidants and protease inhibitors (Blech 1983, Kono 1983), and might also convert the nitric oxide free radical (the so-called endothelial-derived relaxing factor) from a protective, vasodilatory agent that enhances microvascular perfusion to peroxynitrite, a toxic oxidant that can exacerbate microvascular injury

(Beckman 1990). Moreover, the release of circulating mediators such as tumor necrosis factor (TNF) and complement (C'5A), as well as superoxide itself, appears to mediate further D to O conversion, possibly generating a positive feedback loop (Friedl 1989).

PRECLINICAL STUDIES IN ANIMAL MODELS

Antioxidant agents have been used in many models of reperfusion injury. SOD, SOD-derivatives, and SOD-mimics have been shown to ameliorate aspects of oxidant-induced injury, either alone or in combination with other antioxidants in many organs, including the heart (Shlafer 1982, Naslund 1990, Klein 1988, Gelvan 1990, Ohkubo 1990, Sjoquist 1990, Bolli 1990, Zelck 1990, Ely 1992), brain (Davis 1987, Liu 1989, Inoue 1990, Simovic 1990), intestine (Haglund 1988, Ozasa 1990), kidney (Hoshino 1988), and lung (Flick 1981). Other antioxidants, such as mannitol (England 1986, Blasig 1987, Ferreira 1989) and deferoxamine alone (Bolli 1990, Ely 1992) or conjugated to hetastarch (Hallaway 1989, Hedlund 1990), DMSO (Punch 1991) have been shown to ameliorate injury in other models, suggesting that a beneficial effect may be achieved by interrupting the free radical cascade at any of several levels. Other studies, however, have failed to demonstrate a significant benefit (Miura 1988, Marzella1988, Redl 1990). The heterogeneity of these results has led to understandable skepticism over the clinical applicability of free radical ablation. These apparent discrepancies may be understood on at least two accounts. In the first place, different antioxidant compounds will, of course, have differing therapeutic concentrations and serum half-lives, and heterogeneous results should be expected. However, in few situations does the combination of agents acting on different portions of the cascade yield a more favorable response (Zelck 1990, Marinkovix 1990).

Of greater importance, however, is the variable importance of the oxidant-mediated reperfusion component of the total injury sustained in each model of ischemia-reperfusion. This is illustrated by studies of frostbite in a rabbit ear model (Manson 1991). Frostbite can be seen as an analogy to reperfusion injury, where thawing represents an equivalent of reperfusion. After freezing for 30 seconds, the ear survived whether or not antioxidants (SOD or allopurinol) were administered. Similarly, after 2 minutes of freezing, the ear ultimately became necrotic regardless of the administration of antioxidants. However, after 60 or 90 seconds of freezing most ears treated with antioxidants remained viable, whereas most untreated ears became necrotic. The value of this study is that it points to the fact that the treatment of reperfusion injury is exquisitely time dependent. While the quantitative impact of antioxidant therapy appears large within this **therapeutic window**, the extreme narrowness of the window in this case greatly limits the potential of this approach for application to the clinical problem of frostbite.

Similarly, in the cat intestine subjected to partial ischemia, antioxidants administered at reperfusion clearly prevent the increase in capillary permeability seen after one hour, or the superficial necrosis of the villus epithelium seen after 3 hours of partial ischemia (Granger 1981, Parks 1982, Schoenberg 1984). However, longer periods of partial ischemia, or more complete ischemia, produce a more severe injury that is largely unaffected by antioxidant therapy (Haglund 1987, Park 1990). These results suggests that this approach will probably have its greatest potential for the preservation of the intestinal epithelial barrier, the loss of which is probably an important factor in the development of multiple system organ failure (Bulkley 1989, Reilly 1991), rather than for the treatment of the classical clinical syndrome of intestinal ischemia.

This variable quantitative impact of antioxidant therapy was quantitated in a porcine model of human cadaveric renal transplantation [Figure 3]. Here, free radical ablation with either SOD or allopurinol significantly ameliorated the injury seen in kidneys after either 24 or 48 hours of cold ischemic preservation, but had no effect on renal function in kidneys exposed to either shorter or longer periods of cold ischemia (Hoshino 1988). Again, there was a defined ischemic period following which free radical ablation at reperfusion provided a measurable beneficial effect.

This is explained by the fact that the total injury sustained as a consequence of ischemia must necessarily reflect the combined effects of

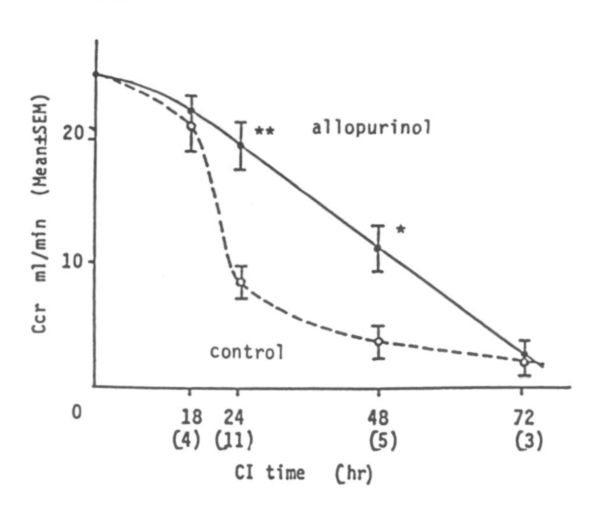

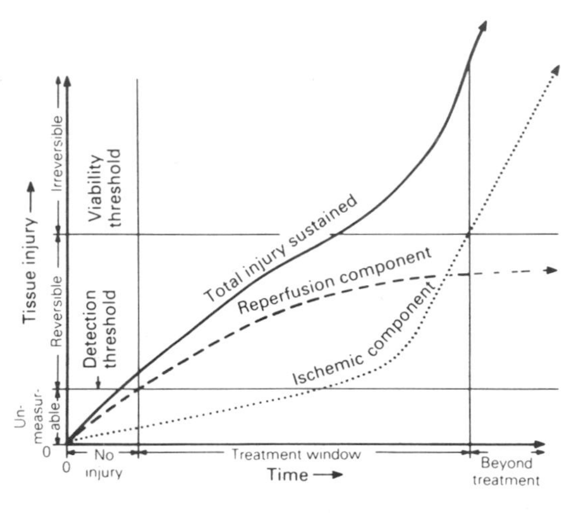

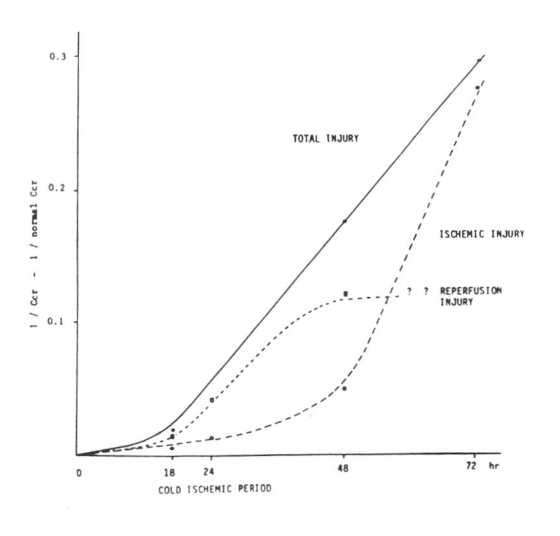

Figure 3. Therapeutic Windo for the Treatment of Fre Radical-Mediated Reperfusic Injury.

A: Posttransplant renal functio expressed as creatine clearance wa studied in allopurinol-treated ar control pig kidneys after variou periods of cold ischemia. (Number c experiments in parentheses; *p<0.0 vs controls; **p < 0.01 vs controls.)

B: The total injury sustained b an organ during a period c hypoperfusion can be divided into tw components; that which is caused b ischemia <u>per se</u>, and that which i caused by toxic oxygen metabolite generated during reperfusion. Eac component begins at the onset c ischemia. Initially, neithe component is large enough to caus measurable injury. As ischemi progresses, the reperfusion componen itself often accounts for as much a two-thirds of the total injur sustained. Eventually, however, point is reached where the ischemi component becomes so large that i precludes viability and therefor overwhelms the component due reperfusion. Only within the centra therapeutic window will free radica ablation have a substanti beneficial effect on the total post ischemic injury sustained.

C: Here, the data from (**A**) ha been transformed to show empiri confirmation of the theoretica relationship shown in (**B**). Tissu injury has been defined as 1/Ccr 1/Ccr(normal). The ischemic injur curve represents the results in th allopurinol-treated kidneys. Th total injury curve represents th results in the control kidneys. Th reperfusion injury curve wa calculated empirically by subtractin the ischemic injury curve from th total injury curve. (From: Hoshin 1988) ischemia <u>per se,</u> as well as of reperfusion. Thus, after short periods of ischemia, the degree of injury from either mechanism is so small that the ablation of reperfusion injury has no measurable nor clinically important effect. (While this point seems to be self-evident, it is important because clinical trials will invariably include patients in this category.) On the other hand, after very long periods of ischemia, the injury due to the ischemic component itself appears to be so great that it overwhelms that component due to reperfusion. The **therapeutic window** for antioxidant therapy may be defined as that period following which the major portion of the total injury sustained is caused by the preventible reperfusion component [Figure 3] (Hoshino 1988). The clinical relevance of the ablation of reperfusion injury is fundamentally dependent on the width of this window. The wide variation in the tolerance of different organs to ischemia suggests that the size and timing of this window varies greatly. This factor, far more than the choice or dose of antioxidant agent, appears to account for most of the striking discrepancies that have been reported for the efficacy of free radical ablation. Moreover, the size of this window appears to be somewhat organ-specific, and relatively unaffected by therapeutic manipulations. Perhaps it most closely reflects the degree to which the primary site of reperfusion injury, the microvasculature, is manifest as parenchymal organ dysfunction. Thus, free radical ablation for the treatment of reperfusion injury seems to have its greatest impact in organs such as the lung, brain, and kidney, where the consequences of a microvascular injury have the most profound functional impact.

ANTIOXIDANT THERAPY IN CLINICAL TRIALS

GENERAL REQUIREMENTS OF TRIAL DESIGN

Clinical trials of free radical ablation for the treatment of reperfusion injury therefore should include not only the conventional requirement for adequate trial design, but also should be designed to adequately account for the effect of this therapeutic window. This includes quantitation of the "dose" (duration and/or degree) of ischemia to allow stratification of a

Table I

Minimal Standards for Clinical Trials of Free Radical Ablation for the Treatment of Reperfusion Injury

I. Conventional Requirements of Good Clinical Trial Design
   A. Unbiased: Prospective, Randomized, Controlled, Blinded.
   B. Adequate numbers and prospective statistical design to allow the recognition of ineffectiveness, as well as of effectiveness (type II error).

II. Account Taken of The Varying Proportional Contribution of the Reperfusion Component to the Sum Total of Post-ischemic Injury
   A. Adequate numbers to allow stratification.
   B. Quantitation of the "dose" (duration and/or degree) of ischemia sustained to allow stratification on this basis.
   C. Indication of a relatively large **therapeutic window** in preclinical animal studies.
   D. Indication that the conditions of the clinical study fall within this **therapeutic window.**

III. Discrimination of Transient from Permanent Effects.

IV. Primary Academic Control of the Study.
   A. Design.
   B. Reporting of negative as well as positive results.

sufficiently large study population on this basis [Table I].

RENAL TRANSPLANTATION

Based upon the encouraging results of the above-described pre-clinical studies in porcine renal transplants, a randomized, double-blind, paired trial was performed in Munich, in which 100 cadaveric renal transplant recipients were randomized to receive either SOD or a placebo into the renal artery at the moment of implantation (reperfusion) (Schneeberger 1989). While early graft function was improved overall, this difference was small and not statistically significant for the group as a whole. Those kidneys preserved for short periods of cold ischemia functioned well without treatment, while those preserved for a very long periods functioned poorly despite treatment. However, that subgroup of kidneys that were cold-preserved for periods of from 25-28 hours (43% of the kidneys studied) showed a dramatic and statistically significant improvement in early function with SOD [Figure 4]. Moreover, in this group, renal function returned in a median of six (instead of 13) days, and the need for postoperative dialysis was reduced from a median of three to one treatments. Thus, in those kidneys preserved for a time that fell within the therapeutic window, where a major component of the injury was mediated at reperfusion by

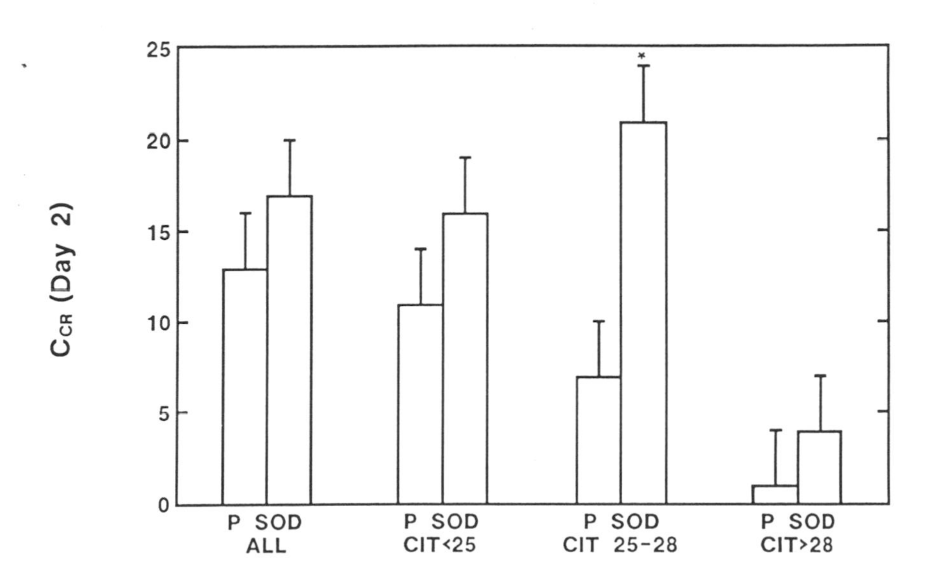

Figure 4. Quantitative Importance of Free Radical Ablation in Human Renal Transplantation. In a prospective, double-blind, paired study of renal transplantation, 100 patients were randomized to receive either SOD or placebo into the renal artery at implantation (reperfusion). When these patients were considered as a whole, there was a small, but statistically insignificant improvement in renal function, assessed here as creatinine clearance [C_{Cr}] at day two after transplantation. However, when the patients were stratified by graft cold ischemia time (CIT), the applicability of the concept of the therapeutic window illustrated in figure 3 became evident. SOD infusion at implantation produced a significant improvement in renal function only in those kidneys subjected to 25 to 28 hours of CIT. This effect was less significant in kidneys subjected to longer or shorter periods of ischemia prior to transplantation. $^{*}p < 0.05$ versus placebo (P). (Modified from Schneeberger 1989. Reproduced with permission from Sussman,M.S., Bulkley,G.B. (1990) In *Methods in Enzymology.* Volume 186. Academic Press, Orlando, pp 723)

free radical generation, the effect of SOD was highly significant statistically and important clinically.

Another randomized, double-blind, placebo-controlled, clinical trial involving 131 evaluable patients examined the effect of recombinant human SOD (rh-SOD) given intravenously minutes prior to graft reperfusion. No differences in graft function were discernible between placebo controls and rh-SOD in the groups receiving grafts preserved for less than 30 hours (short-term preservation). However, in the group receiving grafts preserved long-term, rh-SOD significantly reduced the incidence of acute renal failure and improved early graft function (Schneeberger 1990). Extremely encouraging preliminary data from a similarly designed multicenter, 400 patient trial just completed suggests that the use of allopurinol in the preservation solution dramatically and significantly reduced the incidence of post-transplant renal failure and even patient death [unpublished data].

As a consequence, it is probably no longer ethical to transplant a cadaveric kidney without some form of free radical ablation; indeed, the most commonly used kidney preservation solution, developed by Southard and his colleagues at the University of Wisconsin (UW solution), contains both allopurinol and glutathione (Southard 1991). A report of a European Mulicenter Trial suggested that UW solution resulted in a more rapid reduction in postopertative serum creatinine, higher creatinine clearance rate, and less postoperative dialysis when compared with Euro-Collins solution (Ploeg 1990). This UW solution also allows for significantly large periods of ischemic preservation of cadervaric livers for allotransplantation (Cooper 1990, Olthoff 1990).

CARDIOPULMONARY BYPASS (GLOBAL MYOCARDIAL ISCHEMIA)

A randomized, double-blind clinical study of patients undergoing complex coronary bypass surgery, demonstrated a beneficial effect from allopurinol administered prior to surgery (Johnson 1991). Post-operative mortality decreased from 18% in the control group to just 4% in the allopurinol treated group. Cardiac performance, assessed by cardiac index and the need for inotropic support, was significantly better in the allopurinol group. Another similar, prospective study from Gothenburg on 90 patients undergoing open-heart surgery appeared to show a beneficial effect from allopurinol with respect to post-operative complications (Rashid 1991). Although not a blinded study, the 45 patients receiving allopurinol had significantly lower rates of post-operative arrhythmias and cardiac failure requiring inotropic agents. Another recent clinical trial of 90 patients undergoing coronary artery bypass, or repair or replacement of cardiac valves suggested that xanthine oxidase inhibition with allopurinol resulted in dose-related myocardial protection (Tabayashi 1991). Although no difference was observed in the ventricular stroke work index, allopurinol significantly decreased postoperative serum levels of enzymes associated with myocardial injury.

MYOCARDIAL INFARCTION (REGIONAL MYOCARDIAL ISCHEMIA)

Despite the fact that numerous trials of antioxidants for the treatment of myocardial infarction are said to be underway, the paucity of published results is striking. This is rumored to be due to their negative results, which are not surprising in light of the heterogeneity of the patient populations, the apparent narrowness of the therapeutic window for regional ischemia in the heart (Gallagher 1986, Uraizee 1987, Naslund 1989) and naive trial designs that have failed to take this window into account (Flaherty 1991).

CONCLUSIONS

Toxic oxygen metabolites generated at reperfusion have been implicated as a major source of post-ischemic tissue injury. The free radical-generating enzyme xanthine oxidase, strategically localized in the microvascular endothelium, apparently serves as the trigger for a focal endothelial injury and/or activation that is secondarily amplified by adherent neutrophils, resulting in a substantial microvascular injury which, in many organs, is manifest as clinically important parenchymal organ dysfunction. Therapeutically important modification of this cascade may be achieved by many antioxidants, with varying efficiency, depending primarily upon the relationship of the clinical condition to the therapeutic window. Clearly, the therapeutic potential for such agents as inexpensive and nontoxic as allopurinol is enormous. On the other hand, trials should be prospective, randomized, double-blind, and well designed with respect to population heterogeneity and the width of the therapeutic window. Nevertheless, the early successes of a few well-designed clinical trials suggest that the therapeutic potential for the successful treatment of post-ischemic injury by free radical ablation at reperfusion is substantial, and has already begun to be realized.

REFERENCES

Adkinson,D., Hollwarth,M.E., Benoit,J.N., Parks,D.A., McCord,J.M., Granger,D.N. (1986) Role of free radicals in ischemia-reperfusion injury to the liver. *Acta Physiol. Scand. Suppl.*, 548, 101-107.

Atalla, S.L., Toledo-Pereyra,L.H., MacKenzie,G.H., Cederna,J.P. (1985) Influence of oxygen-derived free radical scavengers on ischemic livers. *Transplantation,* 40, 584-590.

Aust,S.D., Morehouse,L.A., Thomas,C.E. (1985) Role of metals in oxygen radical reactions. *J. Free Radical Biol. Med.*, 1, 3-25.

Beckman,J.S., Beckman,T.W., Chen,J., Marshall,P.A., Freeman,B.A. (1990) Apparent hydroxyl radical production by peroxynitrite: implications for endothelial injury from nitric oxidase and superoxide. *Proc. Natl. Acad. Sci. U.S.A.,* 87, 1620-1624.

Blasig,I.E., Lowe,H., Ebert,B. (1987) Radical trapping and lipid peroxidation during myocardial reperfusion injury: radical scavenging by troxerutin in comparison to mannitol. *Biomed. Biochim. Acta.,* 46, S539-44.

Blech,D.M., Borders,C.L. (1983) Hydroperoxide anion, HO_2^-, is an affinity reagent for the inactivation of yeast Cu, Zn superoxide dismutase: modification of one histidine per subunit. *Arch. Biochem. Biophys*, 224, 579-586.

Bolli,R. (1990) Role of oxygen radicals in postischemic myocardial dysfunction ("Stunning"). *Free Radical Biol. Med.*, 9 (Suppl 1), 153.

Bruder,G., Heid,H.W., Jarasch,E.D., Mather,I.H. (1983) Immunological Identification and determination of xanthine oxidase in cells and tissues. *Differentiation,* 23, 218-225.

Bulkley,G.B. (1989) Mediators of splanchnic organ injury: overview and perspective. in *Splanchnic Ischemia and Multiple Organ Failure* (Marston,A., Bulkley,G.B., Fiddian-Green,R.G., Haglund,U.H., eds.) Edward Arnold Ltd., London, Chapter 16, pp183-191.

Cooper,J., Rettke,S.R., Ludwig,J., Ayalon,A., Sterioff,S., Wiesner,R.H., Krom,R.A.F. (1990) UW solution improves duration and quality of clinical liver transplantation. *Transplantation Proc.*,22, 477-479.

Davis,R.J., Bulkley,G.B., Traystman,R.J. (1987) Role of oxygen free radicals in focal brain ischemia. *Fed. Proc.*, 46, 799.

Eddy,L.J., Stewart,J.R., Jones,H.P., Engerson,T.D., McCord,J.M., Downey,J.M. (1987) Free radical-producing enzyme, xanthine oxidase, is undetectable in human hearts. *Am. J. Physiol.*, 253, H709-H711.

Ely,D., Dunphy,G., Dollwet,H., Richter H., Sellke,F., Azodi,M. (1992) Maintenance of left ventricular function (90%) after twenty-four-hour heart preservation with deferoxamine. *Free radical Biol. Med.*, 12, 479-485.

England,M.D., Cavarocchi,N.C., O'Brien,J.F., Solis,E., Pluth,J.R., Orszulak,T.A., Kaye,M.P., Schaff,H.V. (1986) Influence of antioxidants (mannitol and allopurinol) on oxygen free radical generation during and after cardiopulmonary bypass. (1986) *Circulation,* 74, Suppl. III, 134-137.

Ferreira,R., Burgos,M., Llesuy,S., Molteni,L., Milei,J., Flecha,B.G., Boveris,A. (1989) Reduction of reperfusion injury with mannitol cardioplegia. *Ann. Thoracic Surgery,* 48, 77-84.

Flaherty,J.T., Zweier,J.L. (1991) Role of oxygen radicals in myocardial reperfusion injury: experimental and clinical evidence. *Klin. Wochenschr.,* 69, 1061-1065.

Flick,M.R., Hoeffel,J., Staub,N.C. (1981) Superoxide dismutase prevents increased lung vascular permeability after microemboli. *Fed. Proc.,* 40, 405.

Forman,H.J., Fisher,A.B. (1981) Antioxidant defenses. in *Oxygen and Living Processes an Interdisciplinary Approach* (D.L.Gilbert,ed.), Springer-Verlag, New York, pp.235-249.

Friedl,H.P., Till,G.O., Ryan,U.S., Ward,P.A. (1989) Mediator-induced activation of xanthine oxidase in endothelial cells. *FASEB J.,* 3, 2512-2518.

Gallagher,K.P., Buda,A.J., Pace,D., Gerren,R.A., Shlafer,M. (1986) Failure of superoxide dismutase and catalase to alter size of infarction in conscious dogs after 3 hours of occlusion followed by reperfusion. *Circulation* 73, 1065-1076.

Gelvan,D., Saltman,P., Powell,S. (1990) Prevention of cardiac reperfusion damage by a nitroxide superoxide dismutase mimic. *Free Radical Biol. Med.,* 9 (Suppl 1), 153.

Granger, D.N., Rutili G., McCord J.M. (1981) Superoxide radicals in feline intestinal ischemia. *Gastroenterology,* 81, 22-29.

Granger,D.N., Hollwarth,M.E., Parks,D.A. (1986) Ischemia-reperfusion injury: role of oxygen-derived free radicals. *Acta Physiol. Scand. Suppl.,* 548, 47-63.

Granger,D.N., McCord,J.M., Parks,D.A., Hollwarth,M.E. (1986) Xanthine oxidase inhibitors attenuate ischemia-induced vascular permeability changes in the cat intestine. *Gastroenterology,* 90, 80-84.

Grisham, M.B., Hernandez,L.A., Granger,D.N. (1986) Xanthine oxidase and neutrophil infiltration in intestinal ischemia. *Am. J. Physiol.,* 251, G567-G574.

Haglund,U., Bulkley,G.B., Granger,D.N. (1987) On the pathophysiology of intestinal ischemic injury. *Acta Chir. Scand.,* 153, 321-324.

Haglund,U.H., Morris,J.B., Bulkley,G.B. (1988) Haemodynamic characterization of the isolated (denervated) parabiotically perfused rat jejunum. *Acta. Physiol. Scand.,* 132, 151-158.

Hallaway,P.E., Eaton,J.W., Panter,S.S., Hedlund,B.E. (1989) Modulation of deferoxamine toxicity and clearance by covalent attachment to biocompatible polymers. *Proc. Natl. Acad. Sci. U.S.A.,* 86, 10108-10112.

Hearse,D.J., Manning,A.S., Downey,J.M., Yellon,D.M. (1986) Xanthine oxidase: a critical mediator of myocardial injury during ischemia and reperfusion?. *Acta Physiol.Scand.Suppl.,* 548, 65-78.

Hedlund,B.E., Hallaway,P.E. (1990) Fluid resuscitation with a deferoxamine-colloid conjugate: applications to burn injury and hemorrhagic shock. *Free Radical. Biol. Med.,* 9 (Suppl 1), 153.

Heffner,J.E., Repine,J.E. (1989) Pulmonary strategies of antioxidant defense. *Am. Rev. Respir. Dis.,* 140, 531-554.

Hernandez,L.A., Grisham,M.B., Granger,D.N. (1987a) A role for iron in oxidant-mediated ischemic injury to intestinal microvasculature. *Am. J. Physiol.*, 253, G49-G53.

Hernandez,L.A., Grisham,M.B., Twohig,B., Arfors,K.E., Harlan,J.M., Granger,D.N. (1987b) Role of neutrophils in ischemia-reperfusion-induced microvascular injury. *Am. J. Physiol.*, 253, H699-H703.

Hoshino,T., Maley,W.R., Bulkley,G.B., Williams,G.M. (1988) Ablation of free radical-mediated reperfusion injury for the salvage of kidneys taken from non-heart-beating donors: A quantitative evaluation of the proportion of injury caused by reperfusion following periods of warm, cold, and combined warm and cold ischemia. *Transplantation*, 45, 284-289.

Im,M.J., Manson,P.N., Bulkley,G.B., Hoopes,J.E. (1985) Effects of superoxide dismutase and allopurinol on the survival of acute island skin flaps. *Ann. Surg.*, 201, 357-359.

Inauen,W., Granger,D.N., Meininger,C.J., Schelling,M.E., Granger,H.J., Kvietys,P.R. (1990) Anoxia-reoxygenation-induced, neutrophil-mediated endothelial cell injury: role of elastase. *Am. J. Physiol.*, 259, H925-H931.

Inoue,M., Takeda,Y., Hashimoto,H., Kosaka,F., Sasaki,J. (1990) Inhibition of postishemic reperfusion injury of dog brain by a site directed SOD derivative. *Free Radical Biol. Med.*, 9 (Suppl 1), 154.

Itoh,M., Guth,P.H. (1985) Role of oxygen-derived free radicals in hemorrhagic shock-induced gastric lesions in the rat. *Gastroenterology*, 88, 1162-1167.

Jarasch,E.D., Bruder,G., Heid.H.W. (1986) Significance of xanthine oxidase in capillary endothelial cells. *Acta Physiol. Scand.Suppl.*, 548, 39-46.

Johnson,W.D., Kayser,K.L., Brenowitz,J.B., Saedi,S.F. (1991) A randomized controlled trial of allopurinol in coronary bypass surgery. *Am. Heart J.*, 121, 20-24.

Klein,H.H., Pich,S., Lindert,S., Buchwald,A., Nebendahl,K., Kreuzer,H. (1988) Intracornary superoxide dismutase for the treatment of "reperfusion injury": a blind randomized placebo-controlled trial in ischemic, reperfused porcine hearts. *Basic Res. Cardiol.*, 83, 141-148.

Kono,Y., Fridovich,I. (1983) Inhibition and reactivation of Mn-catalase. *J. Biol. Chem.*, 258, 13646-13648.

Koppenol,W.H. (1985) The reaction of ferrous EDTA with hydrogen peroxide: evidence against hydroxyl radical formation. *J. Free Radical Biol. Med.*, 1, 281-285.

Korthius,R.J., Granger,D.N., Townsley,M.I., Taylor,A.E. (1985) The role of oxygen-derived free radicals in ischemia-induced increases in canine skeletal muscle vascular permeability. *Circ. Res.*, 57, 599-609.

Koyama,I., Bulkley,G.B., Williams,G.M., Im,M.J. (1985) The role of oxygen free radicals in mediating the reperfusion injury of cold-preserved ischemic kidneys. *Transplantation*, 40, 590-595.

Kvietys,P.R., Perry,M.A., Gaginella,T.S., Granger,D.N. (1990) Ethanol enhances leukocyte-endothelial cell interactions in mesenteric venules. *Am. J. Physiol.*, 259, G578-583.

Liu,T.H., Beckman,J.S, Freeman,B.A., Hogan,E.L., Hsu,C. (1989) Polyethylene glycol-conjugated superoxide dismutase and catalase reduce ischemic brain injury. *Am. J. Physiol.*, 256, H589-593.

Manson,P.N., Anthenelli,R.M., Im,M.J., Bulkley,G.B., Hoopes,J.E. (1983) The role of oxygen-free radicals in ischemic tissue injury in island skin flap. *Ann. Surg.*, 198, 87-90.

Manson,P.N., Jesudass,R., Marzella,L., Bulkley,G.B., Im,M.J., Narayan,K.K. (1991) Evidence for an early free radical-mediated reperfusion injury in frostbite. *Free Radical Biol. Med.*, 10, 7-11.

Marinkovix,D., Frederiks,W.M., Maas,A. (1990) Ischemia/Reperfusion Rat Liver Damage - Prevention by Anti-Oxidative/Anti-Inflammatory Combination Therapy. *Free Radical Biol. Med.*, 9 (Suppl 1), 154.
Marzella,L., Jesudass,R.R., Manson,P.N., Myers,R.A.M., Bulkley,G.B. (1988) Functional and structural evaluation of the vasculature of skin flaps after ischemia and reperfusion. *Plast. and Reconstr. Surg.*, 81, 742-750.
Minotti,G., Aust,S.D. (1987) The role of iron in the initiation of lipid peroxidation. *Chem. & Phys. Lipids.*, 44, 191-208.
Miura,T., Yellon,D.M., Kingma,J., Downey,J.,M. (1988) Protection afforded by allopurinol in the first 24 hours of coronary occlusion diminished after 48 hours. *Free Radical Biol. Med.*, 4, 25-30.
Miyachi,M., Vickers,S., Schiller,H.J., Patel,P., Mather,I., Hildreth,J., Kuhajda,F., Bulkley,G.B. (1991) Immunoaffinity localization of the free radical-generating enzyme, xanthine oxidase in the microvascular endothelium of the porcine and human brain. *Circulatory Shock*, 34, A436.
Naslund,U., Haggmark,S., Johansson,G., Marklund,SL, Reiz,S. (1990) Limitation of myocardial infarct size by superoxide dismutase as an adjunct to reperfusion after different durations of coronary occlusion in the pig. *Circ. Research.*, 66, 1294-1301.
Ohkubo,S., Murakami,E., Takekoshi,N., Matsui,S., Nakato,H., Kanemitsu,S., Kitayama,M., Matsuda,T. (1990) Comparison of the effect of r-h-SOD and polyethylene glycol conjugated SOD on myocardial infarct size in a canine of ischemia/reperfusion. *Free Radical Biol. Med.*, 9 (Suppl 1), 155.
Olthoff,K.M., Mills,M.J., Imagawa,D.K., Nuesse,B.J., Derus,L.J., Rosenthal,J.T., Milewicz,A.L., Busuttil,R.W. (1990) Comparison of UW solution and Euro-Collins solution for cold preservation of human liver grafts. *Transplantation* 49, 284-290.
Ozasa,T., Kubota,S., Kimura,M., Hamaguchi,M., Ohwada,S., Watanabe,H. (1990) Effects of superoxide dismutase treatment on ischemia reperfusion injury in total small bowel. *Free Radical Biol. Med.*, 9 (Suppl 1), 156.
Paller,M.S., Hoidal,J.R., Ferris,T.F. (1984) Oxygen free radicals in ischemic acute renal failure in the rat. *J. Clin. Invest.*, 74, 1156-1164.
Park,P.O., Haglund,U., Bulkley,G.B., Flat,K. (1990) The sequence of development of intestinal tissue injury after strangulation ischemia and reperfusion. *Surgery*, 107, 574-580.
Parks,D.A., Bulkley,G.B., Granger,D.N., Hamilton,S.R., McCord,J.M. (1982) Ischemic injury in the cat small intestine: role of superoxide radicals. *Gastroenterology*, 82, 9-15.
Parks,D.A., Granger,D.N. (1983) Ischemia-induced vascular changes: role of xanthine oxidase and hydroxyl radicals. *Am.J. Physiol.*, 245, G285-G289.
Parks,D.A., Granger,D.N. (1986) Xanthine oxidase: biochemistry, distribution, and physiology. *Acta Physiol. Scand. Suppl.* 548, 87-99.
Perler,B.A., Tohmeh,A.C., Bulkley,G.B. (1988) Inhibition of the compartment syndrome in post-ischemic skeletal muscle by free radical ablation at reperfusion. *FASEB J.*, 2, A1875.
Pillai,R., Bando,K., Schueler,S., Zebly,M., Reitz,B.A., Baumgartner,W.A. (1990) Leukocyte depletion results in excellent heart-lung function after 12 hours of storage. *Ann. Thorac. Surgery.*, 50, 211-214.
Ploeg,R.J. (1990) Kidney preservation with the UW and Euro-Collins solution- A preliminary report of a clinical comparison-. *Transplantaion* 49,281-284.
Punch,J., Rees,R., Cashmer,B., Oldham,K., Wilkins,E., Smith,D.J. (1991) Acute lung injury following reperfusion after ischemia in the hind limbs of rats. *J. of Trauma*, 31, 760-765.
Rashid,M.A., William-Olsson,G. (1991) Influence of allopurinol on cardiac complications in open hart operations. *Ann. Thorac. Surg.*, 52, 127-130.

Ratych,R.E., Chuknyiska,R.S., Bulkley,G.B. (1987) The primary localization of free radical generation after anoxia/reoxygenation in isolated endothelial cells. *Surgery,* 102, 122-131.

Redl,H., Schlag,G., Gasser,H., Dinges,H.P., Radmore,K., Davies,J. (1990) hrSOD does not prevent bacterial translocation (BT) in a baboon polytrauma model. *Free Radical Biol. Med.,* 9 (Suppl 1), 156.

Reilly,P.M., Schiller,H.J., Bulkley,G.B. (1991) Reactive oxygen metabolites in shock. in *Care of the Surgical Patient* (Wilmore,D.W., Brennan,M.F., Harken,A.H., et al, eds.), Scientific Am., Inc., New York, Section IV Trauma, Chapter 8, pp1-30.

Romson,J.L., Hook,B.G., Kunkel,S.L., Abrams,G.D., Schork,M.A., Lucchesi,B.R. (1983) Reduction of the extent of ischemic myocardial injury by neutrophil depletion in the dog. *Circulation,* 67, 1016-1023.

Ryan,U.S., Schultz,D.R., Goodwin,J.D., Vann,J.M., Selvaraj,M.P., Hart,M.A. (1989) Role of C1q in phagocytosis of Salmonella minnesota by pulmonary endothelial cells. (1989) *Infection and Immunity,* 57, 1356-1362.

Salim,A.S. (1991) Role of oxygen-derived free radical scavengers in the treatment of recurrent pain produced by chronic pancreatitis -a new approach-. *Arch. Surg.* 126, 1109-1114.

Sanfey,H., Bulkley,G.B., Cameron,J.L. (1985) The pathogenesis of acute pancreatitis: The source and role of oxygen-derived free radicals in three different experimental models. *Ann. Surg.,* 201, 633-639.

Sanfey,H., Sarr,M.G., Bulkley,G.B., Cameron,J.L. (1986) Oxygen-derived free radicals and acute pancreatitis: a review. *Acta Physiol. Scand. Suppl.,* 548,109-118.

Schiller,H.J., Vickers,S., Hildreth,J.J, Mather,I., Kuhajda,F., Bulkley,G.B. (1991) Immunoaffinity localization of xanthine oxidase on the outside surface of the endothelial cell plasma membrane. *Circ. Shock.,* 34, A435.

Schneeberger,H., Illner,W.D., Abendroth,D., Bulkley,G., Rutili,F., Williams,M., Thiel,M., Land,W. (1989) First clinical experiences with superoxide dismutase in kidney transplantation-Results of a double-blind randomized study. *Transplantation Proc.,* 21, 1245-1246.

Schneeberger,H., Schleibner,S., Schilling,M., Illner,W.D., Abendroth,D., Hancke,E., Janicke,U., Land,W. (1990) Prevention of acute renal failure after kidney transplantation by treatment with rh-SOD: interim analysis of a double-blind placebo-controlled trial. *Transplantation Proc.,* 22, 2224-2225.

Schoenberg,M.H., Muhl,E., Sellin,D., Younes,M., Schildberg,F.W., Haglund,U. (1984) Posthypotensive generation of superoxide free radicals-possible role in the pathogenesis of the intestinal mucosal damage. *Acta Chir. Scand.,* 150, 301-309.

Shlafer,M., Kane,P.F., Kirsh,M.M. (1982) Superoxide dismutase plus catalase enhances the efficacy of hypothermic cardioplegia to protect the globally ischemic, reperfused heart. *J. Thorac. Cardiovasc. Surg.,* 83, 830-839.

Simovic,M., Spasic,M., Saicic,Z., Stanimirovic,D., Buzadzic,B., Korac,B., Markovic,M. (1990) The role of antioxidative system of brain tissue for survival combined irradiation injury. *Free Radical Biol. Med.,* 9 (Suppl 1), 157.

Sjoquist,P.O., Marklund,S. (1990) Effects of endothelial bound EC-SOD type C on myocardial damage in reperfused ischemic rat hearts. *Free Radical Biol. Med.,* 9 (Suppl 1), 157.

Southard,J.H., van Gulick,T.M., Ametani,M.S., Vreugdenhil,P.K., Lindell,S.L., Pienaar,B.L., Belzer,F.O. (1991) Important components of the UW solution. *Transplantation,* 49, 251-257.

Suzuki,M., Grisham,M.B., Granger,D.N. (1990) Hydrogen peroxide promotes neutrophil adherence in cat mesenteric venules. *Gastroenterology,* 98, A476.

Tabayashi,K., Suzuki,Y., Nagamine,S., Ito,Y., Sekino.Y., Mohri,H. (1991) A clinical trial of allopurinol (Zyloric) for myocardial protection. *J. Thorac. Cardiovasc. Surg.,* 101, 713-718.

Uraizee,A., Reimer,K.A., Murry,C.E., Jennings,R.B. (1987) Failure of superoxide dismutase to limit size of myocardial infarction after 40 minutes of ischemia and 4 days of reperfusion in dogs. *Circulation* 75, 1237-1248.

Vickers,S., Hildreth,J., Kuhajda,F., Madara,P., Mather,I., Baig,M., Bulkley,G.B. (1990a) Immunohistoaffinity localization of xanthine oxidase in the microvascular endothelial cells of procine and human organs. *Circ. Shock,* 31, 87.

Vickers,S., Miyachi,M., Hildreth,J., Baij,M., Kuhajda,F., Madara,P., Mater,I., Bulkley,G.B. (1990b) Immunohistoaffinity localization of xanthine oxidoreductase (XOR) and histochemical discrimination of dehydrogenase (XD) and oxidase (XO) in situ. *FASEB J.,* 4, A895.

Weiss, S.J. (1989) Tissue destruction by neutrophils. *New England J. Med.,* 320, 365-376.

Werns,S.W., Lucchesi,B.R. (1988) Leukocytes, oxygen radicals, and myocardial injury due to ischemia and reperfusion. *Free Radical Biol. Med.,* 4, 31-37.

Zelck,U., Janichen,F., Karnstedt,U. (1990) Influence of superoxide dismutase and iloprost on the ischemia induced arrhythmias and post-ischemic myocardial function. *Free Radical Biol. Med.,* 9 (Suppl 1), 158.

Free Radicals: From Basic Science to Medicine
G. Poli, E. Albano & M. U. Dianzani (eds.)

METAL-CATALYZED FREE RADICAL INJURIES IN CHILDHOOD: DISORDERS AND PHARMACEUTICAL INTERVENTION

Makoto Mino, Masayuki Miki, Hiromi Ogihara, Toru Ogihara, Hiroshi Yasuda and Ryoya Hirano

Department of Pediatrics, Osaka Medical College, Takatsuki, 569, Japan

SUMMARY: The transition metals, copper and iron, are well known to catalyze the generation of free radicals in vitro and may be involved in the pathogenesis of various diseases. Wilson's disease is an inherited disorder of copper metabolism characterized by progressive accumulation of copper in the liver and its subsequent overflow to extrahepatic tissues.
Non-ceruloplasmin copper is increased and ferroxidase activity is decreased in the plasma. The copper released from the liver into the circulation catalyzes the generation of free radicals, which consume radical trapping antioxidants in the plasma, such as urate, vitamin C and vitamin E. The radical trapping capacity in plasma of Wilson's disease is decreased at the early stage of the disease without therapy, and it is improved by copper chelating therapy with d-penicillamine and tirientine.
In LEC rats (a model of Wilson's disease), the plasma antioxidant activity against ferrous salt-stimulated lipid peroxidation is reduced, and addition of zinc inhibited the peroxidation. This finding provides a basis for the improvement of Wilson's disease by zinc therapy.
Homocystinuria is an inborn error of methionine metabolism characterized by elevated levels of homocysteine in the body fluids which cause premature atherosclerosis and thrombosis. Transition metal-chelated oxidation of homocysteine thiols is thought to lead to the generation of superoxide and hydrogen peroxide. Active oxygen species generated in plasma appear to attack lipoproteins to produce oxidatively modified LDL, which may contribute to the onset of atherosclerosis. We examined the homocysteine-induced oxidation of LDL in the presence of metals and the role of alpha-tocopherol enrichment of LDL in preventing such oxidative modification.

INTRODUCTION.

The transition metals, copper and iron, are well known to catalyze the generation of free radicals in the "in vitro" situation. Many pathological roles for free radical-mediated injury have been suggested in various diseases. In this paper, we would like to discuss metal-chelated free radical injuries in inborn errors of metabolism and the appropriate therapeutic interventions. In this paper, Wilson's disease and homocystinuria are covered.

WILSON'S DISEASE:

Wilson disease is characterized by the accumulation of copper in various body tissues, especially in the liver, because of defective copper excretion into the bile and altered incorporation of copper into ceruloplasmin [CP]. Plasma ceruloplasmin levels are decreased in Wilson's disease patients. Ceruloplasmin has ferroxidase activity and physiologically catalyzes the oxidation of ferrous (Fe^{2+}) to ferric (Fe^{3+}) ion employing dioxygen as a terminal electron acceptor without the generation of any partially reduced oxygen species as intermediate, such as superoxide anion, hydrogen peroxide, or hydroxyl radicals (Danks, 1989).

$$CP\text{-}(Cu^{2+})_4 + 4Fe^{2+} \longrightarrow CP\text{-}(Cu^{+})_4 + 4Fe^{3+}$$
$$CP\text{-}(Cu^{+})_4 + O_2 + 4H \longrightarrow CP\text{-}(Cu^{2+})4 + 2H_2O$$

In Wilson's disease, the low plasma level of ceruloplasmin and the release of copper from the liver into the circulation cause an increase of non-ceruloplasmin copper in the plasma, in spite of the overall low plasma copper level.

1. Total free radical trapping capacity of antioxidants in plasma [TRAP] in Wilson's disease.

Although free radical reactions with transition metals and the involvement of plasma non-ceruloplasmin copper have been proposed to explain the development of acute hemolytic crisis and multiple organ failure with fulminant hepatitis in Wilson's disease, the

precise mechanisms have not yet been elucidated. If non-ceruloplasmin copper released from the liver can catalyze the generation of free radicals, then the non-ceruloplasmin copper status should be examined in Wilson's disease. Plasma copper not incorporated into ceruloplasmin exists as either non-ceruloplasmin copper or loosely bound copper. The amount of non-ceruloplasmin copper can be estimated by subtracting from the total serum copper molecules the value obtained by multiplying the number of ceruloplasmin molecules by six, because ceruloplasmin binds 6 atoms of copper molecules (Danks, 1989). Loosely bound copper can be assayed by chelation with phenanthroline, since it undergoes chelation while copper incorporated into ceruloplasmin does not. The phenanthroline-copper complex degrades DNA in the presence of a reducing agent to yield TBA chromogen (Gutteridge, 1984). Loosely bound copper measured by the phenanthroline method is closely correlated with the non-ceruloplasmin copper level ($r=0{,}969$, $p<0.01$). If non-ceruloplasmin copper increases with the progression of Wilson's disease and free radical generation increases, the consumption of free radical-trapping antioxidants in plasma would be a possible initial event. A test for the total radical trapping antioxidants in plasma [TRAP] (Wayner et al., 1987) provides a mean of assessing the total capacity of chain breaking antioxidants to prevent peroxidation in plasma. The TRAP level can be also calculated by the addition of the individual concentration of the major plasma antioxidants after multiplying each by experimentally determined stoichiometric factors such as 1.7 for ascorbate, 0.33 for sulfhydryl groups, and 2.0 for tocopherol (Wayner et al., 1987). These three compounds are considered to be the major plasma antioxidants, but other compounds (bilirubin, beta-carotene and ubiquinol) may also contribute to the TRAP value (Lindeman et al., 1989).
Ceruloplasmin is also included in this category. The contribution of these unidentified compounds can be obtained by subtracting the calculated TRAP value from the measured one. The clinical course of two patients with Wilson's disease is shown in Table I.

Both patients developed hemolytic anemia and hepatic injury.

Table I

Clinical data of two patients with Wilson's disease during the course of therapy.

Course	Hb (g/dl)	Ht (%)	Reticulo-cytes (%)	Bilir (mg/dl)	sGOT (U/ml)	sGPT (U/ml)	Prothrombin time (%)
Patient S.O.							
Before	9.0	27	2.6	1.6	35	103	80
1 m	10.8	34	2.0	0.8	49	84	86
3 m	11.7	36	NT	0.6	63	60	92
6 m	12.3	34	NT	0.7	36	40	75
Patient Y.M.							
Before	7.6	24	4.1	4.0	250	304	64
2 w	8.2	26	2.9	3.3	212	320	63
1 m	7.8	25	5.4	3.0	154	275	70
5 yr	13.2	39	1.0	0.7	38	79	82

NT: not tested, Hb: hemoglobin, Ht: Hematocrit, Bilir: bilirubin, Before: before treatment.

Table II shows the changes in the TRAP level during the disease course. At the time of diagnosis in the early and active stage of the disease, non-ceruloplasmin copper was increased and the TRAP level was decreased. However, both parameters became closer to normal with therapy. A close inverse correlation was observed between the TRAP and non-ceruloplasmin copper levels, indicating that non-ceruloplasmin copper totally consumed the plasma antioxidants in patients with Wilson's disease unless an appropriate therapy is started.
The correlation of the measured TRAP value (TRAPmeas) with various plasma antioxidants is shown in Table 3. TRAPmeas was best correlated with the calculated TRAP value (TRAPcalc), as would be expected. The second closest correlation of TRAPmeas was observed with urate and unidentified antioxidants, since the latter included ceruloplasmin.

Table II

Changes in antioxidants and plasma TRAP values in two Wilson's disease patients during therapy.

Course	S-Cu	Cu/ non-Cp	UA	ASA	SH	Toc	Un identified antioxidants	Measured TRAP value
Patient S.O.								
Before	4.9	3.5	65	5	75	16	176	325
1 m	6.4	1.8	83	10	105	13	326	512
3 m	2.9	1.5	131	30	215	15	302	624
6 m	2.2	0.8	160	58	228	11	432	836
Patient Y.M.								
Before	14.4	9.6	124	11	63	10	177	399
2 w	13.8	12.4	95	19	65	9	102	195
1 m	13.5	12.6	110	15	104	13	258	487
5 yr	1.5	0.1	232	56	328	19	512	1055
Healthy children	16.8	0.0	255	62 (Mean: n=10)	295	18	310	830

[μM]

S-Cu: serum copper, Cu/non-Cp: non-ceruloplasmin copper, UA: uric acid, ASA: ascorbic acid, Toc: tocopherol, SH: sulphydryl groups, Before: before treatment.

Table III

Correlation of TRAPcalc and calculated trapping capacities of various plasma antioxidants in Wilson's patients with TRAPmeas

	Correlation coefficent (r)	p
TRAPcalc	0.915	< 0.01
Ascorbate	0.558	
SH	0.858	< 0.05
Urate	0.915	< 0.01
Tocopherol	0.606	
Un-identified	0.902	< 0.01

n=12

However, no significant correlation was found with tocopherol and ascorbate, which are known to be the most potent chain-breaking antioxidants and which act synergistically. With respect to plasma tocopherol, no significant reduction was observed in the Wilson's disease patients even before treatment. This may suggest that water-soluble chain-breaking antioxidants including ascorbate and thiols are preferentially utilized in Wilson's disease, and that the reducing effect of water-soluble antioxidants on chromanoxyl radicals of tocopherol can prevent the consumption of tocopherol (Frei et al., 1989, Miki et al., 1989, Motoyama et al., 1989). On the other hand, TRAPmeas was inversely correlated with the levels of non-ceruloplasmin copper and loosely bound copper (Fig. 1). In addition, a decrease in TRAPmeas in the Wilson's disease patients reflected the progression of their anemia (Fig. 2). These findings indicates that non-ceruloplasmin copper probably plays a role in free radical injury related to the pathogenesis of Wilson's disease.

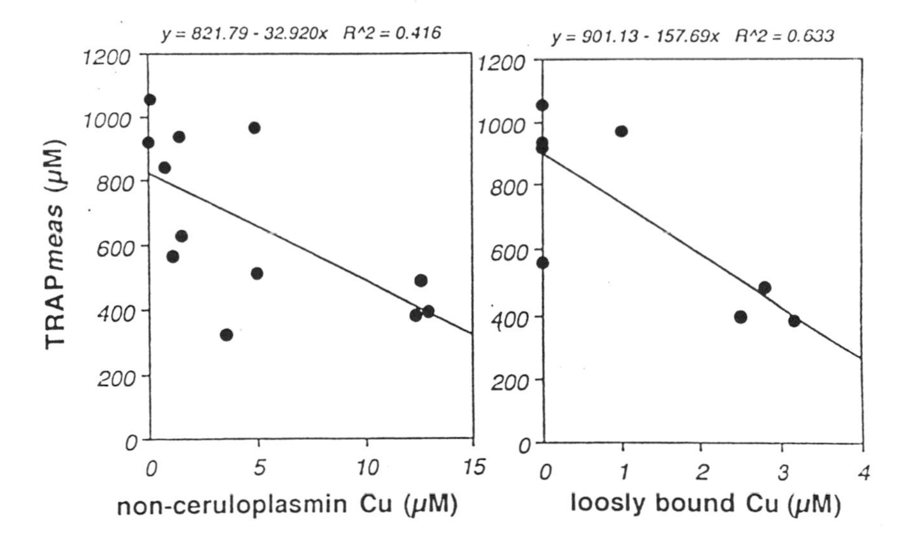

Fig. 1. Correlation of TRAPmeas to non-ceruloplasmin Cu or loosely bound Cu.

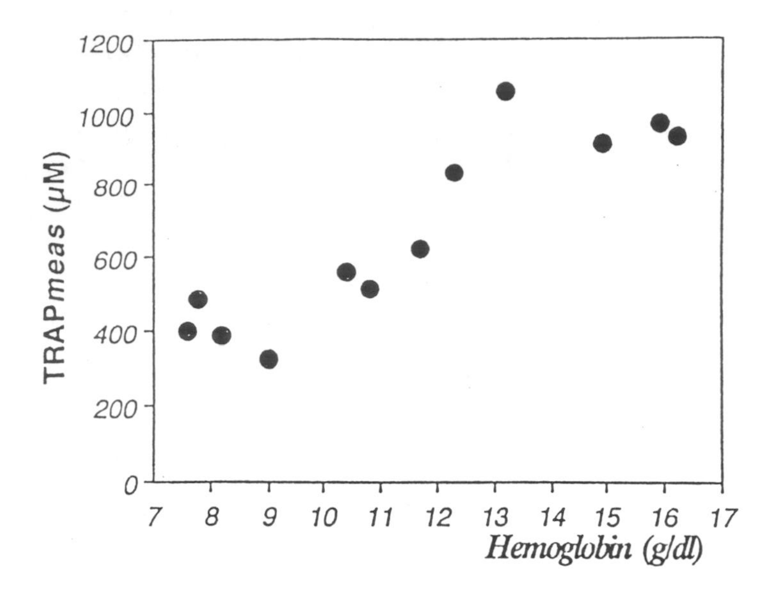

Fig. 2. Relationship between TRAP and Hb in patients with Wilson's disease during therapy course.

2. Free radical generation by copper:

Copper not incorporated into ceruloplasmin readily produces superoxide anions which probably cause free radical injury in various tissues, including the liver, brain, eyes, and kidney. To produce superoxide anion, the presence of monovalent copper (cuprous ion) may be necessary to donate an electron to molecular oxygen, and the thiols involved in proteins seem to play a role in the reduction of cupric to cuprous ion.

Figure 3 shows the relationship between the consumption of thiols and superoxide formation in association with copper sulfate.

In addition, this reaction was also carried out after pretreatment with SDS and iodoacetamide. SDS can expose concealed intrinsic thiols in membrane proteins to increase the amount of reacting thiols, while iodoacetamide leads the surface thiols to be masked and suppresses the reaction (Takanaka et al., 1989). The increase in reacting thiols in erythrocyte ghost membranes with SDS treatment enhanced superoxide production, which was detected by cytochrome C reduction. In contrast, after membrane thiols were masked by iodoacetamide treatment, superoxide production was very slight.

The production of superoxide depends on the consumption of reac-

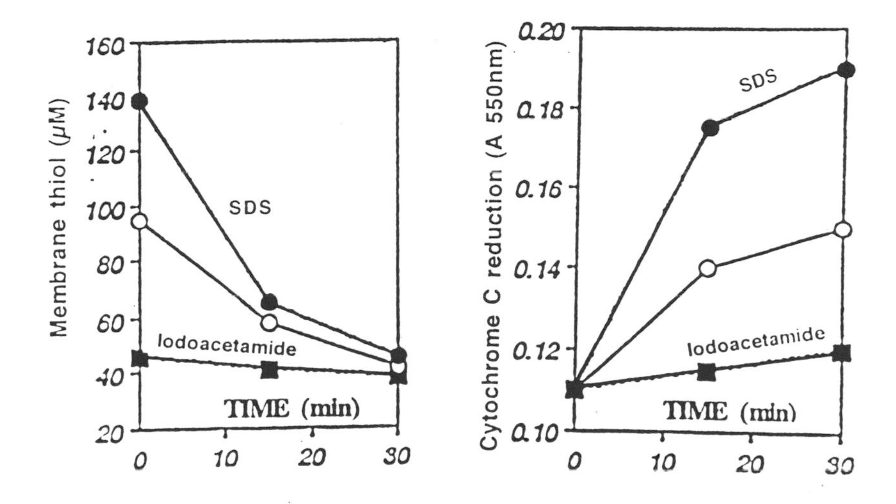

Fig. 3. Membrane thiol composition by Cu and superoxide generation.
SDS treatment: Exposure of concealed intrinsic thiols
Iodoacetamide treatment: masking of the surface thiols.

ting thiols, which might be used for the reduction of cupric to cuprous ion. This may explain the occurrence of haemolysis in Wilson's disease, because an increase in non-ceruloplasmin copper could produce a large amount of free radicals in association with the reduction of copper by membrane thiols. When copper sulfate was reacted with vitamin E-deficient rat erythrocytes, haemolysis developed as early as 60 min after the starting of reaction in parallel with the formation of TBA-reactants. This haemolysis was suppressed by catalase but not by SOD. The superoxide produced by the reaction with cupric ion and thiols is rapidly reduced to hydrogen peroxide either enzymatically or non-enzymatically via the dismutation reaction. The hydrogen peroxide thus produced can penetrate erythrocyte membranes and it may be further reduced to hydroxyl radicals by reacting with the ferrous or perferryl ion that are abundant intracellularly. Then, it can be assumed that the hydroxyl radicals would damage the membrane constituents of the erythrocyte to produce haemolysis. The haemolysis induced by non-ceruloplasmin copper is more evident in vitamin E-deficient erythrocytes as compared to vitamin E-sufficient ones. The above findings indicate that the haemolysis observed in Wilson's disease is probably attributable to a chain of free radical reactions occurring in erythrocyte membranes.

3. Pharmaceutical intervention:

Currently, the administration of copper-chelating agents such as d-penicillamine and trientine (triethylene tetramine dihydrochloride) is performed along with restriction of oral copper intake in order to eliminate accumulated copper via urinary excretion, . In addition, oral zinc sulfate has recently been used to inhibit copper absorption (Danks, 1989). Furthermore, zinc may possibly inhibit the generation of free radicals in Wilson's disease patients. To investigate this hypothesis, a mutant strain of Long-Evans rats (LEC rats) that provide an experimental model very similar to Wilson's disease has been used. LEC rats develop acute hepatitis associated with jaundice at the age of 4 months following progressive accumulation of copper in the liver during the first 4 months of life. These rats have low serum ceruloplasmin levels and massive urinary excretion of copper (Yoshida et al., 1987).

Our working hypothesis has been that the antioxidant effect of plasma from LEC rats would be reduced because of its low ceruloplasmin level. Therefore we have investigated the inhibitory effect of LEC rat plasma upon the stimulation of lipid peroxidation by ferrous ion in phosphatidylcholine (PC) liposomes. As shown in Fig. 4, soybean liposomes have been oxidized by a ferrous ion complex with the chelating agent nitrile triacetate (NTA), and the inhibition of the liposome oxidation has been examined following the addition of plasma from Wistar and LEC rats. The oxidation was assessed from the level of conjugated dienes. In this reaction system, the formation of conjugated

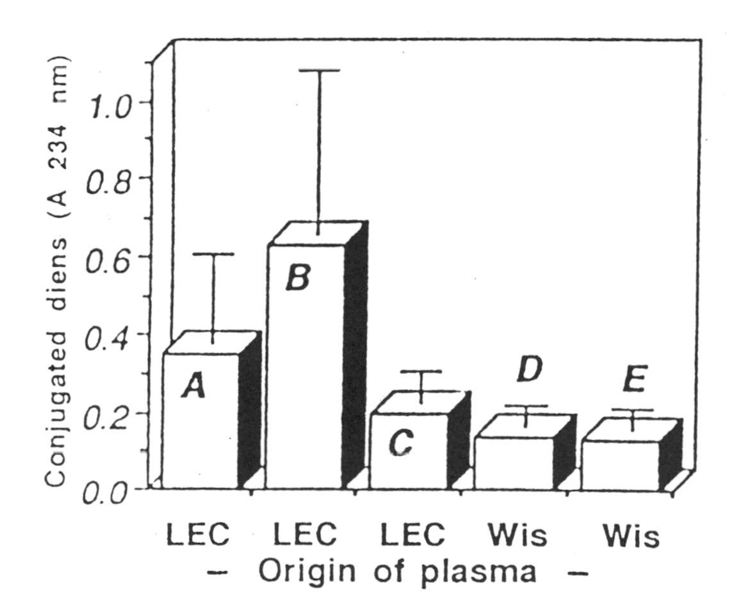

Fig. 4. Effect of plasma on the oxidation of PC-liposomes catalyzed by ferrous ions.
Soybean phosphatidylcholine (PC)-liposomes were oxidized by Fe^{2+}-NTA in the presence of plasma from LEC or Wistar rats.
A: LEC rat plasma only; B: LEC rat plasma pretreated with 1 mM NaN_3; C: LEC rat plasma with 2 µM ceruloplasmin; D: Wistar rat plasma only; E: Wistar rat plasma with 2 µM ceruloplasmin.

dienes is greater in the presence of plasma from LEC rats (A) as compared to that from Wistar rats used as controls (D). The oxidation is enhanced when the ferroxidase activity is initially inhibited by sodium azide (B), while it is inhibited by addition of 2 µM ceruloplasmin (C). The result indicates that the low antioxidant activity of LEC rat plasma may be attributable to their low ceruloplasmin level.

Ferroxidase activity of ceruloplasmin is known to favour the incorporation of iron into transferrin. Therefore, low ferroxidase activity would be expected to accumulate ferrous iron in the liver of LEC rats. The assay of hepatic iron content showed 87 µg/g of wet tissue in LEC rat liver as compared to 32 µg/g in Wistar rats, concomitantly the liver copper content was 238 µg copper/g in LEC and 5 µg copper/g in Wistar rats respectively. The accumulation of iron may also partially contribute to the oxidative injury. In subsequent experiments, the formation of TBA reactants has been examined following the addition to erythrocyte ghost suspensions of NaN_3-pretreated hepatic cytosol in order to suppress intrinsic ferroxidase activity where lipid peroxidation has been initiated by addition of hydrogen peroxide. When the ghost suspension is added to either hepatic cytosol or hydrogen peroxide, the formation of TBA reactants is not so marked using either LEC or Wistar rat liver cytosol. However, following the simultaneous addition of cytosol and hydrogen peroxide, the TBA reactant formation is predominant in the presence of LEC rat liver cytosol which includes high copper and iron.

In this reaction system, the inhibitory effect of chelating agents and zinc on TBA reactant formation has been examined. The addition of zinc, EDTA and desferroxamine (DFO) to this reaction system inhibited the effect of cytosol on TBA formation, whereas d-penicillamine and trientin showed no effects (Table IV). This indicated that copper and iron act as prooxidants under oxidative stress, while zinc can exhibit an antioxidant effect. Since d-penicillamine and trientin do not have any radical scavenging effect, they may act only for elimination of metals via urine.

Table IV

Hepatic cytosol catalized oxidation of RBC ghosts initiated with H_2O_2

Rats	Formation of TBA reactant (A 532 nm x 10^3) LEC	Wistar
ghosts +cytosol + H_2O_2 (complete system)	263 ± 19 (100%)	86 ± 15 (100%)
ghosts + cytosol	60 ± 10	42 ± 7
ghosts + H_2O_2	53 ± 5	
Relative TBA-reactant formation (%) as added to the above complete system		
Zn	63%	56%
EDTA	71%	41%
DFO	78%	32%
PCM	111%	119%
TTM	95%	108%

DFO: desferrioxamine; PCM: d-penicillamine; TTM: triethylene tetramine.

In another study, the production of TBA reactants during the reaction of erythrocyte ghosts with copper sulfate, is inhibited by the addition of zinc (Fig.5, left), while TBA formation resulting from reaction of cupric ion with an azo-compound (ABAP) is not inhibited in the presence of zinc (Fig.5, right). This finding indicated that zinc ions might block the reaction of membrane thiols with cupric ion and thus prevent the formation of cuprous ion. Consistently, the inhibitory effect of zinc occurs in a dose-dependent manner (Fig. 6). The suppressive effect of zinc on copper-mediated free radical injury may contribute to treatment of Wilson's disease.

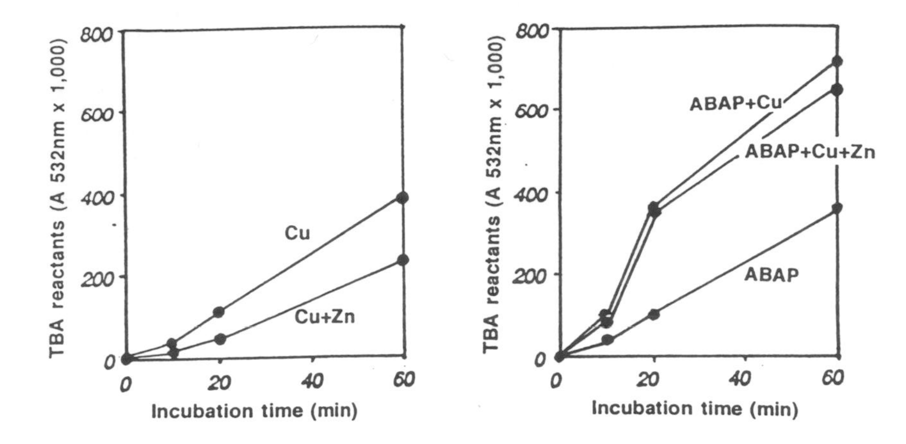

Fig. 5. Antioxidant effect of zinc on oxidation of erythrocyte ghosts by copper and ABAP.

Vitamin E can protect against hemolysis. When vitamin E-deficient erythrocytes are exposed to copper sulfate and ferric chloride, a marked hemolysis and an increase in TBA reactants is observed, while the hemolysis is suppressed in the case of vitamin E-sufficient erythrocytes. Although Wilson's disease patients show no reduction of plasma and erythrocyte tocopherol levels, even when TRAP levels are decreased in the early stage of the disease, a trial of vitamin E administration to prevent ongoing hemolysis may be worthwhile.

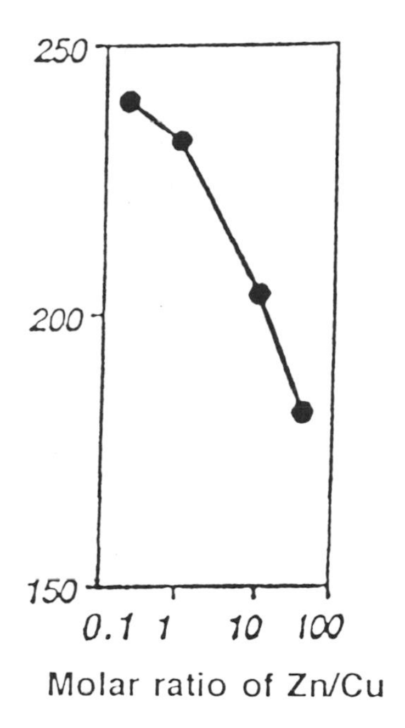

Fig. 6. Suppression of the formation of TBA reactants in the presence of zinc salts when erythrocyte ghosts were reacted with copper salts.

HOMOCYSTINURIA:

Homocystinuria is due to an inborn defect of the enzyme cystathionine synthase. Homocysteine accumulates in the body fluids and causes the clinical symptoms by producing abnormalities of the connective tissues and blood vessels. Among these abnormalities, this article will discuss the occurrence of premature atherosclerosis in this disease.

1. Free radical generation from a high concentration of homocysteine:

We have observed that the oxygen consumption in a solution of homocysteine is increased after the addition of ferric chloride

(Fig. 7). Such an increase in oxygen consumption is accompanied by the rapid decrease of homocysteine thiol groups. The reaction between homocysteine and ferric ion generates superoxide anion, as confirmed by the reduction of cytochrome C and by inhibition following the addition of SOD. These findings indicate that the decrease in thiols of homocysteine may play role in superoxide generation. However, at the moment it is not known whether free ferric ions are present in living tissues even under disease conditions, since tissue iron is mainly incorporated into ferritin. When various concentrations of homocysteine are added to a ferritin solution, we have observed that, as detected with phenanthroline, ferrous ion is released in a dose dependent manner from ferritin, indicating that homocysteine induces the release of ferrous ion from ferritin (Fig. 8).

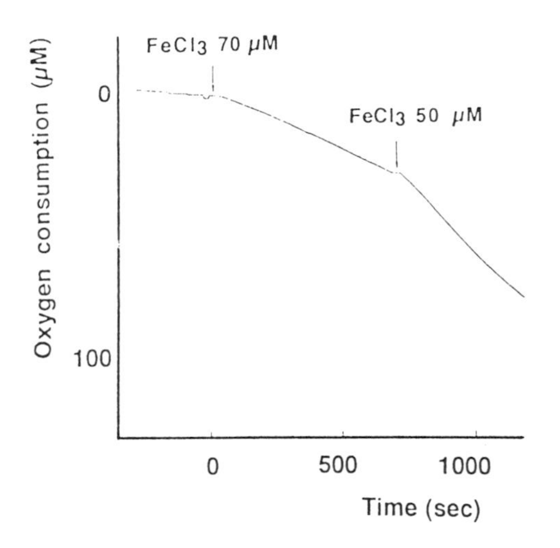

Fig. 7. Oxygen consumption induced by the reaction of homocysteine and ferric chloride.
Basal solution: 2 mM homocysteine in Tris buffer (pH 7.4) 50 μM EDTA at 37° C.

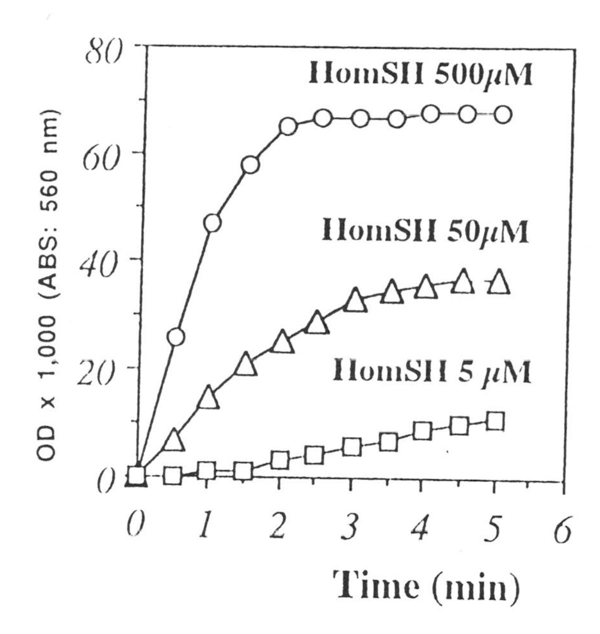

Fig. 8. Ferrous ion production from ferritin reacted with homocysteine
Ferritin (final concentration of 0.1 mg/ml) was added to phosphate buffer (pH 7.8) containing 1-10-phenanthrolin. The production of ferrous ion was monitored at 560 nm.

2. Development of oxidatively modified low density lipoproteins (LDL):

According to the hypothesis of the modified LDL-scavenger receptor as a cause of atherosclerosis (Steinburg et al., 1989) human LDL have been applied to a reaction system with homocysteine and ferric chloride. During the reaction, the tocopherol content of LDL decreases concomitantly with an increase of TBA reactant formation. Furthermore, the PUFA content of LDL decreased rapidly in the order of the degree of unsaturation (Fig. 9).

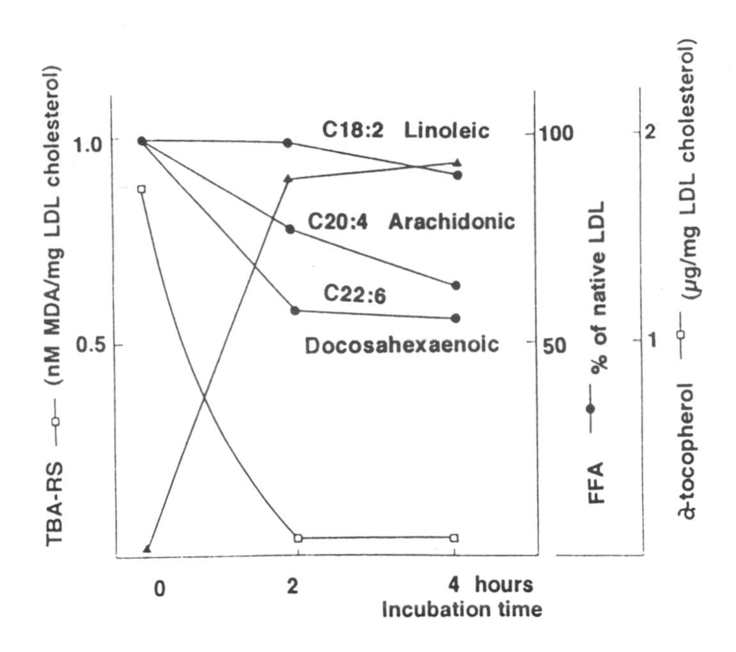

Fig. 9. Changes in the polyunsaturated fatty acid content of LDL oxidized by reaction with homocysteine and ferric chloride.

SDS-PAGE electrophoresis of LDL apoproteins after reaction with homocysteine show protein fragmentation and ferric ion (Fig. 10), and this effect is suppressed in vitamin E-rich LDL.

Agarose gel electrophoresis of LDL oxidized by reaction with homocysteine and ferric chloride shows an increased anodal mobility of LDL as the reaction time is prolonged (Fig. 11).

On this basis, LDL prepared from two patients with homocystinuria have been analyzed by agarose gel electrophoresis. The LDL from the patient with poor therapeutic control shows an abnormal anodal mobility and is decreased in amount as compared with LDL from the well controlled patient or from a healthy adult (Fig. 12), indicating that in homocystinuria "in vivo" oxidative modification of LDL occurs. The cytotoxicity, which is known to be expressed by oxidatively modified LDL (Morel et al., 1984, Hes-

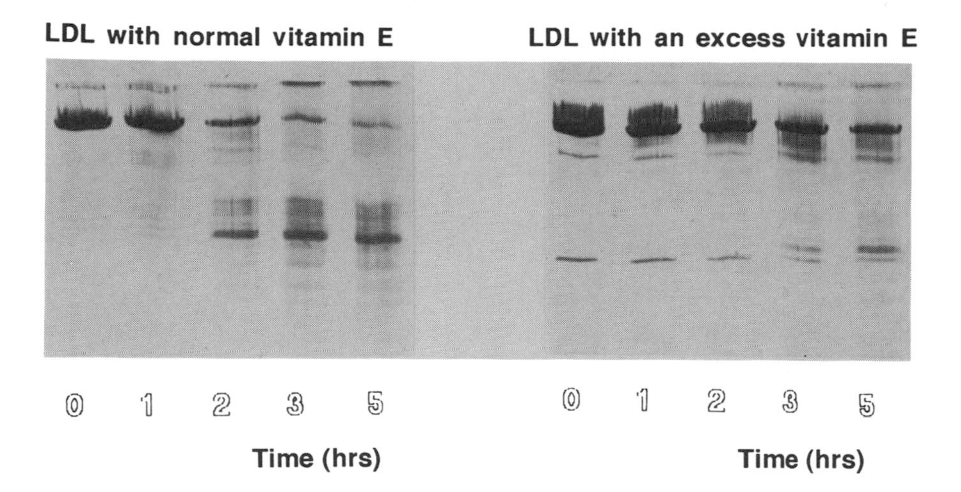

Fig. 10. Changes in the electrophoretic pattern of apo-protein-B in LDL reacted with homocysteine and $FeCl_3$.
LDL (vitamin E-rich and normal native LDL) was obtained from healthy human adults with or without vitamin E administration. It was reacted with a concentration of 0.6 mg total cholesterol/ml with 1 mM homocysteine and in the presence of 200 µM $FeCl_3$-NTA for 5 hrs at 37° C. SDS-PAGE of LDL proteins was carried out using a linear gradient gel (4-20%) at a constant current of 3 mA for 10 hrs at 4° C. The gel was stained with silver for visualization.

sler et al., 1979), has been also confirmed in LDL modified by the reaction with homocysteine and ferric chloride. A decrease in lactate dehydrogenase (LDH) activity and thiols occurs when fibroblasts are cultured with LDL that had been reacted with homocysteine and ferric ion (Fig 13). All the above findings suggest that superoxide generated by the reaction of homocysteine and transition metals might play an important role in LDL modification which may then cause premature atherosclerosis in homocystinuria.

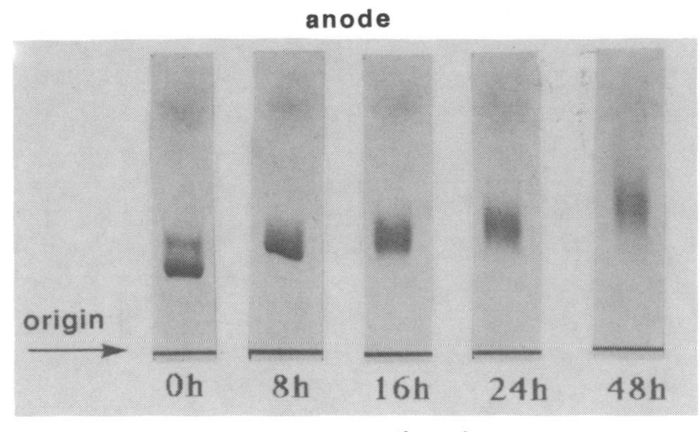

Fig. 11. Changes in the agarose gel electrophoretic pattern of LDL when reacted with homocysteine and $FeCl_3$.

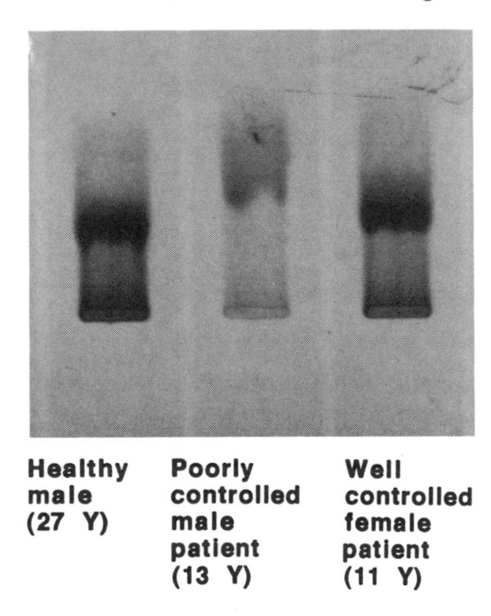

Fig. 12. Agarose gel electrophoretic pattern of LDL in two patients with homocystinuria and a healthy adult control.

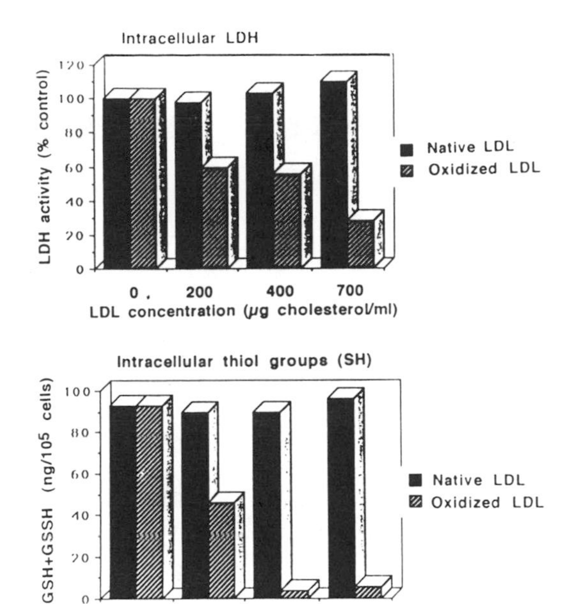

Fig. 13. Cytotoxicity of LDL reacted with homocysteine and $FeCl_3$ towards fibroblasts.
Fibroblasts (4×10^6) at confluence in 6-well plate were incubated with oxidized LDL and native LDL for 8 hrs. Oxidized LDL was produced by the reaction with homocysteine and $FeCl_3$. GSH + GSSG were determined by Tratz's method. The value represents as the mean of three measurements. Cytotoxicity was expressed as a decrease in LDH activity and total GSH + GSSG in cells.

3. Intervention of oxidative changes of LDL reacted with homocysteine and ferric ion:

Vitamin E enrichment of LDL suppresses and/or delays the formation of TBA reactants and the apoprotein changes as compared to LDL that have a normal vitamin E content, indicating the possibility that fortification with vitamin E might suppress the progression of premature atherosclerosis in homocystinuria. However, the efficacy of vitamin E for prevention of the clinical manifestaions mediated by free radicals remains to be confirmed in this disease.

Acknowledgements

We wish to thank Dr E. Niki for his kind assistance and advice. We are also grateful to Eisai Co Ltd. for financial assistance.

REFERENCES

Danks,D.M. Disorders of copper tranport. In the "Metabolc Basis of Inherited Disease, I" Eds Scriber,C.S., Beaudet,A.L., Sly,W.L., and Valle,D. McGraw-Hill, Co., N.Y. (1989) pp. 1411-1422

Frei,B., England,L., and Ames,B.N. (1989) Proc. Natl. Acad. Sci. 86, 6377-6381

Gutteridge,J.M.C. (1984) Biochem. J. 218, 983-985

Hessler,T., Robertson,A.L.Jr., and Chisholm,G.M. (1979) Atherosclerosis, 32, 213-229

Lindeman,J.H.N., Van Zoeren-Gobben,D., Schrijver,J., Speek,A.J., Poorthuis,J.H.M., and Berger,H.M. (1989) Pediatr. Res. 26, 20-24

Miki,M., Motoyama,T., and Mino,M. (1989) Ann. N.Y. Acad. Sci. 570, 474-477

Morel,D.M., DiCorleto,P.E., and Chisolm,G.M. (1984) Arteriosclerosis, 4, 357-364

Motoyama,T., Miki,M., Mino,M., Takahashi,M., and Niki,E. (1989) Arch. Biochem. Biophys. 270, 655-661

Steinburg,D., Parthasarathy,S., Carew,T.E., Khoo.J.C., and Witztum,J.L. (1989) New Engl. J. Med. 320, 915-924

Takanaka,Y., Miki,M., Yasuda,H., and Mino,M. (1991) Arch. Biochem. Biophys. 285, 344-350

Wayner,D.D.M., Burton,G.W., Ingold,K.U., Barclay,L.R.C., and Locke, S.J. (1987) Biochem. Biophys. Acta 924, 408-419

Yoshida,M.C., Masuda,R., and Sasaki,M. (1987) J. Hered., 78, 361-365

Free Radicals: From Basic Science to Medicine
G. Poli, E. Albano & M. U. Dianzani (eds.)

EFFECT OF VITAMIN E AND OTHER ANTIOXIDANTS ON THE OXIDATION RESISTANCE OF HUMAN LOW DENSITY LIPOPROTEINS

H. Puhl, G. Waeg, F. Tatzber and H. Esterbauer

Institute of Biochemistry, University of Graz, Schubertstr. 1, A-8010 Graz, Austria

SUMMARY: Low density lipoprotein (LDL) can be oxidatively modified by vascular cells or transition metal ions like Cu^{2+}. Oxidatively modified LDL (oLDL) exhibits cytotoxic and chemotactic properties and leads to foam cell formation, which are critical steps in atherogenesis. Oxidative modification is a free radical induced lipid peroxidation process, in which the polyunsaturated fatty acids (PUFAs) in LDL are decomposed to a variety of reactive products such as aldehydes. The PUFAs are protected against oxidation by endogenous antioxidants, which consists of mainly tocopherols and to a lesser extend of ubiquinol-10 and carotenoids. The kinetics of oxidation shows three consecutive phases. In the lag phase the antioxidants are consumed, thereafter the PUFAs are oxidized to hydroperoxides (propagation phase). The hydroperoxides are only intermediates which are readily decomposed to aldehydes, accompanied by several changes in the LDL particle (decomposition phase). The lag phase can be used to define the oxidation resistance (OR) of different LDL samples. Interestingly, only a weak correlation was found between antioxidant content and OR. Increasing the α-tocopherol content in LDL by in vitro loading or in vivo supplementation, however always lead to a linear increase in the OR. Yet strong individual variations were observed for the efficacy of α-tocopherol to protect LDL, which probably explain that the vitamin E content alone is not predictive for the OR of an individual LDL. Other antioxidants not contained in LDL were also tested in vitro. All of them prolonged the lag phase, with probucol as the most efficient one.

INTRODUCTION

Several lines of evidence suggest, that oxidative modification of low density lipoprotein (LDL) plays a crucial role on the formation of atherosclerotic plaques (Steinberg et al. 1989). It is mainly based on the findings that vascular endothelial cells, smooth muscle cells or monocytes-macrophages can oxidatively modify LDL to a form which leads to formation of lipid-laden foam cells and exhibits cytotoxic as well as chemotactic properties (for review see Esterbauer et al. 1990). Oxidative modification of LDL is principally a free radical induced lipid peroxidation process, in which the polyunsaturated fatty acids in LDL are oxidized to hydroperoxides in a self-sustaining chain reaction. These hydroperoxides are decomposed to a great variety of products in secondary and tertiary transition metal ion catalyst reactions, including reactive aldehydes such as hexanal, 4-hydroxynonenal (HNE) or malonaldehyde (MDA). From several findings it is likely, that such reactive lipid peroxidation products derivatize amino acid residues of the apolipoprotein B (apo B), as for example ϵ-amino groups of lysine-residues, thereby creating new epitopes which are recognized by the macrophage scavenger receptor (Fig. 1). Ultimately, these modification are then responsible for the uncontrolled degradation of LDL and thus for the formation of foam cells (Jürgens et al. 1987, Steinbrecher et al. 1990).

Immunohistochemical methods indeed showed the presence of MDA or HNE conjugated LDL in lesions of WHHL rabbits (Palinski et al. 1990) and LDL extracted from arterial wall exhibits chemical and biological properties similar to oxidized LDL (Hoff et al. 1982). If lipid peroxidation and its products play an important role in the initial processes of atherogenesis, antioxidants should play a crucial role in preventing LDL from oxidation and its conversion into an atherogenic form.

THE OXIDATION RESISTANCE OF LDL

LDL with an average molecular weight of 2.5 million consists of a core with mainly cholesterylesters and triglycerides surrounded by a monolayer of phospholipids and free cholesterol. Embedded in the monolayer is the apoprotein B 100, which contributes to about 20% of the particle. Roughly half of the fatty acids in the LDL-lipids are polyunsaturated (Table I) and thus prone to free

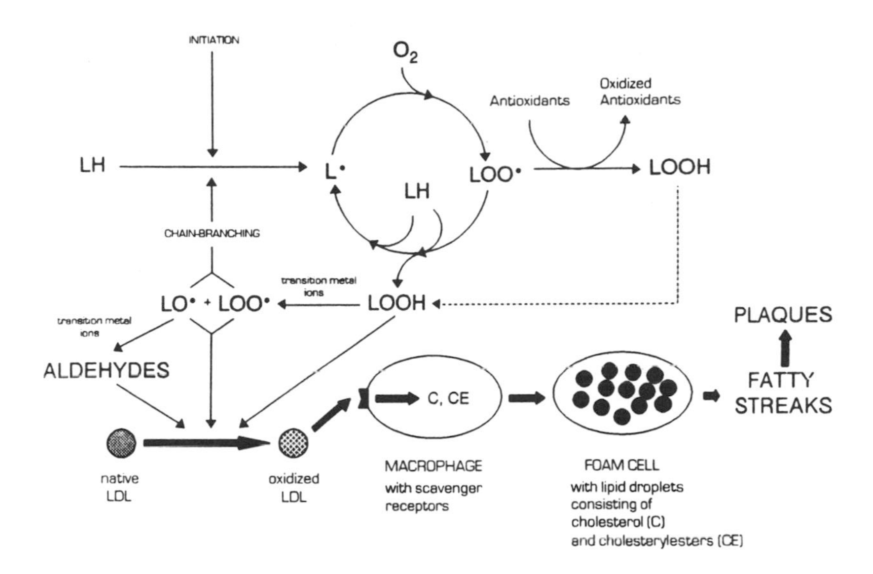

Fig. 1. Scheme showing the major events of LDL oxidation and formation of foam cells. LH = an LDL lipid containing PUFAs; LOO· = lipid peroxyl radical; LO· = lipid alkoxy radical.

radical attack. Protection of these PUFAs is conferred by a great variety of lipophilic antioxidants, which are in a ratio of about 1:150 to the PUFAs. By far the major one is α-tocopherol, with an average content of 6 mol/mol LDL. Other substances with potential antioxidant activity are present in amounts of only about 1/20 to 1/300 of α-tocopherol (Table I). In vitro oxidation of LDL by incubation with macrophages (Jessup et al. 1990), exposure to copper ions or free radical generators like AAPH is always pre-ceeded by a lag phase, during which the endogenous antioxidants of LDL are consumed with ubiquinol-10 followed by α-tocopherol as the first and β-carotene as the last one (Esterbauer et al. 1989, Stocker et al. 1991). When the antioxidants in LDL are exhausted, lipid peroxidation rapidly commences (propagation phase). The kinetics of in vitro oxidation induced by copper ions or free radical generators can be followed as the relative increase in conjugated dienes at 234nm (Esterbauer et al. 1989). From the obtained diene versus time profile the lag phase can be easily determined and used as a marker for the specific oxidation resistance of a particular LDL sample. The mean lag phase in our system of copper-induced LDL oxidation (with a molar ratio of LDL to copper-ions of 1:16) is 69±19.4 minutes, with a range of 33 to 138 minutes (n=121). The sequence of events occurring in the early stage of oxidation suggests, that the oxidation resistance of LDL is dependent on its antioxidant content.

Table 1. Basal values of polyunsaturated fatty acids, antioxidants and lag phases determined for human LDL samples.
This is an updated version of a previous report (Esterbauer et al. 1991a) with a larger number of n. The molecular mass of LDL was assumed to be 2.5 million. All values are in mol/mol LDL, except the lag phase is given in minutes.

	n	mean	SD	range
linoleic acid	31	1101	±298	680 - 1832
arachidonic acid	31	153	±55	48 - 250
docosahexaenoic acid	15	29	±17	15 - 62
alpha-tocopherol	149	7.26	±2.52	2.90 -15.74
gamma-tocopherol	149	0.56	±0.24	0.07 - 1.70
beta-carotene	122	0.29	±0.26	0.03 - 1.87
alpha-carotene	28	0.12	±0.14	0.02 - 0.52
lycopene	136	0.16	±0.11	0.03 - 0.70
cryptoxanthin	114	0.14	±0.13	0.03 - 0.70
cantaxanthin	53	0.02	±0.04	0.01 - 0.24
lutein+zeaxanthin	113	0.04	±0.03	0.01 - 0.16
phytofluene	10	0.05	±0.03	0.02 - 0.11
ubiquinol-10	7	0.10	±0.10	0.03 - 0.35
lag phase	121	69	±19	33 - 138

EFFECT OF α-TOCOPHEROL ON THE OXIDATION RESISTANCE

Since α-tocopherol is quantitatively the most prominent antioxidant in LDL, its relationship to the lag phase was investigated.

A statistical treatment of the α-tocopherol values with the corresponding lag phases by linear regression analysis revealed no correlation between these parameters (r^2=0.043, n=78). The same was found using total antioxidants content instead of α-tocopherol. This finding is in accordance to the results obtained by Babiy et al., who found no correlation between α-tocopherol content and TBARS formation in gamma-irradiated LDL (Babiy et al. 1990). In order to investigate the relation of α-tocopherol to the oxidation resistance in individual LDL samples, LDL was loaded in vitro with α-tocopherol. For that, plasma of individual donors was incubated with different amounts of α-tocopherol (up to 1mM, for 3 hours at 37°C under nitrogen atmosphere) before isolation of LDL (Esterbauer et al 1991b). Thus the α-tocopherol content of LDL from given individuals could be raised several-fold. The oxidation resistance, measured as lag phase of the LDL samples increased always linearely with the α-tocopherol content (Fig. 2).

For all donors investigated so far, this increase was linear depending on the amount of α-tocopherol but varied in its extent. The correlation for each individual LDL can be described by the equation:

$$\mathbf{y = kx+a},$$

where **y** is the oxidation resistance of LDL measured as lag phase in minutes and **x** is the α-tocopherol content of LDL in mol/mol. The slope **k** gives the efficiency of α-tocopherol, i.e. the lag phase produced by one molecule α-tocopherol per LDL particle, whereas the ordinate intercept **a** is an α-tocopherol independent parameter.

A similar relation was found in an ex vivo study after oral supplementation with different doses of 150, 225 800 and 1200 IU RRR-α-tocopherol (Dieber-Rotheneder et al. 1991). Compared to baseline values before and after supplementation, the oxidation resistance increased closely related to the change in α-tocopherol content during three weeks of supplementation. For single probands, the changes were linear correlated according to the

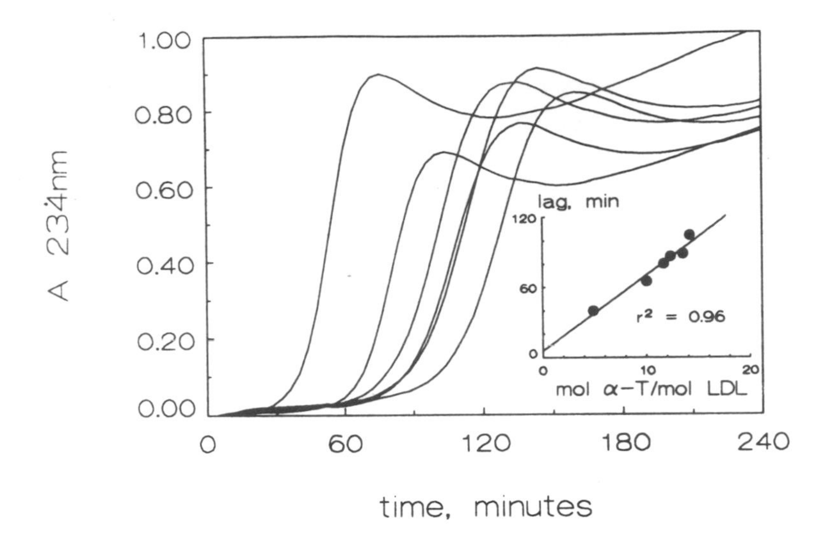

Fig. 2. Determination of the efficacy of α-tocopherol to increase the oxidation resistance of LDL.
The curves show the diene vs. time profiles of 0.25mg total LDL/ml in oxygen saturated PBS oxidized by addition of 1.67μM Cu^{2+}. The insert shows the linear relation between lag phase and α-tocopherol in LDL as estimated by HPLC.

equation given above, with r^2 ranging from 0.56 to 0.95 (Dieber-Rotheneder et al. 1991). Independently how the α-tocopherol content in LDL was increased, whether by in vitro loading or oral supplementation, the values for **k** and **a** showed vary high individual variations with a mean±SD value for **k** of 5.51±4.75 (range 0.7 to 24.8, n=65), and **a** of 35.6±33.3 minutes (range -68.6 to 108.6, n=65). Due to the strong variations of these parameters, the α-tocopherol content of a distinct LDL sample is per se not predictive for the oxidation resistance. **k** and **a** seem to be subject specific parameters, LDL from persons with a large value for **k** but a low or even negative value for **a** for instance, is solely protected by α-tocopherol, whereas low **k** and high **a** values are indicative for additional protection mechanisms. Nevertheless, statistical treatment of all data estimated for α-tocopherol (basal values, in vitro loaded and in vivo supplemented) and the associated lag phases gave a linear correlation with r^2 of 0.46 (n=206, p<0.001), as shown in Fig. 3.

EFFECT OF OTHER ANTIOXIDANTS ON THE OXIDATION RESISTANCE

Other, water and lipid soluble antioxidants not contained in LDL, were also tested. Lipid soluble antioxidants like the vitamin E

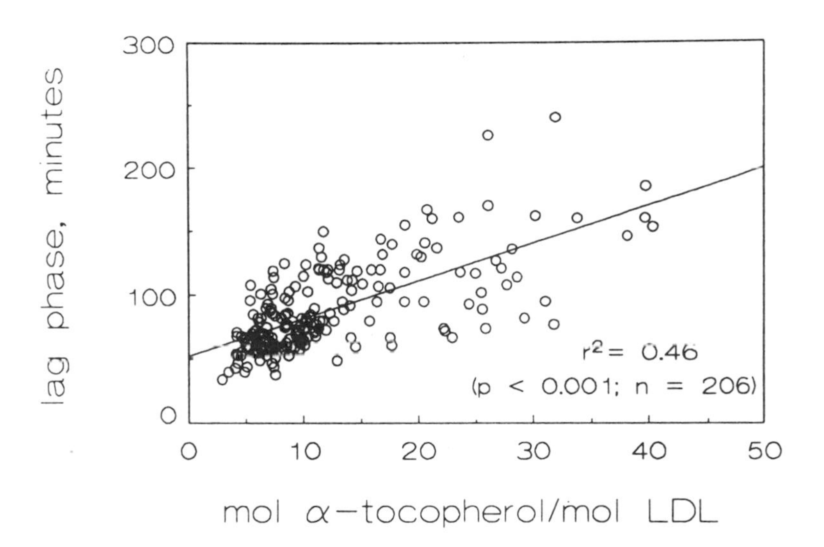

Fig. 3. Statistical relation between the oxidation resistance of LDL and the α-tocopherol content.
In this plot all estimated data for α-tocopherol (basal values, in vitro and in vivo supplemented) and their corresponding lag phases are included.

analogues α- and gamma-tocotrienol or the cholesterol lowering drug probucol were added by incubation of plasma with different concentrations of 0.1 to 1mM before isolation of LDL, as for α-tocopherol. The water soluble antioxidants ascorbate, urate, glutathione and the water soluble vitamin E analogue trolox C were added in various concentrations (up to 100μM) directly to the LDL solution before oxidation.

All of them increased the lag phase dependent on their concentration but with different efficiency, compared on a molar base. The efficiency was calculated in analogy to α-tocopherol from the slope of a linear regression (Table II).

Table 2. Effectiveness of different antioxidants to prolong the lag phase.
The values represent the slope of the regression lines from the correlation of antioxidant concentration against their corresponding lag phases.

Antioxidant	prolongation of lag phase in minutes per 1mol/mol LDL
<u>water soluble</u>	
ascorbate	2.1
urate	6.9
glutathione	3.0
(up to 100mol/mol LDL)	
trolox C	4.2
<u>lipid soluble</u>	
alpha-tocopherol	4.5
alpha-tocotrienol	5.9
gamma-tocotrienol	4.8
probucol	11.7

From the lipid soluble antioxidants probucol was about twice as effective as α-tocopherol to prolong the lag phase, whereas the tocotrienols showed comparable efficiency to α-tocopherol. It must be noted however, that also these antioxidants should be investigated for a possible different, donor specific effectiveness. In case of probucol, analysis of the kinetics of antioxidant disappearance revealed, that in distinction to ascorbate (Esterbauer et al. 1991b) or trolox C vitamin E in LDL is not spared by probucol. Only after vitamin E is markedly reduced also probucol is lost, so that the LDL is protected solely by probucol (Fig. 4). This is in accordance to the findings of Jialal et al. (Jialal et al. 1991), who also showed that addition of probucol did not prevent oxidation of α-tocopherol in LDL.

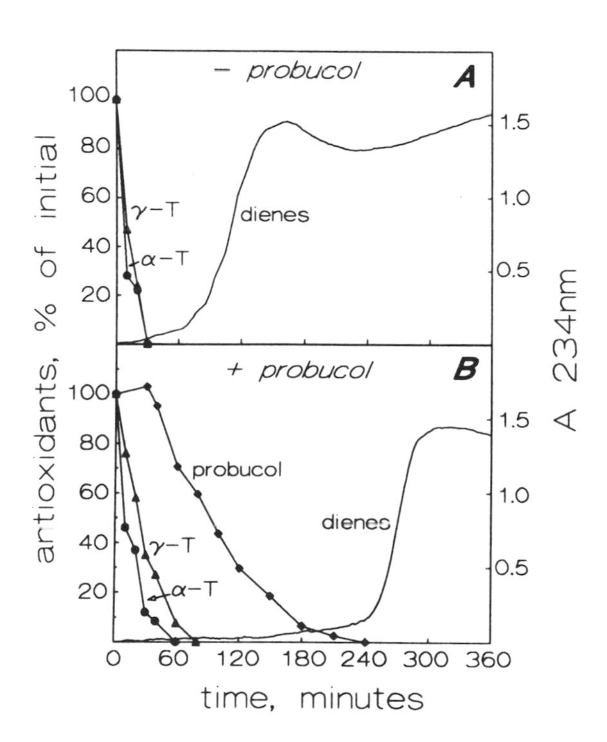

Fig. 4. Kinetics of the consumption of vitamin E and probucol and formation of conjugated dienes.
0.5mg total LDL/ml oxygen saturated PBS were oxidized by addition of 3.37 μM Cu^{2+}. At the indicated time points an aliquot was withdrawn and analyzed for antioxidants by HPLC. A) control, B) after in vitro loading with probucol.

CONCLUSIONS

Although the endogenous antioxidants of LDL are consumed during the lag phase, the length of the lag phase (and thus the oxidation resistance) cannot be predicted from the antioxidant content. If however the content of α-tocopherol, the quantitatively major antioxidant in LDL, is increased either by in vitro loading or oral supplementation, for each individual LDL a highly significant linear correlation between oxidation resistance and α-tocopherol content is obtained. Nevertheless, there are strong individual variations regarding the efficiency of α-tocopherol to increase the oxidation resistance (**k**) as well as in respect to the vitamin E independent component **a** of the oxidation resistance. Preliminary results suggests, that these parameters are characteristic individual constants, which need further investigations to elucidate there nature. The contribution of the other potential antioxidants in LDL to the factor **a** seems to be rather low, since their total concentration is less than one third of α-tocopherol (Table I), provided that there are no additional mechanisms, which increase the capacity of an individual antioxidant. Addition of antioxidants not contained in LDL leads also to an increase in oxidation resistance, whether they were water or lipid soluble. There mode of action however is different, since probucol for instance did not spare vitamin E in LDL.

ACKNOWLEDGEMENT: The author's work has in part been supported by the Association for International Cancer Research, AICR, U.K.

REFERENCES

Babiy, A., Gebicki, J.M., and Sullivan, D.R. (1990) Atherosclerosis 81, 175-182.
Dieber-Rotheneder, M., Puhl, H., Waeg, G., Striegl, G., and Esterbauer, H. (1991) J. Lipid Res. 8, 1325-1332.
Esterbauer, H., Dieber-Rotheneder, M., Waeg, G., Striegl, G., and Jürgens, G. (1990) Chem. Res. Toxicol. 3, 77-92.
Esterbauer, H., Puhl, H., Dieber-Rotheneder, M., Waeg, G., and Rabl, H. (1991a) Ann. Med. 23, 573-581.
Esterbauer, H., Rotheneder, M., Striegl, G., and Waeg, G. (1991b) Am. J. Clin. Nutr. 53, 314S-321S.
Esterbauer, H., Striegl, G., Puhl, H., and Rotheneder, M. (1989) Free Rad. Res. Comms. 6, 67-75.

Hoff, H.F., and Gaubatz, J.W. (1982) Atherosclerosis 42, 273-297.
Jessup, W., Rankin, S.M., DeWhalley, C.V., Hoult, J.R.S., Scott, J., and Leake, D.S. (1990) Biochem. J. 265, 399-405.
Jialal, I., Grundy, S.M. (1991) J. Clin. Invest. 87, 597-601.
Jürgens, G., Hoff, H., Chisholm, G.M., and Esterbauer, H. (1987) Chem. Phys. Lipids 45, 315-336.
Palinski, W., Ylä-Herttuala, S., Rosenfeld, M.E., Butler, S.W., Socher, S.A., Parthasarathy, S., Curitss, L.K., and Witztum, J.L. (1990) Arteriosclerosis 10, 325-335.
Steinberg, D., Parthasarathy, S., Carew, T.E., Khoo, J.C., and Witztum, J.L. (1989) N. Engl. J. Med. 320, 915-924.
Steinbrecher, U.P., Zhang, H., and Lougheed, M. (1990) Free Rad. Biol. Med. 9, 155-168.
Stocker, R., Bowry, V.W., and Frei, B. (1991) Proc. Natl. Acad. Sci. 88, 1646-1650.

Free Radicals: From Basic Science to Medicine
G. Poli, E. Albano & M. U. Dianzani (eds.)

APPROACHES TO THE THERAPY OF GLUTATHIONE DEFICIENCY

A. Meister

Cornell University Medical College, 1300 York Avenue, New York, N.Y. 10021, USA.

SUMMARY The glutathione levels of tissues are tipically decreased in response to oxidative stress induced by treatment with radiation, hyperoxia and various chemical compounds (including drugs) but these modalities are generally non-specific and thus affect many cell structures as well as glutathione. An essentially "pure" glutathione deficiency may be produced by administration to animals of a selective transition-state inhibitor of gamma-glutamylcysteine synthetase (such as buthionine sulfoximine). Inhibition of this enzyme, which catalyzes the first step in the sinthesis of glutathione, produces markedly decreased cellular levels of glutathione because export of glutathione continues in the absence of intracellular glutathione synthesis. Such glutathione deficiency, which leads to oxidative tissue damage, may be ascribed to the very substantial normal physiological endogenous formation of reactive oxygen intermediates and free radicals. This model of oxidative stress thus differs significantly from those produced by exogenous application of stress.
The effects of glutathione deficiency induced in this way are invariably associated with major destruction of mitochondria, which do not sinthesize glutathione but which transport it from cytosol. It has long been known that mitochondria (which lack catalase) normally convert a significant fraction of the oxygen that they use to hydrogen peroxide. This is destroyed by the action of glutathione peroxidases. Severe glutathione deficiency in adult mice is associated with significant damage to skeletal muscle, lung type 2 cells, lymphocytes and epithelial cells of the jejunum and colon. In newborn rats and mice glutathione deficiency leads to formation of cataracts and brain dysfunction associated with striking enlargement and degeneration of cerebral cortex mitochondria. Proximal renal tubular degeneration and hepatic damage also occur. A diastereoisomer of buthionine sulfoximine that does not inhibit glutathione synthesis (L-buthionine-R-sulfoximine) does not produce toxicity, rendering it unlikely that the effects observed after giving L-buthionine-SR-sulfoximine are produced by the sulfoximine moiety itself.
It is important to note that these effects are not prevented or reversed by administration of glutathione (orally or parenterally). However they are prevented by administration of glutathione

monoesters, which are readily transported into virtually all cells and split intracellularly to glutathione. Glutathione deficiency leads to marked decrease in tissue ascorbate levels, and administration of ascorbate, which spares glutathione, decreases mortality in glutathione deficient newborn rats and prevents tissue damage in newborn and older animals. There is linkage between the antioxidant actions of glutathione and ascorbate; thus the oxidative damage associated with glutathione deficiency can be prevented by either ascorbate or glutathione (administered as glutathione monoester).
Other approaches to the modulation of glutathione metabolism have led to potentially useful therapies that increase cellular glutathione levels. Human disorders, such as those associated with ischemia in its various forms, drug-induced toxicities, certain virus-induced and respiratory conditions, and prematurity, may be associated with glutathione deficiency. Application of approaches derived from the study of glutathione metabolism to the therapy of such disorders is worth serious consideration.

INTRODUCTION

Because several human afflictions seem to be associated with oxidative stress there has been interest in the nature of this phenomenon, and especially in the factors that normally control oxidative processes. It is unlikely that conditions such as ageing, cancer, atherosclerosis, arthritis, etc., are solely the result of oxidative stress. Rather it is more probable that the various undelying pathological processes involved in these conditions affect the normal cellular oxidative balance so as to produce oxidative stress. Oxidative damage produced by various mechanisms may lead to extensive morbidity involving abnormalities which further increase oxidative damage. These considerations suggest that suitable antioxidant therapy may be useful for 0the control of the oxidative component of various disease processes. Glutathione is the dominant antioxidant of most mammalian cells (Meister & Anderson, 1983). Therapeutic approaches based on our knowledge of glutathione metabolism are considered in this paper.

Several methods have been used for the production of experimental oxidative stress. These include application of increased oxygen (hyperoxia), radiation, and various agents that lead to oxidation. The last mentioned includes compounds such as hydroperoxides (e.g., t-butyl hydroperoxide), thiol oxidizing agents (e.g., diamide), and thiol reactive agents (e.g., diethylmaleate, phorone) (Plummer et al., 1981). Use of such compounds generally leads to markedly decreased cellular levels of glutathione, but because these compounds are non-specific in their action, many cell constituents (thiol containing and other) are affected (reviewed in Meister, 1991). In the course of research in our laboratory on glutathione biochemistry, we have developed a way of producing oxidative stress that is associated with a essentially "pure" form of glutathione deficiency in which other cell components are, at least initially, largely unaffected (reviewed in Meister, 1991).

In this model of glutathione deficiency experimental animals (or cell suspensions) are treated with buthionine sulfoximine, a selective transition state inhibitor of gamma-glutamylcysteine sinthetase. [Certain other aminoacid sulfoximine may also be used, but methionine sulfoximine is unsuitable because, although inhibits gamma-glutamylcysteine sinthetase effectively, methionine sulfoximine also inhibits glutamine synthetase and inhibition of this enzyme in the brain leads to convulsions and death (Meister, 1978)]. Inhibition of gamma-glutamylcysteine synthetase, which catalyses the first and controlling step of glutathione synthesis, produces markedly decreased levels of cellular glutathione because export of glutathione continues in the absence of intracellular glutathione synthesis. Such glutathione deficiency often leads to oxidative tissue damage, which may be ascribed to the very substantial normal physiological endogenous formation of reactive oxygen intermediates and free radicals. This model of oxidative stress (Martensson et al., 1991), which differs significantly from those produced by exogenous application of stress, may constitute a good mimic of certain human degenerative conditions and diseases.

EFFECTS OF GLUTATHIONE DEFICIENCY PRODUCED BY INHIBITION OF GAMMA-GLUTAMYLCYSTEINE SYNTHETASE.

Of the several known selective inhibitors of gamma-glutamylcysteine synthetase, L-buthionine-SR-sulfoximine (Griffith et al., 1979; Griffith & Meister, 1979; Griffith, 1982) has been the most widely used; it is commercially available. Buthionine sulfoximine is itself not highly reactive chemically. Of the four diastereoisomers of buthionine sulfoximine, only the L-S-isomer is active in inhibiting gamma-glutamylcysteine synthetase. The L-R-isomers of buthionine sulfoximine has little or no effect on gamma-glutamylcysteine synthetase and has therefore been useful as a control compound. Administration of buthionine sulfoximine to experimental animals such as adult rats and mice leads to glutathione deficiency in most tissues (Meister, 1991). Depletion of cellular glutathione by treatment with buthionine sulfoximine sensitizes cells to the toxic effect of substances such as mercuric ions, cadmium ions, melphalan, cisplatin, morphin, and others, as well as to radiation. Such effects of buthionine sulfoximine treatment are potentially useful for the treatment of certain drug- and radiation- resistant tumors, which have high capacity for glutathione synthesis and high levels of this tripeptide (Meister & Griffith, 1979; Meister, 1988; Vistica & Ahmad, 1989; Ozols et al., 1988; Meister, 1986). Depletion of glutathione is therapeutically effective because the tumors have an increased requirement of glutathione as compared to normal tissues which usually have a large excess of glutathione. In recent clinical trials, relatively low doses of buthionine sulfoximine were found to produce significant depletion of glutathione in the peripheral circulating leukocytes of cancer patients without producing toxicity (Hamilton, 1990). It is hoped that depletion of glutathione in tumor cells will be achieved thus resensitizing these tumor cells to anticancer agents, as it has been observed in experimental models.

When larger doses of buthionine sulfoximine are given to

adult mice and newborn rats, severe tissue glutathione deficiency is produced and this leads to cellular oxidative damage. These effects are invariably associated with substantial damage to mitochondria. Mitochondria do not synthetize glutathione, but transport it from the cytosol (Griffith & Meister, 1985; Martensson et al., 1990). It has long been known that mitochondria, which lack catalase, normally convert a significant fraction of the oxygen that they use to hydrogen peroxide (Boveris & Chance, 1973). This is destroyed by the action of glutathione peroxidases. Severe glutathione deficiency in adult mice is associated with significant damage to skeletal muscle (myofiber degeneration), lung type 2 cells (lamellar body disintegration), lymphocytes, and the epithelial cells of the jejunum and colon (Martensson & Meister, 1989; Martensson et al., 1989; Martensson et al., 1990). In newborn rats and mice glutathione deficiency leads to formation of cataracts (Calvin et al., 1986; Martensson et al., 1989). It also produces brain disfunction associated with striking enlargement and degeneration of cerebral cortex mitochondria (Jain et al., 1991), proximal tubular degeneration and hepatic damage (focal necrosis)(Martensson et al., 1991; Martensson & Meister, 1991). L-buthionine-R-sulfoximine, which does not inhibit glutathione synthesis, does not produce toxicity (Martensson et al., 1991; Suthanthiran et al., 1990) rendering it unlikely that the effects observed after giving L-buthionine-SR-sulfoximine are produced by the sulfoximine moiety itself.

THERAPY THAT PREVENTS OR REVERSES THE EFFECTS OF GLUTATHIONE DEFICIENCY.

Severe glutathione deficiency and the associated oxidative stress produced as described above by administration of buthionine sulfoximine, can be prevented or reversed by the administration of glutathione mono(glycyl)esters such as monoethylesters. Such glutathione esters are readily transported into virtually all cells and are split intracellularly to form glutathione (Puri & Meister, 1983; Wellner et al., 1984; Anderson et al., 1985; Anderson and Meister, 1989). In striking contrast, glutathione itself is poorly transported into cells (Meister, 1991). Administration of glutathione, orally or parenterally, does not lead to significantly increased levels of glutathione in tissues, provided that glutathione synthesis is blocked. Nor does it prevent the destructive cellular effect of glutathione deficiency found after giving buthionine sulfoximine. In adult mice and rats, oral administration of glutathione does not lead to significant increased levels of glutathione in the portal blood (Martensson et al., 1990) but cysteine levels are greatly increased (Viña et al., 1989). Intraperitoneal administration of glutathione to adult mice treated with buthionine sulfoximine increases blood levels of glutathione substantially (i.e., from 20-40 µM to 5-20 mM) without significant effect on tissue glutathione levels (Meister, 1983). Although there appears to be little uptake of intact glutathione from the intestinal lumen into the blood plasma under normal conditions, it is conceivable that some uptake of glutathione might occur after giving larger oral doses of glutathione. Although parenteral administration of glutathione increases plasma glutathione levels, the potential practical importance of this approach is limited by the poor transport of glutathione from plasma into cells. The physiological utilization of plasma glutathione involves extracellular degradation of glutathione followed by transport of the resulting aminoacids and dipeptides into cells followed by intracellular glutathione synthesis. Dietary glutathione is degraded within the alimentary tract leading to formation of dipeptide and aminoacids which are absorbed. Although dietary glutathione can thus be used by animals, it is doubtful that significant amounts of intact glutathione are taken up.

Glutathione provides a major source of reducing power for the conversion of dehydroascorbate to ascorbate (Fig. 1). Glutathione itself, or via ascorbate, appears to be the major physiological agent that maintains α-tocopherol and probably other tissue antioxidants in their reduced forms. This has been clearly shown in the case of ascorbic acid.

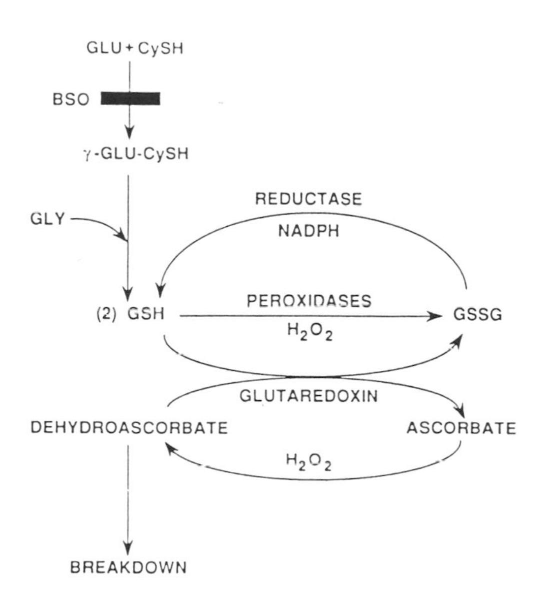

Fig. 1. Interactions involving the antioxidant activities of glutathione (GSH) and ascorbate (Martensson, 1990). Both glutathione and ascorbate act to destroy hydrogen peroxide (and related active oxygen forms). Enzymatic reduction of GSSG (glutathione disulfide) restores glutathione. Dehydroascorbate is reduced to ascorbate by glutathione-dependent thiol transferases (glutaredoxin and protein disulfides isomerase activities). Dehydroascorbate, if not reduced, is readily degraded irreversibly. Glutathione synthesis requires the activity of the two ATP-dependent enzymes, gamma-glutamylcysteine synthetase and glutathione synthetase. The first and rate-controlling step of glutathione synthesis is selectively inhibited by buthionine sulfoximine (BSO).

Thus, glutathione deficiency in newborn rats and adult mice is associated with markedly decreased levels of ascorbate (Martensson et al., 1991; Martensson & Meister, 1991). Oxidative stress, as evaluated by mithocondrial damage (electron microscopy) can be prevented by treatment with glutathione esters or ascorbate. Newborn rats, which like guinea pig and humans, do not appear to synthetize ascorbate, are very sensitive to the effects of glutathione deficiency. Severe glutathione deficiency induced in newborn rats and guinea pigs by giving buthionine sulfoximine leads to high mortality; such severe manifestations of oxidative stress is prevented by administration of glutathione esters or by ascorbate. Notably the levels of glutathione in the tissue (and their mitochondria) of newborn rats treated with buthionine sulfoximine were substantially increased when ascorbate was also given, indicating that ascorbate has a significant sparing effect on glutathione. These antioxidants, whose actions are closely associated (Fig. 1), function together in preventing oxidative stress. The findings indicate that glutathione and ascorbate have action in common in the destruction of reactive oxygen species (Meister, 1992). Glutathione itself reacts directly with reactive oxygen compounds thus functioning in the reduction of hydrogen peroxide and organic peroxides in reactions catalysed by the glutathione peroxidases. Ascorbate can also interact with the peroxides and with other reactive oxygen forms. Some reactions of this type may occur nonenzymatically, but the interaction of hydrogen peroxide with ascorbate is enzyme catalysed in chloroplast, cyanobacteria and soybean nodules (Dalton et al., 1986); an analougous reaction catalysed by ascorbate peroxidase may occur in mammalian tissues. It is of related interest that development of scurvy in guinea pig fed a scorbutic diet is significantly delayed by administration of glutathione monoethyl esters (Han et al., 1992). Thus, the appearance of scorbutic bone changes and hematomas in guinea pigs fed a scorbutic diet was greatly retarded by administration of glutathione esters, and the tissue levels of both ascorbate and glutathione were increased. The metabolic redundancy involved in the overlapping functions of glutathione and ascorbate emphasizes the physiological importance of this antioxidant system.

HUMAN DISORDERS THAT INVOLVE OR MAY INVOLVE GLUTATHIONE DEFICIENCY. POSSIBLE THERAPEUTIC INTERVENTION.

The experimental studies discussed above in which ascorbic acid was found to protect newborn rats and adult mice against the effects of glutathione deficiency suggest that therapy with ascorbic acid would be useful in the treatment of humans that have inborn errors associated with deficiency of glutathione synthesis. Such patients may have inherited deficiency of gamma-glutamylcysteine synthetase or inherited "mild or severe" deficiency of glutathione synthetase (Meister & Larsson, 1989). Treatment of patients with compounds such as glutathione monoesters would also appear to be appropriate, but this has not yet been attempted. However, studies have been carried out with cultured fibroblasts obtained from the patients (Wellner et al., 1984); in these studies, suspension of the cells in media containing glutathione monoesters led to substantial increases in the cellular level of glutathione.

Therapy with ascorbate or glutathione monoesters or L-2-oxothiazolidine-4-carboxylate (see below) may also be useful in the treatment of premature infants who have deficiency of glutathione associated with lack of cystathionase (Sturman et al., 1970). It has recently found that pre-term infants have much lower levels of peripheral blood plasma total glutathione than do full-term infants (Jain et al., 1992). The possibility that glutathione deficiency may also occurs in newborns needs to be further investigated; such deficiency is associated with respiratory distress or related symptoms may appropriately treated with ascorbate glutathione esters or precursors of cysteine.

The ascorbic acid status of patients receiving buthionine sulfoximine therapy for tumors should be monitored. It would be expected that treatment with ascorbate would reverse the effects produced by buthionine sulfoximine. Ascorbate treatment would appear to be a logical therapy for patients exhibiting signs of glutathione deficiency produced by treatment with buthionine sulfoximine. Ascorbate is therefore a potentially rescue therapy.

Only a few data are now available about the glutathione status in various human disease. It seems possible that glutathione deficiency may occur in liver disease and in other conditions, "for example", adult respiratory distress syndrome. It has been found that glutathione deficiency in adult mice leads to lung type 2 cell lamellar body and mitochondrial damage (Jain et al., 1992). These effects are associated with marked decrease of the levels of phosphatidylcholine in the lung and in the bronchoalveolar lining fluids. Treatment of these animals with ascorbate prevented damage to lamellar bodies and the decline of phosphatidylcholine levels in lung and alveolar lining fluids. These experimental findings indicate that glutathione deficiency leads to depletion of lung surfactant and also that such depletion can be prevented by administration of ascorbate. The finding that premature infant who have respiratory distress often have a deficiency of surfactant (Adams et al., 1979), when considered in relation to these experimental studies, suggest that treatment with ascorbate in relatively high doses may have clinical usefulness.

Apart form conditions in which there is a clear indications of glutathione deficiency, treatment with compounds that increase cellular glutathione levels (or with reducing agents) may also be effective for conditions or diseases that are considered to be associated with the effects of the oxidative stress or free radical damage; such conditions include ageing, cancer, atherosclerosis, viral infections such as AIDS, ischemic phenomena, and others. The findings of decreased levels of glutathione in the venous plasma and lung epithelial lining fluid of symptom-free HIV seropositive individuals as compared to normal individuals (Buhl et al., 1989), and similar earlier findings on peripheral blood monocytes and lymphocytes and on plasma (Eck et al., 1979) have attracted attention because of the possibility that there may be a generalized deficiency of glutathione in AIDS. There is currently much interest and activity in this field, and several clinical trials of compounds that increase cellular glutathione levels (e.g. L-2-oxothiazolidine-4-carboxilate and N-acetyl-L-cysteine) are in progress. Treatment with very large doses of ascorbate is also being tried.

It is relevant to mention the finding (Meister, 1983; Suthanthiran et al., 1990) that treatment of peripheral blood monocytes with buthionine sulfoximine decreases lecitine-activated proliferation of T-cells. These and related findings (Suthanthiran et al., 1990; Hamilos et al., 1985) indicate that glutathione is required for proliferation and support the suggestion (Suthanthiran et al., 1990, Buhl et al., 1989) that treatment of patients with AIDS with compounds that increase cellular glutathione may be beneficial.

Utilization of administered glutathione for formation of intracellular glutathione, as discussed above, involves degradation of glutathione, transport of the product formed, and intracellular synthesis of glutathione. More effective approaches involve treatment with compounds that directly supply cysteine and gamma-glutamylcysteine, which are substrates, respectively, of gamma-glutamylcysteine synthetase and glutathione synthetase. L-2-Oxothiazolidine-4-carboxylate and N-acetyl-L-cysteine (Anderson & Meister, 1987) are cysteine delivery compounds that are effective in experimental systems; the former appears to be more effective then the latter (Williamson & Meister, 1987; Williamson et al., 1982). L-2-Oxothiazolidine-4-carboxylate is converted to L-cysteine by the widely distributed intracellular enzyme 5-oxoprolinase. Conversion of N-acetyl-cysteine to cysteine occurs *in vivo*; the site or sites at which the conversion occurs needs further study. Treatment with glutathione monoesters [or other glutathione delivery agents] are also potentially useful and may be more effective than treatment with precursors of the amino acid constituents of glutathione. Thus, treatment with glutathione monoesters leads to direct increase of intracellular glutathione levels without expenditure of cellular energy or requirement for the synthetases.

In addition to the studies which suggest that there is a deficiency of glutathione in AIDS, several in vitro studies suggest that cellular thiols, especially glutathione, might play an important role in regulation of expression of the gene controlled by human immunodeficiency virus (HIV LTR)(Duh et al., 1989; Staal et al., 1990; Roederer et al., 1990; Toledano & Leonard, 1991; Kalebich et al., 1991). For example, stimulation of expression of HIV by tumor necrosis factor α was inhibited by thiols. In studies on a cronologically infected human monocyte cell lyne, the enhancement of virus formation by various inducers was suppressed by glutathione, glutathione monoethylester and N-acetyl-cysteine. Tumor necrosis factor α stimulates transcription of HIV by activating the nuclear factor kB (NF-kB), and such stimulation is inhibited by N-acetyl-cysteine (Toledano & Leonard, 1991). It has been suggested that modulation of the redox state of this factor might represent a post-translational control mechanism. Although these in vitro studies may not necessarily reflect phenomena that can be similarly influenced in vivo, they provide additional support for the idea of testing thiols and related compounds in AIDS.

It has been found that glutathione monoesters protect animals against toxicity due to cadmium ions (Singhal et al., 1987), mercuric ions (Naganuma et al., 1990), and cisplatin (Anderson et al., 1990). Glutathione monoethylester almost completely protected animals against toxicity due to monocrotaline (Teicher et al., 1988), and good protection was found against toxicity to 1,3-bis(2-chloroethyl)-1-nitrosourea and to cyclophosphamide. These and similar observations led to the suggestion that glutathione esters might protect against the toxicity of drugs commonly used for therapy of patients with AIDS (Sunthanthiran et al., 1990). It is of much potential significance that L-2-oxothiazolidine-4-carboxylate was found in experimental studies on mice to protect against AZT-induced hematotoxicity (Willson et al., 1992). Since AZT-induced anemia is a common and difficult clinical problem, approaches involving use of thiols as protectants are clearly of great interest, and should be pursued.

Another potentially valuable use of compounds that increase cellular glutathione levels to tissue ischemia, a complication of atherosclerosis and of surgery that may affect the brain, heart, or other organs. Some preliminary studies (see for example, Maeno et al., 1989) have been reported.

Recent studies suggest that oxidation of low density lipoprotein (LDL) is a key step in the development of atherosclerosis (Parthasarathy et al., 1990; Sparrou et al., 1988; Quinn et al., 1987; Hylä-Herttuala et al., 1989). Although the oxidized LDL hypothesis needs further exploration clinical and animal studies have led to the finding of several compounds whose administration leads to decreased oxidized LDL formation; these include probucol (Schwartz, 1988), N,N'-diphenyl-1,4-phenylenediamine (Sparrow et al., 1992) and ascorbic acid (Jialal & Grundy, 1991). There is evidence from in vitro studies suggesting that glutathione may play a significant role in the protection of cells from the effect oxidized LDL (Kuzuya et al., 1989).

DISCUSSION

Glutathione, which has many functions in cellular metabolism (Meister & Anderson, 1983; Meister, 1991; Larsson et al., 1983; Dolphin et al., 1989; Taniguchi et al., 1989), protects cell from the toxic effects of oxygen and is an important components of the system that uses reduced pyridine nucleotide to provide cells with their reducing properties thus promoting, for example, intracellular formation of cysteine (from cystine) and the thiol forms of proteins. Glutathione is an antioxidant of major significance because it not only can function in the destruction of reactive oxygen compounds, but because it also functions to maintain the reduced forms of such compounds as ascorbate, α-tocopherol and a number of others (Fig. 2).

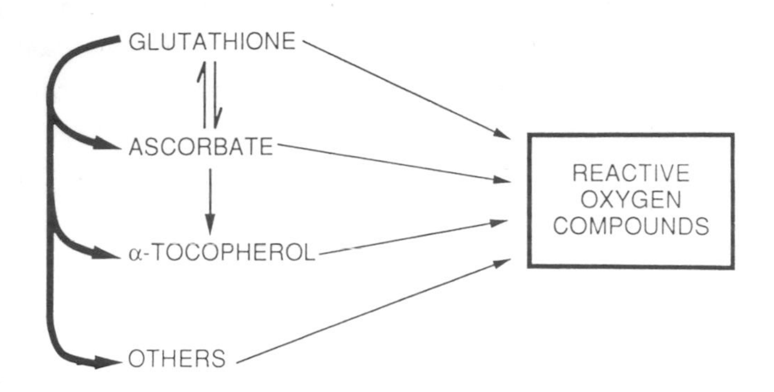

Fig. 2. The Glutathione Antioxidant System. Glutathione functions directly as an antioxidant as substrate for the glutathione peroxidases (selenium-containing ant other), and also by reducing dehydroascorbate to ascorbate. Glutathione functions itself, or via ascorbate,in the mainteinance in reduced forms of α-tocopherol and other antioxidant compounds.

The model of oxidative stress discussed here is based on inhibition of glutathione synthesis by buthionine sulfoximine (Meister, 1991; Martensson et al., 1991).

In this model, oxidative stress is of endogenous origin. Thus, the observed tissue damage is produced by reactive oxygen compounds that are normally formed in metabolism and which are normally destroyed by glutathione and its related antioxidants. In the absence of marked decrease of glutathione synthesis, many cells undergo substantial damage and this is a reflection of the very large amount of oxidative stress with which the cell must normally overcome. In the presence of marked glutathione deficiency, cellular anti-oxidants such as ascorbate and α-tocopherol cannot be restored by reduction and would therefore be of little value to the cell.

The buthionine sulfoximine model seems, in a number of ways, to mimic effects that occur in certain human degenerative states and diseases. Undoubtedly these pathological processes also involve other functions that are required for prevention of oxidative stress, including mechanisms for detection of damaged molecules, removal of these molecules, and their resynthesis in repair processes. When pathological processes damage a component or components of the normal cellular antioxidant systems, the reactive oxygen compounds formed normally in metabolism accumulate leading to further damage. Oxidative stress is thus of endogenous origin, as it is in the buthionine sulfoximine model described here.

As discussed above there are various ways of increasing cellular glutathione and still others may be developed, including genetic approaches involving increase of the synthesis of the two enzymes required for glutathione formation. The model of oxidative stress here, using cataract formation, other types of tissue damage, and production of mortality in newborn rats as indicators, may be expected to be useful for the evaluation of other compounds that may be effective in the prevention of oxidative stress.

REFERENCES

Anderson, M.E, Powrie, F., Puri, R.N. and Meister, A. (1985) Arch. Biochem. Biophys. 239, 538-548.

Anderson, M.E., and Meister, A. (1989) Anal. Biochem. 183, 16-20.

Boveris, A., and Chance, B. (1973) Biochem. J. 134, 707-716.

Calvin, H.I., Medvedowsky, C., and Worgul, B.V. (1986) Science 233, 553-555.

Dalton, D.A., Russel, S.A., Hanus, F.J., Pascoe, G.A., and Evans, H.J. (1986) Proc. Natl. Acad. Sci. USA 83, 3811-3815.

Griffith, O.W., and Meister, A. (1979) J. Biol. Chem. 254, 7558-7560.

Griffith, O.W. (1982) J. Biol. Chem. 257, 13704-13712.
Griffith, O.W., and Meister, A. (1985) Proc. Natl. Acad. Sci. USA 82, 4668-4672.
Hamilton, T., O'Dwyer, P., Young, R., Tew, K., Padavic, K., Comis, R., and Ozols, R. (1990) Proc. An. Meet. Am. Soc. Clin. Oncol. 9, A281.
Han, J., Martensson, J., Meister, A., and Griffith, O.W. (1992) FASEB J: 5, 5631 (Abstract).
Jain, A., Martensson, J., Stole, E., Auld, A., and Meister, A. (1991) Proc. Natl. Acad. Sci. USA 88, 1913-1917.
Martensson, J., and Meister, A. (1989) Proc. Natl. Acad. Sci. USA 86, 471-475.
Martensson, J., Steinherz, R., Jain, A., and Meister, A. (1989) Proc. Natl. Acad. Sci. USA 86, 8727-8731.
Martensson, J., Lai, J.C.K., and Meister,A. (1990a) Proc. Natl. Acad. Sci. USA 87, 7185-7189.
Martensson, J., Jain, A., and Meister, A. (1990b) Proc. Natl. Acad. Sci. USA 87, 1715-1719.
Martensson, J., and Meister, A. (1991) Proc. Natl. Acad. Sci. USA 88, 4656-4660.
Martensson, J., Jain, A., Stole, E., Fryer, W., Auld, P.A.M., and Meister, A. (1991) Proc. Natl. Acad. Sci. USA 88, 9360-9364.
Meister, A. (1978) In: Enzyme-Activated Irreversible Inhibitors (Seiler, N., Jung, M.J., and Koch-Weser, J., eds.) Elsevier-North Holland Biomedical Press, Amsterdam, pp. 187-211.
Meister, A., and Griffith, O.W. (1979) Cancer Treat. Rep. 63, 1115-1121.
Meister, A., and Anderson, M.E. (1983) Ann. Rev. Biochem. 52, 711-760.
Meister, A. (1986) In: Biochemical Modulation of Anticancer Agents: Experimental & Clinical Approaches (Valeriote, F., and Baker, L., eds.) Martinus Nijoff, Boston, Mass., pp. 245-275.
Meister, A. (1988) In: Mechanisms of Drug Resistance in Neoplastic Cells (Tew, K.D., and Woolley, P.V., eds.) Bristol Myers Symposium N°9, Academic Press, New York, pp. 99-126.
Meister, A., and Larsson, A. (1989) In: The Metabolic Bases of Inherited Disease (Scriver, C.R., Beaudet, A.L., Sly, W.S., and Valle, D., eds.) McGraw-Hill, 6th edition, pp. 855-868.
Meister, A. (1991) Pharmacol. Therap. 51, 155-194.
Meister, A. (1992) Biochem. Pharmacol., in press.
Ozols, R.F., Hamilton, T.C., Masuda, H., and Young, R.C. (1988) In: Mechanisms of Drug Resistance in Neoplastic Cells (Tew, K.D., and Woolley, P.V., eds.) Bristol Myers Symposium N°9, Academic Press, New York, pp. 289-305.
Plummer, J.L., Smith, B.R., Sies, H., and Bend, J.R. (1981) Methods Enzymol. 77, 50-51.
Puri, R.N., and Meister, A. (1983) Proc. Natl. Acad. Sci. USA 80, 5258-5260.
Suthanthiran, M., Anderson, M.E., Sharma, V.K., and Meister, A. (1990) Proc. Natl. Acad. Sci. USA 87, 3343-3347.
Viña, J., Perez, C., Furukawa, T., Palacin, M., and Viña, J.R.

(1989) Br. J. Nutr. 62, 683-691.
Vistica, D.T., and Ahmad, S. (1989) In: Glutathione Centennial Molecular Perspective and Clinical Implications (Taniguchi, N., Higashi, T., Sakamoto, Y., and Meister, A., eds.) Academic Press, New York & Tokyo, pp. 301.315.
Wellner, V.P., Anderson, M.E., Puri, R.N., Jensen, G.L., and Meister, A. (1984) Proc. Natl. Acad. Sci. USA 81, 4732-4735.

Free Radicals: From Basic Science to Medicine
G. Poli, E. Albano & M. U. Dianzani (eds.)

UTILIZATION OF ORAL GLUTATHIONE

Lawrence J. Dahm[1], Paula S. Samiec[1], John W. Eley[2,3], Elaine W. Flagg[3], Ralph J. Coates[2,3], and Dean P. Jones[1,2]

[1]Department of Biochemistry, [2]Winship Cancer Center, and [3]Division of Epidemiology, Emory University, Atlanta, GA 30322 USA

Summary: The intestine is exposed continually to dietary toxicants, and it utilizes GSH and GSH-dependent enzyme systems in detoxication of these agents. Recent studies show that enterocytes take up GSH intact and use it to increase degradation of dietary lipid hydroperoxides, enhance GSH S-conjugation of electrophiles, and prevent oxidative injury. The intestine also has a mechanism (i.e., mucosal cysteine secretion) to reduce luminal GSSG to GSH, thereby making more GSH available to enterocytes. Additionally, GSH is transported from lumen to blood to increase plasma GSH concentration. Given the ability of various organ systems to use extracellular (i.e., plasma) GSH for chemical detoxication, increased daily intake of GSH in the diet or from oral therapy may be useful in disease prevention.

Introduction

In mammalian cells, glutathione (L-gamma-glutamyl-L-cysteinylglycine, GSH) is synthesized intracellularly from constituent amino acids and is the most abundant soluble thiol. Its role in detoxication of reactive oxygen species and electrophiles in reactions involving GSH peroxidases and GSH S-transferases, respectively, has been well characterized. Measures to modify intracellular GSH content have underscored the physiological importance of GSH in cellular protection. For example, methods to deplete intracellular GSH content via treatment of cells or animals with electrophilic agents (e.g., diethyl maleate, phorone) or GSH synthesis inhibitors (e.g., buthionine sulfoximine) exacerbate the injury caused by a variety of toxicants. Similarly, treatments which maintain intracellular GSH content (e.g., administration of constituent amino acids or cysteine delivery agents [L-2-oxothiazolidine-4-carboxylate]) afford protection against many of these toxicants. In certain pathologies (e.g., severe oxidative stress), synthesis may not be able to maintain intracellular GSH content. As a result, GSH is depleted and toxicity may result from lipid peroxidation, thiol oxidation, alkylation of critical proteins, as well as other mechanisms.

Certain epithelial cell types including those of intestine, lung (i.e., alveolar type II cells), and kidney have the ability to transport extracellular GSH for use in chemical detoxication (Lash et al., 1986; Hagen et al., 1986, 1988). In recent years, a major focus of our research has been devoted to studying the relationship between dietary GSH and disease prevention. We have found that intestinal utilization of extracellular GSH plays a key role in prevention of oxidative injury to intestine, elimination of peroxidized fats and electrophiles in the diet, and interorgan supply of GSH (see Fig. 1).

Exogenous GSH Protects Cells from Oxidative Injury

Intestinal epithelium is in contact with two pools of GSH. The basolateral membrane of enterocytes is exposed to plasma, which contains about 10-15 µM GSH in rats (Lash and Jones, 1985; Hagen et al., 1990a). The brush-border membrane faces the intestinal lumen, which receives biliary GSH in millimolar concentrations (Eberle et al., 1981). Additionally, GSH enters the intestinal lumen from foods rich in GSH. Recent analyses of GSH content in over 170 foods in the typical Western diet have shown that fresh meats, fruits, and vegetables contain relatively high amounts of GSH (e.g., 3-25 mg/100 g wet weight), whereas cereals, breads, and dairy products are relatively deficient in GSH (e.g., 0-3 mg/100 g wet weight) (Wierzbicka et al., 1989; Jones et al., 1992). Frozen foods generally contain similar amounts of GSH as their fresh counterparts, but other forms of processing as well as cooking result in GSH loss. From GSH contents of 35 common foods, Wierzbicka et al. (1989) estimated daily intake of dietary GSH ranges from 3-130 mg. More recent calculations based upon a more complete survey of foods (Jones et al., 1992) indicate that higher daily consumption (e.g., 150-200 mg) may be common.

Intestinal epithelial cells can elevate their intracellular GSH content by transporting GSH intact across brush-border and basolateral membranes (Lash et al., 1986; Vincenzini et al., 1988). Characteristics of intestinal GSH transport systems recently have been reviewed (Vincenzini et al., 1992). Physiologically, transport of extracellular GSH may be important in preventing cell injury. For example, treatment of intestinal epithelial cells with tert-butylhydroperoxide or menadione causes oxidative injury and progressive loss of cell viability. Administration of exogenous GSH (20 µM-1 mM) in the presence of acivicin and buthionine sulfoximine to inhibit GSH degradation and synthesis, respectively, prevents cell injury, whereas treatment with the constituent amino acids even in the absence of buthionine sulfoximine provides little or no protection (Lash et al., 1986). These results indicate that transport of intact GSH is responsible for the protection and not GSH degradation followed by intracellular resynthesis. Additionally, inclusion of agents (e.g., ophthalmic acid or probenecid) that inhibit uptake of GSH by enterocytes attenuate the protection afforded by extracellular GSH, indicating that GSH does not produce its beneficial effect extracellularly. Because ophthalmic acid and probenecid inhibit intestinal GSH transport in basolateral membranes but not brush-border membranes (Lash et al., 1986; Vincenzini et al., 1992), the results suggest that basolateral uptake of GSH is primarily responsible for protection. However, because probenecid only partially reverses the protective effect afforded by 1 mM GSH, brush-border uptake of GSH may also be involved.

Martensson et al. (1990) have suggested that GSH is critical for intestinal function, because treatment of mice with buthionine sulfoximine for 7 days depletes GSH content in gastrointestinal mucosa and causes degenerative changes in jejunal and colonic epithelium. Epithelial injury is characterized by mitochondrial swelling and degeneration, microvillar desquamation, and vacuolation. Administration of exogenous GSH orally partially restores GSH content in jejunum and colon, and it prevents epithelial injury. Treatment of mice with constituent amino acids (i.e., combination of glutamate, cysteine, and glycine) is without effect. Because of the results from several laboratories demonstrating transport of extracellular GSH across the intestinal brush-border (Linder et al., 1984; Hagen and Jones, 1987; Vincenzini et al., 1988), the most straightforward interpretation of their results is that GSH is transported across the intestinal brush-border to elevate mucosal GSH content and afford protection.

Gastric mucosa also utilizes extracellular GSH for detoxication (Hirota et al., 1989; Stein et al., 1990). Gastric mucosal injury following hemorrhagic shock and reperfusion is thought to be mediated at least in part by toxic oxygen species (Von Ritter et al., 1988). In a rat model, exogenous administration of GSH prevents depletion of gastric mucosal GSH content and gastric injury caused by hemorrhagic shock with reperfusion (Stein et al., 1990). Presumably, the mechanism of protection involves maintenance of intracellular GSH content to eliminate toxic oxygen species generated intracellularly during reperfusion. Because there is little information regarding gastric GSH transport, it is unclear whether restoration of gastric GSH content in this model is due to transport of extracellular GSH or extracellular degradation of GSH followed by intracellular resynthesis. More work is needed to address the issue of GSH transport along the gastrointestinal tract.

Extracellular GSH also provides protection against oxidative injury to epithelial cells of lung (alveolar type II cells) (Hagen et al., 1986; Brown et al., 1992) and kidney (Hagen et al., 1988). As in intestine, transport of extracellular GSH is involved, because protection is afforded under conditions in which GSH synthesis and degradation are inhibited by buthionine sulfoximine and acivicin, respectively. However, unlike the intestine, treatment with equimolar concentrations of constituent amino acids in the absence of buthionine sulfoximine provides significant protection in kidney and complete protection in alveolar type II cells. Thus, these epithelial cells can maintain intracellular GSH content for detoxication by transport of extracellular GSH as well as through GSH synthesis intracellularly. Both mechanisms probably operate physiologically; however, the transport rate for GSH is several-fold higher than the synthesis rate in kidney and alveolar type II cells, suggesting that transport may be more critical in maintaining intracellular GSH levels under conditions threatening cell survival.

In addition to these epithelial cells (i.e., intestinal, renal, alveolar type II) where uptake of GSH appears to provide protection, studies on other cell types indicate that exogenous GSH can provide protection against oxidative injury without uptake. Exposure of alveolar macrophages to hyperoxia causes intracellular GSH depletion and loss of cell function, which are prevented by exogenous administration of GSH (Forman and Skelton, 1990). This protection is abolished by co-treatment with the gamma-glutamyltransferase inhibitor serine-borate, indicating that extracellular GSH degradation, transport of constituent amino acids, and intracellular resynthesis of GSH are involved. In isolated perfused livers from rats, supplementing the perfusion buffer with GSH prevents the decrease in hepatic GSH content and liver injury caused by cyanide or hypoxia/reoxygenation (Younes and Strubelt, 1990). Because hepatic parenchymal cells do not appear to take up GSH intact, the mechanism of protection likely involves extracellular GSH degradation and intracellular resynthesis. This mechanism also is consistent with the protective effect of exogenous GSH on oxidative injury to retinal pigment epithelial cells (Davidson et al., 1992; Sternberg et al., 1992).

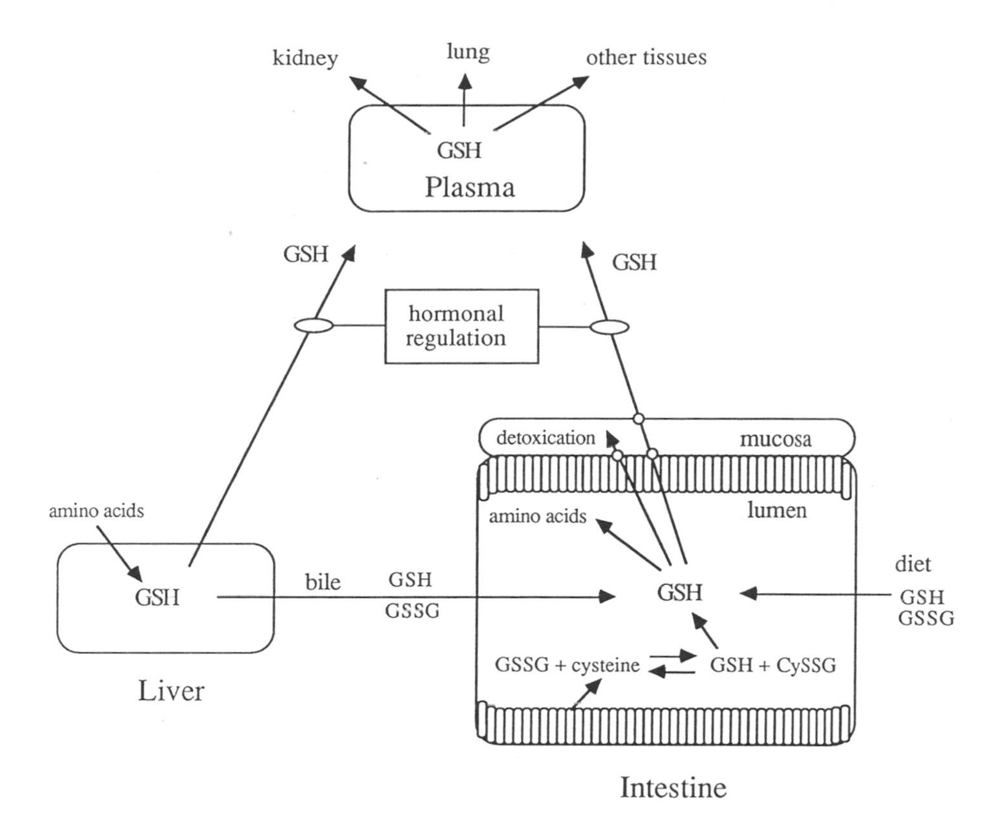

Fig. 1. Scheme showing fates of dietary GSH and GSSG and inte-rorgan supply of GSH. Liver and intestine both supply GSH to blood. In liver, GSH is synthesized from precursor amino acids and is released into sinusoidal blood by a hormonally regulated process (Lauterburg et al., 1984; Sies and Graf, 1985). In the intestinal lumen, dietary and biliary GSH may undergo several fates including degradation to constituent amino acids by -glutamyltransferase and dipeptidases located on the brush-border, uptake into enterocytes to support chemical detoxication, or transepithelial transport from lumen into mesenteric blood. The latter is hormonally regulated (Hagen et al., 1991). Additionally, GSH is synthesized in enterocytes for detoxication and export into blood (not shown). Dietary and biliary GSSG also can supply GSH to mucosa and ultimately to blood, because GSSG is reduced to GSH within the intestinal lumen. The mechanism likely involves thiol/disulfide exchange between GSSG and cysteine released from intestinal mucosa. Plasma GSH can be utilized by kidney, lung, and other tissues to maintain intracellular GSH stores either by transport of intact GSH or by degradation of GSH extracellularly, transport of constituent amino acids, and resynthesis of GSH intracellularly.

Intestinal Use of Extracellular GSH for Elimination of Dietary Peroxides and Electrophiles

Dietary intake of polyunsaturated fats results in exposure of the gut lumen to lipid hydroperoxides. Detoxication of these agents is of critical importance because high fat intake and generation of lipid hydroperoxides is associated with increased risk of cancers (Carrol and Khor, 1975; Reddy, 1983). Glutathione peroxidase is present throughout the small intestinal mucosa and is likely involved in elimination of lipid hydroperoxides (Manohar and Balasubramanian, 1986; Kowalski et al., 1990; Aw et al., 1992). Studies by Aw et al. (1992) demonstrate the importance of intestinal mucosal GSH in detoxication of luminal lipid hydroperoxides. Under control conditions, infusion of peroxidized fish oil intraduodenally over a 6 hr period results in a basal amount of hydroperoxide accumulation in intestinal mucosa and lymph, as well as liver and kidney. This observation indicates that intestinal mucosa could not completely detoxify the luminally administered hydroperoxide. Treatment of rats with agents that decrease mucosal GSH content (e.g., buthionine sulfoximine, diethyl maleate, or phorone) prior to lipid infusion further elevates hydroperoxide accumulation in these tissues compared to control. Inhibition of intestinal GSSG reductase with N,N-bis(2-chloroethyl)-N-nitrosourea (BCNU) produces similar effects, underscoring the importance of GSH maintenance to provide reducing equivalents to GSH peroxidase for continued hydroperoxide detoxication.

Extracellular GSH can be utilized by intestine in detoxication of luminal lipid hydroperoxides. Using everted sacs of rat small intestine, Kowalski et al. (1990) showed that peroxidized methyl linoleate (PML) added to the luminal side (i.e., incubation buffer) appeared on the contraluminal side (i.e., inside sac), which, as described above, indicates that intestine did not completely detoxify PML. Addition of 1 mM GSH to the luminal side has only a slight effect in reducing the appearance of PML

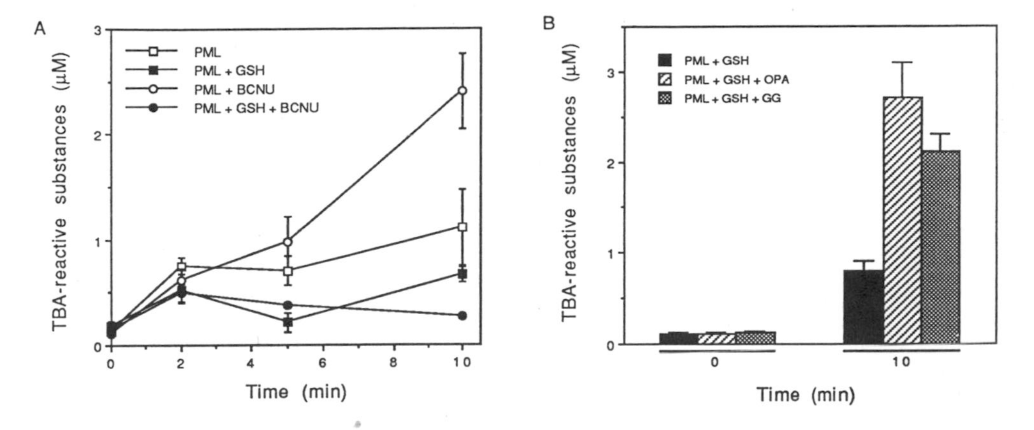

Fig. 2. Effect of luminal GSH on the appearance of peroxidized methyl linoleate (PML) into everted sacs of rat small intestine (panel A). Everted sacs (5 cm) containing oxygenated Krebs Henseleit buffer (pH 7.4) were incubated at 37°C in buffer containing PML emulsified with deoxycholate (8 mM). At 0-10 min, sacs were removed, and the presence of PML within sacs was determined by measuring thiobarbituric acid (TBA)-reactive substances as described by Tappel and Zalkin (1985). Luminal incubation buffer initially contained 75-100 μM TBA-reactive substances. GSH (1 mM) and/or BCNU (10 μM) were included in some luminal incubations as indicated. In panel B, the ability of GSH transport inhibitors to inhibit the effect of added GSH on PML appearance in everted sacs was examined. All sacs were treated with BCNU (10 μM), acivicin (250 μM), and GSH (1 mM). Ophthalmic acid (OPA) and -glutamylglutamate (GG, 10 mM each) were included where indicated. All values are means ± SE from at least three sacs. Data from Kowalski et al. (1990).

on the contraluminal side (Fig. 2A). Inhibition of GSSG reductase with BCNU enhances the appearance of PML into the everted sac, indicating that mucosal PML degradation is reduced. This effect of BCNU is greatly attenuated by luminal addition of GSH (Fig. 2A). Transport of intact GSH is likely responsible for the enhanced degradation of PML, because addition of GSH transport inhibitors (e.g., gamma-glutamylglutamate or ophthalmic acid) in the presence of acivicin abolishes the effect caused by added GSH (Fig. 2B).

Recent preliminary results indicate that intestine also can use extracellular GSH to enhance GSH S-conjugation of electrophiles via GSH S-transferases. When incubated with 1-chloro-2,4-dinitrobenzene (CDNB), freshly isolated epithelial cells from rat intestine generate the GSH S-conjugate of CDNB, and co-treatment with exogenous GSH (1 mM) greatly increases the conjugate formation. Although nonenzymic formation of the GSH S-conjugate does occur, it cannot account for the effect of added GSH. In addition, incubation of cells with equimolar concentrations of constituent amino acids results in an insignificant increase in formation of the GSH S-conjugate of CDNB. Taken together, these observations suggest that intestinal epithelial cells transport extracellular GSH intracellularly and increase the availability of GSH to GSH S-transferases for S-conjugation.

Intraluminal Reduction of Glutathione Disulfide to GSH in Rat Small Intestine

Bile delivers glutathione disulfide (GSSG) as well as GSH to the intestinal lumen, and about 20% of total GSH equivalents in bile are in the form of GSSG (Eberle et al., 1981). In addition, certain foods contain GSSG in amounts several-fold greater than GSH (Wierzbicka et al., 1989; Jones et al., 1992). The fate of GSSG in the intestinal lumen has not been studied extensively, but a study by Grafstrom et al. (1980) indicates that GSSG can be metabolized by gamma-glutamyltransferase and dipeptidases associated with the intestinal brush-border. The preferred route of degradation apparently is initial removal of the gamma-glutamyl moieties of GSSG to yield cystinyl-bis-glycine followed by cleavage of the peptide bonds to ultimately produce cystine (Grafstrom et al., 1980).

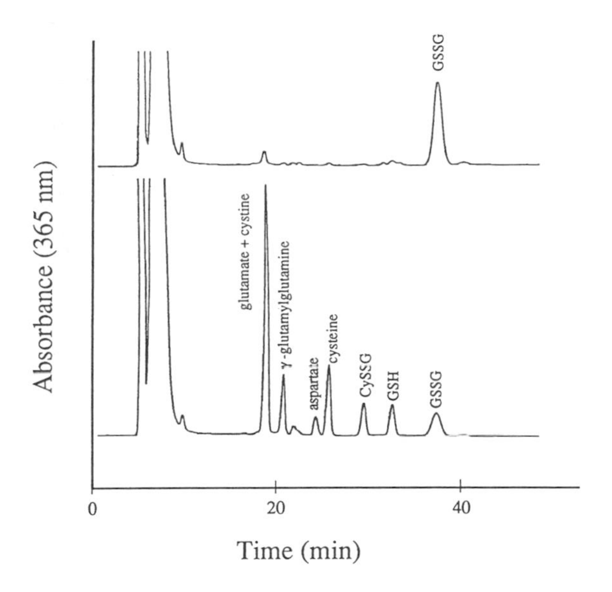

Fig. 3. High performance liquid chromatographic (HPLC) tracings of GSSG incubation buffer (top) and products formed intraluminally in rat jejunum (bottom). A vascularly perfused rat jejunal preparation was used as described previously (Hagen and Jones, 1987). Nine ml of Krebs Henseleit buffer (pH 7.4) containing GSSG and acivicin (250 µM each) were instilled into the jejunal lumen. Twenty min later, luminal contents were removed, spun in a microcentrifuge, acidified with perchloric acid (10% final concentration), and derivatized for HPLC analysis of thiols, disulfides, and other compounds (Reed et al., 1980; Fariss and Reed, 1987). Compounds were identified by retention times of authentic standards, except for the mixed disulfide of GSH and cysteine (CySSG), which was identified after generation via thiol/disulfide exchange between GSSG and cysteine (Jocelyn, 1967). Retention times (min) for luminal products are given in parentheses; gamma-glutamylglutamine (20.80), aspartate (24.53), cysteine (25.93), CySSG (29.86), GSH (33.10), and GSSG (38.00). Cystine and glutamate were not resolved from each other and eluted at 18.81 min. Luminal incubations without GSSG yielded a similar HPLC profile, except that peaks for GSSG and CySSG were absent. Additionally, the GSH peak area in control experiments was <15% of its counterpart after intraluminal GSSG incubation, indicating that mucosal GSH efflux could not account for the presence of GSH in lumen with GSSG incubation.

Studies by Hagen et al. (1990b) indicate that reduction of GSSG to GSH also occurs within the intestinal lumen. They fed rats a diet supplemented with GSSG and observed that the ratio of intraluminal concentrations of GSH to GSSG increased in jejunum but not in other regions of the gastrointestinal tract, suggesting that a mechanism was present to reduce GSSG in duodenum or proximal jejunum. Intraluminal reduction of GSSG to GSH was confirmed in the vascularly perfused rat jejunum (see Fig. 3) and in isolated intestinal segments from duodenum and jejunum (Dahm and Jones, 1991). The mechanism for intraluminal reduction of GSSG appears to be through thiol/disulfide exchange between GSSG and cysteine that is released from intestinal mucosa (see Fig. 1). GSH generated by this mechanism would be available to enterocytes for chemical detoxication or transepithelial transport (see below).

Absorption of Intact GSH from the Intestinal Lumen

As described, luminal GSH can be taken up by enterocytes for detoxication of electrophiles and hydroperoxides. Using everted sacs of rat small intestine, Hunjan and Evered (1985) provided data suggesting that GSH could undergo transepithelial transport. Hagen and Jones (1987) confirmed and characterized transepithelial transport using the vascularly perfused rat small intestine. They found that luminal transport of radiolabelled GSH into the vasculature is Na^+-dependent, inhibited by other gamma-glutamyl compounds, stimulated by α-adrenergic agonists (e.g., phenylephrine), and localized primarily to the midjejunum (Hagen and Jones, 1987; Hagen et al., 1990b; Hagen et al., 1991). Under physiological conditions (i.e., no inhibition of intestinal gamma-glutamyltransferase or GSH synthesis), HPLC analysis of vascular perfusate shows that transport of intact ^{35}S-GSH predominates over luminal degradation and transport of cysteine or cystine when luminal GSH concentrations exceed 0.1 mM. Because luminal duodenal GSH concentration in rats is about 0.5 mM under fasted conditions (Hagen et al., 1990b), these results suggest that transepithelial transport of luminal GSH is favored over degradation in vivo.

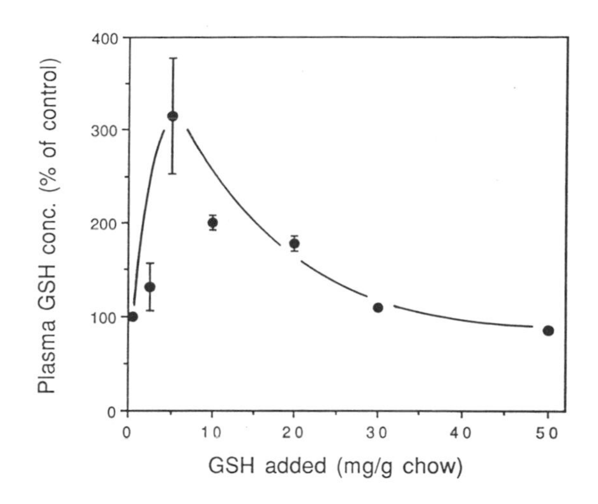

Fig. 4. Concentration dependence of GSH absorption from the gastrointestinal tract. Rats were placed on a semisynthetic powdered diet (AIN-76) for 7 days prior to start of the experiment and were conditioned to eat their chow within 1 hr after darkness. At the start of the experiment, they received AIN-76 alone or the diet supplemented with 2.5, 5, 10, 20, 30, or 50 mg GSH/g chow. The AIN-76 diet itself contained about 0.03 mg GSH/g chow. At 90 min after the eating period ended, rats were anesthetized with ketamine/xylazine, and a sample of blood was obtained by cardiac puncture for plasma measurement of GSH concentration as described by Lash and Jones (1985). This timepoint shows maximal elevation of plasma GSH concentration in time course experiments. All values are expressed as a percentage of zero time control (12 µM). Data are from at least 3 rats (means ± SE) except for 30 and 50 mg GSH/g treatment groups, which are from single rats. Feeding rats 5 mg constituent amino acids/g AIN-76 chow did not result in elevated plasma GSH concentration. Data from Hagen et al. (1990a).

Studies in vivo in rats have confirmed that absorption of GSH from the intestinal tract occurs. After oral gavage with GSH or supplementation with GSH in the diet, plasma GSH concentration is elevated 2 to 5-fold (Hagen and Jones, 1987; Hagen et al., 1990a). Administration of equivalent amounts of constituent amino acids by either method fails to increase plasma GSH concentration over the same 3 hr time course, indicating that transport of intact GSH occurs. Analysis of time course data shows that plasma GSH concentration peaks at about 90 min after consumption of a GSH-supplemented diet (5 mg/g chow) and is followed by a rapid decline to baseline plasma GSH values. We have not identified the mechanism for this decline, but it could relate to reduced GSH efflux from liver and/or enhanced clearance of plasma GSH by other organ systems. Increasing the dietary load of GSH does not further elevate plasma GSH concentration measured 90 min later. Instead, it results in a blunted elevation, and plasma GSH values decline to baseline (i.e., with no dietary GSH supplementation) at the highest dietary load (Fig. 4). Because plasma GSH concentration apparently is not elevated at all with high dietary GSH loads, regulation may be at the level of mucosal transport and not hepatic GSH efflux/extraintestinal clearance. However, if the time course for GSH absorption from the intestinal tract is dependent upon dietary GSH load, measurement of plasma GSH concentration at a single time point could miss any elevation. Thus, although definitive conclusions regarding regulation of plasma GSH concentration cannot be made, the results suggest that it is under tight control. Additionally, they indicate that the intestine likely plays an important role in interorgan transport of GSH (Fig. 1).

Dietary GSH Consumption and Blood Plasma GSH in Humans

Few studies of plasma glutathione levels have been conducted in humans, and in those studies sample sizes have been very small and not analyzed for potential diurnal variation or association with dietary habits. In an exploratory analysis, the population distribution and determinants of plasma total glutathione (i.e., GSH plus all mixed disulfide forms, GSH_t) are being investigated in a group of 100 volunteers aged 18-61 years, sampled in Atlanta, Georgia, U.S.A. during June and July, 1989. Data on demographic and health-related factors were collected by interview, and plasma GSH_t concentration was measured by HPLC. The statistical distribution (mean, median, range, and variance) of plasma GSH_t is being characterized in the entire study group and for demographic subgroups defined by characteristics such as sex, age, race, education, marital status, and religious affiliation. Variation in plasma GSH_t level by health-related factors such as body mass index, consumption of alcohol, former use of cigarettes, usual amount of sleep and exercise, and past or current medical conditions is also being examined.

A second series of analyses using dietary history information for this study population will be used to investigate the association between dietary glutathione intake and plasma glutathione level. GSH and GSH_t, methionine, and cystine intakes have been calculated from food frequency questionnaire data, and concentrations of plasma GSH_t and of other serum antioxidants have been obtained. The data are being analyzed for potential correlation of intakes of GSH and GSH_t with plasma GSH_t level, with adjustment for potentially confounding demographic health-related factors. Additional confounding or effect modification by dietary intake of glutathione precursors and serum antioxidant levels is also be investigated.

With this information on potential demographic and health-related factors that may affect plasma GSH concentration, we expect to be able to use dietary history data to explore the relationship between glutathione intake and risk of specific types of cancer and other chronic diseases.

Comments on Clinical Trials to Assess the Efficacy of GSH Supplementation

Our previous and ongoing research with humans has focused on characterizing the determinants of blood plasma GSH and potential associations between dietary GSH consumption, plasma GSH concentration, and risk of chronic disease. Given that GSH is a well known antioxidant and anticarcinogen, that orally supplied GSH can be absorbed and increase blood plasma GSH concentration, that purified GSH suitable for human consumption is commercially available, and that oral doses of several grams can be taken without serious untoward effects, it is natural to consider carefully controlled clinical trials to scientifically assess the utility of GSH therapy. For such a purpose, it is necessary to establish dose, dosing frequency, and duration of study; the following comments are based upon our experience with studies of dietary GSH and determinants of plasma GSH with the hope of providing reasonable starting points for these types of studies.

Dietary consumption of 50-70 g of protein per day provides about 1.5 to 2 g of cysteine. With dietary consumption of up to 200 mg of GSH, it appears that GSH normally accounts for less than 5% of the cysteine consumed (GSH is only about 33% cysteine). Thus, one can supplement with 200 mg GSH with each meal and only increase total daily cysteine consumption by about 10%. To increase cysteine consumption by about 50%, it is necessary to provide 1 g GSH with each of 3 meals/day. These simple calculations show that two ranges of GSH supply can be achieved readily, one range which provides a several-fold increase in dietary GSH without a substantial increase in total sulfur amino acid supply and a second range where GSH supply is associated with a significant increase in total sulfur amino acid supply.

The lower range (i.e., 200 mg GSH supplement/meal), is sufficient to raise luminal GSH by about 0.6 mM assuming dilution in 1 liter of fluid. From our data in animals, this amount is sufficient to support detoxication of dietary lipid hydroperoxides and electrophiles even if no additional GSH is present in the meal or in the diet. Thus, supply of 200 mg with each meal would appear to be sufficient to reduce any potential health risks associated with consumption of these types of reactive compounds.

For studies to test whether increased oral GSH can decrease risk of disease in tissues other than the gastrointestinal tract, a dose of 200 mg is not likely to have much of an effect. If 200 mg of GSH were introduced into blood and distributed instantaneously, it would only increase plasma GSH concentration to about

100 µM. In fact, absorption would occur over a long period and could not yield such a high concentration. For such a dose to have an effect other than in the intestine, it would have to be cleared from the blood almost exclusively by specific cells that represent only a small fraction of the body. At high concentrations (i.e., 1-2 mM), GSH is cleared largely by the kidney, but comparable information on clearance of lower doses is not available. A more reasonable dose for studies addressing tissues other than the gastrointestinal tract would be in the range of 1 g/meal for 3 meals per day. This would provide a substantial increase in total sulfur amino acids (i.e., 50-70%) and would be expected to increase systemic GSH both by direct supply and by provision of increased availability of cysteine.

As suggested above, the difference between the low and high supplementation ranges may be more than a simple quantitative difference. Low range supplements may have a very important role in reducing risk from dietary toxicants but not provide any potential benefit derived from increasing GSH concentration in other tissues. Numerous animal studies have shown that provision of sulfur amino acids in amounts greater than that required for normal growth and maintenance of nitrogen balance can result in increased tissue GSH concentrations. Cysteine is too reactive to be maintained at high concentrations in cells and is therefore stored in the form of GSH. Evolution appears to have provided multiple benefits from maintaining high concentrations of this "storage" form of cysteine. These benefits include mechanisms to remove toxic peroxides, aldehydes, epoxides, quinones, and other reactive electrophiles. Thus, assessment of a daily requirement for cysteine in terms of growth or nitrogen balance may overlook the real benefits from availability of higher supply of cysteine and GSH. Ample evidence is available showing that GSH is decreased in advanced aging (Kretzschmar et al., 1991), human immunodeficiency viral infection (Buhl et al., 1989), vitamin C deficiency (Henning et al., 1991), hypoxia (Bai, 1992), cirrhosis (Chawla et al., 1984), and other pathophysiological conditions. Thus, controlled studies of sulfur amino acid supplementation, such as by providing GSH, will be of great importance in determining whether this is a useful approach for improving human health.

Acknowledgements

This work has been supported by NIH Grant HL39968

References

Aw, T.Y., Williams, M.W. and Gray, L. (1992) Am. J. Physiol. 262, G99-G106.
Bai, C. (1992) Ph.D. Dissertation, Emory University, Atlanta,GA.
Brown, L.A., Bai, C., and Jones, D.P. (1992) Am. J. Physiol. 262 L305-L312.
Buhl, R., Holroyd, K.J., Mastrangeli, A., Cantin, A.M., Jaffe, H.A., Wells, F.B., Saltini, C., and Crystal, R.G. (1989). Lancet ii, 1294-1298.
Carrol, K.K., and Khor, H.T. (1975) Prog. Biochem. Pharmacol. 10, 308-353.
Chawla, R.K., Lewis, F.W., Kutner, M.H., Bate, D.M., Roy, R.G., and Rudman D. (1984) Gastroenterol. 87, 770-776.
Dahm, L.J., and Jones, D.P. (1991) (abstract) Toxicologist 11, 55.
Davidson, P.C., Sternberg, P., Jones, D.P., and Reed, R.L. (1992) (abstract) Invest. Ophthamol. Vis. Sci. 33, 920.
Eberle, D., Clarke, R., and Kaplowitz, N. (1981) J. Biol. Chem. 256, 2115-2117.
Fariss, M.W., and Reed, D.J. (1987) Methods Enzymol. 143, 101-109.
Forman, H.J., and Skelton, D.C. (1990) Am. J. Physiol. 259, L102-L107.
Grafstrom, R., Stead, A.H., and Orrenius, S. (1980) Eur. J. Biochem. 106, 571-577.
Hagen, T.M., and Jones, D.P. (1987) Am. J. Physiol. 252, G607-G613.
Hagen, T.M., Bai, C., and Jones D.P. (1991) FASEB J. 5, 2721-2727.
Hagen, T.M., Wierzbicka, G.T., Sillau, A.H., Bowman, B.B., and Jones, D.P. (1990a) Am. J. Physiol. 259, G524-G529.
Hagen, T.M., Wierzbicka, G.T., Bowman, B.B., Aw, T.Y., and Jones, D.P. (1990b) Am. J. Physiol. 259, G530-G535.
Hagen, T.M., Brown, L.A., and Jones, D.P. (1986) Biochem. Pharmacol. 35, 4537-4542.
Hagen, T.M., Aw, T.Y. and, Jones, D.P. (1988) Kidney Int. 34, 74-81.
Henning, S.M., Zhang, J.Z., McKee, R.W., Swendseid, M.E., and Jacob R.A. (1991) J. Nutr. 121, 1969-1975.
Hirota, M., Inoue, M., Ando, Y., Hirayama, K., Morino, Y., Sakamoto, K., Mori, K., and Akagi, H. (1989) Gastroenterol. 97, 853-859.
Hunjan, M.K., and Evered, D.F. (1985) Biochim. Biophys. Acta 815,

184-188.
Jocelyn, P.C. (1967) Eur. J. Biochem. 2, 327-331.
Jones, D.P., Coates, R.J., Flagg, E.W., Eley, J.W., Block, G., Greenberg, R.S., Gunter, E.W., and Jackson, B. (1992) Nutr. Cancer 17, 57-75.
Kowalski, D.P., Feeley, R.M., and Jones, D.P. (1990) J. Nutr. 120, 1115-1121.
Kretzschmar, M., Muller, D., Hubscher, H., Marin, E. and Klinger, W. (1991) Int. J. Sports Med. 12, 218-222.
Lash, L.H., and Jones, D.P. (1985) Arch. Biochem. Biophys. 240, 583-592.
Lash, L.H., Hagen, T.M., and Jones, D.P. (1986) Proc. Natl. Acad. Sci. USA. 83, 4641-4645.
Lauterburg, B.H., Adams, J.D., and Mitchell, J.R. (1984) Hepatol. 4, 586-590.
Linder, M., De Burlett, G., and Sudaka, P. (1984) Biochem. Biophys. Res. Commun. 123, 929-936.
Manohar, M., and Balasubramanian, K.A. (1986) Indian J. Biochem. Biophys. 23, 274-278.
Martensson, J., Jain, A., and Meister, A. (1990) Proc. Natl. Acad. Sci. USA. 87, 1715-1719.
Reddy, B.S. (1983) In: Experimental Colon Carcinogenesis (H. Antrup and G.M. Williams eds). C.R.C. Boca Raton, p. 225-239.
Reed, D.J., Babson, J.R., Beatty, P.W., Brodie, A.E., Ellis, W.W., and Potter, D.W. (1980) Anal. Biochem. 106, 55-62.
Sies, H., and Graf, P. (1985) Biochem. J. 226, 545-549.
Stein, H.J., Hinder, R.A., and Oosthuizen, M.M.J. (1990) Surgery 108, 467-474.
Sternberg, P., Reed, R.L., Davidson, P.C., Jones, D.P., Drews-Botsch, C., and Hagen, T.M. (1992) (abstract) Invest. Ophthamol. Vis. Sci. 33, 920.
Tappel, A.L., and Zalkin, H. (1985) Arch. Biochem. Biophys. 80, 326-332.
Vincenzini, M.T., Favilli, F., and Iantomasi, T. (1988) Biochim. Biophys. Acta 942, 107-114.
Vincenzini, M.T., Favilli, F., and Iantomasi, T. (1992) Biochim. Biophys. Acta 1113, 13-23.
Von Ritter, C., Hinder, R.A., Oosthuizen, M.M.J., Svenson, L.G., Hunter, S.J.S., and Lambrecht, H. (1988) Dig. Dis. Sci. 33, 857-864.
Wierzbicka, G.T., Hagen, T.M., and Jones, D.P. (1989) J. Food Comp. and Anal. 2, 327-337.
Younes, M., and Strubelt, O. (1990) Toxicol. Letts. 50, 229-236.

Author Index

Subject Index